HUTCHINS PRICED SCHEDULES 2021

The Small Works and Maintenance Construction Book

71st Edition

Compiled by Griffiths & Hutchins Publications Ltd

Compiled by Griffiths & Hutchins Publications Ltd, Norwich.

© 2020 Griffiths & Hutchins Publications Ltd
Consultant technical editor, Colin Fleming MRICS MCIOB

First published 1945
71st Edition

ISBN 978-1-913293-07-9

A catalogue record for this book is available from the British Library.

Contents

Acknowledgements

As this latest edition of Hutchins Priced Schedules has been produced in a year of particularly unusual circumstances, we are ever more grateful to the people who have helped in making it possible.

These many contributors include manufacturers, suppliers, merchants, contractors, estimators and surveyors, based around the UK, who have generously given their time, energy and resources to bring it together.

Gratitude is again due to the readers, some of whom have a history with Hutchins Priced Schedules going back decades, who have also taken the time to help us review previous editions and consider improvements. Our guides are intended to be used on a day-to-day basis, and so practical insight from the industry is invaluable, and always welcome.

We would also like to pay tribute to a long-time and much valued contributor to this and other price books, André Holden. André had a long history with Hutchins Priced Schedules, including as an editor and technical consultant. He assisted in the early stages of updating this edition, but sadly passed away in August 2020.

Griffiths & Hutchins Publications Ltd

Preface

This price book sets out a reliable and consistent approach to measuring and pricing new build and alterations works for smaller scale projects up to a value in the region of £250,000.

The last edition, published at the end of 2019, sought to assist users pricing work against a backdrop of uncertainty created by ongoing negotiations for the UK's exit from the European Union.

Of course we had not foreseen the sudden and devastating impact of a pandemic on the construction industry and wider economy, that took effect during 2020 and for which the consequences will be long-lasting.

In light of the effect of the Covid-19 virus on costs and working practices, a note has been added to this edition's introduction (see page 6) to help factor in these exceptional circumstances to any estimates and bills of quantities being undertaken in the next 12 months.

This new edition has, as ever, been updated with the latest available prices for plant and material resources in the UK, as well as taking account of wage rates agreed by the Building & Allied Trades Joint Industrial Council (BATJIC) and the Joint Industry Board for Plumbing and Mechanical Engineering Services (JIB-PMES) in the calculation of labour rates.

Due to disruption caused by Covid-19, at the time of publishing this edition JIB-PMES had not yet announced how wage rates would be changed, if at all, for 2021. Our calculations therefore continue to use the current rates, but if and when any changes are announced prior to the publication of the 2022 edition of this book, we will publish our updated calculations at **www.hutchins.guide**

With upheaval caused by Covid-19 as well as ongoing factors surrounding the UK's exit from the European Union, including potential supply and labour issues, all prices provided are of course prefaced with the cautionary note that any changes to the market should be factored into the production of estimates and bills of quantities accordingly.

For experienced users of Hutchins Priced Schedules and new readers alike, we encourage you to read the guidance information that is included in the introductory sections to this edition, some of which has been revised in recent editions to help improve understanding of how best to use the schedules.

The format of each schedule follows that of previous editions with items generally measured in accordance with the trade format found within SMM6. Although this is not the latest version of the Standard Method of Measurement, we know it is one with which many of our readers are familiar and at ease using in their work.

In the last year we have received invaluable feedback from readers to help inform changes to this and future editions of Hutchins. This is never a completed task and we always welcome new comments; the easiest way to get in touch is through our website at **www.hutchins.guide**

Introduction

Background and general use

With its origins in the post-WW2 period of reconstruction, much of the materials, technologies and machinery being used across trades today would be unrecognisable to our Hutchins predecessors. This guide has, however, from its very beginning remained a trusted and reliable source of pricing information across the trades.

If the reader is unfamiliar with the use of price books, reading this introductory section is recommended to help understand the approach that has been taken in compiling the schedules.

The measured rates sections are generally trade-based, ensuring that the task of locating prices is made simpler. As well as a basic contents page, a more detailed table of what is contained in each schedule can be found from page 363 to find specific items more easily.

Measured unit rates are presented at both net and gross price levels; our gross rates allow for 10% profit, which can of course be altered according to requirements. The Memoranda section provides general information, including required quantities for selected work elements, specification information generally and other useful data.

Allowance for Covid-19 impact on working practices

The introduction of special measures to limit the spread of Covid-19 on sites has had a significant impact on productivity. With the possibility that these measures will continue to be implemented, and / or return in future, it is important to factor these in when creating estimates and bills of quantities where necessary.

To ensure the health and safety of construction workers and others who are in the vicinity, considerations including extra cleaning and preparation, reduced gang sizes and one-way systems, must all be accounted for. With this in mind, suitable adjustments should be made to the rates in this book, and we recommend considering a range of 10 to 30%.

Understanding the schedules

To help readers understand how the price for each unit of work is calculated, explanations of the columns included in the schedules are set out below. **All prices in this publication exclude VAT.**

(a) **Unit column**
The unit of measure for the item, e.g. m^3 - the cost per cubic metre.

(b) **Labour hours column**
The time allocated to complete a unit of work as described, expressed as a decimal fraction of an hour, e.g. 0.75 hours equals 45 minutes. The times given are so-called productive hours (see **Labour constants** below for more information). They are intended as an average and include provision for unloading, distribution and fixing where applicable.

(c) **Labour net column**
The cost of labour, calculated by multiplying the labour hours per unit of measure by the all-in labour rate (inclusive of 6.75% site on-costs and 13.25% establishment charges).

(d) **Plant net column**
The cost of the hire of plant per unit of measure, not including fuel or any delivery charges, but including a driver where relevant.

(e) **Materials net column**
The cost of materials per unit of measure, after deduction of discounts if appropriate, plus an allowance for waste.

(f) Unit rate net column
The total net cost of the labour, plant and materials columns.

(g) Labour gross column
The net labour cost plus 10% profit.

(h) Plant gross column
The net plant hire cost plus 10% profit.

(i) Materials gross column
The net materials cost plus 10% profit.

(j) Unit rate gross column
The total of the labour, plant and materials gross columns.

A profit level of 10% of the total of the net labour, plant and materials columns in both the new work and the repair and alterations sections has been assumed as a reasonable return on expenditure, but can of course be adjusted accordingly.

Definition of terms and symbols used

A – amp
BS – British Standard
Btu – British thermal unit
deg – degree
dia – diameter
DTp – Department of Transport
ErP – Energy related Product
G – gauge
g – gram
in – inch
kg – kilogram

kN – kilonewton
kW – kilowatt
l – litre
m – linear metre
m^2 – square metre
m^3 – cubic metre
nr – number
mm – millimetre
PC – prime cost; an allowance earmarked for an item. The cost of some items can vary significantly depending on the specification required and discount available to the purchaser, and should be adjusted accordingly for the estimate.
t – tonne
v – volt
W – watt

Labour constants

It is important to understand how the figure in the labour hours column is calculated in order to apply it to the reader's own situation, whether the work will be carried out by a single person or by a gang. It can also be useful for working out bonus targets.

The figure given in the labour hours column indicates the average number of 'productive' hours required for the unit of work to be carried out. If the labour is carried out by one person and the figure given is four hours, 100% of the labour is attributed to the one person.

If, for example with a 2+1 bricklaying gang, there are two skilled bricklayers and one labourer who is assisting the bricklayers, the four hours of productive labour would be divided by the two bricklayers, so two hours each. There is of course still a cost attached to the labourer for the work they carry out so they are not ignored in the net calculation, but by knowing

Introduction

the productive labour hours, it is possible to apply the average given in this publication to any gang build-up that is actually carrying out the work, whatever the ratio that makes up the labour gang.

Using the bricklaying gang example, the following calculations show how to apply a figure from the labour hours column to any number of operatives completing the unit of work, and how the net labour cost is calculated.

Figure given in labour hours column: 1
Person(s) carrying out the work: 1 (skilled)
Cost of labour: £37.63 x 1 = £37.63
Cost per productive person hour: £37.63 / 1 (skilled) = **£37.63**

Figure given in labour hours column: 1
Person(s) carrying out the work: 1+1 (one skilled and one labourer)
Cost of labour: £37.63 x 1 + £29.58 x 1 = £67.21
Cost per productive person hour: £67.21 / 1 (skilled) = **£67.21**

Figure given in labour hours column: 1
Person(s) carrying out the work: 2+1 (two skilled and one labourer)
Cost of labour: £37.63 x 2 + £29.58 x 1 = £104.84
Cost per productive person hour: £104.84 / 2 (skilled) = **£52.42**

Figure given in labour hours column: 1
Person(s) carrying out the work: 3+1 (three skilled and one labourer)
Cost of labour: £37.63 x 3 + £29.58 x 1 = £142.47
Cost per productive person hour: £142.47 / 3 (skilled) = **£47.49**

Labour costs

Labour outputs used in this publication take into account the latest wage agreements from both the Building & Allied Trades Joint Industrial Council (BATJIC) and the Joint Industry Board for Plumbing and Mechanical Engineering Services (JIB-PMES). At the time of publication, due to continued disruption caused by Covid-19, JIB-PMES had not announced any changes to cover 2021. Current rates have therefore been retained in calculations. If changes are announced, updated calculations will be published at **www.hutchins.guide**

Wage rates agreed by BATJIC, applicable from Monday 22 June 2020 to Sunday 20 June 2021

Adult general operative - £392.73 per 39-hour week; £10.07 per hour
Advanced craftsman - £513.63 per 39-hour week; £13.17 per hour

Calculation of hourly cost	Adult general operative	Advanced craftsman
Labour	£	£
Basic weekly wage	347.42	454.37
Overtime, say 9 hours	120.26	157.28
Non-productive overtime, say 4.5 hours	60.13	78.64
Public holidays, 8 days	18.36	24.01
Annual holidays, 22 days	50.49	66.03
Plus rate, say 50%	298.33	390.17
	894.99	**1,170.50**
Contributions, levies and expenses	£	£
National insurance	100.19	113.55
CITB levy	1.57	2.05
Sickness and injury benefit	4.52	4.52
Workplace pension	23.25	31.52
Death benefit	2.62	2.62
Trade supervision, say	44.05	44.05
Travel allowance, say	23.22	23.22
	199.42	**221.53**

	Adult general operative	Advanced craftsman
Total hourly cost	£	£
Labour rate, as above	894.99	1,170.50
Contributions, etc. as above	199.42	221.53
	1,094.41	**1,392.03**
Inclement weather, say 3%	32.83	41.76
Redundancy reserve, say 3%	32.83	41.76
	1,160.07	**1,475.55**
Employers' liability insurance, say 2%	23.20	29.51
	1,183.27	**1,505.06**
Site on-costs, say 6.75%	79.87	101.59
Overheads, say 13.25%	156.78	199.42
Total weekly cost	**1,419.92**	**1,806.07**
Cost per hour	**29.58**	**37.63**

Wage rates agreed by JIB-PMES, applicable from Monday 6 January 2020

Trained plumber - £513.38 per 37.5-hour week; £13.69 per hour
Advanced plumber - £597.75 per 37.5-hour week; £15.94 per hour

Calculation of hourly cost	Trained plumber	Advanced plumber
Labour	£	£
Basic weekly wage, per 37.5 hours	450.19	524.18
Additional 1.5 hours	27.01	31.45
Overtime, say 9 hours	162.06	188.70
Non-productive overtime, say 4.5 hours	81.03	94.35
Public holidays, 8 days	25.27	29.43
Annual holidays, 24 days	75.82	88.28
Additional holiday pay, say 64 credits	3.15	3.15
Plumbers' welding supplement	29.40	29.40
Plus rate, say 60%	512.36	593.36
	1,366.29	**1,582.30**
Contributions, levies and expenses	£	£
National insurance	117.47	121.79
CITB levy	2.39	2.77
Industry pension scheme	102.47	118.67
Trade supervision, say	68.31	79.12
Trade allowance, say	13.35	13.35
	303.99	**335.70**

Introduction

	Trained plumber	Advanced plumber
Total hourly cost	£	£
Weekly rate, as above	1,366.29	1,582.30
Contributions, etc. as above	303.99	335.70
	1,670.28	**1,918.00**
Inclement weather, say 2%	33.41	38.36
Redundancy reserve, say 2%	33.41	38.36
	1,737.10	**1,994.72**
Employers liability insurance at 2%	34.74	39.89
	1,771.84	**2,034.61**
Site on-costs, say 6.75%	119.60	137.34
Overheads, say 13.25%	234.77	269.59
Total weekly cost	**2,126.21**	**2441.54**
Cost per hour	**44.30**	**50.87**

Establishment charges and profit

20% has been included in the labour net cost to cover site on-costs and establishment charges, for both new and repairs and alteration work. This is made up of:

Site on-costs	6.75%
Establishment charges	13.25%

It is considered that these percentages represent average overhead costs on competitive small works contracts, but they will of course vary according to the size and scope of the contract and the type of work involved.

The extra payments to which operatives are entitled under the Working Rule Agreement for executing certain types of work have been considered in the appropriate labour cost columns.

The following covers the main items included in the assessment of the respective percentages.

Site on-costs: 6.75%

Setting out; cartage
Protection and timekeeping
Sanitary conveniences
Welfare arrangements and safety precautions
Watching and lighting
Sheds and site offices, and record keeping
Deterioration of non-chargeable plant
Removal of debris and other cartage
Site telephone; attendance; incidentals
Gauge boxes; profiles; screeds; templates; samples; trial holes
Replacements, making good and maintenance

Establishment charges: 13.25%

Third Party (about 0.4% of wages)	
Fire insurance	0.25%
Bond fees (if any)	
Estimating costs	
Rent, rates and taxes	
Lighting, heating, electricity	
Other office and overhead costs	13.00%
	13.25%

Converting labour hour decimals into hours and minutes

The time taken to complete a unit of work is given in the Labour hours column as a decimal portion of an hour, e.g. 0.50 hours is 30 minutes. The times are an average and include unloading, distributing, and subsequent fixing in position, if applicable.

The following table shows how to convert decimal hours to hours and minutes. Figures along the top and down the left side of table represent decimal hours to two decimal places; other figures are minutes.

DH*	0.00	0.01	0.02	0.03	0.04	0.05	0.06	0.07	0.08	0.09
0.00	-	1	2	2	3	3	4	5	5	6
0.10	6	7	7	8	9	9	10	11	11	12
0.20	12	12	13	13	15	15	16	17	17	18
0.30	18	19	19	20	21	21	22	23	23	24
0.40	24	25	25	26	27	28	28	29	29	30
0.50	30	31	31	32	33	34	34	35	35	36
0.60	36	37	37	38	38	39	40	41	41	42
0.70	42	43	43	44	44	45	46	47	47	48
0.80	48	49	49	50	50	51	52	52	53	54
0.90	54	55	55	56	56	57	58	58	59	60

*decimal hours

Worked example 1
Labour hours column shows 0.75.
To obtain equivalent minutes: read down the left-hand column to 0.70, then across the line to the column headed 0.05. Answer: 45 minutes.

Worked example 2
Labour hours column shows 1.44
Read down the left-hand column to 0.40, then across the line to the column headed 0.04 = 27 minutes. Add 1 hour. Answer: 1 hour 27 minutes.

The following table shows the reverse calculation; how to present hours and minutes as decimal hours. Figures along the top and down the left side of table represent minutes; other figures are decimal hours.

Mins	0	1	2	3	4	5	6	7	8	9
0	-	0.02	0.03	0.05	0.07	0.08	0.10	0.12	0.13	0.15
10	0.17	0.18	0.20	0.22	0.23	0.25	0.27	0.28	0.30	0.32
20	0.33	0.35	0.37	0.38	0.40	0.42	0.43	0.45	0.47	0.48
30	0.50	0.52	0.53	0.55	0.57	0.58	0.60	0.62	0.63	0.65
40	0.67	0.68	0.70	0.72	0.73	0.75	0.77	0.78	0.80	0.82
50	0.83	0.85	0.87	0.88	0.90	0.92	0.93	0.95	0.97	0.98

Worked example 1
Person takes 45 minutes to complete a unit of work. To obtain decimal hours, read down the left-hand column to 40, then across the line to the column headed 5. Answer: 0.75 hours.

Worked example 2
Person takes 1 hour 27 minutes to complete a unit of work. To obtain decimal hours, read down the left-hand column to 20 then across the line to the column headed 7. This = 0.45 hours. Add 1 hour. Answer: 1.45 hours.

Introduction

Plant and tool hire costs

Plant hire rates used in this book reflect locally hired machines at third quarter 2020. They are calculated on the basis of a 39-hour week and include the cost of a driver where appropriate but do not include VAT, fuel or any delivery charges. Possible savings from longer hire periods or using the contractor's own plant and tools should be considered in estimates accordingly. Common items are included here for reference.

Access and support

	£ per week
Alloy access towers:	
2.2m platform height	65.00
4.2m platform height	123.00
6.2m platform height	183.00
Lightweight staging:	
2.4m long	45.35
3.6m long	53.06
Steel trestles; each	6.50
Scaffold board; each	8.73
Lightweight push-up ladder	
3.5m closed, 6.2m extended	49.47
3.6m closed, 9.1m extended	60.00
Combination ladder; 3.6m	75.00
Extending roof ladder; 3.0m closed, 4.6m extended	46.75
Alloy platform steps:	
6 tread	35.00
8 tread	42.27
10 tread	48.00

General building

	£ per week
Steel props; each	
1.1m closed, 1.8m extended	7.00
1.8m closed, 3.0m extended	7.00
2.6m closed, 3.9m extended	7.00
3.2m closed, 4.9m extended	7.00
Wheelbarrow, tyred, heavy-duty	16.00

Concrete and compaction

	£ per week
Surface scaler, heavy duty; petrol	267.86
Concrete floor grinder; 270mm	302.11
Light-duty needle scaler	102.58
Poker vibrator, 39mm head; electric	106.52
Poker vibrator, 50mm head; petrol	61.34
Beam screed, petrol	250.16
Bulk concrete mixer; diesel	105.19
Light vibrating plate; petrol	50.00
Medium vibrating plate; petrol	69.00
Vibrating roller; 560mm wide	93.83
Rammer, 4-stroke; 6in shoe	128.00
Rammer, 4-stroke; 9in shoe	141.12

Plumbing, pumping and drain clearing

	£ per week
Pipe vice and stand	46.91
Steel pipe bender, manual	68.00

Heating, cooling and drying

	£ per week
Building dryer, portable; 20ltr	105.00
Building dryer, portable; 52ltr	136.50

Introduction

	£ per week
Welding, lighting and power	
Plasterers' light	55.10
Telescopic floodlight; twin head	42.90
Breaking and drilling	
Hydraulic breaker; medium duty	192.00
Air breaker; medium duty	81.58
Fixing, carpentry and sanding	
Rapid-fire cartridge hammer	85.17
Floor edging sander	43.00
Floor sander	56.00
Sawing and cutting	
Chasing machine	128.03
Door trimming saw	44.00
Chainsaw, petrol, including safety kit	136.50
Painting and decorating	
Air compressor, 9cfm; electric	136.00
Air compressor, 15cfm; electric or petrol	156.90
Tyrolean roughcast machine	23.00
Cleaning and floor maintenance	
Compact hot pressure washer	247.19
Electric steam cleaner	115.00

	£ per week
Carpet cleaner	36.00
Carpet cleaner upholstery tool	5.00
Floor scrubber-dryer; single	90.00
Lifting and materials handling	
Rubbish chute section	9.36
Rubbish chute hopper	10.76
Rubbish chute Y-section	11.96
Gardening	
Turf cutter	176.25
Petrol strimmer brush cutter	70.00
Rotary lawn mower, 500mm; petrol	45.00
Electric lawn raker	28.76
Lawn scarifier; petrol	107.03
Lawn aerator; petrol	183.35
Hedge trimmer; electric	28.13
Light duty tiller	72.00
Power digger rotavator	68.00
Post hole borer; manual	22.50
Post hole borer; one-person, petrol	63.84
Stump grinder and chipper; petrol	140.00
Leaf sucker/blower; petrol	93.07
Garden shredder; 6.5hp	101.00

Pricing of scaffolding

As a guide for pricing scaffolding and access equipment, the following rates may be considered likely costs involved for hired equipment, erected and dismantled by a specialist company, and for weekly hire of both putlog and

Introduction

independent scaffolds. Erection, dismantling and initial 4-week period of hire for scaffolding up to 8.00m above ground level:

Putlog scaffold	£12.86/m²
Independent scaffold	£13.55/m²
Chimney access	£296.63

Rental per additional week:

Putlog scaffold	£1.84/m²
Independent scaffold	£1.99/m²
Chimney access	£31.50

Removal of existing

For removal of items, the cost of hiring a skip, delivery to site, wheeling and loading debris into, and the removal when full, disposing of the contents and paying any charges, has been priced separately within the schedule. One instance of this can be found at item 1.01.07.04.

Landfill tax

Landfill tax applies to all waste disposed of by way of landfill at a licensed landfill site on or after 1 October 1996, unless the waste is specifically exempt. The tax, with figures for England and Northern Ireland for 2020/21 and 2021/22 shown in the table, is charged by weight and there are two rates. Inert or inactive waste is subject to the lower rate. Full details and exemptions can be found at www.gov.uk

	Rate from 1 April 2020	Rate from 1 April 2021
Standard rate	£94.15 / tonne	£96.70 / tonne
Lower rate	£3.00 / tonne	£3.10 / tonne

Landfill tax is a devolved issue in Scotland and Wales, though rates for 2020/21 were set at the same as those shown in the table. Rates from April 2021 will be published at www.revenue.scot/scottish-landfill-tax and www.gov.wales/landfill-disposals-tax-rates respectively.

Material costs

Material prices used in this publication have been obtained from manufacturers and suppliers as of third quarter 2020.

All prices exclude VAT and include delivery to site where appropriate. Many small and low value material resources do not include provision for delivery where they would be readily available from local merchants for collection. Allowance has also been made for an average trade discount.

Considerable variation can occur in material prices arising from load size, geographical area and the trading status of the builder. Imported materials are particularly vulnerable due to currency volatility.

Further, prices can be affected by the availability of some materials, for example those in the landscaping schedule, dependent on the season in which the work is taking place.

It is vitally important, therefore, that before entering into contractual situations, confirmation is obtained from suppliers and merchants of material prices relevant to the project.

Reuse of materials

Where materials can be reused, for example in formwork or temporary shoring, the item description includes any assumptions that have been made. For example, if an item says a material is used four times, the material price has been divided by four. In items where formwork is priced by first and subsequent uses, the price of the material is the same per use, but an allowance is made in subsequent uses in case of damage or repurposing. Similarly, the provision for labour is less for subsequent uses to account for less making from scratch being necessary.

The material price can easily be adjusted to suit individual circumstances and whether there is a preference for covering the cost of materials entirely at first use. An example of this would be item 1.01.13.01.

Cost of water

When water has to be priced in bills of quantities, the rate is typically about 0.24% of the contract figure, but district charges vary, and the rate can also vary with type of work.

General guidance: repairs and alterations section

Important notes concerning establishment charges etc.

Most repair and alterations work is estimated on a specification only, or plan and specification without bills of quantities. The schedules in this price book adhere in some cases to measurements and trade headings that are established and familiar but, where suitable, standard measurement rules have been adopted.

Cost depends very much on the size of job and distance from workshop.

The repair sections of this publication are designed for the pricing of repair, adaptation and conversion work, and prices are based on the average conditions under which such work may be carried out.

Adjustments to prices should be made for exceptional working conditions of any work, for a very small or large job, simplicity or complexity of work, accessibility and weather conditions, especially in respect of external work.

The circumstances of a job will also present an obvious need for adjustments. For example, if a tradesperson has to travel specifically to a job, such as re-hanging a door or repairing a burst pipe, it will cost more by comparison than if the tradesperson was already on the site for other work.

Percentages where shown in the work section should be added for pricing work in occupied premises. This is for working in furnished rooms where the necessary extra care and protection affects the speed of execution; but no allowance has been made for the removal and replacement of furniture, carpets, etc.

A basis for the estimate without quantities

It is not usual practice for tenders for small jobs or alteration work to be priced on a bill of quantities. The absence of measured details and contract rates may not be of great importance at the estimating stage (except perhaps on the occasion of making an application for a renovation grant).

However, it can become complicated at the final account stage, especially when there is a long list of variations to be agreed.

Contractors using these schedules as a basis for their estimating could save time and expense when agreeing an estimate, or settling the final account, if the following simple procedure is adopted when submitting the estimate. The following or similar wording could be printed or written, as a footnote to the estimate:

This estimate is based on 'Hutchins Priced Schedules 2021' and rates and prices used are those contained in that edition, and the following percentage adjustments shall apply to this job only.

Introduction

General building work   Add or Deduct* [X]%

Specialist sub-contractors listed below
Trade   Add or Deduct* [X]%
Trade   Add or Deduct* [X]%
Trade   Add or Deduct* [X]%

Delete whichever is not applicable.

These percentage adjustments are subject to the order to commence work being given within [X] months from the date of this estimate.

If the order to commence work is given after the expiry of the stated period, the Contractor reserves the right to revise the percentage adjustments.

The adjustments do not prevent the Contractor from making any subsequent claim for increased cost arising from either national wage agreements, or agreed increases in the cost of materials occurring after the date of the estimate.

Advisory notes

Note 1: The number of months is left to individual contractors' discretion; six months is a reasonable interval, three or four months perhaps on small jobs.

Note 2: The percentage adjustment decisions are the absolute prerogative of the contractor. This will depend on extra over payments, extra travel time, the cost of materials, profit margin and a number of other factors.

Note 3: It is essential to enter the year of the edition of the price book. The fact that a new edition might be published during the course of a contract, could cause misunderstanding if the edition year was not specified.

Note 4: Percentage adjustments stated are binding only to the contract to which they apply and are not obligatory on any other contracts.

Note 5: Percentage adjustments on specialist sub-contractors' work rates included in the various sections are a general guide; quotations received from specialist firms may be substantially different from those quoted. It is, therefore, advisable to consider each specialist trade separately.

Note 6: Contractors must never lose sight of the fact that at some time they may, through no fault of their own, find that a contract is the subject of dispute in a court of law.

If the dispute revolved around the final account and/or the cost of building work, it is quite probable that both the judge and barristers will use the rates in the sections as a 'guiding light'.

If the contractor used certain percentage additions in the estimate but didn't state the additions in writing they stand to lose what may have been a justifiable claim for extra cost.

If, on the other hand, the percentage adjustments were written on the estimate, then such adjustments would have to be taken into consideration by all concerned. In this event the contractor would not have to be subjected to a long and difficult cross-examination on estimating.

This, in turn, would save litigation time and attendant costs, and could ensure a more generous settlement to the contractor than might otherwise have been possible.

Daywork – advisory notes for use when there is no standard form of contract

From time to time, building contractors may carry out either part or the whole of the contract on a daywork basis. This may be due to extreme urgency (e.g. repair after fire, flood or storm damage), complexity of the work or the impatience of the owner, when there is not enough time to prepare an estimate.

These notes are primarily intended for jobs in the £500 to £25,000 range for private owners; too small to justify the services (and fees) of an architect and/or quantity surveyor, and any suggestion of a standard form of contract would be superfluous.

It is, however, essential that the contractors ensure that there will be no misunderstanding about payment when the final account stage is reached.

A letter to the owner along the following lines should suffice.

I/we confirm your written/verbal instructions to carry out sundry repairs and reinstatement work. The cost of the work shall be based on an inclusive hourly rate of £[X] per hour for each operative engaged on the work. The cost of all materials shall be current local costs.

To the sum total of the foregoing wages and material costs shall be added [X]% to cover overheads and profit. Unless we receive your instructions to the contrary it will be assumed that our terms and conditions are acceptable.

If the total value of the work is expected to exceed £1,000 and/or more than one month's duration, add the following paragraph.

On the last day of each month a detailed account will be forwarded to you and payment shall be made within 15 days. In the event of any inadmissibility or otherwise of any charges the whole matter shall be referred to [X], quantity surveyor, whose decision shall be final and binding.

The nomination of a quantity surveyor does not mean that the contractor

has to pay the cost of a retaining fee for a quantity surveyor's services that may never be called on.

It does mean that, if necessary, there is a quantity surveyor who would be prepared to examine all relevant documents for the contractor for an appropriate fee.

The hourly rate for operatives may be based on the cost to employ, that is standard wage rates, plus extra over rate, plus bonus, plus all employers' levies and contributions.

The percentage to be charged is a matter for individual firms; taking into consideration the overall cost of wages and materials is recoverable in full. Something between 120% and 150% would typically suffice.

Wage rates and percentage additions chargeable are the subject of a decision by the contractor and must be stated at the outset of the contract. The recommendations regarding hourly wage rates and percentage additions are not mandatory and may vary from contract to contract.

Preliminaries and general conditions

Preliminaries

The prices given in this publication do not include any provision for preliminaries, but every tender submission needs to include additional costs to cover these.

Unlike measured rates that are directly affected by quantity changes, certain costs in a project are not as affected and require a different approach to recovery. Examples might be scaffolding hire or the provision of a toilet, neither of which are affected by small changes in work quantity. These costs are recovered by adding a sum in the Preliminaries section of tender documents.

The following are examples of the range of items that may be included in the Preliminaries section. The list is not exhaustive but will identify those common items that, if not allowed for in the tender elsewhere, should be included under the Preliminaries section.

Conditions of contract

Careful notice should be taken of the contract conditions in order to establish onerous obligations, restrictions or liabilities that may be imposed by the employer. Particular care should be given to non-standard clauses. Many client bodies have developed contract forms for their own use. If presented with such contract forms the contractor should take legal advice on the significance of the conditions under which the works will be let.

Items that should be evaluated with all contract forms can be considered under the following headings:

1) Access to and possession or use of the site.
2) Limitations of working space.
3) Limitations of working hours.
4) Maintenance of existing services on, under or over the site.
5) Carrying out work in a specific order.
6) The requirement to provide bonds.
7) The requirement to provide insurance cover.

Building contractors' administrative arrangements

An assessment of the builders' site management requirements will normally include allowances for the following:

1) Site supervision and administration.
2) Safety, health and welfare of work people.
3) Transport of work people.

Plant hire

A bar chart programme may be prepared in order to establish the quantity and type of plant required on site, as well as the hire periods involved. Items of plant required for general usage and not already included in the measured rates can be included under the following headings, together with an allowance for transport and maintenance costs.

1) Small plant and tools.
2) Scaffolding.
3) Cranes and lifting plant.
4) Site transport.
5) Plant required for specific trades.

Employers' facilities

The contract may require the contractor to provide facilities, including:

1) Temporary accommodation (e.g. offices).

2)   Telephones.
3)   Signboards.

Contractors' facilities

The contractor should establish the extent of facilities required for their own operations, and the cost of providing them, against the following headings:

1)   Office, compounds, messrooms.
2)   Hoardings and guard rails.
3)   Temporary roads, hard-standings and crossings.
4)   Water for the works; temporary plumbing and distribution.
5)   Lighting and power for the works; temporary installations.
6)   Temporary telephones and the cost of calls.

Temporary works

Temporary works can best be considered under the following headings:

1)   Traffic diversion.
2)   Access roads.
3)   Pumping and dewatering.

Sundry items

Items that do not fall easily under the above headings can be included under this heading, for example:

1)   Testing of materials.
2)   Testing of the works.
3)   Protecting the works from inclement weather.
4)   Removing rubbish and cleaning.
5)   Maintenance of roads and services.

6)   Drying the works.
7)   Control of noise and pollution.
8)   Statutory requirements.

Preliminaries example

Every building contractor will treat the requirement for preliminaries differently, depending upon a number of factors. In addition, each project will have its own special requirements in respect of preliminaries and should be carefully assessed. To assist the reader in calculating preliminaries, a sample calculation is shown below.

Preliminary costs for a small building project of 12 weeks' duration, for example the construction of a domestic extension, approximate value £75,000, might be priced as follows:

Supervision and administration	£1,870.00
Site accommodation	£400.00
Light and power	£105.00
Insurance	£600.00
Water for the works	£65.00
Scaffolding	£3,770.00
Small plant	£310.00
Transport and travelling costs	£380.00
Total preliminaries	**£7,500.00**

This represents 10% of the tender sum of £75,000.

Preliminaries and general conditions

Project overheads and profit

Every contractor needs to recover sufficient income from their activities to fund overheads. In addition, the builder will look to receive a fair return (profit) on the capital invested in the project. Therefore, in addition to the measured rates net cost and the preliminaries mentioned, a sum of money will need to be added to recover profit.

Recovery of overheads and profit can be achieved in a number of ways. Some contractors prefer to add a lump sum to the tender; others prefer to spread the value across the range of measured work items using a percentage adjustment to the rates.

The method adopted in these schedules is to add overheads and profit equally on each rate by the addition of a percentage amount to the rate shown. Net rates include an addition of 20% on labour for site on-costs and establishment charges. Gross rates include a further 10% addition on labour, plant and materials net rates for profit.

Location factors

The prices in this publication are intended as an average for the UK, but of course the location of a project can hugely affect the costs involved. This can be because of a whole range of variables, including market conditions like the supply and demand of labour and materials, levels of competition between contractors tendering for work, accessibility and environmental conditions.

Even within a small geographical area, such as a single town, the same project completed twice could be priced differently dependent on the individual circumstances – how easy it is to access the site, the time of year the project is being completed, the availability of labour at the time, the level of profit required, etc.

With this in mind, estimates should always be put together with all of the relevant factors considered.

However, location factors, here presented regionally, can be used to help make estimates more appropriate, and take into consideration many of the variables mentioned. Because a location factor is not just based on an individual element such as materials or labour, when applying a factor, it is important to apply it to the whole estimate or tender, and not an individual unit rate.

As an example, the average for the UK is 1.00 while the North East is 0.91. This means the overall pricing used in that area should be 91% of that given in the schedules.

Hutchins indicative location factors

Region	Location factor
UK average	*1.00*
North East	0.91
North West	0.95
Yorkshire and the Humber	0.93
East Midlands	1.00
West Midlands	0.94
East of England	1.01
London	1.16
South East	1.06
South West	0.99
Wales	0.93
Scotland	0.93
Northern Ireland	0.69

Hutchins Priced Schedules 2021	Unit	Labour Hours	Labour Net £	Plant Net £	Materials Net £	Unit Rate Net £	Labour Gross £	Plant Gross £	Materials Gross £	Unit Rate Gross £
							(Gross rates include 10% profit)			

1.01.   NEW WORK

1.01.01.   SITE PREPARATION

1.01.01.01.   **Form temporary site road; 150mm hardcore; maintain during period of contract**

Hutchins Priced Schedules 2021	Unit	Labour Hours	Labour Net £	Plant Net £	Materials Net £	Unit Rate Net £	Labour Gross £	Plant Gross £	Materials Gross £	Unit Rate Gross £
1.01.01.01.A 3.00m wide................	m	2.70	79.87	—	24.27	104.13	87.85	—	26.69	114.55

1.01.01.02.   **Break up and remove temporary site road; 150mm hardcore**

1.01.01.02.A 3.00m wide................	m	2.25	66.56	—	—	66.56	73.21	—	—	73.21

1.01.01.03.   **Temporarily enclose site; fencing up to 20 times used**

1.01.01.03.A 1.35m chestnut fencing...........	m	0.25	7.40	—	0.05	7.44	8.13	—	0.05	8.19
1.01.01.03.B 1.80m chain link fencing..........	m	0.45	30.24	—	1.23	31.47	33.27	—	1.35	34.62

1.01.01.04.   **Clear site of bushes, scrub and undergrowth; cut down small trees and grub up roots; burn or deposit in skip**

1.01.01.04.A average 1.50m high............	m²	0.30	8.87	—	—	8.87	9.76	—	—	9.76

1.01.01.05.   **Cut down hedging and grub up roots; burn or deposit in skip; hedge height up to**

1.01.01.05.A 600mm....................	m	1.97	58.27	—	—	58.27	64.10	—	—	64.10
1.01.01.05.B 900mm....................	m	2.63	77.80	—	—	77.80	85.57	—	—	85.57
1.01.01.05.C 1200mm...................	m	3.12	92.29	—	—	92.29	101.52	—	—	101.52
1.01.01.05.D 1500mm...................	m	4.27	126.31	—	—	126.31	138.94	—	—	138.94
1.01.01.05.E 1800mm...................	m	5.74	169.79	—	—	169.79	186.77	—	—	186.77

1.01.01.06.   **Cut down trees, lop off branches and grub up roots; burn or deposit in skip; fill hole with excavated material; size up to**

1.01.01.06.A 450mm girth 140mm dia...........	nr	16.00	473.28	—	—	473.28	520.61	—	—	520.61
1.01.01.06.B 900mm girth 290mm dia...........	nr	28.00	828.24	—	—	828.24	911.06	—	—	911.06
1.01.01.06.C 1350mm girth 430mm dia..........	nr	42.00	1242.36	—	—	1242.36	1366.60	—	—	1366.60
1.01.01.06.D 1800mm girth 570mm dia..........	nr	56.00	1656.48	—	—	1656.48	1822.13	—	—	1822.13
1.01.01.06.E 2250mm girth 720mm dia..........	nr	69.00	2041.02	—	—	2041.02	2245.12	—	—	2245.12
1.01.01.06.F 2700mm girth 860mm dia..........	nr	81.00	2395.98	—	—	2395.98	2635.58	—	—	2635.58
1.01.01.06.G 3150mm girth 1000mm dia.........	nr	92.00	2721.36	—	—	2721.36	2993.50	—	—	2993.50
1.01.01.06.H 3600mm girth 1150mm dia.........	nr	102.00	3017.16	—	—	3017.16	3318.88	—	—	3318.88

Excavation, Earthwork and Concrete Work

Hutchins Priced Schedules 2021		Unit	Labour Hours	Labour Net £	Plant Net £	Materials Net £	Unit Rate Net £	Labour Gross £	Plant Gross £	Materials Gross £	Unit Rate Gross £
									(Gross rates include 10% profit)		
1.01.	**NEW WORK**										
1.01.01.	**SITE PREPARATION**										
1.01.01.07.	**Excavate top soil to be preserved; by hand; average depth**										
1.01.01.07.A	150mm	m²	0.64	18.93	—	—	18.93	20.82	—	—	20.82
1.01.01.07.B	225mm	m²	0.93	27.51	—	—	27.51	30.26	—	—	30.26
1.01.01.07.C	300mm	m²	1.26	37.27	—	—	37.27	41.00	—	—	41.00
1.01.01.08.	**Excavate top soil to be preserved; by machine; average depth**										
1.01.01.08.A	150mm	m²	—	—	0.22	—	0.22	—	0.25	—	0.25
1.01.01.08.B	225mm	m²	—	—	0.35	—	0.35	—	0.39	—	0.39
1.01.01.08.C	300mm	m²	—	—	0.45	—	0.45	—	0.49	—	0.49
1.01.02.	**EXCAVATION BY HAND**										
1.01.02.01.	**Excavate to reduce levels; maximum depth not exceeding 0.25m**										
1.01.02.01.A	loose soil	m³	2.63	77.80	—	—	77.80	85.57	—	—	85.57
1.01.02.01.B	firm soil; sand	m³	3.15	93.18	—	—	93.18	102.49	—	—	102.49
1.01.02.01.C	light clay; compact soil; gravel	m³	3.94	116.55	—	—	116.55	128.20	—	—	128.20
1.01.02.01.D	stiff heavy clay	m³	5.25	155.30	—	—	155.30	170.82	—	—	170.82
1.01.02.01.E	soft chalk	m³	7.88	233.09	—	—	233.09	256.40	—	—	256.40
1.01.02.02.	**Excavate to reduce levels; maximum depth not exceeding 1.00m**										
1.01.02.02.A	loose soil	m³	2.71	80.16	—	—	80.16	88.18	—	—	88.18
1.01.02.02.B	firm soil; sand	m³	3.25	96.14	—	—	96.14	105.75	—	—	105.75
1.01.02.02.C	light clay; compact soil; gravel	m³	4.06	120.09	—	—	120.09	132.10	—	—	132.10
1.01.02.02.D	stiff heavy clay	m³	5.42	160.32	—	—	160.32	176.36	—	—	176.36
1.01.02.02.E	soft chalk	m³	8.13	240.49	—	—	240.49	264.53	—	—	264.53
1.01.02.03.	**Excavate to reduce levels; maximum depth not exceeding 2.00m**										
1.01.02.03.A	loose soil	m³	2.88	85.19	—	—	85.19	93.71	—	—	93.71
1.01.02.03.B	firm soil; sand	m³	3.46	102.35	—	—	102.35	112.58	—	—	112.58
1.01.02.03.C	light clay; compact soil; gravel	m³	4.33	128.08	—	—	128.08	140.89	—	—	140.89
1.01.02.03.D	stiff heavy clay	m³	5.77	170.68	—	—	170.68	187.74	—	—	187.74
1.01.02.03.E	soft chalk	m³	8.65	255.87	—	—	255.87	281.45	—	—	281.45
1.01.02.05.	**Excavate for basement; maximum depth not exceeding 0.25m**										
1.01.02.05.A	loose soil	m³	2.63	77.80	—	—	77.80	85.57	—	—	85.57
1.01.02.05.B	firm soil; sand	m³	3.15	93.18	—	—	93.18	102.49	—	—	102.49
1.01.02.05.C	light clay; compact soil; gravel	m³	3.94	116.55	—	—	116.55	128.20	—	—	128.20

Excavation, Earthwork and Concrete Work

Hutchins Priced Schedules 2021		Unit	Labour Hours	Labour Net £	Plant Net £	Materials Net £	Unit Rate Net £	Labour Gross £	Plant Gross £	Materials Gross £	Unit Rate Gross £
								(Gross rates include 10% profit)			
1.01.	**NEW WORK**										
1.01.02.	**EXCAVATION BY HAND**										
1.01.02.05.	**Excavate for basement; maximum depth not exceeding 0.25m**										
1.01.02.05.D	stiff heavy clay	m³	5.25	155.30	—	—	155.30	170.82	—	—	170.82
1.01.02.05.E	soft chalk	m³	7.88	233.09	—	—	233.09	256.40	—	—	256.40
1.01.02.06.	**Excavate for basement; maximum depth not exceeding 1.00m**										
1.01.02.06.A	loose soil	m³	2.71	80.16	—	—	80.16	88.18	—	—	88.18
1.01.02.06.B	firm soil; sand	m³	3.25	96.14	—	—	96.14	105.75	—	—	105.75
1.01.02.06.C	light clay; compact soil; gravel	m³	4.06	120.09	—	—	120.09	132.10	—	—	132.10
1.01.02.06.D	stiff heavy clay	m³	5.42	160.32	—	—	160.32	176.36	—	—	176.36
1.01.02.06.E	soft chalk	m³	8.13	240.49	—	—	240.49	264.53	—	—	264.53
1.01.02.07.	**Excavate for basement; maximum depth not exceeding 2.00m**										
1.01.02.07.A	loose soil	m³	3.37	99.68	—	—	99.68	109.65	—	—	109.65
1.01.02.07.B	firm soil; sand	m³	4.04	119.50	—	—	119.50	131.45	—	—	131.45
1.01.02.07.C	light clay; compact soil; gravel	m³	5.05	149.38	—	—	149.38	164.32	—	—	164.32
1.01.02.07.D	stiff heavy clay	m³	6.73	199.07	—	—	199.07	218.98	—	—	218.98
1.01.02.07.E	soft chalk	m³	10.10	298.76	—	—	298.76	328.63	—	—	328.63
1.01.02.09.	**Excavate pit to receive bases of stanchions, isolated piers, etc.; maximum depth not exceeding 0.25m**										
1.01.02.09.A	loose soil	m³	2.71	80.16	—	—	80.16	88.18	—	—	88.18
1.01.02.09.B	firm soil; sand	m³	3.25	96.14	—	—	96.14	105.75	—	—	105.75
1.01.02.09.C	light clay; compact soil; gravel	m³	4.06	120.09	—	—	120.09	132.10	—	—	132.10
1.01.02.09.D	stiff heavy clay	m³	5.42	160.32	—	—	160.32	176.36	—	—	176.36
1.01.02.09.E	soft chalk	m³	8.13	240.49	—	—	240.49	264.53	—	—	264.53
1.01.02.10.	**Excavate pit to receive bases of stanchions, isolated piers, etc.; maximum depth not exceeding 1.00m**										
1.01.02.10.A	loose soil	m³	2.97	87.85	—	—	87.85	96.64	—	—	96.64
1.01.02.10.B	firm soil; sand	m³	3.56	105.30	—	—	105.30	115.84	—	—	115.84
1.01.02.10.C	light clay; compact soil; gravel	m³	4.45	131.63	—	—	131.63	144.79	—	—	144.79
1.01.02.10.D	stiff heavy clay	m³	5.93	175.41	—	—	175.41	192.95	—	—	192.95
1.01.02.10.E	soft chalk	m³	8.90	263.26	—	—	263.26	289.59	—	—	289.59
1.01.02.11.	**Excavate pit to receive bases of stanchions, isolated piers, etc.; maximum depth not exceeding 2.00m**										
1.01.02.11.A	loose soil	m³	3.72	110.04	—	—	110.04	121.04	—	—	121.04
1.01.02.11.B	firm soil; sand	m³	4.46	131.93	—	—	131.93	145.12	—	—	145.12
1.01.02.11.C	light clay; compact soil; gravel	m³	5.58	165.06	—	—	165.06	181.56	—	—	181.56

Excavation, Earthwork and Concrete Work

Hutchins Priced Schedules 2021	Unit	Labour Hours	Labour Net £	Plant Net £	Materials Net £	Unit Rate Net £	Labour Gross £	Plant Gross £	Materials Gross £	Unit Rate Gross £
							(Gross rates include 10% profit)			
1.01.	**NEW WORK**									
1.01.02.	**EXCAVATION BY HAND**									
1.01.02.11.	Excavate pit to receive bases of stanchions, isolated piers, etc.; maximum depth not exceeding 2.00m									
1.01.02.11.D	stiff heavy clay	m³	7.43	219.78	—	—	219.78	241.76	—	—	241.76
1.01.02.11.E	soft chalk	m³	11.15	329.82	—	—	329.82	362.80	—	—	362.80
1.01.02.13.	**Excavate trenches to receive foundations; exceeding 0.30m in width; maximum depth not exceeding 0.25m**									
1.01.02.13.A	loose soil	m³	2.71	80.16	—	—	80.16	88.18	—	—	88.18
1.01.02.13.B	firm soil; sand	m³	3.25	96.14	—	—	96.14	105.75	—	—	105.75
1.01.02.13.C	light clay; compact soil; gravel	m³	4.06	120.09	—	—	120.09	132.10	—	—	132.10
1.01.02.13.D	stiff heavy clay	m³	5.42	160.32	—	—	160.32	176.36	—	—	176.36
1.01.02.13.E	soft chalk	m³	8.13	240.49	—	—	240.49	264.53	—	—	264.53
1.01.02.14.	**Excavate trenches to receive foundations; exceeding 0.30m in width; maximum depth not exceeding 1.00m**									
1.01.02.14.A	loose soil	m³	2.97	87.85	—	—	87.85	96.64	—	—	96.64
1.01.02.14.B	firm soil; sand	m³	3.56	105.30	—	—	105.30	115.84	—	—	115.84
1.01.02.14.C	light clay; compact soil; gravel	m³	4.45	131.63	—	—	131.63	144.79	—	—	144.79
1.01.02.14.D	stiff heavy clay	m³	5.93	175.41	—	—	175.41	192.95	—	—	192.95
1.01.02.14.E	soft chalk	m³	8.90	263.26	—	—	263.26	289.59	—	—	289.59
1.01.02.15.	**Excavate trenches to receive foundations; exceeding 0.30m in width; maximum depth not exceeding 2.00m**									
1.01.02.15.A	loose soil	m³	3.72	110.04	—	—	110.04	121.04	—	—	121.04
1.01.02.15.B	firm soil; sand	m³	4.46	131.93	—	—	131.93	145.12	—	—	145.12
1.01.02.15.C	light clay; compact soil; gravel	m³	5.58	165.06	—	—	165.06	181.56	—	—	181.56
1.01.02.15.D	stiff heavy clay	m³	7.43	219.78	—	—	219.78	241.76	—	—	241.76
1.01.02.15.E	soft chalk	m³	11.15	329.82	—	—	329.82	362.80	—	—	362.80
1.01.02.17.	**Excavate trenches to receive foundations; not exceeding 0.30m in width; maximum depth not exceeding 0.25m**									
1.01.02.17.A	loose soil	m³	0.28	8.28	—	—	8.28	9.11	—	—	9.11
1.01.02.17.B	firm soil; sand	m³	0.34	10.06	—	—	10.06	11.06	—	—	11.06
1.01.02.17.C	light clay; compact soil; gravel	m³	0.43	12.72	—	—	12.72	13.99	—	—	13.99
1.01.02.17.D	stiff heavy clay	m³	0.57	16.86	—	—	16.86	18.55	—	—	18.55
1.01.02.17.E	soft chalk	m³	0.85	25.14	—	—	25.14	27.66	—	—	27.66

Hutchins Priced Schedules 2021	Unit	Labour Hours	Labour Net £	Plant Net £	Materials Net £	Unit Rate Net £	Labour Gross £	Plant Gross £	Materials Gross £	Unit Rate Gross £
							(Gross rates include 10% profit)			
1.01. NEW WORK										
1.01.02. EXCAVATION BY HAND										
1.01.02.18. Excavate trenches to receive foundations; not exceeding 0.30m in width; maximum depth not exceeding 0.50m										
1.01.02.18.A loose soil	m³	0.54	15.97	—	—	15.97	17.57	—	—	17.57
1.01.02.18.B firm soil; sand	m³	0.65	19.23	—	—	19.23	21.15	—	—	21.15
1.01.02.18.C light clay; compact soil; gravel	m³	0.81	23.96	—	—	23.96	26.36	—	—	26.36
1.01.02.18.D stiff heavy clay	m³	1.08	31.95	—	—	31.95	35.14	—	—	35.14
1.01.02.18.E soft chalk	m³	1.63	48.22	—	—	48.22	53.04	—	—	53.04
1.01.02.19. Excavate trenches to receive foundations; not exceeding 0.30m in width; maximum depth not exceeding 0.75m										
1.01.02.19.A loose soil	m³	0.78	23.07	—	—	23.07	25.38	—	—	25.38
1.01.02.19.B firm soil; sand	m³	0.93	27.51	—	—	27.51	30.26	—	—	30.26
1.01.02.19.C light clay; compact soil; gravel	m³	1.16	34.31	—	—	34.31	37.74	—	—	37.74
1.01.02.19.D stiff heavy clay	m³	1.55	45.85	—	—	45.85	50.43	—	—	50.43
1.01.02.19.E soft chalk	m³	2.33	68.92	—	—	68.92	75.81	—	—	75.81
1.01.02.20. Excavate trenches to receive foundations; not exceeding 0.30m in width; maximum depth not exceeding 1.00m										
1.01.02.20.A loose soil	m³	1.00	29.58	—	—	29.58	32.54	—	—	32.54
1.01.02.20.B firm soil; sand	m³	1.20	35.50	—	—	35.50	39.05	—	—	39.05
1.01.02.20.C light clay; compact soil; gravel	m³	1.50	44.37	—	—	44.37	48.81	—	—	48.81
1.01.02.20.D stiff heavy clay	m³	2.00	59.16	—	—	59.16	65.08	—	—	65.08
1.01.02.20.E soft chalk	m³	3.00	88.74	—	—	88.74	97.61	—	—	97.61
1.01.03. EXCAVATION BY MACHINE										
1.01.03.01. Excavate to reduce levels; maximum depth not exceeding 0.25m										
1.01.03.01.A loose soil	m³	—	—	5.42	—	5.42	—	5.96	—	5.96
1.01.03.01.B firm soil; sand	m³	—	—	6.73	—	6.73	—	7.40	—	7.40
1.01.03.01.C light clay; compact soil; gravel	m³	—	—	8.30	—	8.30	—	9.13	—	9.13
1.01.03.01.D stiff heavy clay	m³	—	—	11.19	—	11.19	—	12.30	—	12.30
1.01.03.01.E soft chalk	m³	—	—	16.92	—	16.92	—	18.61	—	18.61
1.01.03.02. Excavate to reduce levels; maximum depth not exceeding 1.00m										
1.01.03.02.A loose soil	m³	—	—	4.45	—	4.45	—	4.90	—	4.90
1.01.03.02.B firm soil; sand	m³	—	—	5.42	—	5.42	—	5.96	—	5.96
1.01.03.02.C light clay; compact soil; gravel	m³	—	—	6.73	—	6.73	—	7.40	—	7.40
1.01.03.02.D stiff heavy clay	m³	—	—	8.94	—	8.94	—	9.84	—	9.84
1.01.03.02.E soft chalk	m³	—	—	13.40	—	13.40	—	14.74	—	14.74

Excavation, Earthwork and Concrete Work

Hutchins Priced Schedules 2021	Unit	Labour Hours	Labour Net £	Plant Net £	Materials Net £	Unit Rate Net £	Labour Gross £	Plant Gross £	Materials Gross £	Unit Rate Gross £
						(Gross rates include 10% profit)				
1.01.	**NEW WORK**									
1.01.03.	**EXCAVATION BY MACHINE**									
1.01.03.03.	**Excavate to reduce levels; maximum depth not exceeding 2.00m**									
1.01.03.03.A	loose soil	m³	—	—	5.42	—	5.42	—	5.96	—	5.96
1.01.03.03.B	firm soil; sand	m³	—	—	6.73	—	6.73	—	7.40	—	7.40
1.01.03.03.C	light clay; compact soil; gravel	m³	—	—	8.30	—	8.30	—	9.13	—	9.13
1.01.03.03.D	stiff heavy clay	m³	—	—	11.19	—	11.19	—	12.30	—	12.30
1.01.03.03.E	soft chalk	m³	—	—	16.92	—	16.92	—	18.61	—	18.61
1.01.03.05.	**Excavate for basement; maximum depth not exceeding 0.25m**									
1.01.03.05.A	loose soil	m³	—	—	6.38	—	6.38	—	7.02	—	7.02
1.01.03.05.B	firm soil; sand	m³	—	—	7.66	—	7.66	—	8.43	—	8.43
1.01.03.05.C	light clay; compact soil; gravel	m³	—	—	9.58	—	9.58	—	10.54	—	10.54
1.01.03.05.D	stiff heavy clay	m³	—	—	12.79	—	12.79	—	14.07	—	14.07
1.01.03.05.E	soft chalk	m³	—	—	19.49	—	19.49	—	21.44	—	21.44
1.01.03.06.	**Excavate for basement; maximum depth not exceeding 1.00m**									
1.01.03.06.A	loose soil	m³	—	—	6.06	—	6.06	—	6.66	—	6.66
1.01.03.06.B	firm soil; sand	m³	—	—	7.34	—	7.34	—	8.07	—	8.07
1.01.03.06.C	light clay; compact soil; gravel	m³	—	—	8.94	—	8.94	—	9.84	—	9.84
1.01.03.06.D	stiff heavy clay	m³	—	—	11.79	—	11.79	—	12.97	—	12.97
1.01.03.06.E	soft chalk	m³	—	—	17.88	—	17.88	—	19.67	—	19.67
1.01.03.07.	**Excavate for basement; maximum depth not exceeding 2.00m**									
1.01.03.07.A	loose soil	m³	—	—	6.38	—	6.38	—	7.02	—	7.02
1.01.03.07.B	firm soil; sand	m³	—	—	7.66	—	7.66	—	8.43	—	8.43
1.01.03.07.C	light clay; compact soil; gravel	m³	—	—	9.58	—	9.58	—	10.54	—	10.54
1.01.03.07.D	stiff heavy clay	m³	—	—	12.79	—	12.79	—	14.07	—	14.07
1.01.03.07.E	soft chalk	m³	—	—	19.49	—	19.49	—	21.44	—	21.44
1.01.03.09.	**Excavate pit to receive bases of stanchions, isolated piers, etc.; maximum depth not exceeding 0.25m**									
1.01.03.09.A	loose soil	m³	—	—	8.94	—	8.94	—	9.84	—	9.84
1.01.03.09.B	firm soil; sand	m³	—	—	10.83	—	10.83	—	11.92	—	11.92
1.01.03.09.C	light clay; compact soil; gravel	m³	—	—	13.40	—	13.40	—	14.74	—	14.74
1.01.03.09.D	stiff heavy clay	m³	—	—	17.88	—	17.88	—	19.67	—	19.67
1.01.03.09.E	soft chalk	m³	—	—	27.15	—	27.15	—	29.86	—	29.86

Hutchins Priced Schedules 2021	Unit	Labour Hours	Labour Net £	Plant Net £	Materials Net £	Unit Rate Net £	Labour Gross £	Plant Gross £	Materials Gross £	Unit Rate Gross £
						(Gross rates include 10% profit)				

1.01.   NEW WORK

1.01.03.   EXCAVATION BY MACHINE

1.01.03.10.   Excavate pit to receive bases of stanchions, isolated piers, etc.; maximum depth not exceeding 1.00m

1.01.03.10.A	loose soil	m³	—	—	7.02	—	7.02	—	7.72	—	7.72
1.01.03.10.B	firm soil; sand	m³	—	—	8.62	—	8.62	—	9.48	—	9.48
1.01.03.10.C	light clay; compact soil; gravel	m³	—	—	10.83	—	10.83	—	11.92	—	11.92
1.01.03.10.D	stiff heavy clay	m³	—	—	14.36	—	14.36	—	15.79	—	15.79
1.01.03.10.E	soft chalk	m³	—	—	21.73	—	21.73	—	23.90	—	23.90

1.01.03.11.   Excavate pit to receive bases of stanchions, isolated piers, etc.; maximum depth not exceeding 2.00m

1.01.03.11.A	loose soil	m³	—	—	8.94	—	8.94	—	9.84	—	9.84
1.01.03.11.B	firm soil; sand	m³	—	—	10.83	—	10.83	—	11.92	—	11.92
1.01.03.11.C	light clay; compact soil; gravel	m³	—	—	13.40	—	13.40	—	14.74	—	14.74
1.01.03.11.D	stiff heavy clay	m³	—	—	17.88	—	17.88	—	19.67	—	19.67
1.01.03.11.E	soft chalk	m³	—	—	27.15	—	27.15	—	29.86	—	29.86

1.01.03.13.   Excavate trenches to receive foundations; exceeding 0.30m in width; maximum depth not exceeding 0.25m

1.01.03.13.A	loose soil	m³	—	—	5.42	—	5.42	—	5.96	—	5.96
1.01.03.13.B	firm soil; sand	m³	—	—	6.73	—	6.73	—	7.40	—	7.40
1.01.03.13.C	light clay; compact soil; gravel	m³	—	—	8.30	—	8.30	—	9.13	—	9.13
1.01.03.13.D	stiff heavy clay	m³	—	—	11.19	—	11.19	—	12.30	—	12.30
1.01.03.13.E	soft chalk	m³	—	—	16.92	—	16.92	—	18.61	—	18.61

1.01.03.14.   Excavate trenches to receive foundations; exceeding 0.30m in width; maximum depth not exceeding 1.00m

1.01.03.14.A	loose soil	m³	—	—	4.45	—	4.45	—	4.90	—	4.90
1.01.03.14.B	firm soil; sand	m³	—	—	5.42	—	5.42	—	5.96	—	5.96
1.01.03.14.C	light clay; compact soil; gravel	m³	—	—	6.73	—	6.73	—	7.40	—	7.40
1.01.03.14.D	stiff heavy clay	m³	—	—	8.94	—	8.94	—	9.84	—	9.84
1.01.03.14.E	soft chalk	m³	—	—	13.40	—	13.40	—	14.74	—	14.74

1.01.03.15.   Excavate trenches to receive foundations; exceeding 0.30m in width; maximum depth not exceeding 2.00m

1.01.03.15.A	loose soil	m³	—	—	5.42	—	5.42	—	5.96	—	5.96
1.01.03.15.B	firm soil; sand	m³	—	—	6.73	—	6.73	—	7.40	—	7.40
1.01.03.15.C	light clay; compact soil; gravel	m³	—	—	8.30	—	8.30	—	9.13	—	9.13
1.01.03.15.D	stiff heavy clay	m³	—	—	17.88	—	17.88	—	19.67	—	19.67

Excavation, Earthwork and Concrete Work

Hutchins Priced Schedules 2021	Unit	Labour Hours	Labour Net £	Plant Net £	Materials Net £	Unit Rate Net £	Labour Gross £	Plant Gross £	Materials Gross £	Unit Rate Gross £
							(Gross rates include 10% profit)			
1.01.	**NEW WORK**									
1.01.03.	**EXCAVATION BY MACHINE**									
1.01.03.15.	Excavate trenches to receive foundations; exceeding 0.30m in width; maximum depth not exceeding 2.00m									
1.01.03.15.E	soft chalk	m³	—	—	27.15	—	27.15	—	29.86	—	29.86
1.01.03.17.	Excavate trenches to receive foundations; not exceeding 0.30m in width; maximum depth not exceeding 0.25m									
1.01.03.17.A	loose soil	m³	—	—	3.81	—	3.81	—	4.20	—	4.20
1.01.03.17.B	firm soil; sand	m³	—	—	4.78	—	4.78	—	5.25	—	5.25
1.01.03.17.C	light clay; compact soil; gravel	m³	—	—	5.77	—	5.77	—	6.35	—	6.35
1.01.03.17.D	stiff heavy clay	m³	—	—	7.66	—	7.66	—	8.43	—	8.43
1.01.03.17.E	soft chalk	m³	—	—	11.51	—	11.51	—	12.66	—	12.66
1.01.03.18.	Excavate trenches to receive foundations; not exceeding 0.30m in width; maximum depth not exceeding 0.50m									
1.01.03.18.A	loose soil	m³	—	—	6.38	—	6.38	—	7.02	—	7.02
1.01.03.18.B	firm soil; sand	m³	—	—	7.66	—	7.66	—	8.43	—	8.43
1.01.03.18.C	light clay; compact soil; gravel	m³	—	—	9.58	—	9.58	—	10.54	—	10.54
1.01.03.18.D	stiff heavy clay	m³	—	—	12.79	—	12.79	—	14.07	—	14.07
1.01.03.18.E	soft chalk	m³	—	—	19.49	—	19.49	—	21.44	—	21.44
1.01.03.19.	Excavate trenches to receive foundations; not exceeding 0.30m in width; maximum depth not exceeding 0.75m									
1.01.03.19.A	loose soil	m³	—	—	8.94	—	8.94	—	9.84	—	9.84
1.01.03.19.B	firm soil; sand	m³	—	—	10.83	—	10.83	—	11.92	—	11.92
1.01.03.19.C	light clay; compact soil; gravel	m³	—	—	13.40	—	13.40	—	14.74	—	14.74
1.01.03.19.D	stiff heavy clay	m³	—	—	17.88	—	17.88	—	19.67	—	19.67
1.01.03.19.E	soft chalk	m³	—	—	27.15	—	27.15	—	29.86	—	29.86
1.01.03.20.	Excavate trenches to receive foundations; not exceeding 0.30m in width; maximum depth not exceeding 1.00m									
1.01.03.20.A	loose soil	m³	—	—	11.51	—	11.51	—	12.66	—	12.66
1.01.03.20.B	firm soil; sand	m³	—	—	14.04	—	14.04	—	15.44	—	15.44
1.01.03.20.C	light clay; compact soil; gravel	m³	—	—	17.56	—	17.56	—	19.32	—	19.32
1.01.03.20.D	stiff heavy clay	m³	—	—	23.33	—	23.33	—	25.67	—	25.67
1.01.03.20.E	soft chalk	m³	—	—	34.81	—	34.81	—	38.29	—	38.29

Excavation, Earthwork and Concrete Work

Hutchins Priced Schedules 2021	Unit	Labour Hours	Labour Net £	Plant Net £	Materials Net £	Unit Rate Net £	Labour Gross £	Plant Gross £	Materials Gross £	Unit Rate Gross £
						(Gross rates include 10% profit)				
1.01. **NEW WORK**										
1.01.04. **BREAKING UP BY HAND**										
1.01.04.01. Extra over excavation for breaking up										
1.01.04.01.A brickwork in lime mortar	m³	3.00	88.74	—	—	88.74	97.61	—	—	97.61
1.01.04.01.B brickwork in cement mortar	m³	4.00	118.32	—	—	118.32	130.15	—	—	130.15
1.01.04.01.C concrete	m³	6.00	177.48	—	—	177.48	195.23	—	—	195.23
1.01.04.01.D reinforced concrete	m³	7.50	221.85	—	—	221.85	244.04	—	—	244.04
1.01.04.01.E semi-hard rock (sandstone, etc.)	m³	6.63	195.97	—	—	195.97	215.56	—	—	215.56
1.01.05. **BREAKING UP BY MACHINE**										
1.01.05.01. Extra over excavation for breaking up using compressed air equipment										
1.01.05.01.A concrete; average thickness 150mm . . .	m²	0.55	16.30	8.25	—	24.55	17.93	9.07	—	27.00
1.01.05.01.B concrete; average thickness 300mm . . .	m²	1.10	32.63	16.51	—	49.14	35.89	18.16	—	54.05
1.01.05.01.C reinforced concrete; average thickness 150mm	m²	0.73	21.56	10.91	—	32.48	23.72	12.00	—	35.72
1.01.05.01.D reinforced concrete; average thickness 300mm	m²	1.46	43.16	21.84	—	65.00	47.47	24.03	—	71.50
1.01.05.01.E brickwork in lime mortar	m³	2.00	59.04	29.88	—	88.92	64.95	32.87	—	97.81
1.01.05.01.F brickwork in cement mortar	m³	2.66	78.68	39.82	—	118.50	86.55	43.80	—	130.35
1.01.05.01.G concrete	m³	3.68	108.77	55.04	—	163.81	119.64	60.55	—	180.19
1.01.05.01.H reinforced concrete	m³	4.86	143.88	72.81	—	216.69	158.26	80.10	—	238.36
1.01.05.01.I semi-hard rock (sandstone, etc.)	m³	4.22	124.86	63.19	—	188.05	137.34	69.51	—	206.85
1.01.06. **EARTHWORK SUPPORT**										
1.01.06.01. Earthwork support (four uses assumed) in firm ground to opposing faces not exceeding 2.00m apart; maximum depth not exceeding										
1.01.06.01.A 1.00m	m²	0.64	18.93	—	2.35	21.28	20.82	—	2.59	23.41
1.01.06.01.B 2.00m	m²	0.70	20.71	—	2.60	23.31	22.78	—	2.86	25.64
1.01.06.02. Earthwork support (four uses assumed) in loose ground to opposing faces not exceeding 2.00m apart; maximum depth not exceeding										
1.01.06.02.A 1.00m	m²	4.98	147.31	—	16.85	164.15	162.04	—	18.53	180.57
1.01.06.02.B 2.00m	m²	5.45	161.12	—	18.64	179.76	177.23	—	20.50	197.74
1.01.07. **DISPOSAL OF EXCAVATED MATERIAL**										
1.01.07.01. Fill barrows; deposit										
1.01.07.01.A wheel up to 20m	m³	1.00	29.58	—	—	29.58	32.54	—	—	32.54
1.01.07.01.B add; wheel each additional 20m	m³	0.45	13.31	—	—	13.31	14.64	—	—	14.64

Excavation, Earthwork and Concrete Work

Hutchins Priced Schedules 2021	Unit	Labour Hours	Labour Net £	Plant Net £	Materials Net £	Unit Rate Net £	Labour Gross £	Plant Gross £	Materials Gross £	Unit Rate Gross £
							(Gross rates include 10% profit)			
1.01. NEW WORK										
1.01.07. DISPOSAL OF EXCAVATED MATERIAL										
1.01.07.02. Excavated material moved by hand from spoil heap or side of excavations; deposit in skip; average distance from spoil heap or excavation										
1.01.07.02.A 25m	m³	3.78	111.81	—	—	111.81	122.99	—	—	122.99
1.01.07.02.B 50m	m³	4.40	130.15	—	—	130.15	143.17	—	—	143.17
1.01.07.03. Hand loading; transporting; depositing in spoil heaps; average distance from excavation										
1.01.07.03.A 25m	m³	3.78	111.81	—	—	111.81	122.99	—	—	122.99
1.01.07.03.B 50m	m³	4.40	130.15	—	—	130.15	143.17	—	—	143.17
1.01.07.04. Hire of skip; delivery to site; removing when full; disposal of contents; payment of tipping charges skip size										
1.01.07.04.A 4.50m3	m³	—	—	43.60	—	43.60	—	47.96	—	47.96
1.01.07.05. Excavated material loaded from spoil heaps or side of excavation into lorry and cart to contractor's tip										
1.01.07.05.A by hand	m³	1.75	51.77	23.75	—	75.52	56.94	26.13	—	83.07
1.01.07.05.B by machine	m³	—	—	26.96	—	26.96	—	29.65	—	29.65
1.01.08. FILLING										
1.01.08.01. Excavated material as filling to excavations; by hand; deposited; compacted										
1.01.08.01.A in 250mm layers	m³	1.50	44.37	—	—	44.37	48.81	—	—	48.81
1.01.08.02. Excavated material as filling in making up levels; by hand; wheeling average 25m; deposited; compacted; thickness										
1.01.08.02.A over 250mm	m³	2.20	65.08	—	—	65.08	71.58	—	—	71.58
1.01.08.02.B average 100mm	m²	0.36	10.65	—	—	10.65	11.71	—	—	11.71
1.01.08.02.C average 150mm	m²	0.46	13.61	—	—	13.61	14.97	—	—	14.97
1.01.08.02.D average 200mm	m²	0.55	16.27	—	—	16.27	17.90	—	—	17.90
1.01.08.03. Economy topsoil (PC sum £30 per m3) filling to make up levels by hand; wheel average 25m; deposited; compacted; thickness										
1.01.08.03.A over 250mm	m³	1.30	38.45	—	39.75	78.20	42.30	—	43.73	86.02
1.01.08.03.B average 100mm	m²	0.17	5.03	—	3.99	9.02	5.53	—	4.39	9.92
1.01.08.03.C average 150mm	m²	0.26	7.69	—	5.97	13.66	8.46	—	6.57	15.03

Hutchins Priced Schedules 2021	Unit	Labour Hours	Labour Net £	Plant Net £	Materials Net £	Unit Rate Net £	Labour Gross £	Plant Gross £	Materials Gross £	Unit Rate Gross £
								(Gross rates include 10% profit)		

1.01.   **NEW WORK**

1.01.08.   **FILLING**

1.01.08.03.   Economy topsoil (PC sum £30 per m3) filling to make up levels by hand; wheel average 25m; deposited; compacted; thickness

	Unit	Labour Hours	Labour Net £	Plant Net £	Materials Net £	Unit Rate Net £	Labour Gross £	Plant Gross £	Materials Gross £	Unit Rate Gross £
1.01.08.03.D average 200mm.................	m²	0.35	10.35	—	7.95	18.30	11.39	—	8.75	20.13

1.01.08.04.   Imported hardcore filling to make up levels; by hand; wheel average 25m; deposited; compacted; thickness

	Unit	Labour Hours	Labour Net £	Plant Net £	Materials Net £	Unit Rate Net £	Labour Gross £	Plant Gross £	Materials Gross £	Unit Rate Gross £
1.01.08.04.A over 250mm.................	m³	2.70	79.87	—	64.68	144.55	87.85	—	71.15	159.00
1.01.08.04.B average 75mm................	m²	0.27	7.99	—	4.88	12.87	8.79	—	5.37	14.16
1.01.08.04.C average 100mm................	m²	0.36	10.65	—	6.48	17.13	11.71	—	7.13	18.84
1.01.08.04.D average 150mm................	m²	0.54	15.97	—	9.72	25.69	17.57	—	10.69	28.26
1.01.08.04.E average 200mm................	m²	0.72	21.30	—	12.96	34.25	23.43	—	14.25	37.68

1.01.08.05.   Hand packing hardcore to form vertical or battering faces; thickness

	Unit	Labour Hours	Labour Net £	Plant Net £	Materials Net £	Unit Rate Net £	Labour Gross £	Plant Gross £	Materials Gross £	Unit Rate Gross £
1.01.08.05.A over 250mm.................	m²	1.26	37.27	—	—	37.27	41.00	—	—	41.00
1.01.08.05.B average 100mm................	m	0.20	5.92	—	—	5.92	6.51	—	—	6.51
1.01.08.05.C average 150mm................	m	0.30	8.87	—	—	8.87	9.76	—	—	9.76
1.01.08.05.D average 200mm................	m	0.34	10.06	—	—	10.06	11.06	—	—	11.06

1.01.09.   **SURFACE TREATMENTS**

1.01.09.01.   Level and compact

	Unit	Labour Hours	Labour Net £	Plant Net £	Materials Net £	Unit Rate Net £	Labour Gross £	Plant Gross £	Materials Gross £	Unit Rate Gross £
1.01.09.01.A bottoms of excavation...........	m²	0.12	3.55	—	—	3.55	3.90	—	—	3.90

1.01.09.02.   Grade and compact bottom of excavation or surface of filling to

	Unit	Labour Hours	Labour Net £	Plant Net £	Materials Net £	Unit Rate Net £	Labour Gross £	Plant Gross £	Materials Gross £	Unit Rate Gross £
1.01.09.02.A falls........................	m²	0.15	4.44	—	—	4.44	4.88	—	—	4.88
1.01.09.02.B crossfalls...................	m²	0.27	7.99	—	—	7.99	8.79	—	—	8.79

1.01.09.03.   Blind surfaces of soil or hardcore filling with sand; thickness

	Unit	Labour Hours	Labour Net £	Plant Net £	Materials Net £	Unit Rate Net £	Labour Gross £	Plant Gross £	Materials Gross £	Unit Rate Gross £
1.01.09.03.A 25mm......................	m²	0.10	2.96	—	0.97	3.92	3.25	—	1.06	4.32
1.01.09.03.B 50mm......................	m²	0.14	4.14	—	1.93	6.07	4.56	—	2.12	6.68

1.01.09.04.   Blind surfaces of soil or hardcore filling with ash; thickness

	Unit	Labour Hours	Labour Net £	Plant Net £	Materials Net £	Unit Rate Net £	Labour Gross £	Plant Gross £	Materials Gross £	Unit Rate Gross £
1.01.09.04.A 25mm......................	m²	0.12	3.55	—	4.88	8.43	3.90	—	5.37	9.28
1.01.09.04.B 50mm......................	m²	0.16	4.73	—	9.99	14.72	5.21	—	10.99	16.20

1.01.10.   **CONCRETE WORK**

1.01.10.01.   Concrete GEN1 in

	Unit	Labour Hours	Labour Net £	Plant Net £	Materials Net £	Unit Rate Net £	Labour Gross £	Plant Gross £	Materials Gross £	Unit Rate Gross £
1.01.10.01.A foundation trench...............	m³	4.50	133.11	—	92.26	225.37	146.42	—	101.49	247.91
1.01.10.01.B bed, spread over site and levelled 100mm thick..................	m³	7.50	221.85	—	92.26	314.11	244.04	—	101.49	345.52
1.01.10.01.C bed, spread over site and levelled 150mm thick..................	m³	7.00	207.06	—	92.26	299.32	227.77	—	101.49	329.25

Excavation, Earthwork and Concrete Work

Hutchins Priced Schedules 2021	Unit	Labour Hours	Labour Net £	Plant Net £	Materials Net £	Unit Rate Net £	Labour Gross £	Plant Gross £	Materials Gross £	Unit Rate Gross £
						(Gross rates include 10% profit)				
1.01.	**NEW WORK**									
1.01.10.	**CONCRETE WORK**									
1.01.10.01.	**Concrete GEN1 in**									
1.01.10.01.D	bed, spread over site and levelled 300mm thick	m³	5.00	147.90	—	92.26	240.16	162.69	—	101.49	264.18
1.01.10.02.	**Concrete GEN2 in**									
1.01.10.02.A	foundation trench	m³	4.50	133.11	—	94.79	227.90	146.42	—	104.26	250.68
1.01.10.02.B	treads, risers and landings (formwork measured separately)	m³	5.25	155.30	—	94.79	250.08	170.82	—	104.26	275.09
1.01.10.02.C	isolated pier holes	m³	5.40	159.73	—	94.79	254.52	175.71	—	104.26	279.97
1.01.10.02.D	small quantities to hearths (including formwork) 125mm thick	m²	3.00	88.74	—	12.34	101.08	97.61	—	13.57	111.19
1.01.10.03.	**Fill cavity with fine concrete**									
1.01.10.03.A	50mm	m²	0.65	19.23	—	5.19	24.41	21.15	—	5.70	26.85
1.01.10.04.	**Labour tamped finish surface of unset concrete to**									
1.01.10.04.A	levels	m²	0.15	4.44	—	—	4.44	4.88	—	—	4.88
1.01.10.04.B	falls	m²	0.16	4.73	—	—	4.73	5.21	—	—	5.21
1.01.10.04.C	crossfalls	m²	0.18	5.32	—	—	5.32	5.86	—	—	5.86
1.01.10.04.D	cambers	m²	0.19	5.62	—	—	5.62	6.18	—	—	6.18
1.01.10.04.E	slopes	m²	0.16	4.73	—	—	4.73	5.21	—	—	5.21
1.01.10.05.	**Labour spade finish surface of unset concrete to**									
1.01.10.05.A	levels	m²	0.22	6.51	—	—	6.51	7.16	—	—	7.16
1.01.10.05.B	falls	m²	0.25	7.40	—	—	7.40	8.13	—	—	8.13
1.01.10.05.C	crossfalls	m²	0.29	8.58	—	—	8.58	9.44	—	—	9.44
1.01.10.05.D	cambers	m²	0.32	9.47	—	—	9.47	10.41	—	—	10.41
1.01.10.05.E	slopes	m²	0.25	7.40	—	—	7.40	8.13	—	—	8.13
1.01.10.06.	**Labour trowelled finish surface of unset concrete to**									
1.01.10.06.A	levels	m²	0.29	8.58	—	—	8.58	9.44	—	—	9.44
1.01.10.06.B	falls	m²	0.34	10.06	—	—	10.06	11.06	—	—	11.06
1.01.10.06.C	crossfalls	m²	0.41	12.13	—	—	12.13	13.34	—	—	13.34
1.01.10.06.D	cambers	m²	0.45	13.31	—	—	13.31	14.64	—	—	14.64
1.01.10.06.E	slopes	m²	0.34	10.06	—	—	10.06	11.06	—	—	11.06
1.01.10.07.	**Labour power floating finish surface of unset concrete to**									
1.01.10.07.A	levels	m²	0.24	7.10	0.94	—	8.04	7.81	1.03	—	8.84
1.01.10.07.B	falls	m²	0.27	7.99	1.05	—	9.04	8.79	1.16	—	9.95
1.01.10.07.C	crossfalls	m²	0.32	9.47	1.25	—	10.72	10.41	1.38	—	11.79
1.01.10.07.D	cambers	m²	0.35	10.35	1.37	—	11.72	11.39	1.50	—	12.89

Hutchins Priced Schedules 2021	Unit	Labour Hours	Labour Net £	Plant Net £	Materials Net £	Unit Rate Net £	Labour Gross £	Plant Gross £	Materials Gross £	Unit Rate Gross £
							(Gross rates include 10% profit)			
1.01. **NEW WORK**										
1.01.10. **CONCRETE WORK**										
1.01.10.07. Labour power floating finish surface of unset concrete to										
1.01.10.07.E slopes	m²	0.27	7.99	1.05	—	9.04	8.79	1.16	—	9.95
1.01.10.08. Extra for working concrete around										
1.01.10.08.A pipes or cables	m²	0.20	5.92	—	—	5.92	6.51	—	—	6.51
1.01.10.10. Carborundum non-slip grain surfacing to										
1.01.10.10.A concrete steps, etc.	m²	0.25	7.40	—	3.31	10.70	8.13	—	3.64	11.77
1.01.10.11. Sisalkraft building sheets and laying										
1.01.10.11.A under concrete floors	m²	0.04	1.18	—	2.14	3.33	1.30	—	2.36	3.66
1.01.10.12. Polythene building film and laying										
1.01.10.12.A under concrete floors	m²	0.04	1.18	—	0.37	1.55	1.30	—	0.40	1.71
1.01.10.13. Treat concrete floors with three applications of										
1.01.10.13.A silicate of soda solution	m²	0.30	8.87	—	0.57	9.45	9.76	—	0.63	10.39
1.01.10.14. Groove in concrete for and including galvanised steel water bar										
1.01.10.14.A 25mm x 6mm	m	0.75	22.19	—	12.30	34.48	24.40	—	13.53	37.93
1.01.10.15. Groove including pinning lugs for										
1.01.10.15.A sliding door track (track fittings measured separately)	m	0.55	16.27	—	—	16.27	17.90	—	—	17.90
1.01.10.16. Hack face of concrete for key										
1.01.10.16.A by hand	m²	0.50	14.79	—	—	14.79	16.27	—	—	16.27
1.01.10.16.B by machine	m²	0.25	7.40	0.50	—	7.89	8.13	0.55	—	8.68
1.01.10.17. Bitumen expansion joint 9mm										
1.01.10.17.A 100mm deep seal top edge	m	0.50	14.79	—	3.28	18.07	16.27	—	3.61	19.88
1.01.10.18. Form holes for pipes through 100mm concrete; make good										
1.01.10.18.A not exceeding 55mm dia	nr	0.22	6.51	—	—	6.51	7.16	—	—	7.16
1.01.10.18.B 55mm - 110mm dia	nr	0.31	9.17	—	—	9.17	10.09	—	—	10.09
1.01.10.19. Form holes for pipes through 150mm concrete; make good										
1.01.10.19.A not exceeding 55mm dia	nr	0.37	10.94	—	—	10.94	12.04	—	—	12.04
1.01.10.19.B 55mm - 110mm dia	nr	0.46	13.61	—	—	13.61	14.97	—	—	14.97
1.01.10.20. Form mortices in concrete for iron dowels and ragbolts and grout in										
1.01.10.20.A 100mm deep	nr	0.50	14.79	—	0.10	14.89	16.27	—	0.11	16.38

Excavation, Earthwork and Concrete Work

Hutchins Priced Schedules 2021	Unit	Labour Hours	Labour Net £	Plant Net £	Materials Net £	Unit Rate Net £	Labour Gross £	Plant Gross £	Materials Gross £	Unit Rate Gross £
									(Gross rates include 10% profit)	
1.01.	**NEW WORK**									
1.01.10.	**CONCRETE WORK**									
1.01.10.21.	**Grouting in**									
1.01.10.21.A	foundation bolts and stanchion bases . . .	nr	0.50	14.79	—	0.20	14.99	16.27	—	0.22	16.49
1.01.11.	**READY-MIXED PLAIN CONCRETE**									
1.01.11.01.	**Foundations**									
1.01.11.01.A	exceeding 300mm thick	m³	3.00	88.74	—	92.26	181.00	97.61	—	101.49	199.10
1.01.11.01.B	not exceeding 300mm thick	m³	3.30	97.61	—	92.26	189.88	107.38	—	101.49	208.86
1.01.11.02.	**Floors or oversite concrete**									
1.01.11.02.A	100mm thick	m³	6.00	177.48	—	92.26	269.74	195.23	—	101.49	296.72
1.01.11.02.B	150mm thick	m³	5.50	162.69	—	92.26	254.95	178.96	—	101.49	280.45
1.01.11.02.C	300mm thick	m³	4.80	141.98	—	92.26	234.25	156.18	—	101.49	257.67
1.01.12.	**READY-MIXED REINFORCED CONCRETE (REINFORCEMENT AND FORMWORK MEASURED SEPARATELY)**									
1.01.12.01.	**Foundations**									
1.01.12.01.A	exceeding 300mm thick	m³	4.00	118.32	—	92.26	210.58	130.15	—	101.49	231.64
1.01.12.02.	**Suspended floors or roofs**									
1.01.12.02.A	100mm thick	m³	7.50	221.85	—	92.26	314.11	244.04	—	101.49	345.52
1.01.12.02.B	150mm thick	m³	7.00	207.06	—	92.26	299.32	227.77	—	101.49	329.25
1.01.12.02.C	300mm thick	m³	6.00	177.48	—	92.26	269.74	195.23	—	101.49	296.72
1.01.12.03.	**Walls**									
1.01.12.03.A	100mm thick	m³	8.00	236.64	—	92.26	328.90	260.30	—	101.49	361.79
1.01.12.03.B	150mm thick	m³	7.50	221.85	—	92.26	314.11	244.04	—	101.49	345.52
1.01.12.03.C	300mm thick	m³	6.50	192.27	—	92.26	284.53	211.50	—	101.49	312.99
1.01.12.04.	**Columns; sectional area**									
1.01.12.04.A	not exceeding 0.05m2	m³	10.00	295.80	—	92.26	388.06	325.38	—	101.49	426.87
1.01.12.04.B	0.05m2 - 0.10m2	m³	9.00	266.22	—	92.26	358.48	292.84	—	101.49	394.33
1.01.12.04.C	exceeding 0.10m2	m³	8.00	236.64	—	92.26	328.90	260.30	—	101.49	361.79
1.01.12.05.	**Beams; sectional area**									
1.01.12.05.A	not exceeding 0.05m2	m³	9.00	266.22	—	92.26	358.48	292.84	—	101.49	394.33
1.01.12.05.B	0.05m2 - 0.10m2	m³	8.00	236.64	—	92.26	328.90	260.30	—	101.49	361.79
1.01.12.05.C	exceeding 0.10m2	m³	7.00	207.06	—	92.26	299.32	227.77	—	101.49	329.25

Hutchins Priced Schedules 2021		Unit	Labour Hours	Labour Net £	Plant Net £	Materials Net £	Unit Rate Net £	Labour Gross £	Plant Gross £	Materials Gross £	Unit Rate Gross £
								(Gross rates include 10% profit)			
1.01.	**NEW WORK**										
1.01.13.	**SAWN SOFTWOOD FORMWORK**										
1.01.13.01.	**Horizontal soffit to floors, landings and the like (up to four uses)**										
1.01.13.01.A	first use	m²	2.70	101.60	—	11.27	112.87	111.76	—	12.40	124.16
1.01.13.01.B	each subsequent use	m²	2.30	86.36	—	12.25	98.61	95.00	—	13.47	108.47
1.01.13.02.	**Sloping soffit of floors, roofs, staircases and the like (up to four uses)**										
1.01.13.02.A	first use	m²	3.60	135.47	—	11.27	146.74	149.01	—	12.40	161.41
1.01.13.02.B	each subsequent use	m²	3.06	115.15	—	12.25	127.40	126.66	—	13.47	140.13
1.01.13.03.	**Vertical or battering sides of foundations, ground beams, large machine bases and the like (up to four uses)**										
1.01.13.03.A	first use	m²	3.45	129.82	—	3.96	133.79	142.81	—	4.36	147.16
1.01.13.03.B	each subsequent use	m²	2.93	110.37	—	4.19	114.55	121.41	—	4.60	126.01
1.01.13.04.	**Vertical or battering sides of walls, solid balustrades and the like (up to four uses)**										
1.01.13.04.A	first use	m²	3.45	129.82	—	11.27	141.09	142.81	—	12.40	155.20
1.01.13.04.B	each subsequent use	m²	2.93	110.37	—	12.25	122.62	121.41	—	13.47	134.88
1.01.13.05.	**Vertical or battering sides of stanchion casings, columns, piers, pilasters and the like (up to four uses)**										
1.01.13.05.A	first use	m²	3.60	135.47	—	10.89	146.36	149.01	—	11.98	160.99
1.01.13.05.B	each subsequent use	m²	3.06	115.15	—	11.74	126.89	126.66	—	12.92	139.58
1.01.13.06.	**Sides and soffits of openings in walls, recesses in walls, projecting panels on walls and the like (up to four uses)**										
1.01.13.06.A	first use	m²	3.60	135.47	—	11.27	146.74	149.01	—	12.40	161.41
1.01.13.06.B	each subsequent use	m²	3.06	115.15	—	12.25	127.40	126.66	—	13.47	140.13
1.01.13.08.	**Sides and soffits of horizontal beam casings, beams, lintels and the like (up to four uses)**										
1.01.13.08.A	first use	m²	3.90	146.76	—	11.27	158.03	161.43	—	12.40	173.83
1.01.13.08.B	each subsequent use	m²	3.32	124.74	—	12.25	136.99	137.22	—	13.47	150.69
1.01.13.09.	**Sides and soffits of sloping beam casings, staircase strings and the like (up to four uses)**										
1.01.13.09.A	first use	m²	4.65	174.98	—	11.27	186.25	192.48	—	12.40	204.87
1.01.13.09.B	each subsequent use	m²	3.95	148.75	—	12.25	161.00	163.63	—	13.47	177.10

Excavation, Earthwork and Concrete Work

Hutchins Priced Schedules 2021	Unit	Labour Hours	Labour Net £	Plant Net £	Materials Net £	Unit Rate Net £	Labour Gross £	Plant Gross £	Materials Gross £	Unit Rate Gross £
						(Gross rates include 10% profit)				
1.01.	**NEW WORK**									
1.01.13.	**SAWN SOFTWOOD FORMWORK**									
1.01.13.10.	**Sloping upper surface of beam casings, beams, staircase strings and the like; exceeding 15 deg from the horizontal (up to four uses)**									
1.01.13.10.A	first use	m²	3.90	146.76	—	11.27	158.03	161.43	—	12.40	173.83
1.01.13.10.B	each subsequent use	m²	3.32	124.74	—	12.25	136.99	137.22	—	13.47	150.69
1.01.13.11.	**Isolated beam casings and isolated beams (up to four uses)**									
1.01.13.11.A	first use	m²	4.20	158.05	—	11.27	169.31	173.85	—	12.40	186.25
1.01.13.11.B	each subsequent use	m²	3.57	134.34	—	12.25	146.59	147.77	—	13.47	161.25
1.01.13.12.	**Edges or faces of beds and the like; not exceeding 250mm high (up to four uses)**									
1.01.13.12.A	first use	m	0.38	14.30	—	0.96	15.26	15.73	—	1.05	16.78
1.01.13.12.B	each subsequent use	m	0.32	12.15	—	1.05	13.20	13.37	—	1.15	14.52
1.01.13.13.	**Edges of suspended floors, landings, roofs and the like; not exceeding 250mm wide (up to four uses)**									
1.01.13.13.A	first use	m	0.75	28.22	—	2.93	31.15	31.04	—	3.22	34.26
1.01.13.13.B	each subsequent use	m	0.64	24.01	—	3.10	27.11	26.41	—	3.41	29.82
1.01.13.14.	**Sides of kerbs and upstands and the like; not exceeding 250mm high (up to four uses)**									
1.01.13.14.A	first use	m	0.68	25.59	—	1.04	26.62	28.15	—	1.14	29.29
1.01.13.14.B	each subsequent use	m	0.58	21.75	—	1.08	22.84	23.93	—	1.19	25.12
1.01.13.15.	**Risers of steps and staircases not exceeding 250mm wide (up to four uses)**									
1.01.13.15.A	first use	m	0.60	22.58	—	2.93	25.50	24.84	—	3.22	28.05
1.01.13.15.B	each subsequent use	m	0.51	19.19	—	3.10	22.29	21.11	—	3.41	24.52
1.01.13.16.	**Edges and soffits of projecting eaves not exceeding 600mm girth (up to four uses)**									
1.01.13.16.A	first use	m	2.25	84.67	—	6.83	91.50	93.13	—	7.51	100.65
1.01.13.16.B	each subsequent use	m	1.91	71.99	—	7.38	79.37	79.18	—	8.12	87.31
1.01.13.17.	**Projecting or sunk cornices, bands and the like; not exceeding 250mm girth (up to four uses)**									
1.01.13.17.A	first use	m	0.60	22.58	—	2.93	25.50	24.84	—	3.22	28.05

Hutchins Priced Schedules 2021	Unit	Labour Hours	Labour Net £	Plant Net £	Materials Net £	Unit Rate Net £	Labour Gross £	Plant Gross £	Materials Gross £	Unit Rate Gross £
										(Gross rates include 10% profit)

1.01. NEW WORK

1.01.13. SAWN SOFTWOOD FORMWORK

1.01.13.17. Projecting or sunk cornices, bands and the like; not exceeding 250mm girth (up to four uses)

	Unit	Labour Hours	Labour Net £	Plant Net £	Materials Net £	Unit Rate Net £	Labour Gross £	Plant Gross £	Materials Gross £	Unit Rate Gross £
1.01.13.17.B each subsequent use	m	0.51	19.19	—	3.10	22.29	21.11	—	3.41	24.52
1.01.13.18. Throats, grooves, chases, rebates, chamfers and the like; not exceeding 100mm wide (up to four uses)										
1.01.13.18.A first use	m	0.20	7.53	—	1.29	8.81	8.28	—	1.42	9.70
1.01.13.18.B each subsequent use	m	0.17	6.40	—	1.34	7.73	7.04	—	1.47	8.51
1.01.13.19. Labours on formwork										
1.01.13.19.A raking cutting	m	0.20	7.53	—	—	7.53	8.28	—	—	8.28
1.01.13.19.B curved cutting	m	0.65	24.46	—	—	24.46	26.91	—	—	26.91
1.01.14. PLYWOOD FORMWORK										
1.01.14.01. Horizontal soffit of floors, landings and the like (up to three uses)										
1.01.14.01.A first use	m²	2.84	106.87	—	6.69	113.55	117.56	—	7.35	124.91
1.01.14.01.B each subsequent use	m²	2.41	90.84	—	6.73	97.56	99.92	—	7.40	107.32
1.01.14.02. Sloping soffit of floors, roofs, staircases and the like (up to three uses)										
1.01.14.02.A first use	m²	3.74	140.74	—	6.69	147.42	154.81	—	7.35	162.16
1.01.14.02.B each subsequent use	m²	3.18	119.63	—	6.73	126.35	131.59	—	7.40	138.99
1.01.14.03. Vertical or battering sides of foundations, ground beams, large machine bases and the like (up to three uses)										
1.01.14.03.A first use	m²	3.59	135.09	—	5.86	140.95	148.60	—	6.44	155.04
1.01.14.03.B each subsequent use	m²	3.05	114.85	—	6.73	121.57	126.33	—	7.40	133.73
1.01.14.04. Vertical or battering sides of walls, solid balustrades and the like (up to three uses)										
1.01.14.04.A first use	m²	3.59	135.09	—	5.78	140.87	148.60	—	6.36	154.96
1.01.14.04.B each subsequent use	m²	3.05	114.85	—	6.69	121.54	126.33	—	7.36	133.69
1.01.14.05. Vertical or battering sides of stanchion casings, columns, piers, pilasters and the like (up to three uses)										
1.01.14.05.A first use	m²	3.74	140.74	—	5.48	146.22	154.81	—	6.03	160.84
1.01.14.05.B each subsequent use	m²	3.18	119.63	—	6.54	126.16	131.59	—	7.19	138.78

Excavation, Earthwork and Concrete Work

Hutchins Priced Schedules 2021	Unit	Labour Hours	Labour Net £	Plant Net £	Materials Net £	Unit Rate Net £	Labour Gross £	Plant Gross £	Materials Gross £	Unit Rate Gross £
							(Gross rates include 10% profit)			

1.01. **NEW WORK**

1.01.14. **PLYWOOD FORMWORK**

1.01.14.06. Sides and soffits of openings in walls, recesses in walls, projecting panels on walls and the like (up to three uses)

1.01.14.06.A	first use	m²	3.74	140.74	—	5.97	146.71	154.81	—	6.57	161.38
1.01.14.06.B	each subsequent use	m²	3.18	119.63	—	6.88	126.50	131.59	—	7.56	139.15
1.01.14.08.	**Sides and soffits of horizontal beam casings, beams, lintels and the like (up to three uses)**										
1.01.14.08.A	first use	m²	4.04	152.03	—	5.86	157.88	167.23	—	6.44	173.67
1.01.14.08.B	each subsequent use	m²	3.43	129.22	—	6.73	135.95	142.14	—	7.40	149.54
1.01.14.09.	**Sides and soffits of sloping beam casings, staircase strings and the like (up to three uses)**										
1.01.14.09.A	first use	m²	4.79	180.25	—	5.86	186.10	198.27	—	6.44	204.71
1.01.14.09.B	each subsequent use	m²	4.07	153.23	—	6.73	159.96	168.55	—	7.40	175.95
1.01.14.10.	**Sloping upper surface of beam casings, beams, staircase strings and the like, exceeding 15 deg from the (up to three uses)**										
1.01.14.10.A	first use	m²	4.04	152.03	—	5.86	157.88	167.23	—	6.44	173.67
1.01.14.10.B	each subsequent use	m²	3.43	129.22	—	6.73	135.95	142.14	—	7.40	149.54
1.01.14.11.	**Isolated beam casings and isolated beams (up to three uses)**										
1.01.14.11.A	first use	m²	4.34	163.31	—	5.86	169.17	179.65	—	6.44	186.09
1.01.14.11.B	each subsequent use	m²	3.69	138.82	—	6.73	145.54	152.70	—	7.40	160.10
1.01.14.12.	**Edges or faces of beds and the like; not exceeding 250mm high (up to three uses)**										
1.01.14.12.A	first use	m	0.38	14.30	—	1.31	15.61	15.73	—	1.44	17.17
1.01.14.12.B	each subsequent use	m	0.32	12.15	—	1.76	13.92	13.37	—	1.94	15.31
1.01.14.13.	**Edges of suspended floors, landings, roofs and the like; not exceeding 250mm wide (up to three uses)**										
1.01.14.13.A	first use	m	0.75	28.22	—	1.53	29.75	31.04	—	1.68	32.73
1.01.14.13.B	each subsequent use	m	0.64	24.01	—	1.74	25.75	26.41	—	1.91	28.32
1.01.14.14.	**Sides of kerbs and upstands and the like; not exceeding 250mm high (up to three uses)**										
1.01.14.14.A	first use	m	0.68	25.59	—	1.32	26.91	28.15	—	1.45	29.60
1.01.14.14.B	each subsequent use	m	0.58	21.75	—	1.74	23.49	23.93	—	1.91	25.84

Hutchins Priced Schedules 2021	Unit	Labour Hours	Labour Net £	Plant Net £	Materials Net £	Unit Rate Net £	Labour Gross £	Plant Gross £	Materials Gross £	Unit Rate Gross £
							(Gross rates include 10% profit)			
1.01. **NEW WORK**										
1.01.14. **PLYWOOD FORMWORK**										
1.01.14.15. **Risers of steps and staircases not exceeding 250mm wide (up to three uses)**										
1.01.14.15.A first use	m	0.60	22.58	—	1.62	24.20	24.84	—	1.78	26.62
1.01.14.15.B each subsequent use	m	0.51	19.19	—	1.85	21.04	21.11	—	2.04	23.15
1.01.14.16. **Edges and soffits of projecting eaves not exceeding 600mm girth (up to three uses)**										
1.01.14.16.A first use	m	2.25	84.67	—	3.76	88.43	93.13	—	4.14	97.27
1.01.14.16.B each subsequent use	m	1.91	71.99	—	4.38	76.37	79.18	—	4.82	84.00
1.01.14.17. **Projecting or sunk cornices, bands and the like; not exceeding 250mm girth (up to three uses)**										
1.01.14.17.A first use	m	0.60	22.58	—	1.62	24.20	24.84	—	1.78	26.62
1.01.14.17.B each subsequent use	m	0.51	19.19	—	1.85	21.04	21.11	—	2.04	23.15
1.01.14.18. **Throats, grooves, chases, rebates, chamfers and the like; not exceeding 100mm wide (up to three uses)**										
1.01.14.18.A first use	m	0.20	7.53	—	0.64	8.17	8.28	—	0.70	8.98
1.01.14.18.B each subsequent use	m	0.17	6.40	—	0.71	7.11	7.04	—	0.78	7.82
1.01.14.19. **Labours on formwork**										
1.01.14.19.A raking cutting	m	0.20	7.53	—	—	7.53	8.28	—	—	8.28
1.01.14.19.B curved cutting	m	0.65	24.46	—	—	24.46	26.91	—	—	26.91
1.01.15. **REINFORCEMENT**										
1.01.15.01. **Deformed high yield steel bar reinforcement; supplied, cut, bent, labelled and fixed; including tying wire, distance blocks and ordinary spacers**										
1.01.15.01.A 6mm	m	0.12	4.52	—	0.15	4.66	4.97	—	0.16	5.13
1.01.15.01.B 8mm	m	0.10	3.76	—	0.25	4.02	4.14	—	0.28	4.42
1.01.15.01.C 10mm	m	0.09	3.39	—	0.40	3.79	3.73	—	0.44	4.16
1.01.15.01.D 12mm	m	0.08	3.01	—	0.58	3.59	3.31	—	0.64	3.95
1.01.15.01.E 16mm	m	0.07	2.63	—	1.02	3.65	2.90	—	1.12	4.02
1.01.15.01.F 20mm	m	0.07	2.63	—	1.60	4.23	2.90	—	1.76	4.65
1.01.15.01.G 25mm	m	0.07	2.63	—	2.49	5.12	2.90	—	2.74	5.63

Excavation, Earthwork and Concrete Work

Hutchins Priced Schedules 2021		Unit	Labour Hours	Labour Net £	Plant Net £	Materials Net £	Unit Rate Net £	Labour Gross £	Plant Gross £	Materials Gross £	Unit Rate Gross £
								(Gross rates include 10% profit)			
1.01.	**NEW WORK**										
1.01.15.	**REINFORCEMENT**										
1.01.15.02.	**Deformed high yield steel bar links, stirrups and binders; cut, bent, labelled and fixed; including tying wire and special spacers**										
1.01.15.02.A	6mm	m	0.14	5.27	—	0.15	5.42	5.80	—	0.16	5.96
1.01.15.02.B	8mm	m	0.12	4.52	—	0.25	4.77	4.97	—	0.28	5.24
1.01.15.03.	**Fabric reinforcement; BS 4483:2005; in slabs including tying wire and distance blocks, with allowance for 200mm laps**										
1.01.15.03.B	A142	m²	0.14	5.27	—	1.72	6.99	5.80	—	1.89	7.69
1.01.15.03.C	A193	m²	0.14	5.27	—	2.34	7.61	5.80	—	2.58	8.37
1.01.15.03.D	A252	m²	0.16	6.02	—	3.02	9.04	6.62	—	3.33	9.95
1.01.15.03.E	A393	m²	0.18	6.77	—	4.73	11.50	7.45	—	5.20	12.65
1.01.15.03.F	B283	m²	0.14	5.27	—	2.97	8.24	5.80	—	3.27	9.06
1.01.15.03.G	B385	m²	0.16	6.02	—	3.61	9.63	6.62	—	3.97	10.60
1.01.15.03.H	B503	m²	0.18	6.77	—	4.67	11.45	7.45	—	5.14	12.59
1.01.15.03.I	B785	m²	0.20	7.53	—	6.58	14.11	8.28	—	7.24	15.52
1.01.15.03.J	C283	m²	0.14	5.27	—	3.83	9.10	5.80	—	4.22	10.01
1.01.15.03.K	C385	m²	0.14	5.27	—	5.01	10.28	5.80	—	5.51	11.30
1.01.15.03.L	C503	m²	0.16	6.02	—	6.38	12.40	6.62	—	7.02	13.65
1.01.15.04.	**Fabric reinforcement; BS 4483:2005; in casings to steel columns and beams; including bending tying wire and distance blocks, with allowance for 200mm laps**										
1.01.15.04.A	D49	m²	0.30	11.29	—	2.91	14.20	12.42	—	3.20	15.62
1.01.15.04.B	D98	m²	0.35	13.17	—	1.34	14.51	14.49	—	1.48	15.97
1.01.16.	**PRECAST CONCRETE**										
1.01.16.01.	**Copings; weathered and throated; bedded in gauged mortar; pointed**										
1.01.16.01.A	300mm x 50mm	m	0.50	33.74	—	13.11	46.85	37.11	—	14.42	51.54
1.01.16.01.B	350mm x 50mm	m	0.85	57.40	—	15.58	72.98	63.14	—	17.14	80.28
1.01.16.02.	**Jambs and heads; bedded in gauged mortar; pointed**										
1.01.16.02.A	100mm x 75mm	m	0.20	13.51	—	12.16	25.67	14.86	—	13.38	28.24
1.01.16.02.B	225mm x 75mm	m	0.28	18.55	—	13.35	31.90	20.40	—	14.69	35.09
1.01.16.02.C	350mm x 75mm	m	0.38	25.34	—	19.18	44.52	27.87	—	21.10	48.97

Hutchins Priced Schedules 2021	Unit	Labour Hours	Labour Net £	Plant Net £	Materials Net £	Unit Rate Net £	Labour Gross £	Plant Gross £	Materials Gross £	Unit Rate Gross £
						(Gross rates include 10% profit)				
1.01.	**NEW WORK**									
1.01.16.	**PRECAST CONCRETE**									
1.01.16.03.	**Sills; weathered; throated; grooved; bedded in gauged mortar; pointed**									
1.01.16.03.A	225mm x 75mm	m	0.30	20.23	—	13.35	33.58	22.25	—	14.69	36.94
1.01.16.03.B	275mm x 100mm	m	0.33	21.91	—	16.08	37.99	24.10	—	17.69	41.79
1.01.16.03.C	275mm x 150mm	m	0.38	25.34	—	19.06	44.39	27.87	—	20.96	48.83
1.01.16.04.	**Thresholds; weathered; bedded in gauged mortar; pointed**									
1.01.16.04.A	275mm x 75mm	m	0.28	18.55	—	19.06	37.61	20.40	—	20.96	41.37
1.01.16.06.	**Lintels; rectangular; reinforced; hoisted; bedded in gauged mortar 100mm x 65mm; length**									
1.01.16.06.A	450mm	nr	0.05	3.36	—	5.00	8.36	3.70	—	5.50	9.20
1.01.16.06.B	600mm	nr	0.07	4.37	—	6.63	11.00	4.81	—	7.29	12.10
1.01.16.06.C	750mm	nr	0.09	5.71	—	8.25	13.96	6.28	—	9.07	15.35
1.01.16.06.D	900mm	nr	0.10	6.72	—	9.87	16.59	7.39	—	10.86	18.25
1.01.16.06.E	1050mm	nr	0.12	7.80	—	11.49	19.29	8.58	—	12.64	21.21
1.01.16.06.F	1200mm	nr	0.13	8.80	—	13.12	21.92	9.68	—	14.43	24.11
1.01.16.06.G	1350mm	nr	0.15	10.15	—	14.74	24.89	11.16	—	16.22	27.38
1.01.16.06.H	1500mm	nr	0.17	11.16	—	16.36	27.52	12.27	—	18.00	30.27
1.01.16.07.	**Lintels; rectangular; reinforced; hoisted; bedded in gauged mortar; 140mm x 100mm; length**									
1.01.16.07.A	450mm	nr	0.12	7.80	—	10.07	17.87	8.58	—	11.08	19.66
1.01.16.07.B	600mm	nr	0.15	10.15	—	13.39	23.54	11.16	—	14.73	25.89
1.01.16.07.C	750mm	nr	0.19	12.84	—	16.70	29.53	14.12	—	18.37	32.49
1.01.16.07.D	900mm	nr	0.23	15.19	—	20.02	35.21	16.71	—	22.02	38.73
1.01.16.07.E	1050mm	nr	0.27	17.88	—	23.32	41.20	19.67	—	25.66	45.32
1.01.16.07.F	1200mm	nr	0.30	20.23	—	26.64	46.87	22.25	—	29.31	51.56
1.01.16.07.G	1350mm	nr	0.34	22.99	—	29.96	52.95	25.28	—	32.96	58.24
1.01.16.07.H	1500mm	nr	0.38	25.34	—	33.40	58.73	27.87	—	36.73	64.61
1.01.16.08.	**Lintels; rectangular; reinforced; hoisted; bedded in gauged mortar; 150mm x 65mm; length**									
1.01.16.08.A	450mm	nr	0.08	5.04	—	6.87	11.91	5.54	—	7.56	13.10
1.01.16.08.B	600mm	nr	0.10	6.72	—	9.12	15.84	7.39	—	10.03	17.42
1.01.16.08.C	750mm	nr	0.13	8.47	—	11.35	19.82	9.32	—	12.49	21.80
1.01.16.08.D	900mm	nr	0.15	10.15	—	13.60	23.75	11.16	—	14.96	26.12
1.01.16.08.E	1050mm	nr	0.18	11.83	—	15.85	27.68	13.01	—	17.43	30.44

Excavation, Earthwork and Concrete Work

Hutchins Priced Schedules 2021		Unit	Labour Hours	Labour Net £	Plant Net £	Materials Net £	Unit Rate Net £	Labour Gross £	Plant Gross £	Materials Gross £	Unit Rate Gross £
								(Gross rates include 10% profit)			
1.01.	**NEW WORK**										
1.01.16.	**PRECAST CONCRETE**										
1.01.16.08.	Lintels; rectangular; reinforced; hoisted; bedded in gauged mortar; 150mm x 65mm; length										
1.01.16.08.F	1200mm	nr	0.20	13.51	—	18.09	31.60	14.86	—	19.90	34.76
1.01.16.08.G	1350mm	nr	0.23	15.19	—	20.34	35.53	16.71	—	22.38	39.08
1.01.16.08.H	1500mm	nr	0.25	16.87	—	22.59	39.46	18.56	—	24.85	43.40
1.01.16.09.	Lintels; rectangular; reinforced; hoisted; bedded in gauged mortar; 215mm x 65mm; length										
1.01.16.09.A	450mm	nr	0.11	7.12	—	10.54	17.67	7.84	—	11.60	19.44
1.01.16.09.B	600mm	nr	0.14	9.48	—	14.01	23.49	10.42	—	15.41	25.83
1.01.16.09.C	750mm	nr	0.18	11.83	—	17.49	29.31	13.01	—	19.23	32.25
1.01.16.09.D	900mm	nr	0.21	14.18	—	20.95	35.13	15.60	—	23.05	38.64
1.01.16.09.E	1050mm	nr	0.25	16.53	—	24.43	40.96	18.19	—	26.87	45.06
1.01.16.09.F	1200mm	nr	0.28	18.89	—	27.89	46.78	20.77	—	30.68	51.45
1.01.16.09.G	1350mm	nr	0.32	21.57	—	31.37	52.94	23.73	—	34.50	58.23
1.01.16.09.H	1500mm	nr	0.36	23.99	—	34.83	58.83	26.39	—	38.31	64.71
1.01.16.10.	Lintels; rectangular; reinforced; hoisted; bedded in gauged mortar 255mm x 65mm; length										
1.01.16.10.A	450mm	nr	0.13	8.47	—	12.48	20.95	9.32	—	13.72	23.04
1.01.16.10.B	600mm	nr	0.17	11.16	—	16.59	27.75	12.27	—	18.25	30.52
1.01.16.10.C	750mm	nr	0.21	13.85	—	20.71	34.55	15.23	—	22.78	38.01
1.01.16.10.D	900mm	nr	0.25	16.87	—	24.82	41.69	18.56	—	27.31	45.86
1.01.16.10.E	1050mm	nr	0.29	19.56	—	28.94	48.50	21.51	—	31.83	53.35
1.01.16.10.F	1200mm	nr	0.33	22.25	—	33.06	55.30	24.47	—	36.36	60.83
1.01.16.10.G	1350mm	nr	0.37	25.00	—	37.17	62.17	27.50	—	40.89	68.39
1.01.16.10.H	1500mm	nr	0.42	28.03	—	41.29	69.32	30.83	—	45.42	76.25
1.01.16.13.	Precast concrete padstones; bedded in gauged mortar										
1.01.16.13.A	215mm x 140mm x 102mm	nr	0.12	8.20	—	9.23	17.43	9.02	—	10.16	19.18
1.01.16.13.B	300mm x 140mm x 102mm	nr	0.17	11.43	—	11.56	22.99	12.57	—	12.72	25.29
1.01.16.13.C	215mm x 215mm x 102mm	nr	0.19	12.57	—	13.84	26.41	13.83	—	15.23	29.05
1.01.16.13.D	440mm x 140mm x 102mm	nr	0.25	16.74	—	17.15	33.88	18.41	—	18.86	37.27
1.01.16.13.E	440mm x 215mm x 102mm	nr	0.38	25.67	—	19.28	44.95	28.24	—	21.21	49.45
1.01.16.13.F	440mm x 140mm x 215mm	nr	0.53	35.29	—	24.12	59.41	38.81	—	26.53	65.35
1.01.16.14.	Pier caps; bedded in gauged mortar										
1.01.16.14.A	300mm x 300mm x 75mm for 225mm piers	nr	0.50	33.74	—	12.68	46.42	37.11	—	13.94	51.06
1.01.16.14.B	400mm x 400mm x 75mm for 337mm piers	nr	0.65	43.89	—	12.82	56.71	48.28	—	14.11	62.38

Hutchins Priced Schedules 2021	Unit	Labour Hours	Labour Net £	Plant Net £	Materials Net £	Unit Rate Net £	Labour Gross £	Plant Gross £	Materials Gross £	Unit Rate Gross £
							(Gross rates include 10% profit)			
1.02.	**REPAIRS AND ALTERATIONS**									
1.02.01.	**EXCAVATION**									
1.02.01.01.	**Excavate by hand over site area; wheel 18.00m; deposit in skip**									
1.02.01.01.A	average 300mm deep	m²	1.86	55.02	—	—	55.02	60.52	—	—	60.52
1.02.01.02.	**Excavate by hand for trenches to receive foundations; wheel 18.00m; deposit in skip**									
1.02.01.02.A	not exceeding 1.00m deep	m³	6.52	192.86	—	—	192.86	212.15	—	—	212.15
1.02.01.02.B	exceeding 1.00m deep and not exceeding 2.00m deep	m³	8.02	237.23	—	—	237.23	260.95	—	—	260.95
1.02.01.03.	**Excavate by hand for basement; wheel 18.00m deposit in skip**									
1.02.01.03.A	not exceeding 1.00m deep	m³	6.00	177.48	—	—	177.48	195.23	—	—	195.23
1.02.01.03.B	exceeding 1.00m deep and not exceeding 2.00m deep	m³	6.72	198.78	—	—	198.78	218.66	—	—	218.66
1.02.01.04.	**Excavated material as filling to excavations deposited and compacted by hand in**									
1.02.01.04.A	250mm layers	m³	2.00	59.16	—	—	59.16	65.08	—	—	65.08
1.02.01.05.	**Extra over excavation for breaking up by hand brickwork in**									
1.02.01.05.A	old foundations	m³	7.09	209.72	—	—	209.72	230.69	—	—	230.69
1.02.01.06.	**Hire of skip, delivery to site; removing when full, disposal of contents, payment of tipping charges skip size**									
1.02.01.06.A	4.50m3	m³	—	—	43.60	—	43.60	—	47.96	—	47.96
1.02.01.07.	**Earthwork support (four uses assumed) in firm ground to opposing faces not exceeding 2.00m apart; maximum depth not exceeding**									
1.02.01.07.A	1.00m	m²	0.85	25.14	—	2.35	27.49	27.66	—	2.59	30.24
1.02.01.07.B	2.00m	m²	0.93	27.51	—	2.60	30.11	30.26	—	2.86	33.12
1.02.01.08.	**Earthwork support (four uses assumed) in loose ground to opposing faces not exceeding 2.00m apart; maximum depth not exceeding**									
1.02.01.08.A	1.00m	m²	6.05	178.96	—	16.85	195.80	196.85	—	18.53	215.39

Excavation, Earthwork and Concrete Work

Hutchins Priced Schedules 2021	Unit	Labour Hours	Labour Net £	Plant Net £	Materials Net £	Unit Rate Net £	Labour Gross £	Plant Gross £	Materials Gross £	Unit Rate Gross £
							(Gross rates include 10% profit)			

1.02.   REPAIRS AND ALTERATIONS

1.02.01.   EXCAVATION

1.02.01.08.   Earthwork support (four uses assumed) in loose ground to opposing faces not exceeding 2.00m apart; maximum depth not exceeding

1.02.01.08.B 2.00m	m²	6.62	195.82	—	18.64	214.46	215.40	—	20.50	235.90

1.02.01.09.   Imported hardcore compacted to receive concrete to finished thickness

1.02.01.09.A 100mm	m²	0.48	14.20	—	6.48	20.68	15.62	—	7.13	22.74
1.02.01.09.B 150mm	m²	0.72	21.30	—	9.72	31.01	23.43	—	10.69	34.12
1.02.01.09.C 225mm	m²	1.08	31.95	—	14.60	46.54	35.14	—	16.06	51.20

1.02.02.   CONCRETE WORK

1.02.02.01.   Portland cement concrete in foundations

1.02.02.01.A GEN1 mix	m³	6.00	177.48	—	91.39	268.87	195.23	—	100.53	295.76
1.02.02.01.B GEN2 mix	m³	6.00	177.48	—	93.89	271.37	195.23	—	103.28	298.51

1.02.02.02.   Concrete GEN1 oversite; thickness

1.02.02.02.A 100mm	m³	9.00	266.22	—	91.39	357.61	292.84	—	100.53	393.37
1.02.02.02.B 150mm	m³	8.50	251.43	—	91.39	342.82	276.57	—	100.53	377.10

1.02.02.03.   Concrete GEN1 oversite in patches not exceeding 4.00m2 in area; including jointing to existing; thickness

1.02.02.03.A 100mm	m³	14.00	414.12	—	91.39	505.51	455.53	—	100.53	556.06
1.02.02.03.B 150mm	m³	13.50	399.33	—	91.39	490.72	439.26	—	100.53	539.79

1.02.02.04.   Extra oversite concrete for

1.02.02.04.A preparing to receive asphalt, tiling, etc. including extra cement	m²	0.45	13.31	—	2.53	15.84	14.64	—	2.79	17.43
1.02.02.04.B trowel to smooth surface	m²	0.55	16.27	—	—	16.27	17.90	—	—	17.90

1.02.02.05.   Sprinkle surface of concrete with

1.02.02.05.A coarse carborundum at 1kg/m2 and lightly trowel	m²	0.55	16.27	—	3.28	19.55	17.90	—	3.60	21.50

1.02.02.06.   Clean existing concrete or rendered floors and treat with

1.02.02.06.A application of silicate of soda solution	m²	0.30	8.87	—	0.57	9.45	9.76	—	0.63	10.39

1.02.02.07.   Reinforced concrete lintels cast in situ including reinforcement and formwork (up to three times used); size

1.02.02.07.A 113mm x 150mm	m	0.38	25.34	—	3.92	29.26	27.87	—	4.31	32.18
1.02.02.07.B 113mm x 225mm	m	0.45	30.38	—	4.91	35.29	33.42	—	5.40	38.82
1.02.02.07.C 225mm x 150mm	m	0.65	43.89	—	6.22	50.11	48.28	—	6.85	55.12
1.02.02.07.D 225mm x 225mm	m	0.73	48.93	—	8.12	57.05	53.82	—	8.93	62.75

Hutchins Priced Schedules 2021	Unit	Labour Hours	Labour Net £	Plant Net £	Materials Net £	Unit Rate Net £	Labour Gross £	Plant Gross £	Materials Gross £	Unit Rate Gross £
							(Gross rates include 10% profit)			
1.02. REPAIRS AND ALTERATIONS										
1.02.02. CONCRETE WORK										
1.02.02.09. Needle through brickwork; shore with one pair Acrow or other adjustable struts to every linear metre or part thereof (maximum span 2.70m); cut out and remove defective lintel and supply; hoist and build in reinforced concrete lintel cast in situ (formwork used up to three times); make good all brickwork (reclaimed facing bricks, PC sum £780 per 1000) and plaster disturbed; lintel size										
1.02.02.09.A 225mm x 150mm	m	3.09	207.61	7.45	14.41	229.47	228.37	8.20	15.85	252.42
1.02.02.09.B 225mm x 225mm	m	3.29	221.05	7.45	16.30	244.81	243.16	8.20	17.93	269.29
1.02.02.10. Cut away triangular area of brickwork above lintel; cut out and remove defective lintel and supply, hoist and build in precast reinforced concrete lintel; rebuild brickwork over including facing bricks (PC sum £550 per 1000) to match existing; make good internal plaster; lintel size										
1.02.02.10.A 225mm x 225mm	m	5.85	393.18	—	26.62	419.80	432.50	—	29.28	461.78
1.02.02.11. Take out stone or concrete sill; supply and build in cast concrete sill including all making good (formwork used up to three times); sill size										
1.02.02.11.A 225mm x 75mm	m	0.55	37.10	—	13.22	50.32	40.81	—	14.54	55.35
1.02.02.12. Pier caps										
1.02.02.12.A 300mm x 300mm x 75mm for 225mm piers	nr	0.50	33.74	—	12.68	46.42	37.11	—	13.94	51.06
1.02.02.12.B 400mm x 400mm x 75mm for 338mm piers	nr	0.65	43.89	—	12.82	56.71	48.28	—	14.11	62.38
1.02.02.13. Break up and remove old concrete steps, form new steps in										
1.02.02.13.A concrete GEN1 mix, including wrought formwork to risers and ends (four uses assumed), and surfaces of treads trowelled smooth	m³	9.54	641.25	—	123.89	765.14	705.38	—	136.28	841.66
1.02.02.14. Break up and remove concrete floors, pavings, etc.; at ground level and load into skip										
1.02.02.14.A not exceeding 150mm	m²	2.25	66.56	—	—	66.56	73.21	—	—	73.21

Excavation, Earthwork and Concrete Work

Hutchins Priced Schedules 2021		Unit	Labour Hours	Labour Net £	Plant Net £	Materials Net £	Unit Rate Net £	Labour Gross £	Plant Gross £	Materials Gross £	Unit Rate Gross £
										(Gross rates include 10% profit)	
1.02.	**REPAIRS AND ALTERATIONS**										
1.02.02.	**CONCRETE WORK**										
1.02.02.14.	Break up and remove concrete floors, pavings, etc.; at ground level and load into skip										
1.02.02.14.B	150mm - 225mm	m²	4.00	118.32	—	—	118.32	130.15	—	—	130.15
1.02.02.14.C	225mm - 300mm	m²	6.00	177.48	—	—	177.48	195.23	—	—	195.23
1.02.02.15.	Break up and remove reinforced concrete floors, pavings, etc.; at ground level and load into skip										
1.02.02.15.A	not exceeding 150mm	m²	3.40	100.57	—	—	100.57	110.63	—	—	110.63
1.02.02.15.B	150mm - 225mm	m²	6.00	177.48	—	—	177.48	195.23	—	—	195.23
1.02.02.15.C	225mm - 300mm	m²	9.00	266.22	—	—	266.22	292.84	—	—	292.84
1.02.02.16.	Break up concrete paving 750mm wide for new wall and remove. Excavate trench and part return, fill in and ram and remove remainder. Make good concrete paving (foundation concrete measured separately)										
1.02.02.16.A	100mm thick	m	2.85	84.30	—	5.66	89.96	92.73	—	6.22	98.96
1.02.02.17.	Hack up broken or sunken areas of concrete paving, spread and consolidate hardcore 150mm, lay new concrete to falls, joint to existing including trowel to form smooth surface										
1.02.02.17.A	100mm	m²	1.08	72.59	—	25.75	98.33	79.85	—	28.32	108.17
1.02.02.17.B	150mm	m²	1.23	82.67	—	30.80	113.46	90.94	—	33.88	124.81
1.02.02.18.	Hack surface of existing paving or floors and grout and render in										
1.02.02.18.A	19mm cement mortar (1:2:5)	m²	0.50	33.74	—	2.44	36.18	37.11	—	2.68	39.80
1.02.02.19.	Hack off defective cement rendering to steps (treads and risers) and make out in										
1.02.02.19.A	25mm cement and sand (1:3) trowelled including nosings and arrises	m²	1.26	84.35	—	3.56	87.91	92.78	—	3.92	96.70
1.02.02.20.	Clean and hack existing concrete surface to form key for										
1.02.02.20.A	granolithic paving	m²	0.40	11.83	—	—	11.83	13.02	—	—	13.02
1.02.02.21.	Roughen and grout edge of existing concrete paving to new										
1.02.02.21.A	100mm	m	0.40	11.83	—	—	11.83	13.02	—	—	13.02
1.02.02.21.B	150mm	m	0.50	14.79	—	—	14.79	16.27	—	—	16.27

Hutchins Priced Schedules 2021	Unit	Labour Hours	Labour Net £	Plant Net £	Materials Net £	Unit Rate Net £	Labour Gross £	Plant Gross £	Materials Gross £	Unit Rate Gross £
							(Gross rates include 10% profit)			
1.02.	**REPAIRS AND ALTERATIONS**									
1.02.02.	**CONCRETE WORK**									
1.02.02.22.	**Break up reinforced concrete walls, columns, beams, suspended floors or roofs and loading into**									
1.02.02.22.A	skip......................	m³	27.00	798.66	—	—	798.66	878.53	—	—	878.53
1.02.02.23.	**Cut holes through concrete for pipes, bars, etc.; per 25mm depth of cut; make good; area**									
1.02.02.23.A	not exceeding 0.003m2...........	nr	0.20	5.92	—	0.24	6.15	6.51	—	0.26	6.77
1.02.02.23.B	0.003m2 - 0.023m2..............	nr	0.40	11.83	—	0.24	12.07	13.02	—	0.26	13.28
1.02.02.24.	**Cut holes through reinforced concrete for pipes, bars, etc.; per 25mm depth of cut; make good; area**									
1.02.02.24.A	not exceeding 0.003m2...........	nr	0.30	8.87	—	0.24	9.11	9.76	—	0.26	10.02
1.02.02.24.B	0.003m2 - 0.023m2..............	nr	0.60	17.75	—	0.24	17.99	19.52	—	0.26	19.78
1.02.02.25.	**Form concrete GEN2 kerbs and channels; including all necessary formwork (up to three times used) but excluding excavation**									
1.02.02.25.A	average 0.047m2 sectional area......	m	0.50	33.74	—	5.44	39.18	37.11	—	5.98	43.10

Brickwork and Blockwork

Hutchins Priced Schedules 2021	Unit	Labour Hours	Labour Net £	Plant Net £	Materials Net £	Unit Rate Net £	Labour Gross £	Plant Gross £	Materials Gross £	Unit Rate Gross £
										(Gross rates include 10% profit)

2.01. NEW WORK

2.01.01. CLASS B ENGINEERING BRICKWORK (PC SUM £500 PER 1000) IN CEMENT MORTAR (1:3)

	Unit	Labour Hours	Labour Net £	Plant Net £	Materials Net £	Unit Rate Net £	Labour Gross £	Plant Gross £	Materials Gross £	Unit Rate Gross £
2.01.01.01. Walls										
2.01.01.01.A half brick thick	m²	1.15	60.49	—	34.66	95.15	66.54	—	38.13	104.67
2.01.01.01.B one brick thick	m²	1.96	102.74	—	67.99	170.73	113.02	—	74.79	187.81
2.01.01.02. Skins of hollow walls										
2.01.01.02.A half brick thick	m²	1.15	60.49	—	34.66	95.15	66.54	—	38.13	104.67
2.01.01.02.B one brick thick	m²	1.96	102.74	—	67.99	170.73	113.02	—	74.79	187.81
2.01.01.03. Honeycomb sleeper walls										
2.01.01.03.A half brick thick	m²	0.89	46.44	—	26.31	72.75	51.09	—	28.94	80.02
2.01.01.04. For every £10 per 1000 variation in the price of bricks, add or deduct as follows										
2.01.01.04.A half brick walls	m²	—	—	—	0.62	0.62	—	—	0.68	0.68
2.01.01.04.B one brick walls	m²	—	—	—	1.24	1.24	—	—	1.36	1.36
2.01.01.05. Extra over brickwork in cement mortar (1:3) for fair face and flush pointing one side as the work proceeds										
2.01.01.05.A stretcher bond	m²	0.31	16.46	—	—	16.46	18.11	—	—	18.11
2.01.01.05.B Flemish bond	m²	0.34	17.61	—	—	17.61	19.37	—	—	19.37
2.01.01.05.C margins	m	0.07	3.88	—	—	3.88	4.27	—	—	4.27
2.01.02. COMMON BRICKWORK (PC SUM £450 PER 1000) IN GAUGED MORTAR (1:1:6)										
2.01.02.01. Walls										
2.01.02.01.A half brick thick	m²	1.11	58.40	—	31.50	89.90	64.24	—	34.65	98.89
2.01.02.01.B one brick thick	m²	1.82	95.30	—	61.65	156.95	104.83	—	67.82	172.65
2.01.02.02. Skins of hollow walls										
2.01.02.02.A half brick thick	m²	1.11	58.40	—	31.50	89.90	64.24	—	34.65	98.89
2.01.02.02.B one brick thick	m²	1.82	95.30	—	61.65	156.95	104.83	—	67.82	172.65
2.01.02.03. Honeycomb sleeper walls										
2.01.02.03.A half brick wall	m²	0.86	44.98	—	25.07	70.05	49.47	—	27.58	77.05

Brickwork and Blockwork

Hutchins Priced Schedules 2021	Unit	Labour Hours	Labour Net £	Plant Net £	Materials Net £	Unit Rate Net £	Labour Gross £	Plant Gross £	Materials Gross £	Unit Rate Gross £
							(Gross rates include 10% profit)			
2.01.	**NEW WORK**									
2.01.02.	**COMMON BRICKWORK (PC SUM £450 PER 1000) IN GAUGED MORTAR (1:1:6)**									
2.01.02.04.	**Projections of footings and chimney breasts**									
2.01.02.04.A	half brick thick	m²	1.25	65.42	—	31.89	97.31	71.96	—	35.08	107.04
2.01.02.04.B	one brick thick	m²	2.05	107.25	—	65.51	172.77	117.98	—	72.07	190.04
2.01.02.04.C	one and a half brick thick	m²	2.45	128.32	—	99.08	227.40	141.16	—	108.98	250.14
2.01.02.05.	**Isolated piers and chimney stacks**									
2.01.02.05.A	one brick thick	m²	2.27	118.78	—	65.51	184.30	130.66	—	72.07	202.73
2.01.02.05.B	one and a half brick thick	m²	2.71	142.06	—	99.08	241.13	156.26	—	108.98	265.25
2.01.02.05.C	two bricks thick	m²	3.33	174.77	—	132.12	306.89	192.25	—	145.33	337.58
2.01.02.06.	**Projection of attached piers, plinths, bands, oversailing courses and the like**									
2.01.02.06.A	215mm x 102.5mm	m	0.37	19.40	—	8.16	27.56	21.33	—	8.98	30.32
2.01.02.06.B	215mm x 215mm	m	0.97	50.64	—	15.88	66.52	55.70	—	17.47	73.17
2.01.02.06.C	327.5mm x 102.5mm	m	0.53	27.78	—	11.57	39.35	30.56	—	12.73	43.29
2.01.02.06.D	327.5mm x 215mm	m	1.82	95.30	—	24.17	119.47	104.83	—	26.59	131.42
2.01.02.07.	**For every £10 per 1000 variation in the price of bricks, add or deduct as follows**									
2.01.02.07.A	half brick walls	m²	—	—	—	0.62	0.62	—	—	0.68	0.68
2.01.02.07.B	one brick walls	m²	—	—	—	1.24	1.24	—	—	1.36	1.36
2.01.02.07.C	one and a half brick walls	m²	—	—	—	1.86	1.86	—	—	2.05	2.05
2.01.02.07.D	two brick walls	m²	—	—	—	2.48	2.48	—	—	2.73	2.73
2.01.02.09.	**Extra over common brickwork in gauged mortar (1:1:6) for fair face and flush pointing one side as the work proceeds**									
2.01.02.09.A	stretcher bond	m²	0.27	14.05	—	—	14.05	15.45	—	—	15.45
2.01.02.09.B	Flemish bond	m²	0.29	15.41	—	—	15.41	16.95	—	—	16.95
2.01.02.09.C	margins	m	0.07	3.56	—	—	3.56	3.92	—	—	3.92
2.01.03.	**FACING BRICKWORK (PC SUM £550 PER 1000), IN GAUGED MORTAR (1:1:6)**									
2.01.03.01.	**Walls**									
2.01.03.01.A	half brick thick; stretcher bond	m²	1.37	72.03	—	39.13	111.16	79.23	—	43.04	122.27
2.01.03.01.B	one brick thick; double stretcher bond . .	m²	2.61	136.82	—	79.13	215.95	150.50	—	87.04	237.54
2.01.03.01.C	one brick thick; English bond	m²	2.75	144.16	—	79.52	223.67	158.57	—	87.47	246.04
2.01.03.01.D	one brick thick; English garden wall bond	m²	2.66	139.54	—	79.52	219.06	153.50	—	87.47	240.96
2.01.03.01.E	one brick thick; English cross bond	m²	2.89	151.60	—	81.72	233.31	166.76	—	89.89	256.65
2.01.03.01.F	one brick thick; Flemish bond	m²	2.84	148.66	—	79.39	228.05	163.53	—	87.33	250.85
2.01.03.01.G	one brick thick; Flemish garden wall bond	m²	2.70	141.32	—	79.39	220.71	155.46	—	87.33	242.78

Hutchins Priced Schedules 2021	Unit	Labour Hours	Labour Net £	Plant Net £	Materials Net £	Unit Rate Net £	Labour Gross £	Plant Gross £	Materials Gross £	Unit Rate Gross £
							(Gross rates include 10% profit)			
2.01. **NEW WORK**										
2.01.03. **FACING BRICKWORK (PC SUM £550 PER 1000), IN GAUGED MORTAR (1:1:6)**										
2.01.03.02. **Skins of hollow walls**										
2.01.03.02.A half brick thick; stretcher bond	m²	1.37	72.03	—	39.00	111.03	79.23	—	42.90	122.13
2.01.03.02.B half brick thick; English bond (snapped headers)	m²	1.55	81.25	—	57.28	138.53	89.38	—	63.01	152.39
2.01.03.02.C half brick thick; Flemish bond (snapped headers)	m²	1.59	83.35	—	51.23	134.58	91.68	—	56.35	148.04
2.01.03.04. **For every £10 per 1000 variation in the price of bricks, add or deduct as follows**										
2.01.03.04.A half brick walls; stretcher bond	m²	—	—	—	0.66	0.66	—	—	0.73	0.73
2.01.03.04.B half brick walls; Flemish bond	m²	—	—	—	0.89	0.89	—	—	0.98	0.98
2.01.04. **ARCHES AND COPINGS**										
2.01.04.01. **Extra over brickwork for flat arch, 112mm soffit (common bricks PC sum £450 per 1000; facing bricks PC sum £550 per 1000)**										
2.01.04.01.A commons; 112mm high	m	0.40	21.07	—	0.45	21.52	23.18	—	0.50	23.68
2.01.04.01.B commons; 225mm high	m	0.54	28.20	—	0.45	28.65	31.02	—	0.50	31.52
2.01.04.01.C facings; 112mm high	m	0.50	26.31	—	4.07	30.39	28.95	—	4.48	33.43
2.01.04.01.D facings; 225mm high	m	0.67	35.12	—	7.47	42.59	38.63	—	8.21	46.85
2.01.04.02. **Extra over brickwork for camber arch; 112mm soffit (common bricks PC sum £450 per 1000; facing bricks PC sum £550 per 1000)**										
2.01.04.02.A commons; 225mm high	m	1.14	59.76	—	0.45	60.21	65.73	—	0.50	66.23
2.01.04.02.B facings; 225mm high	m	1.21	63.32	—	7.47	70.79	69.66	—	8.21	77.87
2.01.04.03. **Extra over brickwork for segmental arch; 225mm deep in two half brick rings (common bricks PC sum £450 per 1000; facing bricks PC sum £550 per 1000)**										
2.01.04.03.A commons	m	1.27	66.78	—	0.90	67.68	73.46	—	0.99	74.45
2.01.04.03.B facings	m	1.41	73.81	—	7.47	81.27	81.19	—	8.21	89.40
2.01.04.04. **Extra over brickwork in rough relieving arches (PC sum £450 per 1000)**										
2.01.04.04.A commons; 112mm high	m	0.20	10.59	—	0.45	11.04	11.65	—	0.50	12.14
2.01.04.04.B commons; 225mm high	m	0.34	17.61	—	0.45	18.06	19.37	—	0.50	19.87

Brickwork and Blockwork

Hutchins Priced Schedules 2021	Unit	Labour Hours	Labour Net £	Plant Net £	Materials Net £	Unit Rate Net £	Labour Gross £	Plant Gross £	Materials Gross £	Unit Rate Gross £
									(Gross rates include 10% profit)	
2.01.	**NEW WORK**									
2.01.04.	**ARCHES AND COPINGS**									
2.01.04.05.	**Brick-on-edge coping to one brick wall; double tile creasing course; two small fillets; pointed all round (common bricks PC sum £450 per 1000; facing bricks PC sum £550 per 1000)**									
2.01.04.05.A	commons	m	1.61	84.40	—	13.02	97.41	92.84	—	14.32	107.15
2.01.04.05.B	facings	m	1.68	87.96	—	14.52	102.48	96.76	—	15.97	112.72
2.01.06.	**SILLS AND STEPS**									
2.01.06.01.	**Heather brown quarry tile sill, one tile deep, rounded on one edge, bedded, jointed and pointed in cement mortar**									
2.01.06.01.B	200mm x 200mm x 14mm	m	0.25	13.00	—	12.39	25.39	14.30	—	13.63	27.93
2.01.06.01.C	150mm x 150mm x 12mm	m	0.22	11.64	—	14.15	25.79	12.80	—	15.57	28.37
2.01.06.02.	**Red quarry tile sill, one tile deep, rounded on one edge, bedded, jointed and pointed in cement mortar**									
2.01.06.02.B	200mm x 200mm x 14mm	m	0.25	13.00	—	12.39	25.39	14.30	—	13.63	27.93
2.01.06.02.D	150mm x 150mm x 12mm	m	0.22	11.64	—	14.15	25.79	12.80	—	15.57	28.37
2.01.06.03.	**Two courses roofing tiles set sloping to form**									
2.01.06.03.A	external window sills; including pointing .	m	0.91	47.49	—	6.35	53.85	52.24	—	6.99	59.23
2.01.06.04.	**Brick-on-edge step in hard red paviors (PC sum £1135 per 1000); bedded and pointed**									
2.01.06.04.A	225mm wide	m	0.94	49.27	—	17.50	66.77	54.20	—	19.25	73.45
2.01.06.04.B	338mm wide	m	2.01	105.47	—	25.68	131.15	116.02	—	28.25	144.27
2.01.06.05.	**Brick-on-edge sill; bedded and pointed**									
2.01.06.05.A	225mm in facing bricks (PC sum £550 per 1000)	m	0.87	45.71	—	8.76	54.48	50.28	—	9.64	59.92
2.01.06.05.B	225mm in single bullnosed engineering bricks	m	0.87	45.71	—	26.93	72.64	50.28	—	29.62	79.90
2.01.08.	**DENSE CONCRETE BLOCK WALLING**									
2.01.08.01.	**Blockwork in walls; partitions or skins of hollow walls**									
2.01.08.01.A	75mm solid	m²	0.74	39.00	—	11.11	50.11	42.90	—	12.22	55.12
2.01.08.01.B	100mm solid	m²	0.88	46.13	—	14.37	60.50	50.74	—	15.81	66.55
2.01.08.01.C	140mm solid	m²	1.04	54.52	—	22.87	77.38	59.97	—	25.15	85.12
2.01.08.01.D	140mm hollow	m²	1.10	57.66	—	21.67	79.33	63.43	—	23.83	87.26

Brickwork and Blockwork

Hutchins Priced Schedules 2021		Unit	Labour Hours	Labour Net £	Plant Net £	Materials Net £	Unit Rate Net £	Labour Gross £	Plant Gross £	Materials Gross £	Unit Rate Gross £
								(Gross rates include 10% profit)			
2.01.	**NEW WORK**										
2.01.08.	**DENSE CONCRETE BLOCK WALLING**										
2.01.08.01.	Blockwork in walls; partitions or skins of hollow walls										
2.01.08.01.E	190mm hollow...............	m²	1.24	65.11	—	25.51	90.61	71.62	—	28.06	99.67
2.01.08.01.F	215mm hollow...............	m²	1.37	71.71	—	28.68	100.39	78.88	—	31.55	110.43
2.01.08.01.G	extra for fair face and flush pointing blockwork as the work proceeds; any thickness; one face...........	m²	0.09	4.61	—	0.51	5.13	5.07	—	0.57	5.64
2.01.08.01.H	extra for fair face and flush pointing blockwork as the work proceeds; any thickness; both faces...........	m²	0.23	11.95	—	0.51	12.47	13.15	—	0.57	13.71
2.01.09.	**AIRCRETE CONCRETE BLOCK WALLING**										
2.01.09.01.	Thermalite blockwork in walls; partitions or skins of hollow walls										
2.01.09.01.C	75mm.....................	m²	0.67	35.12	—	12.32	47.44	38.63	—	13.55	52.18
2.01.09.01.E	100mm....................	m²	0.80	42.15	—	15.95	58.09	46.36	—	17.54	63.90
2.01.09.01.H	150mm....................	m²	1.01	52.73	—	26.42	79.15	58.01	—	29.06	87.07
2.01.09.01.K	215mm....................	m²	1.27	66.78	—	36.77	103.55	73.46	—	40.45	113.91
2.01.09.01.L	extra for fair face and flush pointing blockwork as the work proceeds; any thickness; one face...........	m²	0.09	4.61	—	1.03	5.64	5.07	—	1.13	6.21
2.01.09.01.M	extra for fair face and flush pointing blockwork as the work proceeds; any thickness; both faces...........	m²	0.23	11.95	—	1.03	12.98	13.15	—	1.13	14.28
2.01.09.02.	Celcon Solar grade; high insulation blockwork in walls; partitions or skins of hollow walls										
2.01.09.02.A	100mm....................	m²	1.01	53.05	—	21.47	74.52	58.35	—	23.62	81.97
2.01.09.02.G	215mm....................	m²	1.57	82.30	—	45.60	127.90	90.53	—	50.16	140.69
2.01.09.02.H	extra for fair face and flush pointing blockwork as the work proceeds; any thickness; one face...........	m²	0.09	4.61	—	1.03	5.64	5.07	—	1.13	6.21
2.01.09.02.I	extra for fair face and flush pointing blockwork as the work proceeds; any thickness; both faces...........	m²	0.23	11.95	—	1.03	12.98	13.15	—	1.13	14.28
2.01.09.03.	Celcon Standard grade blockwork in walls; partitions or skins of hollow walls										
2.01.09.03.E	100mm....................	m²	0.80	42.15	—	15.85	58.00	46.36	—	17.44	63.80
2.01.09.03.F	140mm....................	m²	0.94	49.22	—	28.21	77.43	54.14	—	31.03	85.18
2.01.09.03.H	150mm....................	m²	1.01	52.73	—	30.13	82.87	58.01	—	33.15	91.15
2.01.09.03.K	215mm....................	m²	1.27	66.78	—	43.59	110.38	73.46	—	47.95	121.42

Brickwork and Blockwork

Hutchins Priced Schedules 2021	Unit	Labour Hours	Labour Net £	Plant Net £	Materials Net £	Unit Rate Net £	Labour Gross £	Plant Gross £	Materials Gross £	Unit Rate Gross £
								(Gross rates include 10% profit)		
2.01.	**NEW WORK**									
2.01.09.	**AIRCRETE CONCRETE BLOCK WALLING**									
2.01.09.03.	**Celcon Standard grade blockwork in walls; partitions or skins of hollow walls**									
2.01.09.03.L	extra for fair face and flush pointing blockwork as the work proceeds; any thickness; one face	m²	0.09	4.61	—	1.03	5.64	5.07	—	1.13	6.21
2.01.09.03.M	extra for fair face and flush pointing blockwork as the work proceeds; any thickness; both faces	m²	0.23	11.95	—	1.03	12.98	13.15	—	1.13	14.28
2.01.13.	**DAMP PROOF COURSES**									
2.01.13.01.	**Polythene; horizontal; bedded in gauged mortar (1:1:6)**									
2.01.13.01.A	112.5mm wide	m	0.03	1.47	—	0.31	1.78	1.61	—	0.34	1.96
2.01.13.01.B	225mm wide	m	0.05	2.83	—	0.77	3.60	3.11	—	0.84	3.96
2.01.13.01.C	over 225mm wide	m²	0.23	12.27	—	3.18	15.45	13.49	—	3.50	16.99
2.01.13.01.D	over 225mm wide; forming cavity gutter in hollow wall	m²	0.38	19.71	—	3.18	22.89	21.68	—	3.50	25.18
2.01.13.02.	**Polythene; vertical; bedded in gauged mortar (1:1:6)**									
2.01.13.02.A	over 225mm wide	m²	0.70	36.90	—	3.18	40.09	40.59	—	3.50	44.09
2.01.13.03.	**Fibre based bitumen; horizontal; bedded in gauged mortar (1:1:6)**									
2.01.13.03.A	112.5mm wide	m	0.03	1.47	—	1.64	3.11	1.61	—	1.81	3.42
2.01.13.03.B	225mm wide	m	0.05	2.83	—	3.41	6.24	3.11	—	3.75	6.87
2.01.13.03.C	over 225mm wide	m²	0.23	12.27	—	15.51	27.77	13.49	—	17.06	30.55
2.01.13.03.D	over 225mm wide; forming cavity gutter in hollow wall	m²	0.38	19.71	—	15.51	35.22	21.68	—	17.06	38.74
2.01.13.04.	**Fibre based bitumen; vertical; bedded in gauged mortar (1:1:6)**									
2.01.13.04.A	over 225mm wide	m²	0.70	36.90	—	15.51	52.41	40.59	—	17.06	57.65
2.01.13.05.	**Hessian based bitumen; horizontal; bedded in gauged mortar (1:1:6)**									
2.01.13.05.A	112.5mm wide	m	0.03	1.47	—	1.37	2.84	1.61	—	1.51	3.12
2.01.13.05.B	225mm wide	m	0.05	2.83	—	2.98	5.81	3.11	—	3.28	6.39
2.01.13.05.C	over 225mm wide	m²	0.23	12.27	—	13.09	25.36	13.49	—	14.40	27.89
2.01.13.05.D	over 225mm wide; forming cavity gutter in hollow wall	m²	0.38	19.71	—	13.09	32.80	21.68	—	14.40	36.08
2.01.13.06.	**Hessian based bitumen; vertical; bedded in gauged mortar (1:1:6)**									
2.01.13.06.A	over 225mm wide	m²	0.70	36.90	—	13.09	50.00	40.59	—	14.40	55.00

Hutchins Priced Schedules 2021	Unit	Labour Hours	Labour Net £	Plant Net £	Materials Net £	Unit Rate Net £	Labour Gross £	Plant Gross £	Materials Gross £	Unit Rate Gross £
							(Gross rates include 10% profit)			
2.01. **NEW WORK**										
2.01.13. **DAMP PROOF COURSES**										
2.01.13.07. **Hyload pitch polymer; horizontal; 100mm laps sealed with Hyload contact adhesive; bedded in gauged mortar (1:1:6)**										
2.01.13.07.A 112.5mm wide	m	0.03	1.47	—	1.49	2.96	1.61	—	1.64	3.26
2.01.13.07.B 225mm wide	m	0.05	2.83	—	3.35	6.18	3.11	—	3.69	6.80
2.01.13.07.C over 225mm wide	m²	0.23	12.27	—	14.17	26.44	13.49	—	15.59	29.08
2.01.13.07.D over 225mm wide; forming cavity gutter in hollow wall	m²	0.38	19.71	—	14.17	33.88	21.68	—	15.59	37.27
2.01.13.08. **Hyload pitch polymer; vertical; bedded in gauged mortar (1:1:6)**										
2.01.13.08.A over 225mm wide	m²	0.70	36.90	—	7.24	44.15	40.59	—	7.97	48.56
2.01.13.11. **Fibre based lead lined; horizontal; bedded in gauged mortar (1:1:6)**										
2.01.13.11.A 112.5mm wide	m	0.03	1.78	—	3.95	5.73	1.96	—	4.35	6.31
2.01.13.11.B 225mm wide	m	0.07	3.56	—	8.36	11.92	3.92	—	9.19	13.11
2.01.13.11.C over 225mm wide	m²	0.31	16.15	—	36.23	52.38	17.76	—	39.86	57.62
2.01.13.11.D over 225mm wide; forming cavity gutter in hollow wall	m²	0.50	26.00	—	36.23	62.23	28.60	—	39.86	68.46
2.01.13.13. **Fibre based lead lined; vertical; bedded in gauged mortar (1:1:6)**										
2.01.13.13.A over 225mm wide	m²	0.70	36.90	—	36.23	73.14	40.59	—	39.86	80.45
2.01.13.14. **Hessian based lead lined; horizontal; bedded in gauged mortar (1:1:6)**										
2.01.13.14.A 112.5mm wide	m	0.03	1.78	—	4.15	5.93	1.96	—	4.57	6.53
2.01.13.14.B 225mm wide	m	0.07	3.56	—	8.40	11.96	3.92	—	9.24	13.16
2.01.13.14.C over 225mm wide	m²	0.31	16.15	—	38.04	54.19	17.76	—	41.84	59.60
2.01.13.14.D over 225mm wide; forming cavity gutter in hollow wall	m²	0.50	26.00	—	38.04	64.04	28.60	—	41.84	70.44
2.01.13.15. **Hessian based lead lined; vertical; bedded in gauged mortar (1:1:6)**										
2.01.13.15.A over 225mm wide	m²	0.70	36.90	—	38.04	74.94	40.59	—	41.84	82.44
2.01.13.18. **Welsh slates; two courses; horizontal; bedded in cement mortar (1:3)**										
2.01.13.18.A over 225mm wide	m²	1.38	72.44	—	61.34	133.79	79.69	—	67.48	147.17
2.01.13.19. **Welsh slates; two courses; vertical; bedded in cement mortar (1:3)**										
2.01.13.19.A over 225mm wide	m²	2.11	110.71	—	61.34	172.06	121.78	—	67.48	189.26

Brickwork and Blockwork

								(Gross rates include 10% profit)		
2.01.	**NEW WORK**									
2.01.13.	**DAMP PROOF COURSES**									
2.01.13.20.	**IKOpro Synthaprufe; vertical membrane; three coats brushed on; final covering dusted with clean sharp sand**									
2.01.13.20.A	not exceeding 150mm wide	m	0.07	3.88	—	1.01	4.89	4.27	—	1.11	5.37
2.01.13.20.B	150mm - 300mm wide	m	0.12	6.29	—	1.96	8.25	6.92	—	2.16	9.08
2.01.13.20.C	over 300mm wide	m²	0.26	13.73	—	6.37	20.11	15.11	—	7.01	22.12
2.01.15.	**AIR BRICKS AND SOOT DOORS**									
2.01.15.01.	**Air bricks; terracotta; building in**									
2.01.15.01.A	215mm x 65mm; square hole	nr	0.15	7.76	—	4.26	12.02	8.53	—	4.69	13.22
2.01.15.01.B	215mm x 140mm; square hole	nr	0.20	10.59	—	6.22	16.81	11.65	—	6.84	18.49
2.01.15.01.C	215mm x 215mm; square hole	nr	0.22	11.64	—	14.32	25.96	12.80	—	15.75	28.55
2.01.15.01.D	215mm x 65mm; louvred hole	nr	0.15	7.76	—	4.84	12.59	8.53	—	5.32	13.85
2.01.15.01.E	215mm x 140mm; louvred hole	nr	0.20	10.59	—	7.69	18.28	11.65	—	8.46	20.11
2.01.15.02.	**Air bricks; cast iron; building in**									
2.01.15.02.A	225mm x 75mm	nr	0.15	7.76	—	23.44	31.20	8.53	—	25.78	34.32
2.01.15.02.B	225mm x 150mm	nr	0.20	10.59	—	29.21	39.80	11.65	—	32.13	43.78
2.01.15.02.C	225mm x 225mm	nr	0.22	11.64	—	35.11	46.74	12.80	—	38.62	51.42
2.01.15.03.	**Louvred ventilators; aluminium; screw fixed**									
2.01.15.03.A	225mm x 75mm	nr	0.15	7.76	—	6.72	14.48	8.53	—	7.39	15.93
2.01.15.03.B	225mm x 150mm	nr	0.20	10.59	—	8.20	18.79	11.65	—	9.03	20.67
2.01.15.03.C	225mm x 225mm	nr	0.22	11.64	—	9.49	21.13	12.80	—	10.44	23.24
2.01.15.04.	**Air ventilators; plaster; flyproof; set in plastering**									
2.01.15.04.A	225mm x 75mm	nr	0.15	7.76	—	9.22	16.97	8.53	—	10.14	18.67
2.01.15.04.B	225mm x 150mm	nr	0.20	10.59	—	10.39	20.98	11.65	—	11.43	23.08
2.01.15.04.C	225mm x 225mm	nr	0.22	11.64	—	11.69	23.33	12.80	—	12.86	25.66
2.01.15.05.	**Terracotta cavity liners; 200mm long; building in**									
2.01.15.05.A	215mm x 65mm	nr	0.20	10.59	—	7.12	17.71	11.65	—	7.84	19.49
2.01.15.05.B	215mm x 140mm	nr	0.25	13.31	—	8.06	21.38	14.65	—	8.87	23.51
2.01.15.05.C	215mm x 215mm	nr	0.37	19.40	—	19.36	38.76	21.33	—	21.30	42.63
2.01.15.06.	**Soot doors; double cover; frame; cast iron; building in**									
2.01.15.06.A	225mm x 150mm	nr	0.27	14.05	—	29.12	43.17	15.45	—	32.04	47.49
2.01.15.06.B	225mm x 225mm	nr	0.30	15.83	—	30.25	46.08	17.41	—	33.28	50.69

Hutchins Priced Schedules 2021	Unit	Labour Hours	Labour Net £	Plant Net £	Materials Net £	Unit Rate Net £	Labour Gross £	Plant Gross £	Materials Gross £	Unit Rate Gross £
						(Gross rates include 10% profit)				
2.01. **NEW WORK**										
2.01.17. **CHIMNEY FLUE LININGS, BENDS, POTS, ETC.**										
2.01.17.01. **Parge and core flues**										
2.01.17.01.A 225mm x 225mm...............	m	0.67	35.12	—	1.90	37.02	38.63	—	2.09	40.72
2.01.17.02. **Clay flue linings; BS EN 1457-1:2012; bedded and jointed in cement mortar (1:3)**										
2.01.17.02.A 125mm dia...................	m	0.34	17.61	—	6.49	24.11	19.37	—	7.14	26.52
2.01.17.02.B 150mm dia...................	m	0.37	19.40	—	7.04	26.43	21.33	—	7.74	29.08
2.01.17.02.C 185mm dia...................	m	0.40	21.07	—	12.00	33.07	23.18	—	13.20	36.38
2.01.17.02.D 210mm dia...................	m	0.42	22.02	—	13.46	35.47	24.22	—	14.80	39.02
2.01.17.02.E 225mm dia...................	m	0.44	23.06	—	15.48	38.54	25.37	—	17.02	42.39
2.01.17.02.F 250mm dia...................	m	0.47	24.64	—	17.83	42.47	27.10	—	19.61	46.71
2.01.17.02.G 300mm dia...................	m	0.47	24.64	—	28.90	53.54	27.10	—	31.79	58.89
2.01.17.02.H 185mm x 185mm...............	m	0.41	21.49	—	12.54	34.03	23.64	—	13.79	37.43
2.01.17.02.I 225mm x 225mm...............	m	0.42	22.02	—	20.34	42.36	24.22	—	22.38	46.59
2.01.17.02.J 300mm x 300mm...............	m	0.44	23.06	—	38.45	61.51	25.37	—	42.29	67.66
2.01.17.04. **Clay flue bends; BS EN 1457-1:2012; bedded and jointed in cement mortar (1:3)**										
2.01.17.04.A 125mm dia...................	nr	0.39	20.23	—	39.71	59.94	22.26	—	43.68	65.94
2.01.17.04.B 150mm dia...................	nr	0.42	22.02	—	41.13	63.14	24.22	—	45.24	69.46
2.01.17.04.C 185mm dia...................	nr	0.47	24.64	—	44.88	69.52	27.10	—	49.37	76.47
2.01.17.04.D 210mm dia...................	nr	0.49	25.69	—	83.49	109.17	28.25	—	91.84	120.09
2.01.17.04.E 225mm dia...................	nr	0.50	26.31	—	63.25	89.57	28.95	—	69.58	98.53
2.01.17.04.F 250mm dia...................	nr	0.56	29.25	—	104.54	133.79	32.18	—	114.99	147.17
2.01.17.04.G 300mm dia...................	nr	0.67	35.12	—	159.25	194.37	38.63	—	175.17	213.80
2.01.17.04.H 185mm x 185mm...............	nr	0.45	23.59	—	58.02	81.61	25.95	—	63.82	89.77
2.01.17.04.I 225mm x 225mm...............	nr	0.48	24.95	—	106.35	131.31	27.45	—	116.99	144.44
2.01.17.04.J 300mm x 300mm...............	nr	0.53	27.99	—	174.22	202.21	30.79	—	191.64	222.43
2.01.17.14. **Chimney pots; clay; roll top; set and flaunched in cement mortar (1:3); height**										
2.01.17.14.A 300mm.....................	nr	0.77	40.47	—	41.02	81.48	44.52	—	45.12	89.63
2.01.17.14.B 375mm.....................	nr	0.85	44.50	—	49.58	94.09	48.96	—	54.54	103.50
2.01.17.14.C 450mm.....................	nr	0.93	48.54	—	48.43	96.97	53.40	—	53.27	106.67
2.01.17.14.D 600mm.....................	nr	1.09	57.35	—	72.88	130.23	63.08	—	80.17	143.25
2.01.17.14.E 750mm.....................	nr	1.35	70.50	—	101.84	172.35	77.56	—	112.03	189.58

Brickwork and Blockwork

Hutchins Priced Schedules 2021	Unit	Labour Hours	Labour Net £	Plant Net £	Materials Net £	Unit Rate Net £	Labour Gross £	Plant Gross £	Materials Gross £	Unit Rate Gross £
						(Gross rates include 10% profit)				
2.01.	**NEW WORK**									
2.01.17.	**CHIMNEY FLUE LININGS, BENDS, POTS, ETC.**									
2.01.17.14.	**Chimney pots; clay; roll top; set and flaunched in cement mortar (1:3); height**									
2.01.17.14.F	900mm	nr	1.60	83.66	—	135.61	219.27	92.03	—	149.17	241.20
2.01.17.15.	**Chimney pots; clay; cannon head; set and flaunched in cement mortar (1:3); height**									
2.01.17.15.A	300mm	nr	0.77	40.47	—	54.47	94.94	44.52	—	59.92	104.44
2.01.17.15.B	450mm	nr	0.93	48.54	—	66.58	115.12	53.40	—	73.24	126.63
2.01.17.15.C	600mm	nr	1.09	57.35	—	97.85	155.20	63.08	—	107.63	170.72
2.01.17.23.	**Damp course to stacks (measured overall flues)**									
2.01.17.23.A	double slate	m²	2.01	105.47	—	120.91	226.38	116.02	—	133.00	249.02
2.01.18.	**FIREPLACES**									
2.01.18.01.	**Building in only**									
2.01.18.01.A	continuous burning fire	nr	2.41	126.54	—	10.09	136.63	139.20	—	11.09	150.29
2.01.18.01.C	underfloor draught fire	nr	5.37	281.29	—	10.09	291.37	309.41	—	11.09	320.51
2.01.18.01.D	tiled surround and hearth including assembly; joint and set in cement mortar (1:3)	nr	5.70	298.90	—	6.29	305.19	328.79	—	6.92	335.71
2.01.20.	**TURNING PIECES AND CENTRING**									
2.01.20.01.	**Turning pieces to flat arches (four uses assumed); 112mm soffit**									
2.01.20.01.A	first use	m	0.30	15.83	—	2.02	17.85	17.41	—	2.22	19.63
2.01.20.01.B	each subsequent use	m	0.27	14.05	—	2.27	16.32	15.45	—	2.49	17.95
2.01.20.02.	**Turning pieces to flat arches (four uses assumed); 225mm soffit**									
2.01.20.02.A	first use	m	0.37	19.40	—	5.92	25.32	21.33	—	6.51	27.85
2.01.20.02.B	each subsequent use	m	0.30	15.83	—	6.55	22.38	17.41	—	7.21	24.62
2.01.20.03.	**Centring to segmental arches (four uses assumed); 225mm wide soffit**									
2.01.20.03.A	1.20m span; first use	nr	0.84	43.93	—	12.97	56.90	48.32	—	14.27	62.59
2.01.20.03.B	1.50m span; first use	nr	0.87	45.71	—	16.25	61.96	50.28	—	17.88	68.16
2.01.20.03.C	1.80m span; first use	nr	0.97	50.95	—	19.53	70.48	56.05	—	21.48	77.53
2.01.20.03.D	1.20m span; each subsequent use	nr	0.71	37.32	—	14.23	51.56	41.06	—	15.66	56.71
2.01.20.03.E	1.50m span; each subsequent use	nr	0.74	38.84	—	17.89	56.73	42.73	—	19.68	62.40
2.01.20.03.F	1.80m span; each subsequent use	nr	0.83	43.30	—	21.54	64.84	47.63	—	23.69	71.32

Hutchins Priced Schedules 2021	Unit	Labour Hours	Labour Net £	Plant Net £	Materials Net £	Unit Rate Net £	Labour Gross £	Plant Gross £	Materials Gross £	Unit Rate Gross £
							(Gross rates include 10% profit)			
2.01. **NEW WORK**										
2.01.22. **SUNDRIES**										
2.01.22.01. **Forming cavity not exceeding 100mm; between skins of hollow wall**										
2.01.22.01.A no wall ties	m²	0.05	2.83	—	—	2.83	3.11	—	—	3.11
2.01.22.01.C twin triangular ties; stainless steel; 5nr per m2	m²	0.09	4.61	—	0.66	5.28	5.07	—	0.73	5.80
2.01.22.01.E Catnic ties; stainless steel; 5nr per m2 . .	m²	0.09	4.61	—	1.27	5.88	5.07	—	1.40	6.47
2.01.22.02. **Close cavity not exceeding 100mm wide; between skins of hollow wall at ends, jambs or sills of openings**										
2.01.22.02.A common brickwork (PC sum £450 per 1000) half brick thick	m	0.24	12.69	—	3.41	16.09	13.95	—	3.75	17.70
2.01.22.02.B common brickwork (PC sum £450 per 1000) half brick thick; bituminous felt . . .	m	0.32	16.88	—	4.92	21.80	18.57	—	5.41	23.98
2.01.22.02.C common brickwork (PC sum £450 per 1000) half brick thick; slate	m	0.40	21.07	—	9.16	30.24	23.18	—	10.08	33.26
2.01.22.02.D blockwork 100mm thick	m	0.20	10.59	—	5.14	15.73	11.65	—	5.65	17.30
2.01.22.02.E blockwork 100mm thick; bituminous felt .	m	0.24	12.69	—	6.65	19.34	13.95	—	7.32	21.27
2.01.22.02.F blockwork 100mm thick; slate	m	0.32	16.88	—	10.90	27.77	18.57	—	11.98	30.55
2.01.22.02.G Kingspan Thermabate 50; 50-60mm cavity width	m	0.08	4.19	—	6.01	10.20	4.61	—	6.61	11.22
2.01.22.02.H Kingspan Thermabate 65; 65-75mm cavity width	m	0.10	5.24	—	6.45	11.69	5.77	—	7.09	12.86
2.01.22.02.I Kingspan Thermabate 75; 75-85mm cavity width	m	0.11	5.98	—	6.45	12.42	6.57	—	7.09	13.67
2.01.22.02.J Kingspan Thermabate 85; 85-95mm cavity width	m	0.13	7.02	—	6.96	13.99	7.73	—	7.66	15.38
2.01.22.02.K Kingspan Thermabate 90; 90-100mm cavity width	m	0.15	7.76	—	7.62	15.38	8.53	—	8.39	16.92
2.01.22.03. **Close cavity not exceeding 100mm wide; at top of hollow wall with single course of blocks laid flat in gauged mortar (1:1:6)**										
2.01.22.03.A 75mm	m	0.18	9.65	—	2.68	12.32	10.61	—	2.94	13.55
2.01.22.03.B 100mm	m	0.22	11.64	—	3.48	15.12	12.80	—	3.83	16.63
2.01.22.03.C 150mm	m	0.30	15.83	—	7.49	23.32	17.41	—	8.23	25.65
2.01.22.04. **Rake out joints of brickwork**										
2.01.22.04.A to form key for plaster work	m²	0.34	17.61	—	—	17.61	19.37	—	—	19.37
2.01.22.04.B and point in cement mortar	m²	0.80	42.15	—	0.51	42.66	46.36	—	0.57	46.93
2.01.22.05. **Prepare top of brick wall**										
2.01.22.05.A for raising	m²	1.34	70.35	—	—	70.35	77.38	—	—	77.38

Brickwork and Blockwork

Hutchins Priced Schedules 2021		Unit	Labour Hours	Labour Net £	Plant Net £	Materials Net £	Unit Rate Net £	Labour Gross £	Plant Gross £	Materials Gross £	Unit Rate Gross £
								(Gross rates include 10% profit)			
2.01.	**NEW WORK**										
2.01.22.	**SUNDRIES**										
2.01.22.07.	**Setting brickwork up to 50mm forward or backward**										
2.01.22.07.A	raised or sunk panels	m²	0.44	22.86	—	—	22.86	25.14	—	—	25.14
2.01.22.08.	**Jablite expanded polystyrene board; cavity wall insulation; wedging in position between wall ties; fixing with insulation retaining discs**										
2.01.22.08.A	25mm thick	m²	0.13	7.02	—	2.61	9.63	7.73	—	2.87	10.59
2.01.22.08.B	40mm thick	m²	0.13	7.02	—	4.03	11.06	7.73	—	4.44	12.16
2.01.22.08.C	50mm thick	m²	0.15	7.76	—	5.23	12.99	8.53	—	5.75	14.29
2.01.22.08.D	75mm thick	m²	0.15	7.76	—	7.53	15.29	8.53	—	8.28	16.82
2.01.22.08.E	100mm thick	m²	0.16	8.49	—	10.02	18.51	9.34	—	11.02	20.36
2.01.22.09.	**DriTherm 37 cavity wall insulation; wedging in position between wall ties**										
2.01.22.09.A	50mm thick	m²	0.10	5.24	—	3.68	8.92	5.77	—	4.04	9.81
2.01.22.09.B	75mm thick	m²	0.11	5.98	—	3.93	9.90	6.57	—	4.32	10.89
2.01.22.09.C	85mm thick	m²	0.13	6.81	—	4.61	11.42	7.50	—	5.07	12.57
2.01.22.09.D	100mm thick	m²	0.15	7.76	—	4.66	12.42	8.53	—	5.13	13.66
2.01.22.10.	**Rockwool cavity wall insulation; wedging in position between wall ties**										
2.01.22.10.A	50mm thick	m²	0.10	5.24	—	4.42	9.66	5.77	—	4.86	10.63
2.01.22.10.B	75mm thick	m²	0.11	5.98	—	5.40	11.37	6.57	—	5.94	12.51
2.01.22.11.	**Fill bottom of cavity wall with concrete; 50mm cavity**										
2.01.22.11.A	300mm high	m	0.09	4.61	—	1.57	6.18	5.07	—	1.72	6.80
2.01.22.12.	**Double tile creasing course; nibless flat tiles (PC sum £600 per 1000); red; machine-made; projecting 50mm from wall face in cement mortar (1:3)**										
2.01.22.12.A	half brick wall	m	0.64	33.44	—	8.76	42.20	36.79	—	9.63	46.42
2.01.22.12.B	one brick wall	m	0.84	43.93	—	16.91	60.84	48.32	—	18.60	66.92
2.01.22.14.	**Set one course of brickwork**										
2.01.22.14.A	forward or backward (strings)	m	0.11	5.66	—	—	5.66	6.23	—	—	6.23
2.01.22.14.B	dentil course up to 50mm projection	m	0.22	11.64	—	—	11.64	12.80	—	—	12.80
2.01.22.14.C	oversailing per course	m	0.13	7.02	—	—	7.02	7.73	—	—	7.73
2.01.22.14.D	plinth course per course to 50mm projection	m	0.07	3.56	—	—	3.56	3.92	—	—	3.92

Brickwork and Blockwork

Hutchins Priced Schedules 2021		Unit	Labour Hours	Labour Net £	Plant Net £	Materials Net £	Unit Rate Net £	Labour Gross £	Plant Gross £	Materials Gross £	Unit Rate Gross £
							(Gross rates include 10% profit)				
2.01.	**NEW WORK**										
2.01.22.	**SUNDRIES**										
2.01.22.15.	**Wedge and pin up common brickwork (PC sum £450 per 1000) to underside of existing construction with slates in cement mortar (1:3)**										
2.01.22.15.A	half brick thick	m	0.28	14.78	—	6.26	21.04	16.26	—	6.88	23.14
2.01.22.15.B	one brick thick	m	0.57	29.88	—	10.76	40.63	32.87	—	11.83	44.70
2.01.22.15.C	one and a half brick thick	m	0.87	45.71	—	15.71	61.42	50.28	—	17.28	67.56
2.01.22.16.	**Bed wood frames and sills in mortar and point up**										
2.01.22.16.A	one side	m	0.28	14.78	—	0.51	15.30	16.26	—	0.57	16.83
2.01.22.16.B	both sides	m	0.44	23.17	—	0.90	24.07	25.49	—	0.99	26.48
2.01.22.17.	**Pointing in gun-grade polysulphide-based mastic sealant**										
2.01.22.17.A	one side	m	0.15	8.07	—	1.59	9.66	8.88	—	1.75	10.63
2.01.22.17.B	both sides	m	0.31	16.15	—	3.10	19.24	17.76	—	3.41	21.17
2.01.22.18.	**Bed plate in**										
2.01.22.18.A	mortar	m	0.15	7.76	—	1.31	9.06	8.53	—	1.44	9.97
2.01.22.22.	**Beam filling to**										
2.01.22.22.A	one brick wall	m	0.44	23.17	—	—	23.17	25.49	—	—	25.49
2.01.22.23.	**Cut chases in brickwork for**										
2.01.22.23.A	small pipe or conduit	m	0.60	31.66	—	—	31.66	34.83	—	—	34.83
2.01.22.24.	**Cut groove in brick sill for**										
2.01.22.24.A	water bar	m	0.54	28.20	—	—	28.20	31.02	—	—	31.02
2.01.22.25.	**Rake out joints in brickwork for**										
2.01.22.25.A	turn in of asphalt skirting or metal horizontal flashings	m	0.27	14.05	—	—	14.05	15.45	—	—	15.45
2.01.22.25.B	turn in of asphalt skirting or metal stepped flashings	m	0.40	21.07	—	—	21.07	23.18	—	—	23.18
2.01.22.26.	**Cut and bond ends of wall in class B engineering bricks (PC sum £500 per 1000) to existing**										
2.01.22.26.A	half brick thick	m	0.32	16.88	—	1.00	17.88	18.57	—	1.10	19.67
2.01.22.26.B	one brick thick	m	0.47	24.64	—	2.00	26.64	27.10	—	2.20	29.30
2.01.22.26.C	one and a half brick thick	m	0.69	36.17	—	3.00	39.17	39.79	—	3.30	43.09

Brickwork and Blockwork

Hutchins Priced Schedules 2021	Unit	Labour Hours	Labour Net £	Plant Net £	Materials Net £	Unit Rate Net £	Labour Gross £	Plant Gross £	Materials Gross £	Unit Rate Gross £
										(Gross rates include 10% profit)

2.01.   NEW WORK

2.01.22.   SUNDRIES

2.01.22.27.   Cut and bond ends of wall in common brickwork (PC sum £450 per 1000) to existing

2.01.22.27.A	half brick thick	m	0.32	16.88	—	0.90	17.78	18.57	—	0.99	19.56
2.01.22.27.B	one brick thick	m	0.47	24.64	—	1.80	26.44	27.10	—	1.98	29.08
2.01.22.27.C	one and a half brick thick	m	0.69	36.17	—	2.70	38.87	39.79	—	2.97	42.76
2.01.22.28.	**Cut and bond ends of wall in facing bricks (PC sum £550 per 1000) to existing**										
2.01.22.28.A	half brick thick	m	0.32	16.88	—	1.10	17.98	18.57	—	1.21	19.78
2.01.22.28.B	one brick thick	m	0.47	24.64	—	2.20	26.84	27.10	—	2.42	29.52
2.01.22.28.C	one and a half brick thick	m	0.69	36.17	—	3.30	39.47	39.79	—	3.63	43.42
2.01.22.29.	**Quoin up jambs in common brickwork (PC sum £450 per 1000) in gauged mortar (1:1:6)**										
2.01.22.29.A	half brick thick	m	0.80	42.15	—	5.46	47.61	46.36	—	6.01	52.37
2.01.22.29.B	one brick thick	m	1.21	63.32	—	10.35	73.67	69.66	—	11.39	81.04
2.01.22.29.C	one and a half brick thick	m	1.57	82.30	—	15.94	98.24	90.53	—	17.54	108.07
2.01.22.30.	**Quoin up jambs in facing bricks (PC sum £550 per 1000) in gauged mortar (1:1:6)**										
2.01.22.30.A	half brick thick	m	1.01	52.94	—	6.56	59.51	58.24	—	7.22	65.46
2.01.22.30.B	one brick thick	m	1.21	63.32	—	12.45	75.77	69.66	—	13.70	83.35
2.01.22.30.C	one and a half brick thick	m	1.85	97.08	—	19.14	116.23	106.79	—	21.06	127.85
2.01.22.32.	**Mesh reinforcement in walls**										
2.01.22.32.A	63mm wide	m	0.17	9.12	—	0.35	9.47	10.03	—	0.38	10.41
2.01.22.32.B	175mm wide	m	0.23	11.95	—	0.87	12.82	13.15	—	0.96	14.11
2.01.22.33.	**Angle fillets**										
2.01.22.33.A	cement mortar	m	0.27	14.05	—	0.47	14.52	15.45	—	0.52	15.98
2.01.22.34.	**Raking and cutting of facing bricks (PC sum £550 per 1000)**										
2.01.22.34.C	fair raking or splay cutting	m	0.22	11.64	—	6.05	17.69	12.80	—	6.66	19.46
2.01.22.34.D	fair curved cutting	m	0.44	23.17	—	6.05	29.22	25.49	—	6.66	32.14
2.01.22.34.E	fair squint or birdsmouth angle	m	0.22	11.64	—	7.70	19.34	12.80	—	8.47	21.27
2.01.22.34.F	fair chamfered or round angle	m	0.30	15.83	—	6.05	21.88	17.41	—	6.66	24.07
2.01.22.40.	**Catnic stainless steel stronghold wall connector system; plugging to wall with plugs and coach screws provided; twist and sliding arms to 225mm centres and building into joints of walls**										
2.01.22.40.A	60mm - 250mm thick	m	0.21	10.90	—	2.60	13.51	11.99	—	2.86	14.86

Hutchins Priced Schedules 2021	Unit	Labour Hours	Labour Net £	Plant Net £	Materials Net £	Unit Rate Net £	Labour Gross £	Plant Gross £	Materials Gross £	Unit Rate Gross £
							(Gross rates include 10% profit)			
2.01.	**NEW WORK**									
2.01.22.	**SUNDRIES**									
2.01.22.44.	**Unload; hoist; build in metal windows; door frames; pinning lugs; to brickwork; pointing externally**									
2.01.22.44.A	not exceeding 0.50m2	nr	0.74	38.69	—	0.36	39.04	42.55	—	0.39	42.95
2.01.22.44.B	0.50m2 - 1.00m2	nr	0.97	50.95	—	0.71	51.66	56.05	—	0.78	56.83
2.01.22.44.C	1.00m2 - 1.50m2	nr	1.34	70.35	—	1.19	71.53	77.38	—	1.31	78.69
2.01.22.44.D	exceeding 1.50m2	nr	1.51	79.15	—	1.54	80.70	87.07	—	1.70	88.77
2.01.22.45.	**Hole for small pipe (not exceeding 55mm dia) through walls; make good**									
2.01.22.45.A	half brick thick	nr	0.27	14.05	—	0.36	14.40	15.45	—	0.39	15.84
2.01.22.45.B	one brick thick	nr	0.44	23.17	—	0.47	23.64	25.49	—	0.52	26.01
2.01.22.45.C	one and a half brick thick	nr	0.72	37.95	—	0.59	38.55	41.75	—	0.65	42.40
2.01.22.45.D	blockwork 100mm thick	nr	0.24	12.69	—	0.24	12.92	13.95	—	0.26	14.22
2.01.22.46.	**Hole for large pipe (55mm - 110mm dia) through walls; make good**									
2.01.22.46.A	half brick thick	nr	0.32	16.88	—	0.36	17.24	18.57	—	0.39	18.96
2.01.22.46.B	one brick thick	nr	0.56	29.56	—	0.47	30.04	32.52	—	0.52	33.04
2.01.22.46.C	one and a half brick thick	nr	0.89	46.44	—	0.59	47.04	51.09	—	0.65	51.74
2.01.22.46.D	blockwork 100mm thick	nr	0.28	14.78	—	0.36	15.14	16.26	—	0.39	16.65
2.01.22.47.	**Hole for extra large pipe (exceeding 110mm dia) through walls; make good**									
2.01.22.47.A	half brick thick	nr	0.40	21.07	—	0.36	21.43	23.18	—	0.39	23.57
2.01.22.47.B	one brick thick	nr	0.68	35.86	—	0.47	36.33	39.44	—	0.52	39.96
2.01.22.47.C	one and a half brick thick	nr	0.82	42.98	—	0.59	43.58	47.28	—	0.65	47.94
2.01.22.47.D	blockwork 100mm thick	nr	0.36	18.98	—	0.36	19.33	20.87	—	0.39	21.27
2.01.22.50.	**Cut mortices for iron bolts, stays, etc. in brickwork and grout in cement mortar**									
2.01.22.50.A	per 25mm depth of mortice	nr	0.05	2.83	—	0.12	2.95	3.11	—	0.13	3.24
2.01.22.52.	**Cut and pin in brickwork**									
2.01.22.52.A	sink bearers, radiator brackets, holderbats, etc.	nr	0.50	26.31	—	0.47	26.79	28.95	—	0.52	29.47
2.01.22.52.B	end of steel joists; not exceeding 250mm high	nr	0.60	31.66	—	0.71	32.37	34.83	—	0.78	35.61
2.01.22.52.C	end of steel joists; 250mm - 500mm high .	nr	0.91	47.49	—	1.19	48.68	52.24	—	1.31	53.55
2.01.22.53.	**Galvanised frame ties screwed to wood frame and built into brickwork**									
2.01.22.53.A	203mm girth	nr	0.13	7.02	—	0.34	7.36	7.73	—	0.37	8.10

Brickwork and Blockwork

Hutchins Priced Schedules 2021	Unit	Labour Hours	Labour Net £	Plant Net £	Materials Net £	Unit Rate Net £	Labour Gross £	Plant Gross £	Materials Gross £	Unit Rate Gross £
						(Gross rates include 10% profit)				
2.01.	**NEW WORK**									
2.01.22.	**SUNDRIES**									
2.01.22.53.	**Galvanised frame ties screwed to wood frame and built into brickwork**									
2.01.22.53.B	254mm girth	nr	0.15	7.76	—	0.35	8.11	8.53	—	0.38	8.92
2.02.	**REPAIRS AND ALTERATIONS**									
2.02.01.	**DEMOLISHING BRICKWORK**									
2.02.01.01.	**Demolish brickwork, any height, and set aside arisings**									
2.02.01.01.A	half brick walls	m²	0.75	22.19	—	—	22.19	24.40	—	—	24.40
2.02.01.01.B	one brick walls	m²	1.35	39.93	—	—	39.93	43.93	—	—	43.93
2.02.01.01.C	one and a half brick walls	m²	1.85	54.72	—	—	54.72	60.20	—	—	60.20
2.02.01.01.D	two brick walls	m²	2.40	70.99	—	—	70.99	78.09	—	—	78.09
2.02.01.02.	**Sort, clean and stack sound bricks for reuse; removing remaining**									
2.02.01.02.A	lime mortar	1000	14.00	414.12	—	—	414.12	455.53	—	—	455.53
2.02.01.02.B	composite mortar	1000	17.00	502.86	—	—	502.86	553.15	—	—	553.15
2.02.01.02.C	cement mortar	1000	22.00	650.76	—	—	650.76	715.84	—	—	715.84
2.02.01.03.	**Hand load rubble to skip, cart away to tip**									
2.02.01.03.A	and pay all tipping fees	m³	2.00	59.16	43.60	—	102.76	65.08	47.96	—	113.04
2.02.03.	**CHIMNEY STACK REMOVAL**									
2.02.03.01.	**Pull down chimney stacks; clean sound whole bricks for reuse and remove remaining**									
2.02.03.01.A	up to 9.00m high or two storeys; lime mortar	m³	18.78	984.55	—	—	984.55	1083.01	—	—	1083.01
2.02.03.01.B	up to 9.00m high or two storeys; cement mortar	m³	24.15	1265.84	—	—	1265.84	1392.42	—	—	1392.42
2.02.03.01.C	extra over for each additional 3.00m or storey height; lime mortar	m³	4.70	246.16	—	—	246.16	270.78	—	—	270.78
2.02.03.01.D	extra over for each additional 3.00m or storey height; cement mortar	m³	6.04	316.41	—	—	316.41	348.05	—	—	348.05
2.02.05.	**CHIMNEY STACK REBUILDING**									
2.02.05.01.	**Rebuild single flue chimney in common brickwork (PC sum £450 per 1000), in cement mortar (1:3), include building in 185mm dia socketed and rebated clay flue liners, BS EN 13502:2002; up to 9.00m or two storeys high; overall plan dimensions**									
2.02.05.01.A	450mm x 450mm	m	4.02	210.94	—	74.68	285.61	232.03	—	82.14	314.18

Hutchins Priced Schedules 2021		Unit	Labour Hours	Labour Net £	Plant Net £	Materials Net £	Unit Rate Net £	Labour Gross £	Plant Gross £	Materials Gross £	Unit Rate Gross £
								(Gross rates include 10% profit)			
2.02.	**REPAIRS AND ALTERATIONS**										
2.02.05.	**CHIMNEY STACK REBUILDING**										
2.02.05.01.	**Rebuild single flue chimney in common brickwork (PC sum £450 per 1000), in cement mortar (1:3), include building in 185mm dia socketed and rebated clay flue liners, BS EN 13502:2002; up to 9.00m or two storeys high; overall plan dimensions**										
2.02.05.01.B	675mm x 675mm...............	m	6.48	339.68	—	116.16	455.84	373.65	—	127.78	501.43
2.02.05.02.	**Rebuild double flue chimney in common brickwork (PC sum £450 per 1000), in cement mortar (1:3), include building in 185mm dia socketed and rebated clay flue liners, BS EN 13502:2002; up to 9.00m or two storeys high; overall plan dimensions**										
2.02.05.02.A	450mm x 750mm...............	m	3.49	182.84	—	85.85	268.70	201.13	—	94.44	295.56
2.02.05.02.B	675mm x 675mm...............	m	5.84	305.92	—	128.76	434.68	336.52	—	141.63	478.15
2.02.05.03.	**Take off loose chimney pot and reset including flaunching**										
2.02.05.03.A	up to two storeys or 9.00m high......	nr	1.34	70.35	—	0.83	71.18	77.38	—	0.91	78.30
2.02.05.04.	**Take down and remove chimney pot, supply, set and flaunch new pot; up to two storeys or 9.00m high**										
2.02.05.04.A	300mm pot..................	nr	1.68	87.96	—	52.81	140.77	96.76	—	58.09	154.85
2.02.05.04.B	375mm pot..................	nr	1.71	89.69	—	64.92	154.61	98.66	—	71.41	170.07
2.02.05.04.C	450mm pot..................	nr	1.74	91.42	—	64.92	156.34	100.56	—	71.41	171.97
2.02.05.04.D	600mm pot..................	nr	1.81	94.99	—	96.19	191.17	104.48	—	105.81	210.29
2.02.05.04.E	add for each additional storey or 3.00m high; 300mm pot..............	nr	0.18	9.54	—	—	9.54	10.49	—	—	10.49
2.02.05.04.F	add for each additional storey or 3.00m high; 375mm pot..............	nr	0.22	11.43	—	—	11.43	12.57	—	—	12.57
2.02.05.04.G	add for each additional storey or 3.00m high; 450mm pot..............	nr	0.25	13.31	—	—	13.31	14.65	—	—	14.65
2.02.05.04.H	add for each additional storey or 3.00m high; 600mm pot..............	nr	0.34	17.61	—	—	17.61	19.37	—	—	19.37
2.02.06.	**DAMP PROOFING**										
2.02.06.01.	**Treat with silicone or similar damp-proofing liquid**										
2.02.06.01.A	external brick walls.............	m²	0.15	7.76	—	0.06	7.82	8.53	—	0.07	8.60
2.02.07.	**BRICKWORK REPAIRS**										
2.02.07.01.	**Cut out defective brickwork and reface with new facing bricks (PC sum £550 per 1000), in**										
2.02.07.01.A	cement mortar.................	m²	3.82	200.45	—	38.55	239.01	220.50	—	42.41	262.91
2.02.07.01.B	lime mortar..................	m²	3.35	175.82	—	38.75	214.56	193.40	—	42.62	236.02

Brickwork and Blockwork

Hutchins Priced Schedules 2021	Unit	Labour Hours	Labour Net £	Plant Net £	Materials Net £	Unit Rate Net £	Labour Gross £	Plant Gross £	Materials Gross £	Unit Rate Gross £
							(Gross rates include 10% profit)			
2.02. **REPAIRS AND ALTERATIONS**										
2.02.07. **BRICKWORK REPAIRS**										
2.02.07.02. **Cut out single facing bricks and reface with new facing bricks (PC sum £550 per 1000), in**										
2.02.07.02.A cement mortar...............	nr	0.34	17.61	—	0.67	18.28	19.37	—	0.74	20.11
2.02.07.02.B lime mortar...............	nr	0.23	12.27	—	0.68	12.94	13.49	—	0.75	14.24
2.02.07.03. **Rake out mortar and repoint**										
2.02.07.03.A perished mortar............	m²	0.58	30.30	—	0.51	30.81	33.33	—	0.57	33.89
2.02.07.03.B sound mortar.............	m²	1.44	75.59	—	0.51	76.10	83.15	—	0.57	83.71
2.02.07.04. **Cut out fractures in brickwork and lace in common brickwork (PC sum £450 per 1000), approximately 405mm wide, 225mm thick in**										
2.02.07.04.A cement mortar............	m	2.41	126.54	—	29.16	155.70	139.20	—	32.08	171.27
2.02.07.04.B lime mortar.............	m	1.61	84.40	—	29.25	113.65	92.84	—	32.18	125.01
2.02.07.05. **Cut out fractures in brickwork, and lace in facing bricks (PC sum £550 per 1000), approximately 405mm wide, 225mm thick in**										
2.02.07.05.A cement mortar............	m	2.65	138.91	—	35.15	174.07	152.80	—	38.67	191.47
2.02.07.05.B lime mortar.............	m	1.84	96.66	—	35.15	131.82	106.33	—	38.67	145.00
2.02.07.06. **Take down segmental arch and rebuild in facing bricks (PC sum £550 per 1000), 225mm high on face, including centring**										
2.02.07.06.A 113mm wide soffit.............	m	4.02	210.94	—	8.89	219.83	232.03	—	9.78	241.81
2.02.07.06.B 225mm wide soffit.............	m	5.23	274.26	—	17.01	291.28	301.69	—	18.72	320.40
2.02.09. **OPENINGS IN BRICK WALLS**										
2.02.09.01. **Cut opening through brickwork (common brickwork PC sum £450 per 1000, facing brickwork PC sum £550 per 1000) in cement mortar for doors, windows, etc.; including all necessary shoring; make good, in**										
2.02.09.01.A half brick walls...............	m²	1.34	58.72	0.36	20.09	79.17	64.59	0.39	22.10	87.09
2.02.09.01.B one brick walls...............	m²	2.68	117.54	0.72	36.98	155.24	129.29	0.79	40.68	170.77
2.02.09.01.C one and a half brick walls..........	m²	4.02	176.26	0.99	53.99	231.24	193.88	1.09	59.39	254.36
2.02.09.01.D two brick walls...............	m²	4.85	220.19	1.30	70.44	291.92	242.21	1.43	77.48	321.11

Hutchins Priced Schedules 2021	Unit	Labour Hours	Labour Net £	Plant Net £	Materials Net £	Unit Rate Net £	Labour Gross £	Plant Gross £	Materials Gross £	Unit Rate Gross £
							(Gross rates include 10% profit)			
2.02.	**REPAIRS AND ALTERATIONS**									
2.02.09.	**OPENINGS IN BRICK WALLS**									
2.02.09.03.	**Infill openings in brickwork in common brickwork (PC sum £450 per 1000), in gauged mortar; in small areas and bond to existing**									
2.02.09.03.A	half brick walls	m²	1.64	86.18	—	32.66	118.84	94.80	—	35.93	130.72
2.02.09.03.B	one brick walls	m²	3.02	158.20	—	65.00	223.20	174.02	—	71.50	245.52
2.02.09.03.C	one and a half brick walls	m²	4.53	237.36	—	97.66	335.02	261.09	—	107.43	368.52
2.02.10.	**DAMP PROOF COURSES**									
2.02.10.01.	**Damp-proof course in short lengths in existing walls including cutting out brickwork and building in with new common brickwork (PC sum £450 per 1000)**									
2.02.10.01.A	half brick wide; two course slate	m	1.24	65.11	—	7.61	72.71	71.62	—	8.37	79.98
2.02.10.01.B	half brick wide; bitumen felt	m	1.04	54.52	—	6.33	60.85	59.97	—	6.96	66.93
2.02.10.01.C	one brick wide and over; two course slate	m²	4.36	228.55	—	46.89	275.44	251.41	—	51.58	302.98
2.02.10.01.D	one brick wide and over; bitumen felt . . .	m²	3.59	188.08	—	58.99	247.07	206.89	—	64.89	271.78
2.02.15.	**FRAME REBEDDING**									
2.02.15.01.	**Take out and rebed door or window frame including**									
2.02.15.01.A	point externally; make good internally . . .	m	0.37	19.40	—	0.47	19.87	21.33	—	0.52	21.86
2.02.15.02.	**Rake out defective pointing around door or window frame and repoint in**									
2.02.15.02.A	cement mortar	m	0.30	15.83	—	0.13	15.96	17.41	—	0.14	17.56
2.02.15.02.B	mastic sealant	m	0.40	21.07	—	0.15	21.23	23.18	—	0.17	23.35
2.02.16.	**FIRE AND HEARTH REPAIRS**									
2.02.16.01.	**Take out existing fireplace, including surround and hearth**									
2.02.16.01.A	small iron	nr	2.68	140.70	—	—	140.70	154.76	—	—	154.76
2.02.16.01.B	large tiled	nr	3.45	181.06	—	—	181.06	199.16	—	—	199.16
2.02.16.01.C	free standing	nr	2.15	112.49	—	—	112.49	123.74	—	—	123.74
2.02.16.02.	**Take out and reset existing fireplace, including surround and hearth**									
2.02.16.02.A	small iron	nr	6.71	351.63	—	12.58	364.21	386.80	—	13.83	400.63
2.02.16.02.B	large tiled	nr	8.22	430.79	—	16.97	447.75	473.87	—	18.66	492.53
2.02.16.02.C	free standing	nr	5.50	288.31	—	4.98	293.29	317.14	—	5.48	322.62

Brickwork and Blockwork

Hutchins Priced Schedules 2021	Unit	Labour Hours	Labour Net £	Plant Net £	Materials Net £	Unit Rate Net £	Labour Gross £	Plant Gross £	Materials Gross £	Unit Rate Gross £
						(Gross rates include 10% profit)				
2.02.	**REPAIRS AND ALTERATIONS**									
2.02.18.	**VENTILATION**									
2.02.18.01.	**Air ventilator cut out and remove**									
2.02.18.01.A	plaster fly proof 225mm x 75mm	nr	0.20	10.59	—	—	10.59	11.65	—	—	11.65
2.02.18.01.B	plaster fly proof 225mm x 150mm	nr	0.25	13.00	—	—	13.00	14.30	—	—	14.30
2.02.18.01.C	terracotta 225mm x 75mm	nr	0.21	11.22	—	—	11.22	12.34	—	—	12.34
2.02.18.01.D	terracotta 225mm x 150mm	nr	0.28	14.78	—	—	14.78	16.26	—	—	16.26
2.02.18.01.E	galvanised iron 215mm x 65mm	nr	0.21	11.22	—	—	11.22	12.34	—	—	12.34
2.02.18.01.F	galvanised iron 215mm x 140mm	nr	0.28	14.78	—	—	14.78	16.26	—	—	16.26
2.02.18.03.	**Air ventilator cut through brick wall (any thickness); build in; make good**									
2.02.18.03.A	plaster fly proof 225mm x 75mm	nr	0.32	16.88	—	8.38	25.26	18.57	—	9.22	27.79
2.02.18.03.B	plaster fly proof 225mm x 150mm	nr	0.37	19.40	—	9.44	28.84	21.33	—	10.39	31.72
2.02.18.03.C	terracotta 215mm x 65mm; square hole .	nr	0.34	17.61	—	3.55	21.16	19.37	—	3.91	23.28
2.02.18.03.D	terracotta 215mm x 140mm; square hole .	nr	0.40	21.07	—	5.27	26.34	23.18	—	5.80	28.98
2.02.18.03.E	terracotta 215mm x 65mm; louvred hole .	nr	0.34	17.61	—	4.12	21.74	19.37	—	4.54	23.91
2.02.18.03.F	terracotta 215mm x 140mm; louvred hole	nr	0.40	21.07	—	6.74	27.81	23.18	—	7.42	30.60
2.02.18.03.G	galvanised iron 225mm x 75mm	nr	0.34	17.61	—	22.61	40.22	19.37	—	24.87	44.25
2.02.18.03.H	galvanised iron 225mm x 150mm	nr	0.40	21.07	—	28.26	49.33	23.18	—	31.09	54.27

Hutchins Priced Schedules 2021		Unit	Labour Hours	Labour Net £	Plant Net £	Materials Net £	Unit Rate Net £	Labour Gross £	Plant Gross £	Materials Gross £	Unit Rate Gross £
								(Gross rates include 10% profit)			
3.01.	**NEW WORK**										
3.01.01.	**CARCASSING SAWN SOFTWOOD**										
3.01.01.01.	**Floors**										
3.01.01.01.A	47mm x 100mm	m	0.12	4.52	—	3.50	8.01	4.97	—	3.85	8.81
3.01.01.01.B	47mm x 125mm	m	0.14	5.08	—	4.23	9.31	5.59	—	4.65	10.24
3.01.01.01.C	47mm x 150mm	m	0.15	5.53	—	5.07	10.60	6.08	—	5.58	11.66
3.01.01.01.D	47mm x 175mm	m	0.17	6.43	—	5.97	12.41	7.08	—	6.57	13.65
3.01.01.01.E	47mm x 200mm	m	0.20	7.38	—	6.86	14.23	8.11	—	7.54	15.66
3.01.01.01.F	47mm x 225mm	m	0.22	8.28	—	7.58	15.86	9.11	—	8.34	17.45
3.01.01.01.G	63mm x 125mm	m	0.15	5.80	—	5.41	11.20	6.37	—	5.95	12.32
3.01.01.01.H	63mm x 150mm	m	0.19	6.96	—	6.36	13.32	7.66	—	7.00	14.66
3.01.01.01.I	63mm x 175mm	m	0.22	8.13	—	7.47	15.59	8.94	—	8.21	17.15
3.01.01.01.J	63mm x 200mm	m	0.25	9.26	—	8.57	17.82	10.18	—	9.42	19.61
3.01.01.01.K	63mm x 225mm	m	0.28	10.42	—	9.56	19.98	11.47	—	10.51	21.98
3.01.01.01.L	75mm x 125mm	m	0.18	6.89	—	6.44	13.32	7.57	—	7.08	14.66
3.01.01.01.M	75mm x 150mm	m	0.22	8.28	—	7.58	15.86	9.11	—	8.34	17.45
3.01.01.01.N	75mm x 175mm	m	0.26	9.63	—	8.89	18.53	10.60	—	9.78	20.38
3.01.01.01.O	75mm x 200mm	m	0.29	11.03	—	10.21	21.23	12.13	—	11.23	23.35
3.01.01.01.P	75mm x 225mm	m	0.33	12.42	—	11.37	23.79	13.66	—	12.51	26.17
3.01.01.02.	**Partitions**										
3.01.01.02.B	38mm x 100mm	m	0.21	7.90	—	2.82	10.73	8.69	—	3.11	11.80
3.01.01.02.C	47mm x 75mm	m	0.22	8.28	—	2.63	10.90	9.11	—	2.89	11.99
3.01.01.02.D	47mm x 100mm	m	0.27	10.16	—	3.50	13.66	11.18	—	3.85	15.02
3.01.01.03.	**Flat roofs**										
3.01.01.03.A	47mm x 150mm	m	0.29	10.91	—	5.07	15.98	12.00	—	5.58	17.58
3.01.01.03.B	47mm x 175mm	m	0.32	12.04	—	5.97	18.02	13.25	—	6.57	19.82
3.01.01.03.C	47mm x 200mm	m	0.36	13.55	—	6.86	20.40	14.90	—	7.54	22.44
3.01.01.03.D	75mm x 100mm	m	0.29	10.91	—	5.15	16.06	12.00	—	5.66	17.66
3.01.01.04.	**Pitched roofs including ceiling joists**										
3.01.01.04.A	25mm x 100mm	m	0.12	4.52	—	1.86	6.37	4.97	—	2.04	7.01
3.01.01.04.B	25mm x 125mm	m	0.14	5.27	—	2.25	7.52	5.80	—	2.47	8.27
3.01.01.04.C	25mm x 150mm	m	0.14	5.27	—	2.70	7.97	5.80	—	2.97	8.76
3.01.01.04.D	38mm x 75mm	m	0.14	5.27	—	2.12	7.39	5.80	—	2.33	8.13
3.01.01.04.E	38mm x 100mm	m	0.14	5.27	—	2.82	8.09	5.80	—	3.11	8.90
3.01.01.04.F	38mm x 125mm	m	0.14	5.27	—	3.42	8.69	5.80	—	3.77	9.56
3.01.01.04.G	38mm x 150mm	m	0.15	5.64	—	4.11	9.75	6.21	—	4.52	10.73

Woodwork

Hutchins Priced Schedules 2021		Unit	Labour Hours	Labour Net £	Plant Net £	Materials Net £	Unit Rate Net £	Labour Gross £	Plant Gross £	Materials Gross £	Unit Rate Gross £
									(Gross rates include 10% profit)		
3.01.	**NEW WORK**										
3.01.01.	CARCASSING SAWN SOFTWOOD										
3.01.01.04.	Pitched roofs including ceiling joists										
3.01.01.04.H	47mm x 75mm	m	0.14	5.27	—	2.63	7.89	5.80	—	2.89	8.68
3.01.01.04.I	47mm x 100mm	m	0.15	5.64	—	3.50	9.14	6.21	—	3.85	10.06
3.01.01.04.J	47mm x 125mm	m	0.15	5.64	—	4.23	9.88	6.21	—	4.65	10.86
3.01.01.04.K	47mm x 150mm	m	0.15	5.64	—	5.07	10.72	6.21	—	5.58	11.79
3.01.01.04.L	63mm x 100mm	m	0.15	5.64	—	3.24	8.89	6.21	—	3.57	9.78
3.01.01.04.M	63mm x 125mm	m	0.15	5.64	—	5.41	11.05	6.21	—	5.95	12.16
3.01.01.04.N	63mm x 150mm	m	0.16	6.02	—	6.36	12.38	6.62	—	7.00	13.62
3.01.01.04.O	63mm x 175mm	m	0.16	6.02	—	7.47	13.49	6.62	—	8.21	14.83
3.01.01.04.P	75mm x 100mm	m	0.15	5.64	—	5.15	10.79	6.21	—	5.66	11.87
3.01.01.04.Q	75mm x 125mm	m	0.16	6.02	—	6.44	12.46	6.62	—	7.08	13.70
3.01.01.04.R	75mm x 150mm	m	0.17	6.40	—	7.58	13.98	7.04	—	8.34	15.38
3.01.01.04.S	75mm x 175mm	m	0.18	6.77	—	8.89	15.67	7.45	—	9.78	17.23
3.01.01.04.T	100mm x 150mm	m	0.18	6.77	—	10.11	16.88	7.45	—	11.12	18.57
3.01.01.04.U	100mm x 175mm	m	0.20	7.53	—	11.85	19.38	8.28	—	13.04	21.32
3.01.01.04.V	100mm x 200mm	m	0.21	7.90	—	13.13	21.03	8.69	—	14.44	23.13
3.01.01.04.W	100mm x 225mm	m	0.22	8.28	—	14.76	23.04	9.11	—	16.24	25.35
3.01.01.05.	**Kerbs, bearers and the like**										
3.01.01.05.A	25mm x 100mm	m	0.05	1.88	—	1.86	3.74	2.07	—	2.04	4.11
3.01.01.05.B	38mm x 75mm	m	0.05	1.88	—	2.12	4.00	2.07	—	2.33	4.40
3.01.01.05.C	38mm x 100mm	m	0.07	2.63	—	2.82	5.46	2.90	—	3.11	6.00
3.01.01.05.D	47mm x 75mm	m	0.07	2.63	—	2.63	5.26	2.90	—	2.89	5.79
3.01.01.05.E	47mm x 100mm	m	0.09	3.39	—	3.50	6.88	3.73	—	3.85	7.57
3.01.01.05.F	75mm x 100mm	m	0.14	5.27	—	5.15	10.41	5.80	—	5.66	11.45
3.01.01.06.	**Noggins**										
3.01.01.06.A	38mm x 50mm	m	0.30	11.29	—	1.42	12.71	12.42	—	1.56	13.98
3.01.01.06.B	47mm x 50mm	m	0.35	13.17	—	1.75	14.92	14.49	—	1.93	16.42
3.01.01.06.C	47mm x 75mm	m	0.40	15.05	—	1.89	16.94	16.56	—	2.08	18.64
3.01.01.07.	**Herringbone strutting between joists (measured over joists)**										
3.01.01.07.A	38mm x 50mm	m	0.45	16.93	—	1.42	18.35	18.63	—	1.56	20.19
3.01.01.07.B	47mm x 50mm	m	0.50	18.82	—	1.75	20.57	20.70	—	1.93	22.63
3.01.01.09.	**Solid strutting between joists (measured over joists)**										
3.01.01.09.A	47mm x 100mm	m	0.35	13.17	—	2.76	15.93	14.49	—	3.04	17.53
3.01.01.09.B	47mm x 125mm	m	0.40	15.05	—	4.23	19.28	16.56	—	4.65	21.21
3.01.01.09.C	47mm x 175mm	m	0.50	18.82	—	5.97	24.79	20.70	—	6.57	27.27

Hutchins Priced Schedules 2021		Unit	Labour Hours	Labour Net £	Plant Net £	Materials Net £	Unit Rate Net £	Labour Gross £	Plant Gross £	Materials Gross £	Unit Rate Gross £
								(Gross rates include 10% profit)			
3.01.	**NEW WORK**										
3.01.01.	**CARCASSING SAWN SOFTWOOD**										
3.01.01.10.	**Sprocket pieces**										
3.01.01.10.A	47mm x 50mm x 200mm	nr	0.17	6.40	—	0.37	6.77	7.04	—	0.41	7.45
3.01.01.10.B	47mm x 100mm x 600mm	nr	0.20	7.53	—	1.76	9.28	8.28	—	1.93	10.21
3.01.01.11.	**Extra labour trimming to openings**										
3.01.01.11.A	500mm x 1000mm; joist 47mm x 100mm	nr	1.35	50.80	—	—	50.80	55.88	—	—	55.88
3.01.01.11.B	750mm x 1000mm; joist 47mm x 100mm	nr	1.58	59.27	—	—	59.27	65.19	—	—	65.19
3.01.01.11.C	600mm x 1200mm; joist 47mm x 175mm	nr	1.62	60.96	—	—	60.96	67.06	—	—	67.06
3.01.01.11.D	900mm x 1000mm; joist 47mm x 175mm	nr	2.43	91.44	—	—	91.44	100.58	—	—	100.58
3.01.01.11.E	600mm x 1200mm; joist 47mm x 200mm	nr	1.62	60.96	—	—	60.96	67.06	—	—	67.06
3.01.01.11.F	900mm x 1500mm; joist 47mm x 200mm	nr	2.16	81.28	—	—	81.28	89.41	—	—	89.41
3.01.01.11.G	600mm x 1200mm; joist 75mm x 175mm	nr	1.62	60.96	—	—	60.96	67.06	—	—	67.06
3.01.01.11.H	1000mm x 2000mm; joist 75mm x 175mm	nr	2.70	101.60	—	—	101.60	111.76	—	—	111.76
3.01.01.12.	**Trussed rafter; stress graded; sawn softwood; pressure impregnated; raised through two storeys; fixed in position; 450mm eaves overhang fan truss 22.5 deg pitch; span over wall plates**										
3.01.01.12.A	5.00m	nr	0.83	55.72	—	49.24	104.96	61.29	—	54.16	115.45
3.01.01.12.B	6.00m	nr	0.85	56.93	—	60.16	117.08	62.62	—	66.17	128.79
3.01.01.12.C	7.00m	nr	0.86	58.07	—	68.63	126.70	63.88	—	75.49	139.37
3.01.01.12.D	8.00m	nr	0.88	59.28	—	79.55	138.83	65.21	—	87.50	152.71
3.01.01.12.E	9.00m	nr	0.90	60.42	—	108.63	169.05	66.46	—	119.50	185.96
3.01.01.12.F	10.00m	nr	0.92	61.63	—	117.39	179.02	67.79	—	129.13	196.92
3.01.01.13.	**Trussed rafter; stress graded; sawn softwood; pressure impregnated; raised through two storeys; fixed in position; 450mm eaves overhang fan truss 35 deg pitch; span over wall plates**										
3.01.01.13.A	5.00m	nr	0.86	58.07	—	52.88	110.95	63.88	—	58.17	122.04
3.01.01.13.B	6.00m	nr	0.88	59.28	—	67.42	126.70	65.21	—	74.16	139.37
3.01.01.13.C	7.00m	nr	0.90	60.42	—	77.12	137.54	66.46	—	84.83	151.29
3.01.01.13.D	8.00m	nr	0.92	61.63	—	88.02	149.65	67.79	—	96.82	164.62
3.01.01.13.E	9.00m	nr	0.93	35.15	—	119.54	154.68	38.66	—	131.49	170.15
3.01.01.13.F	10.00m	nr	0.95	63.98	—	129.23	193.22	70.38	—	142.16	212.54

Woodwork

|---|---|---|---|---|---|---|---|---|---|---|
							(Gross rates include 10% profit)			
3.01.	**NEW WORK**									
3.01.01.	CARCASSING SAWN SOFTWOOD									
3.01.01.14.	**Trussed rafter; stress graded; sawn softwood; pressure impregnated; raised through two storeys; fixed in position; 450mm eaves overhang fan truss 45 deg pitch; span over wall plates**									
3.01.01.14.A	5.00m	nr	0.95	63.85	—	58.94	122.79	70.23	—	64.83	135.06
3.01.01.14.B	6.00m	nr	0.97	65.19	—	75.91	141.10	71.71	—	83.50	155.21
3.01.01.14.C	7.00m	nr	0.99	66.47	—	91.66	158.13	73.12	—	100.83	173.94
3.01.01.14.D	8.00m	nr	1.01	67.75	—	115.90	183.65	74.52	—	127.49	202.01
3.01.01.14.E	9.00m	nr	1.03	69.02	—	141.35	210.37	75.93	—	155.48	231.41
3.01.01.14.F	10.00m	nr	1.05	70.37	—	154.68	225.05	77.41	—	170.15	247.55
3.01.01.15.	**Trussed rafter; stress graded; sawn softwood; pressure impregnated; raised through two storeys; fixed in position; 450mm eaves overhang fink truss 22.5 deg pitch; span over wall plates**									
3.01.01.15.A	5.00m	nr	0.83	55.72	—	41.71	97.43	61.29	—	45.88	107.17
3.01.01.15.B	6.00m	nr	0.85	56.93	—	50.93	107.86	62.62	—	56.03	118.64
3.01.01.15.C	7.00m	nr	0.86	58.07	—	58.09	116.16	63.88	—	63.90	127.78
3.01.01.15.D	8.00m	nr	0.88	59.28	—	67.31	126.59	65.21	—	74.04	139.25
3.01.01.15.E	9.00m	nr	0.90	60.42	—	91.88	152.30	66.46	—	101.07	167.53
3.01.01.15.F	10.00m	nr	0.92	61.63	—	99.28	160.92	67.79	—	109.21	177.01
3.01.01.16.	**Trussed rafter; stress graded; sawn softwood; pressure impregnated; raised through two storeys; fixed in position; 450mm eaves overhang fink truss 35 deg pitch; span over wall plates**									
3.01.01.16.A	5.00m	nr	0.86	58.07	—	44.79	102.86	63.88	—	49.27	113.15
3.01.01.16.B	6.00m	nr	0.88	59.28	—	57.07	116.35	65.21	—	62.78	127.99
3.01.01.16.C	7.00m	nr	0.90	60.42	—	65.26	125.69	66.46	—	71.79	138.26
3.01.01.16.D	8.00m	nr	0.92	61.63	—	74.47	136.10	67.79	—	81.92	149.72
3.01.01.16.E	9.00m	nr	0.93	62.77	—	101.09	163.86	69.05	—	111.20	180.25
3.01.01.16.F	10.00m	nr	0.95	63.98	—	109.28	173.26	70.38	—	120.21	190.59

Hutchins Priced Schedules 2021	Unit	Labour Hours	Labour Net £	Plant Net £	Materials Net £	Unit Rate Net £	Labour Gross £	Plant Gross £	Materials Gross £	Unit Rate Gross £
								(Gross rates include 10% profit)		
3.01.	**NEW WORK**									
3.01.01.	**CARCASSING SAWN SOFTWOOD**									
3.01.01.17.	**Trussed rafter; stress graded; sawn softwood; pressure impregnated; raised through two storeys; fixed in position; 450mm eaves overhang fink truss 45 deg pitch; span over wall plates**									
3.01.01.17.A	5.00m.	nr	0.95	63.85	—	53.65	117.50	70.23	—	59.01	129.25
3.01.01.17.B	6.00m.	nr	0.97	65.19	—	69.07	134.27	71.71	—	75.98	147.69
3.01.01.17.C	7.00m.	nr	0.99	66.47	—	83.40	149.87	73.12	—	91.74	164.86
3.01.01.17.D	8.00m.	nr	1.01	67.75	—	105.44	173.18	74.52	—	115.98	190.50
3.01.01.17.E	9.00m.	nr	1.03	69.02	—	128.57	197.59	75.93	—	141.42	217.35
3.01.01.17.F	10.00m.	nr	1.05	70.37	—	140.69	211.06	77.41	—	154.76	232.17
3.01.03.	**FIRST FIXINGS CHIPBOARD**									
3.01.03.01.	**Boarding to floors; moisture-resistant; tongued and grooved joints; thickness**									
3.01.03.01.A	18mm.	m²	0.35	13.17	—	5.47	18.64	14.49	—	6.02	20.51
3.01.03.01.B	18mm raking cutting	m	0.54	20.32	—	0.72	21.04	22.35	—	0.79	23.14
3.01.03.01.C	18mm curved cutting	m	0.93	35.00	—	1.66	36.65	38.50	—	1.82	40.32
3.01.03.01.D	22mm.	m²	0.42	15.80	—	6.02	21.82	17.39	—	6.62	24.00
3.01.03.01.E	22mm raking cutting	m	0.60	22.58	—	0.79	23.37	24.84	—	0.87	25.71
3.01.03.01.F	22mm curved cutting	m	0.99	37.25	—	1.82	39.08	40.98	—	2.00	42.98
3.01.05.	**FIRST FIXINGS STERLING BOARD**									
3.01.05.01.	**Boarding to floors; butt joints; thickness**									
3.01.05.01.A	18mm.	m²	0.32	12.04	—	6.51	18.55	13.25	—	7.16	20.41
3.01.05.01.B	18mm raking cutting	m	0.49	18.44	—	1.04	19.47	20.28	—	1.14	21.42
3.01.05.01.C	18mm curved cutting	m	0.84	31.61	—	2.07	33.68	34.77	—	2.28	37.05
3.01.05.01.D	11mm.	m²	0.27	10.16	—	5.11	15.27	11.18	—	5.62	16.80
3.01.05.01.E	11mm raking cutting	m	0.40	15.05	—	0.81	15.87	16.56	—	0.89	17.45
3.01.05.01.F	11mm curved cutting	m	0.63	23.71	—	1.63	25.33	26.08	—	1.79	27.87
3.01.06.	**FIRST FIXINGS PLYWOOD**									
3.01.06.01.	**Boarding to roofs; butt joints; thickness**									
3.01.06.01.A	18mm.	m²	0.44	16.56	—	7.74	24.30	18.21	—	8.51	26.73
3.01.06.01.B	18mm raking cutting	m	0.40	15.05	—	1.23	16.28	16.56	—	1.35	17.91
3.01.06.01.C	18mm curved cutting	m	1.06	39.89	—	2.46	42.35	43.88	—	2.71	46.58
3.01.06.01.D	25mm.	m²	0.50	18.82	—	13.77	32.58	20.70	—	15.14	35.84
3.01.06.01.E	25mm raking cutting	m	0.43	16.18	—	2.19	18.37	17.80	—	2.41	20.21
3.01.06.01.F	25mm curved cutting	m	1.13	42.52	—	4.38	46.90	46.77	—	4.82	51.59
3.01.06.02.	**Boarding to roofs; sloping; butt joints thickness**									
3.01.06.02.A	18mm.	m²	0.46	17.31	—	7.74	25.05	19.04	—	8.51	27.55
3.01.06.02.B	18mm raking cutting	m	0.40	15.05	—	1.23	16.28	16.56	—	1.35	17.91

Woodwork

(Gross rates include 10% profit)

3.01.   **NEW WORK**

3.01.06.   **FIRST FIXINGS PLYWOOD**

3.01.06.02.   **Boarding to roofs; sloping; butt joints thickness**

		Unit	Labour Hours	Labour Net £	Plant Net £	Materials Net £	Unit Rate Net £	Labour Gross £	Plant Gross £	Materials Gross £	Unit Rate Gross £
3.01.06.02.C	18mm curved cutting	m	1.06	39.89	—	2.46	42.35	43.88	—	2.71	46.58
3.01.06.02.D	25mm	m²	0.52	19.57	—	13.77	33.33	21.52	—	15.14	36.67
3.01.06.02.E	25mm raking cutting	m	0.43	16.18	—	2.19	18.37	17.80	—	2.41	20.21
3.01.06.02.F	25mm curved cutting	m	1.13	42.52	—	4.38	46.90	46.77	—	4.82	51.59
3.01.06.05.	**Boarding to dormers; tops or cheeks; butt joints; thickness**										
3.01.06.05.A	18mm; over 300mm wide	m²	0.73	27.47	—	7.74	35.21	30.22	—	8.51	38.73
3.01.06.05.B	18mm; not exceeding 150mm wide	m	0.11	4.14	—	1.23	5.37	4.55	—	1.35	5.91
3.01.06.05.C	18mm; 150mm to 300mm wide	m	0.22	8.28	—	2.46	10.74	9.11	—	2.71	11.81
3.01.06.05.D	18mm; raking cutting	m	0.40	15.05	—	1.23	16.28	16.56	—	1.35	17.91
3.01.06.05.E	18mm; curved cutting	m	1.06	39.89	—	2.46	42.35	43.88	—	2.71	46.58
3.01.06.05.F	25mm; over 300mm wide	m²	0.83	31.23	—	13.77	45.00	34.36	—	15.14	49.50
3.01.06.05.G	25mm; not exceeding 150mm wide	m	0.13	4.89	—	2.19	7.08	5.38	—	2.41	7.79
3.01.06.05.H	25mm; 150mm to 300mm wide	m	0.25	9.41	—	4.38	13.79	10.35	—	4.82	15.16
3.01.06.05.I	25mm; raking cutting	m	0.43	16.18	—	2.19	18.37	17.80	—	2.41	20.21
3.01.06.05.J	25mm; curved cutting	m	1.13	42.52	—	4.38	46.90	46.77	—	4.82	51.59
3.01.06.06.	**Boarding to gutters; bottoms or sides; butt joints; thickness**										
3.01.06.06.A	18mm; over 300mm wide	m²	2.76	103.86	—	7.74	111.60	114.24	—	8.51	122.76
3.01.06.06.B	18mm; not exceeding 150mm wide	m	0.41	15.43	—	1.23	16.66	16.97	—	1.35	18.32
3.01.06.06.C	18mm; 150mm to 300mm wide	m	0.83	31.23	—	2.46	33.69	34.36	—	2.71	37.06
3.01.06.06.D	18mm; raking cutting	m	0.40	15.05	—	1.23	16.28	16.56	—	1.35	17.91
3.01.06.06.E	18mm; curved cutting	m	1.06	39.89	—	2.46	42.35	43.88	—	2.71	46.58
3.01.06.06.F	25mm; over 300mm wide	m²	3.14	118.16	—	13.77	131.92	129.97	—	15.14	145.12
3.01.06.06.G	25mm; not exceeding 150mm wide	m	0.47	17.69	—	2.19	19.88	19.45	—	2.41	21.86
3.01.06.06.H	25mm; 150mm to 300mm wide	m	0.94	35.37	—	4.38	39.75	38.91	—	4.82	43.73
3.01.06.06.I	25mm; raking cutting	m	0.43	16.18	—	2.19	18.37	17.80	—	2.41	20.21
3.01.06.06.J	25mm; curved cutting	m	1.13	42.52	—	4.38	46.90	46.77	—	4.82	51.59
3.01.06.07.	**Boarding to eaves; verges; fascias and the like; butt joints; thickness**										
3.01.06.07.A	18mm; over 300mm wide	m²	1.52	57.20	—	7.74	64.94	62.92	—	8.51	71.43
3.01.06.07.B	18mm; not exceeding 150mm wide	m	0.23	8.65	—	1.23	9.89	9.52	—	1.35	10.87
3.01.06.07.C	18mm; 150mm to 300mm wide	m	0.46	17.31	—	2.46	19.77	19.04	—	2.71	21.75
3.01.06.07.D	18mm; raking cutting	m	0.40	15.05	—	1.23	16.28	16.56	—	1.35	17.91
3.01.06.07.E	18mm; curved cutting	m	1.06	39.89	—	2.46	42.35	43.88	—	2.71	46.58
3.01.06.07.F	25mm; over 300mm wide	m²	1.66	62.47	—	13.77	76.23	68.71	—	15.14	83.85
3.01.06.07.G	25mm; not exceeding 150mm wide	m	0.25	9.41	—	2.19	11.60	10.35	—	2.41	12.76
3.01.06.07.H	25mm; 150mm to 300mm wide	m	0.50	18.82	—	4.38	23.19	20.70	—	4.82	25.51

Hutchins Priced Schedules 2021	Unit	Labour Hours	Labour Net £	Plant Net £	Materials Net £	Unit Rate Net £	Labour Gross £	Plant Gross £	Materials Gross £	Unit Rate Gross £
							(Gross rates include 10% profit)			
3.01.	**NEW WORK**									
3.01.06.	FIRST FIXINGS PLYWOOD									
3.01.06.07.	Boarding to eaves; verges; fascias and the like; butt joints; thickness									
3.01.06.07.I	25mm; raking cutting	m	0.43	16.18	—	2.19	18.37	17.80	—	2.41	20.21
3.01.06.07.J	25mm; curved cutting	m	1.13	42.52	—	4.38	46.90	46.77	—	4.82	51.59
3.01.08.	FIRST FIXINGS SOFTWOOD									
3.01.08.01.	Boarding to roofs; 150mm wide boards; butt joints; thickness									
3.01.08.01.A	19mm	m²	0.70	26.34	—	14.28	40.62	28.98	—	15.71	44.68
3.01.08.01.B	19mm; raking cutting	m	0.20	7.53	—	2.27	9.80	8.28	—	2.50	10.78
3.01.08.01.C	19mm; curved cutting	m	0.60	22.58	—	4.54	27.12	24.84	—	5.00	29.83
3.01.08.01.D	25mm	m²	0.75	28.22	—	14.70	42.92	31.04	—	16.17	47.21
3.01.08.01.E	25mm; raking cutting	m	0.20	7.53	—	2.34	9.86	8.28	—	2.57	10.85
3.01.08.01.F	25mm; curved cutting	m	0.60	22.58	—	4.68	27.25	24.84	—	5.14	29.98
3.01.08.02.	Boarding to roofs; sloping; 150mm wide boards; butt joints; thickness									
3.01.08.02.A	19mm	m²	1.05	39.51	—	14.28	53.79	43.46	—	15.71	59.17
3.01.08.02.B	19mm; raking cutting	m	0.20	7.53	—	2.27	9.80	8.28	—	2.50	10.78
3.01.08.02.C	19mm; curved cutting	m	0.60	22.58	—	4.54	27.12	24.84	—	5.00	29.83
3.01.08.02.D	25mm	m²	1.13	42.52	—	14.70	57.22	46.77	—	16.17	62.94
3.01.08.02.E	25mm; raking cutting	m	0.20	7.53	—	2.34	9.86	8.28	—	2.57	10.85
3.01.08.02.F	25mm; curved cutting	m	0.60	22.58	—	4.68	27.25	24.84	—	5.14	29.98
3.01.08.03.	Boarding to dormers; tops or cheeks; 150mm wide boards; butt joints; thickness									
3.01.08.03.A	19mm	m²	0.99	37.25	—	14.28	51.53	40.98	—	15.71	56.69
3.01.08.03.B	19mm; raking cutting	m	0.20	7.53	—	2.27	9.80	8.28	—	2.50	10.78
3.01.08.03.C	19mm; curved cutting	m	0.60	22.58	—	4.54	27.12	24.84	—	5.00	29.83
3.01.08.03.D	25mm	m²	1.07	40.26	—	14.70	54.96	44.29	—	16.17	60.46
3.01.08.03.E	25mm; raking cutting	m	0.20	7.53	—	2.34	9.86	8.28	—	2.57	10.85
3.01.08.03.F	25mm; curved cutting	m	0.60	22.58	—	4.68	27.25	24.84	—	5.14	29.98
3.01.08.04.	Boarding to gutters; bottoms or sides; sloping; thickness									
3.01.08.04.A	19mm; over 300mm wide	m²	1.51	56.82	—	14.28	71.10	62.50	—	15.71	78.21
3.01.08.04.B	19mm; not exceeding 150mm wide	m	0.76	28.60	—	2.27	30.87	31.46	—	2.50	33.96
3.01.08.04.C	19mm; 150mm to 300mm wide	m	0.98	36.88	—	4.54	41.42	40.57	—	5.00	45.56
3.01.08.04.D	19mm; raking cutting	m	0.20	7.53	—	2.27	9.80	8.28	—	2.50	10.78
3.01.08.04.E	19mm; curved cutting	m	0.60	22.58	—	4.54	27.12	24.84	—	5.00	29.83
3.01.08.04.F	25mm; over 300mm wide	m²	1.51	56.82	—	14.70	71.52	62.50	—	16.17	78.67
3.01.08.04.G	25mm; not exceeding 150mm wide	m	0.76	28.60	—	2.34	30.94	31.46	—	2.57	34.03
3.01.08.04.H	25mm; 150mm to 300mm wide	m	0.98	36.88	—	4.68	41.55	40.57	—	5.14	45.71

Woodwork

Hutchins Priced Schedules 2021	Unit	Labour Hours	Labour Net £	Plant Net £	Materials Net £	Unit Rate Net £	Labour Gross £	Plant Gross £	Materials Gross £	Unit Rate Gross £
								(Gross rates include 10% profit)		
3.01.	**NEW WORK**									
3.01.08.	**FIRST FIXINGS SOFTWOOD**									
3.01.08.04.	Boarding to gutters; bottoms or sides; sloping; thickness									
3.01.08.04.I	25mm; raking cutting	m	0.20	7.53	—	2.34	9.86	8.28	—	2.57	10.85
3.01.08.04.J	25mm; curved cutting	m	0.60	22.58	—	4.68	27.25	24.84	—	5.14	29.98
3.01.08.05.	Boarding to verges; fascias; soffits, thickness									
3.01.08.05.A	19mm; over 300mm wide	m²	1.34	50.42	—	14.28	64.70	55.47	—	15.71	71.17
3.01.08.05.B	19mm; not exceeding 150mm wide	m	0.68	25.59	—	2.27	27.86	28.15	—	2.50	30.65
3.01.08.05.C	19mm; 150mm to 300mm wide	m	0.87	32.74	—	4.54	37.28	36.01	—	5.00	41.01
3.01.08.05.D	19mm; raking cutting	m	0.20	7.53	—	2.27	9.80	8.28	—	2.50	10.78
3.01.08.05.E	19mm; curved cutting	m	0.60	22.58	—	4.54	27.12	24.84	—	5.00	29.83
3.01.08.05.F	25mm; over 300mm wide	m²	1.34	50.42	—	14.70	65.12	55.47	—	16.17	71.64
3.01.08.05.G	25mm; not exceeding 150mm wide	m	0.68	25.59	—	2.34	27.93	28.15	—	2.57	30.72
3.01.08.05.H	25mm; 150mm to 300mm wide	m	0.87	32.74	—	4.68	37.41	36.01	—	5.14	41.16
3.01.08.05.I	25mm; raking cutting	m	0.20	7.53	—	2.34	9.86	8.28	—	2.57	10.85
3.01.08.05.J	25mm; curved cutting	m	0.60	22.58	—	4.68	27.25	24.84	—	5.14	29.98
3.01.08.10.	Firrings; 50mm wide; average depth									
3.01.08.10.A	38mm	m	0.15	5.64	—	0.76	6.40	6.21	—	0.83	7.04
3.01.08.10.B	50mm	m	0.15	5.64	—	0.93	6.58	6.21	—	1.03	7.24
3.01.08.10.C	75mm	m	0.17	6.40	—	1.41	7.80	7.04	—	1.55	8.58
3.01.08.11.	Bearers									
3.01.08.11.A	25mm x 50mm	m	0.12	4.52	—	0.93	5.45	4.97	—	1.03	6.00
3.01.08.11.B	38mm x 50mm	m	0.12	4.52	—	1.42	5.93	4.97	—	1.56	6.53
3.01.08.11.C	47mm x 50mm	m	0.12	4.52	—	1.75	6.27	4.97	—	1.93	6.90
3.01.08.11.D	47mm x 75mm	m	0.14	5.27	—	1.89	7.16	5.80	—	2.08	7.87
3.01.08.12.	Angle fillets									
3.01.08.12.A	38mm x 38mm	m	0.13	4.89	—	0.65	5.54	5.38	—	0.72	6.10
3.01.08.12.B	47mm x 50mm	m	0.13	4.89	—	0.98	5.87	5.38	—	1.07	6.46
3.01.08.12.C	75mm x 75mm	m	0.15	5.64	—	1.49	7.14	6.21	—	1.64	7.85
3.01.08.13.	Tilting fillets									
3.01.08.13.A	19mm x 38mm	m	0.12	4.52	—	0.54	5.05	4.97	—	0.59	5.56
3.01.08.13.B	25mm x 50mm	m	0.12	4.52	—	0.93	5.45	4.97	—	1.03	6.00
3.01.08.13.C	38mm x 50mm	m	0.12	4.52	—	1.42	5.93	4.97	—	1.56	6.53
3.01.08.13.D	47mm x 75mm	m	0.14	5.27	—	1.89	7.16	5.80	—	2.08	7.87
3.01.08.13.E	75mm x 100mm	m	0.16	6.02	—	5.15	11.17	6.62	—	5.66	12.28
3.01.08.14.	Grounds or battens									
3.01.08.14.A	19mm x 38mm	m	0.07	2.63	—	0.54	3.17	2.90	—	0.59	3.49
3.01.08.14.B	19mm x 50mm	m	0.07	2.63	—	0.70	3.34	2.90	—	0.77	3.67

Hutchins Priced Schedules 2021		Unit	Labour Hours	Labour Net £	Plant Net £	Materials Net £	Unit Rate Net £	Labour Gross £	Plant Gross £	Materials Gross £	Unit Rate Gross £
									(Gross rates include 10% profit)		
3.01.	**NEW WORK**										
3.01.08.	**FIRST FIXINGS SOFTWOOD**										
3.01.08.14.	**Grounds or battens**										
3.01.08.14.C	25mm x 50mm	m	0.07	2.63	—	0.93	3.57	2.90	—	1.03	3.93
3.01.08.15.	**Framework to bath panel at 500mm centres both ways**										
3.01.08.15.A	25mm x 50mm	m²	1.09	41.02	—	6.02	47.04	45.12	—	6.62	51.74
3.01.08.18.	**Framework to walls at 300mm centres one way; 600mm centres other way**										
3.01.08.18.A	25mm x 50mm	m²	1.54	57.95	—	4.95	62.90	63.75	—	5.45	69.19
3.01.08.18.B	38mm x 50mm	m²	1.54	57.95	—	7.51	65.46	63.75	—	8.26	72.01
3.01.08.18.C	47mm x 50mm	m²	1.54	57.95	—	9.29	67.24	63.75	—	10.22	73.97
3.01.08.18.D	47mm x 75mm	m²	1.56	58.70	—	10.02	68.72	64.57	—	11.02	75.59
3.01.08.18.E	75mm x 75mm	m²	1.58	59.46	—	20.48	79.93	65.40	—	22.53	87.93
3.01.08.19.	**Framework as bracketing and cradling around steelwork**										
3.01.08.19.A	25mm x 50mm	m²	1.70	63.97	—	7.25	71.23	70.37	—	7.98	78.35
3.01.08.19.B	47mm x 50mm	m²	1.80	67.73	—	13.61	81.35	74.51	—	14.97	89.48
3.01.08.19.C	47mm x 75mm	m²	1.90	71.50	—	14.67	86.17	78.65	—	16.14	94.79
3.01.08.20.	**Blockings wedged between flanges of steelwork**										
3.01.08.20.A	47mm x 50mm x 150mm	nr	0.14	5.27	—	0.28	5.55	5.80	—	0.31	6.10
3.01.08.20.B	47mm x 75mm x 225mm	nr	0.15	5.64	—	0.45	6.09	6.21	—	0.50	6.70
3.01.08.20.C	47mm x 100mm x 300mm	nr	0.16	6.02	—	0.88	6.90	6.62	—	0.97	7.59
3.01.08.22.	**Floor fillets set in concrete**										
3.01.08.22.A	38mm x 50mm	m	0.14	5.27	—	1.42	6.69	5.80	—	1.56	7.35
3.01.08.22.B	47mm x 50mm	m	0.14	5.27	—	1.75	7.02	5.80	—	1.93	7.72
3.01.09.	**FIRST FIXINGS SOFTWOOD BOARDING**										
3.01.09.01.	**Boarding to floors; 100mm wide boards; butt joints; thickness**										
3.01.09.01.A	19mm	m²	0.60	22.58	—	14.28	36.86	24.84	—	15.71	40.54
3.01.09.01.B	19mm raking cutting	m	0.20	7.53	—	2.27	9.80	8.28	—	2.50	10.78
3.01.09.01.C	19mm curved cutting	m	0.60	22.58	—	4.54	27.12	24.84	—	5.00	29.83
3.01.09.01.D	25mm	m²	0.60	22.58	—	14.70	37.28	24.84	—	16.17	41.01
3.01.09.01.E	25mm raking cutting	m	0.20	7.53	—	2.34	9.86	8.28	—	2.57	10.85
3.01.09.01.F	25mm curved cutting	m	0.60	22.58	—	4.68	27.25	24.84	—	5.14	29.98
3.01.09.01.G	32mm	m²	0.66	24.84	—	28.85	53.69	27.32	—	31.74	59.06
3.01.09.01.H	32mm raking cutting	m	0.22	8.28	—	4.59	12.87	9.11	—	5.05	14.15
3.01.09.01.I	32mm curved cutting	m	0.63	23.71	—	9.18	32.89	26.08	—	10.10	36.17

Woodwork

|---|---|---|---|---|---|---|---|---|---|---|
| | | | | | | (Gross rates include 10% profit) | | | | |

3.01.   **NEW WORK**

3.01.09.   **FIRST FIXINGS SOFTWOOD BOARDING**

3.01.09.02.   **Boarding to floors; 150mm wide boards; butt joints; thickness**

3.01.09.02.A	19mm	m²	0.50	18.82	—	14.35	33.17	20.70	—	15.79	36.49
3.01.09.02.B	19mm raking cutting	m	0.20	7.53	—	2.28	9.81	8.28	—	2.51	10.79
3.01.09.02.C	19mm curved cutting	m	0.60	22.58	—	4.57	27.14	24.84	—	5.02	29.86
3.01.09.02.D	25mm	m²	0.50	18.82	—	15.47	34.28	20.70	—	17.01	37.71
3.01.09.02.E	25mm raking cutting	m	0.20	7.53	—	2.46	9.99	8.28	—	2.71	10.98
3.01.09.02.F	25mm curved cutting	m	0.60	22.58	—	4.92	27.50	24.84	—	5.41	30.25
3.01.09.02.G	32mm	m²	0.55	20.70	—	19.60	40.30	22.77	—	21.56	44.33
3.01.09.02.H	32mm raking cutting	m	0.20	7.53	—	3.12	10.64	8.28	—	3.43	11.71
3.01.09.02.I	32mm curved cutting	m	0.60	22.58	—	6.24	28.81	24.84	—	6.86	31.70

3.01.09.03.   **Boarding to floors; 125mm wide boards; tongued and grooved joints, thickness**

3.01.09.03.A	22mm	m²	0.65	24.46	—	13.91	38.37	26.91	—	15.30	42.21
3.01.09.03.B	22mm raking cutting	m	0.20	7.53	—	2.21	9.74	8.28	—	2.43	10.71
3.01.09.03.C	22mm curved cutting	m	0.60	22.58	—	4.43	27.00	24.84	—	4.87	29.70
3.01.09.03.D	25mm	m²	0.65	24.46	—	17.30	41.76	26.91	—	19.03	45.94
3.01.09.03.E	25mm raking cutting	m	0.20	7.53	—	2.75	10.28	8.28	—	3.03	11.31
3.01.09.03.F	25mm curved cutting	m	0.60	22.58	—	5.50	28.08	24.84	—	6.05	30.89

3.01.09.04.   **Boarding to floors; 150mm wide boards; tongued and grooved joints; thickness**

3.01.09.04.A	25mm	m²	0.60	22.58	—	14.35	36.93	24.84	—	15.79	40.62
3.01.09.04.B	25mm raking cutting	m	0.20	7.53	—	2.28	9.81	8.28	—	2.51	10.79
3.01.09.04.C	25mm curved cutting	m	0.60	22.58	—	4.57	27.14	24.84	—	5.02	29.86

3.01.09.05.   **Mitred margin**

3.01.09.05.A	25mm x 75mm	m	0.40	15.05	—	1.37	16.42	16.56	—	1.50	18.06

3.01.09.06.   **Boarding to integral walls; 125mm wide boards; tongued and grooved and V jointed; thickness**

3.01.09.06.A	19mm	m²	0.91	34.24	—	17.91	52.16	37.67	—	19.70	57.37
3.01.09.06.B	19mm raking cutting	m	0.20	7.53	—	2.85	10.38	8.28	—	3.13	11.41
3.01.09.06.C	19mm curved cutting	m	0.60	22.58	—	5.70	28.28	24.84	—	6.27	31.10
3.01.09.06.D	25mm	m²	0.91	34.24	—	20.25	54.50	37.67	—	22.28	59.95
3.01.09.06.E	25mm raking cutting	m	0.20	7.53	—	3.22	10.75	8.28	—	3.54	11.82
3.01.09.06.F	25mm curved cutting	m	0.60	22.58	—	6.44	29.02	24.84	—	7.09	31.92

Woodwork

Hutchins Priced Schedules 2021	Unit	Labour Hours	Labour Net £	Plant Net £	Materials Net £	Unit Rate Net £	Labour Gross £	Plant Gross £	Materials Gross £	Unit Rate Gross £
									(Gross rates include 10% profit)	
3.01.	**NEW WORK**									
3.01.09.	**FIRST FIXINGS SOFTWOOD BOARDING**									
3.01.09.07.	**Boarding to internal ceilings; 125mm wide boards; tongued, grooved and V jointed; thickness**									
3.01.09.07.A	19mm	m²	1.16	43.65	—	17.91	61.56	48.02	—	19.70	67.72
3.01.09.07.B	19mm raking cutting	m	0.20	7.53	—	2.85	10.38	8.28	—	3.13	11.41
3.01.09.07.C	19mm curved cutting	m	0.60	22.58	—	5.70	28.28	24.84	—	6.27	31.10
3.01.09.07.D	25mm	m²	1.16	43.65	—	20.25	63.91	48.02	—	22.28	70.30
3.01.09.07.E	25mm raking cutting	m	0.20	7.53	—	3.22	10.75	8.28	—	3.54	11.82
3.01.09.07.F	25mm curved cutting	m	0.60	22.58	—	6.44	29.02	24.84	—	7.09	31.92
3.01.09.08.	**Boarding to internal walls; knotty pine; 100mm wide boards; tongued and grooved; thickness**									
3.01.09.08.A	13mm	m²	0.97	36.50	—	14.65	51.15	40.15	—	16.11	56.26
3.01.09.08.B	13mm raking cutting	m	0.20	7.53	—	2.33	9.86	8.28	—	2.56	10.84
3.01.09.08.C	13mm curved cutting	m	0.60	22.58	—	4.66	27.24	24.84	—	5.13	29.96
3.01.09.09.	**Boarding to internal ceilings; knotty pine; 100mm wide boards; tongued and grooved; thickness**									
3.01.09.09.A	13mm	m²	1.21	45.53	—	14.65	60.18	50.09	—	16.11	66.20
3.01.09.09.B	13mm raking cutting	m	0.20	7.53	—	2.33	9.86	8.28	—	2.56	10.84
3.01.09.09.C	13mm curved cutting	m	0.60	22.58	—	4.66	27.24	24.84	—	5.13	29.96
3.01.10.	**SECOND FIXINGS SHEET LININGS**									
3.01.10.01.	**3.2mm hardboard linings to walls**									
3.01.10.01.A	over 300mm wide	m²	0.36	13.55	—	1.41	14.95	14.90	—	1.55	16.45
3.01.10.01.B	not exceeding 300mm wide	m	0.15	5.64	—	0.47	6.11	6.21	—	0.52	6.72
3.01.10.02.	**3.2mm hardboard linings to ceilings**									
3.01.10.02.A	over 300mm wide	m²	0.41	15.43	—	1.41	16.84	16.97	—	1.55	18.52
3.01.10.02.B	not exceeding 300mm wide	m	0.17	6.40	—	0.47	6.87	7.04	—	0.52	7.55
3.01.10.02.C	raking cutting	m	0.16	6.02	—	0.08	6.10	6.62	—	0.08	6.71
3.01.10.02.D	curved cutting	m	0.53	19.94	—	0.15	20.09	21.94	—	0.16	22.10
3.01.10.03.	**12mm chipboard lining to walls**									
3.01.10.03.A	over 300mm wide	m²	0.46	17.31	—	2.99	20.30	19.04	—	3.29	22.33
3.01.10.03.B	not exceeding 300mm wide	m	0.19	7.15	—	1.00	8.15	7.86	—	1.10	8.96
3.01.10.04.	**12mm chipboard lining to ceilings**									
3.01.10.04.A	over 300mm wide	m²	0.53	19.94	—	2.99	22.94	21.94	—	3.29	25.23
3.01.10.04.B	not exceeding 300mm wide	m	0.22	8.28	—	1.00	9.28	9.11	—	1.10	10.20
3.01.10.04.C	raking cutting	m	0.33	12.42	—	0.48	12.89	13.66	—	0.52	14.18

Woodwork

	Unit	Labour Hours	Labour Net £	Plant Net £	Materials Net £	Unit Rate Net £	Labour Gross £	Plant Gross £	Materials Gross £	Unit Rate Gross £
										(Gross rates include 10% profit)
3.01.	**NEW WORK**									
3.01.10.	**SECOND FIXINGS SHEET LININGS**									
3.01.10.04.	**12mm chipboard lining to ceilings**									
3.01.10.04.D	curved cutting	m	0.73	27.47	—	0.95	28.42	30.22	—	1.05	31.26
3.01.10.06.	**12mm Celotex insulation board to walls**									
3.01.10.06.A	over 300mm wide	m²	0.29	10.91	—	4.89	15.81	12.00	—	5.38	17.39
3.01.10.06.B	not exceeding 300mm wide	m	0.12	4.52	—	1.63	6.15	4.97	—	1.79	6.76
3.01.10.07.	**12mm Celotex insulation board to ceilings**									
3.01.10.07.A	over 300mm wide	m²	0.33	12.42	—	4.89	17.31	13.66	—	5.38	19.04
3.01.10.07.B	not exceeding 300mm wide	m	0.14	5.27	—	1.63	6.90	5.80	—	1.79	7.59
3.01.10.07.C	raking cutting	m	0.14	5.27	—	0.78	6.05	5.80	—	0.86	6.65
3.01.10.07.D	curved cutting	m	0.40	15.05	—	1.56	16.61	16.56	—	1.71	18.27
3.01.10.08.	**6mm fire retardant MDF board lining to walls**									
3.01.10.08.A	over 300mm wide	m²	0.40	15.05	—	13.85	28.90	16.56	—	15.23	31.79
3.01.10.08.B	not exceeding 300mm wide	m	0.16	6.02	—	4.62	10.64	6.62	—	5.08	11.70
3.01.10.09.	**6mm fire retardant MDF board lining to ceilings**									
3.01.10.09.A	over 300mm wide	m²	0.46	17.31	—	13.85	31.16	19.04	—	15.23	34.28
3.01.10.09.B	not exceeding 300mm wide	m	0.18	6.77	—	4.62	11.39	7.45	—	5.08	12.53
3.01.10.09.C	raking cutting	m	0.33	12.42	—	2.20	14.62	13.66	—	2.42	16.08
3.01.10.09.D	curved cutting	m	0.93	35.00	—	4.41	39.40	38.50	—	4.85	43.34
3.01.10.10.	**9mm fire retardant MDF board lining to walls**									
3.01.10.10.A	over 300mm wide	m²	0.43	16.18	—	16.87	33.05	17.80	—	18.56	36.36
3.01.10.10.B	not exceeding 300mm wide	m	0.18	6.77	—	5.62	12.40	7.45	—	6.19	13.64
3.01.10.11.	**9mm fire retardant MDF board lining to ceilings**									
3.01.10.11.A	over 300mm wide	m²	0.50	18.82	—	16.87	35.69	20.70	—	18.56	39.26
3.01.10.11.B	not exceeding 300mm wide	m	0.21	7.90	—	5.62	13.53	8.69	—	6.19	14.88
3.01.10.11.C	raking cutting	m	0.33	12.42	—	2.68	15.10	13.66	—	2.95	16.61
3.01.10.11.D	curved cutting	m	0.93	35.00	—	5.37	40.36	38.50	—	5.90	44.40
3.01.10.14.	**4mm plywood birch faced lining to walls**									
3.01.10.14.A	over 300mm wide	m²	0.44	16.56	—	3.45	20.01	18.21	—	3.80	22.01
3.01.10.14.B	not exceeding 300mm wide	m	0.18	6.77	—	1.15	7.92	7.45	—	1.27	8.72
3.01.10.16.	**4mm plywood birch faced lining to ceilings**									
3.01.10.16.A	over 300mm wide	m²	0.51	19.19	—	3.45	22.65	21.11	—	3.80	24.91

Hutchins Priced Schedules 2021	Unit	Labour Hours	Labour Net £	Plant Net £	Materials Net £	Unit Rate Net £	Labour Gross £	Plant Gross £	Materials Gross £	Unit Rate Gross £
								(Gross rates include 10% profit)		
3.01.	**NEW WORK**									
3.01.10.	**SECOND FIXINGS SHEET LININGS**									
3.01.10.16.	**4mm plywood birch faced lining to ceilings**									
3.01.10.16.B	not exceeding 300mm wide	m	0.21	7.90	—	1.15	9.05	8.69	—	1.27	9.96
3.01.10.16.C	raking cutting	m	0.27	10.16	—	0.55	10.71	11.18	—	0.60	11.78
3.01.10.16.D	curved cutting	m	0.60	22.58	—	1.10	23.68	24.84	—	1.21	26.04
3.01.10.18.	**5.5mm plywood birch faced lining to walls**									
3.01.10.18.A	over 300mm wide	m²	0.48	18.06	—	4.55	22.61	19.87	—	5.00	24.87
3.01.10.18.B	not exceeding 300mm wide	m	0.20	7.53	—	1.52	9.04	8.28	—	1.67	9.95
3.01.10.19.	**5.5mm plywood birch faced lining to ceilings**									
3.01.10.19.A	over 300mm wide	m²	0.55	20.70	—	4.55	25.24	22.77	—	5.00	27.77
3.01.10.19.B	not exceeding 300mm wide	m	0.23	8.65	—	1.52	10.17	9.52	—	1.67	11.19
3.01.10.19.C	raking cutting	m	0.27	10.16	—	0.72	10.88	11.18	—	0.80	11.97
3.01.10.19.D	curved cutting	m	0.60	22.58	—	1.45	24.02	24.84	—	1.59	26.43
3.01.10.21.	**12mm blockboard lining to walls**									
3.01.10.21.A	over 300mm wide	m²	0.56	21.07	—	11.13	32.20	23.18	—	12.24	35.42
3.01.10.21.B	not exceeding 300mm wide	m	0.23	8.65	—	3.71	12.36	9.52	—	4.08	13.60
3.01.10.22.	**12mm blockboard lining to ceilings**									
3.01.10.22.A	over 300mm wide	m²	0.64	24.08	—	11.13	35.21	26.49	—	12.24	38.73
3.01.10.22.B	not exceeding 300mm wide	m	0.27	10.16	—	3.71	13.87	11.18	—	4.08	15.26
3.01.10.22.C	raking cutting	m	0.33	12.42	—	1.77	14.19	13.66	—	1.95	15.61
3.01.10.22.D	curved cutting	m	0.73	27.47	—	3.54	31.01	30.22	—	3.89	34.11
3.01.10.24.	**18mm blockboard lining to walls**									
3.01.10.24.A	over 300mm wide	m²	0.60	22.58	—	8.91	31.49	24.84	—	9.81	34.64
3.01.10.24.B	not exceeding 300mm wide	m	0.25	9.41	—	2.97	12.38	10.35	—	3.27	13.62
3.01.10.25.	**18mm blockboard lining to ceilings**									
3.01.10.25.A	over 300mm wide	m²	0.69	25.96	—	8.91	34.88	28.56	—	9.81	38.37
3.01.10.25.B	not exceeding 300mm wide	m	0.29	10.91	—	2.97	13.88	12.00	—	3.27	15.27
3.01.10.25.C	raking cutting	m	0.33	12.42	—	1.42	13.84	13.66	—	1.56	15.22
3.01.10.25.D	curved cutting	m	0.85	31.99	—	2.84	34.82	35.18	—	3.12	38.30
3.01.12.	**SECOND FIXINGS SHEET CASINGS**									
3.01.12.01.	**Cupboards; 3.2mm hardboard sides; softwood framing**									
3.01.12.01.A	25mm x 25mm	m²	6.45	242.71	—	6.23	248.94	266.98	—	6.85	273.84
3.01.12.01.B	32mm x 32mm	m²	6.70	252.12	—	7.05	259.17	277.33	—	7.75	285.08
3.01.12.01.C	38mm x 38mm	m²	6.70	252.12	—	7.79	259.91	277.33	—	8.57	285.90

Woodwork

Hutchins Priced Schedules 2021

Hutchins Priced Schedules 2021	Unit	Labour Hours	Labour Net £	Plant Net £	Materials Net £	Unit Rate Net £	Labour Gross £	Plant Gross £	Materials Gross £	Unit Rate Gross £
						(Gross rates include 10% profit)				
3.01.	**NEW WORK**									
3.01.12.	**SECOND FIXINGS SHEET CASINGS**									
3.01.12.02.	**Cupboards; hardboard backs**									
3.01.12.02.A	3.2mm	m²	0.40	15.05	—	1.41	16.46	16.56	—	1.55	18.10
3.01.12.04.	**Cupboards; 3.2mm hardboard doors; softwood framing**									
3.01.12.04.A	25mm x 25mm	m²	7.50	282.23	—	6.23	288.46	310.45	—	6.85	317.30
3.01.12.04.B	32mm x 32mm	m²	7.80	293.51	—	7.05	300.56	322.87	—	7.75	330.62
3.01.12.04.C	38mm x 38mm	m²	7.80	293.51	—	7.79	301.30	322.87	—	8.57	331.43
3.01.12.06.	**Cupboards; 5.5mm plywood sides; softwood framing**									
3.01.12.06.A	25mm x 25mm	m²	6.45	242.71	—	9.37	252.08	266.98	—	10.31	277.29
3.01.12.06.B	32mm x 32mm	m²	6.70	252.12	—	10.19	262.31	277.33	—	11.20	288.54
3.01.12.06.C	38mm x 38mm	m²	6.70	252.12	—	10.93	263.05	277.33	—	12.02	289.35
3.01.12.08.	**Cupboards; plywood backs**									
3.01.12.08.A	6.5mm	m²	0.50	18.82	—	6.32	25.14	20.70	—	6.95	27.65
3.01.12.09.	**Cupboards; 6.5mm plywood doors; softwood framing**									
3.01.12.09.A	25mm x 25mm	m²	7.50	282.23	—	11.14	293.37	310.45	—	12.26	322.71
3.01.12.09.B	32mm x 32mm	m²	7.50	282.23	—	11.96	294.19	310.45	—	13.16	323.60
3.01.12.09.C	38mm x 38mm	m²	7.80	293.51	—	12.70	306.22	322.87	—	13.97	336.84
3.01.12.12.	**Boxed pipe casing; 150mm height x 150mm deep; 19mm x 25mm sawn softwood framing; front fixed with brass screws and cups**									
3.01.12.12.A	3.2mm hardboard front	m	1.15	43.27	—	1.83	45.11	47.60	—	2.01	49.62
3.01.12.12.B	6.5mm plywood front	m	1.30	48.92	—	3.47	52.39	53.81	—	3.82	57.63
3.01.12.13.	**Pelmet casings 225mm girth; 75mm x 25mm sawn softwood framing; 19mm x 100mm softwood top; 6.5mm plywood front 125mm deep**									
3.01.12.13.A	over 1.00mm long	m	1.25	47.04	—	2.13	49.17	51.74	—	2.34	54.08
3.01.12.13.B	short lengths; not exceeding 0.90m long .	nr	1.30	48.92	—	2.13	51.05	53.81	—	2.34	56.15
3.01.12.13.C	short lengths; not exceeding 1.20m long .	nr	1.50	56.45	—	2.54	58.99	62.09	—	2.80	64.89
3.01.12.13.D	extra for boxed end	nr	0.45	16.93	—	0.13	17.06	18.63	—	0.14	18.77
3.01.12.14.	**Pelmet casings 225mm girth; 75mm x 25mm sawn softwood framing; part plugged and screwed to brick or concrete; 19mm x 100mm softwood top; 6.5mm plywood front 125mm deep**									
3.01.12.14.A	over 1.00m long	m	1.50	56.45	—	2.13	58.57	62.09	—	2.34	64.43
3.01.12.14.B	short lengths; not exceeding 0.90m long .	nr	1.60	60.21	—	2.13	62.34	66.23	—	2.34	68.57

Hutchins Priced Schedules 2021		Unit	Labour Hours	Labour Net £	Plant Net £	Materials Net £	Unit Rate Net £	Labour Gross £	Plant Gross £	Materials Gross £	Unit Rate Gross £
								(Gross rates include 10% profit)			
3.01.	**NEW WORK**										
3.01.12.	**SECOND FIXINGS SHEET CASINGS**										
3.01.12.14.	Pelmet casings 225mm girth; 75mm x 25mm sawn softwood framing; part plugged and screwed to brick or concrete; 19mm x 100mm softwood top; 6.5mm plywood front 125mm deep										
3.01.12.14.C	short lengths; not exceeding 1.20m long .	nr	1.80	67.73	—	2.54	70.28	74.51	—	2.80	77.31
3.01.12.14.D	extra for boxed end	nr	0.45	16.93	—	0.13	17.06	18.63	—	0.14	18.77
3.01.12.16.	**Plastic curtain track including**										
3.01.12.16.A	brackets, gliders, stops and the like	m	0.65	24.46	—	9.91	34.37	26.91	—	10.90	37.80
3.01.13.	**SECOND FIXINGS PVCU CLADDING**										
3.01.13.01.	**Swish profiled cladding; fixed in accordance with manufacturer's instructions; 38mm x 25mm sawn softwood framing at 450mm centres**										
3.01.13.01.A	to walls over 300mm wide	m²	0.50	33.54	—	24.17	57.71	36.89	—	26.59	63.48
3.01.13.02.	**Extra over cladding for**										
3.01.13.02.A	edge trim	m	0.11	7.53	—	4.34	11.86	8.28	—	4.77	13.05
3.01.16.	**SECOND FIXINGS WROUGHT SOFTWOOD**										
3.01.16.01.	**Skirtings**										
3.01.16.01.A	19mm x 100mm chamfered or pencil rounded	m	0.12	4.52	—	1.30	5.82	4.97	—	1.43	6.40
3.01.16.01.B	25mm x 150mm Torus/Ovolo; Torus/Ogee	m	0.14	5.27	—	4.44	9.71	5.80	—	4.89	10.68
3.01.16.01.C	25mm x 225mm Torus/Ogee	m	0.17	6.40	—	5.76	12.16	7.04	—	6.34	13.38
3.01.16.01.D	returned ends	nr	0.19	7.15	—	—	7.15	7.86	—	—	7.86
3.01.16.01.E	mitres	nr	0.12	4.52	—	—	4.52	4.97	—	—	4.97
3.01.16.02.	**Picture rail**										
3.01.16.02.A	25mm x 50mm	m	0.12	4.52	—	3.17	7.69	4.97	—	3.49	8.46
3.01.16.03.	**Dado rail**										
3.01.16.03.A	32mm x 63mm	m	0.19	7.15	—	3.09	10.24	7.86	—	3.40	11.26
3.01.16.04.	**Architraves**										
3.01.16.04.A	19mm x 50mm chamfered and rounded .	m	0.14	5.27	—	1.22	6.49	5.80	—	1.34	7.13
3.01.16.04.B	19mm x 63mm chamfered and rounded .	m	0.14	5.27	—	1.53	6.80	5.80	—	1.69	7.48
3.01.16.04.C	25mm x 75mm chamfered and rounded .	m	0.14	5.27	—	2.12	7.39	5.80	—	2.33	8.13
3.01.16.04.D	19mm x 50mm Ogee	m	0.14	5.27	—	1.92	7.19	5.80	—	2.11	7.91
3.01.16.04.E	19mm x 63mm Ogee	m	0.14	5.27	—	2.34	7.61	5.80	—	2.58	8.37
3.01.16.04.F	19mm x 50mm Ovolo	m	0.14	5.27	—	2.11	7.38	5.80	—	2.32	8.12
3.01.16.04.G	25mm x 75mm Torus	m	0.14	5.27	—	2.25	7.52	5.80	—	2.47	8.27
3.01.16.04.H	38mm x 150mm Edwardian	m	0.14	5.27	—	10.19	15.45	5.80	—	11.20	17.00
3.01.16.04.I	19mm x 50mm twice rounded	m	0.14	5.27	—	3.07	8.33	5.80	—	3.37	9.17

Woodwork

Hutchins Priced Schedules 2021	Unit	Labour Hours	Labour Net £	Plant Net £	Materials Net £	Unit Rate Net £	Labour Gross £	Plant Gross £	Materials Gross £	Unit Rate Gross £
						(Gross rates include 10% profit)				
3.01.	**NEW WORK**									
3.01.16.	**SECOND FIXINGS WROUGHT SOFTWOOD**									
3.01.16.04.	**Architraves**									
3.01.16.04.J	returned ends	nr	0.19	7.15	—	—	7.15	7.86	—	—	7.86
3.01.16.04.K	mitres	nr	0.12	4.52	—	—	4.52	4.97	—	—	4.97
3.01.16.06.	**Stops**									
3.01.16.06.B	16mm x 38mm Ovolo	m	0.12	4.52	—	1.07	5.59	4.97	—	1.18	6.15
3.01.16.06.C	16mm x 50mm Ovolo	m	0.12	4.52	—	1.34	5.86	4.97	—	1.48	6.45
3.01.16.06.G	32mm x 38mm fire check	m	0.14	5.27	—	2.40	7.67	5.80	—	2.64	8.44
3.01.16.07.	**Glazing beads**									
3.01.16.07.A	8mm x 12mm	m	0.06	2.26	—	1.05	3.31	2.48	—	1.16	3.64
3.01.16.07.B	8mm x 16mm	m	0.06	2.26	—	1.91	4.17	2.48	—	2.10	4.59
3.01.16.07.C	25mm x 45mm fire check	m	0.12	4.52	—	3.24	7.76	4.97	—	3.57	8.54
3.01.16.08.	**Quadrants**									
3.01.16.08.A	6mm	m	0.07	2.63	—	0.67	3.31	2.90	—	0.74	3.64
3.01.16.08.B	9mm	m	0.07	2.63	—	0.75	3.38	2.90	—	0.82	3.72
3.01.16.08.C	12mm	m	0.07	2.63	—	0.95	3.58	2.90	—	1.04	3.94
3.01.16.08.D	15mm	m	0.07	2.63	—	1.22	3.85	2.90	—	1.34	4.24
3.01.16.08.E	18mm	m	0.07	2.63	—	1.42	4.05	2.90	—	1.56	4.46
3.01.16.08.F	21mm	m	0.07	2.63	—	1.58	4.21	2.90	—	1.73	4.63
3.01.16.09.	**Half rounds**									
3.01.16.09.A	16mm x 4mm	m	0.07	2.63	—	0.75	3.38	2.90	—	0.82	3.72
3.01.16.09.B	18mm x 8mm	m	0.07	2.63	—	0.99	3.62	2.90	—	1.09	3.98
3.01.16.11.	**Scotia**									
3.01.16.11.A	9mm	m	0.07	2.63	—	1.38	4.01	2.90	—	1.51	4.41
3.01.16.11.B	15mm	m	0.07	2.63	—	1.62	4.25	2.90	—	1.78	4.68
3.01.16.11.D	18mm	m	0.07	2.63	—	1.69	4.32	2.90	—	1.86	4.76
3.01.16.11.E	21mm	m	0.07	2.63	—	2.46	5.09	2.90	—	2.70	5.60
3.01.16.11.F	30mm	m	0.07	2.63	—	3.42	6.06	2.90	—	3.77	6.66
3.01.16.15.	**Window boards**									
3.01.16.15.A	32mm x 150mm	m	0.24	9.03	—	3.79	12.82	9.93	—	4.17	14.10
3.01.16.15.C	32mm x 225mm	m	0.32	12.04	—	8.13	20.17	13.25	—	8.94	22.19
3.01.16.17.	**Floating shelves; solid oak**									
3.01.16.17.F	30mm x 200mm	m	0.22	8.28	—	33.83	42.11	9.11	—	37.21	46.32
3.01.16.17.G	40mm x 200mm	m	0.22	8.28	—	37.92	46.19	9.11	—	41.71	50.81
3.01.16.21.	**Shelves; worktops; slatted with 50mm wide slats at 75mm centres; thickness**									
3.01.16.21.A	19mm	m²	1.56	58.70	—	8.76	67.46	64.57	—	9.63	74.20
3.01.16.21.B	25mm	m²	1.56	58.70	—	11.58	70.28	64.57	—	12.73	77.31

Hutchins Priced Schedules 2021	Unit	Labour Hours	Labour Net £	Plant Net £	Materials Net £	Unit Rate Net £	Labour Gross £	Plant Gross £	Materials Gross £	Unit Rate Gross £
						(Gross rates include 10% profit)				

3.01. **NEW WORK**

3.01.16. **SECOND FIXINGS WROUGHT SOFTWOOD**

3.01.16.21. Shelves; worktops; slatted with 50mm wide slats at 75mm centres; thickness

3.01.16.21.C	32mm	m²	1.56	58.70	—	17.51	76.21	64.57	—	19.26	83.84

3.01.16.23. Bearers

3.01.16.23.A	19mm x 38mm	m	0.12	4.52	—	0.40	4.91	4.97	—	0.44	5.41
3.01.16.23.B	25mm x 50mm	m	0.12	4.52	—	0.82	5.33	4.97	—	0.90	5.87
3.01.16.23.C	47mm x 50mm	m	0.12	4.52	—	1.46	5.98	4.97	—	1.61	6.57
3.01.16.23.D	47mm x 75mm	m	0.12	4.52	—	1.94	6.46	4.97	—	2.14	7.10

3.01.16.24. Bearers; framed

3.01.16.24.A	19mm x 38mm	m	0.15	5.64	—	0.40	6.04	6.21	—	0.44	6.65
3.01.16.24.B	25mm x 50mm	m	0.15	5.64	—	0.82	6.46	6.21	—	0.90	7.11
3.01.16.24.C	47mm x 50mm	m	0.15	5.64	—	1.46	7.10	6.21	—	1.61	7.81
3.01.16.24.D	47mm x 75mm	m	0.15	5.64	—	1.94	7.59	6.21	—	2.14	8.35

3.01.16.26. Framing; framed

3.01.16.26.A	19mm x 38mm	m	0.19	7.15	—	0.40	7.55	7.86	—	0.44	8.30
3.01.16.26.B	25mm x 50mm	m	0.19	7.15	—	0.82	7.97	7.86	—	0.90	8.77
3.01.16.26.C	47mm x 50mm	m	0.19	7.15	—	1.46	8.61	7.86	—	1.61	9.47
3.01.16.26.D	47mm x 75mm	m	0.19	7.15	—	1.94	9.09	7.86	—	2.14	10.00

3.01.16.40. Frames; at jambs or heads

3.01.16.40.A	32mm x 75mm	m	0.25	9.41	—	1.90	11.31	10.35	—	2.09	12.44
3.01.16.40.B	32mm x 100mm	m	0.25	9.41	—	2.40	11.81	10.35	—	2.64	12.99
3.01.16.40.C	32mm x 150mm	m	0.25	9.41	—	2.94	12.35	10.35	—	3.23	13.58
3.01.16.40.D	47mm x 75mm	m	0.28	10.54	—	1.94	12.48	11.59	—	2.14	13.73
3.01.16.40.E	47mm x 100mm	m	0.28	10.54	—	2.60	13.14	11.59	—	2.86	14.45
3.01.16.40.F	47mm x 150mm	m	0.28	10.54	—	4.65	15.19	11.59	—	5.12	16.71

3.01.16.41. Frames; once rebated; at jambs or heads

3.01.16.41.A	47mm x 75mm	m	0.28	10.54	—	9.21	19.74	11.59	—	10.13	21.72
3.01.16.41.B	47mm x 100mm	m	0.28	10.54	—	10.16	20.70	11.59	—	11.18	22.77
3.01.16.41.C	47mm x 150mm	m	0.31	11.67	—	14.98	26.65	12.83	—	16.48	29.31
3.01.16.41.D	75mm x 100mm	m	0.35	13.17	—	15.55	28.72	14.49	—	17.11	31.59
3.01.16.41.F	75mm x 150mm	m	0.35	13.17	—	21.83	35.00	14.49	—	24.01	38.50

3.01.16.42. Frames; once rebated; once grooved; at jambs or heads

3.01.16.42.A	47mm x 100mm	m	0.28	10.54	—	10.83	21.36	11.59	—	11.91	23.50
3.01.16.42.B	47mm x 125mm	m	0.28	10.54	—	13.41	23.94	11.59	—	14.75	26.34
3.01.16.42.C	47mm x 150mm	m	0.31	11.67	—	15.70	27.36	12.83	—	17.27	30.10
3.01.16.42.D	75mm x 100mm	m	0.35	13.17	—	16.23	29.40	14.49	—	17.86	32.34
3.01.16.42.F	75mm x 150mm	m	0.35	13.17	—	22.49	35.66	14.49	—	24.74	39.23

Woodwork

Hutchins Priced Schedules 2021	Unit	Labour Hours	Labour Net £	Plant Net £	Materials Net £	Unit Rate Net £	Labour Gross £	Plant Gross £	Materials Gross £	Unit Rate Gross £
							(Gross rates include 10% profit)			
3.01.	**NEW WORK**									
3.01.16.	**SECOND FIXINGS WROUGHT SOFTWOOD**									
3.01.16.43.	**Frames; at mullions or transoms**									
3.01.16.43.A	32mm x 75mm	m	0.19	7.15	—	6.88	14.03	7.86	—	7.57	15.43
3.01.16.43.B	32mm x 100mm	m	0.19	7.15	—	8.55	15.70	7.86	—	9.40	17.27
3.01.16.43.C	32mm x 150mm	m	0.19	7.15	—	11.79	18.94	7.86	—	12.97	20.84
3.01.16.44.	**Frames; twice rebated; at mullions or transoms**									
3.01.16.44.A	38mm x 100mm	m	0.28	10.54	—	9.42	19.95	11.59	—	10.36	21.95
3.01.16.44.B	38mm x 150mm	m	0.28	10.54	—	13.15	23.68	11.59	—	14.46	26.05
3.01.16.44.C	47mm x 100mm	m	0.28	10.54	—	10.82	21.35	11.59	—	11.90	23.49
3.01.16.44.D	75mm x 100mm	m	0.35	13.17	—	16.19	29.36	14.49	—	17.81	32.30
3.01.16.44.E	75mm x 150mm	m	0.35	13.17	—	22.48	35.65	14.49	—	24.73	39.22
3.01.16.45.	**Frames; once sunk weathered; once rebated; three times grooved; at sills**									
3.01.16.45.A	75mm x 150mm	m	0.35	13.17	—	27.86	41.03	14.49	—	30.64	45.13
3.01.16.45.B	75mm x 175mm	m	0.35	13.17	—	35.56	48.73	14.49	—	39.12	53.61
3.01.16.46.	**Linings; tongued at angles**									
3.01.16.46.A	25mm x 75mm	m	0.25	9.41	—	6.27	15.68	10.35	—	6.90	17.24
3.01.16.46.B	25mm x 100mm	m	0.25	9.41	—	7.71	17.11	10.35	—	8.48	18.83
3.01.16.46.C	25mm x 125mm	m	0.25	9.41	—	9.11	18.52	10.35	—	10.03	20.37
3.01.16.46.D	25mm x 150mm	m	0.25	9.41	—	10.82	20.22	10.35	—	11.90	22.24
3.01.16.46.E	32mm x 100mm	m	0.27	10.16	—	8.55	18.71	11.18	—	9.40	20.58
3.01.16.46.F	32mm x 125mm	m	0.27	10.16	—	10.17	20.33	11.18	—	11.19	22.37
3.01.16.46.G	32mm x 150mm	m	0.27	10.16	—	11.79	21.95	11.18	—	12.97	24.15
3.01.16.47.	**Linings; once rebated; tongued at angles**									
3.01.16.47.A	38mm x 100mm	m	0.28	10.54	—	6.47	17.00	11.59	—	7.11	18.70
3.01.16.47.B	38mm x 150mm	m	0.28	10.54	—	8.83	19.37	11.59	—	9.71	21.30
3.01.16.48.	**Threshold**									
3.01.16.48.A	50mm x 150mm sunk, weathered and grooved	m	0.40	15.05	—	9.66	24.71	16.56	—	10.63	27.18
3.01.22.	**SECOND FIXINGS INTERNAL DOORS**									
3.01.22.01.	**Moulded wood fibre face, 6 panel, middleweight cellular core (c.21kg); white primed; includes hinges**									
3.01.22.01.A	1981mm x 762mm x 35mm	nr	0.95	35.75	—	54.28	90.03	39.32	—	59.71	99.03
3.01.22.01.B	1981mm x 838mm x 35mm	nr	0.95	35.75	—	55.07	90.82	39.32	—	60.57	99.90
3.01.22.01.C	2040mm x 726mm x 40mm	nr	1.00	37.63	—	66.23	103.86	41.39	—	72.85	114.24

Hutchins Priced Schedules 2021	Unit	Labour Hours	Labour Net £	Plant Net £	Materials Net £	Unit Rate Net £	Labour Gross £	Plant Gross £	Materials Gross £	Unit Rate Gross £
							(Gross rates include 10% profit)			

3.01. **NEW WORK**

3.01.22. SECOND FIXINGS INTERNAL DOORS

3.01.22.01. Moulded wood fibre face, 6 panel, middleweight cellular core (c.21kg); white primed; includes hinges

3.01.22.01.D	2040mm x 826mm x 40mm	nr	1.00	37.63	—	58.34	95.97	41.39	—	64.18	105.57

3.01.22.02. Moulded wood fibre face; 4 panel; middleweight cellular core (c.21kg); white primed; includes hinges

3.01.22.02.A	1981mm x 762mm x 35mm	nr	0.95	35.75	—	54.28	90.03	39.32	—	59.71	99.03
3.01.22.02.B	1981mm x 838mm x 35mm	nr	0.95	35.75	—	55.07	90.82	39.32	—	60.57	99.90
3.01.22.02.C	2040mm x 726mm x 40mm	nr	1.00	37.63	—	58.88	96.51	41.39	—	64.77	106.16
3.01.22.02.D	2040mm x 826mm x 40mm	nr	1.00	37.63	—	58.34	95.97	41.39	—	64.18	105.57

3.01.22.03. Moulded wood fibre face; 2 panel; middleweight cellular core (c.21kg); white primed; includes hinges

3.01.22.03.A	1981mm x 762mm x 35mm	nr	0.95	35.75	—	59.14	94.89	39.32	—	65.06	104.38
3.01.22.03.B	1981mm x 838mm x 35mm	nr	0.95	35.75	—	59.78	95.53	39.32	—	65.76	105.08
3.01.22.03.C	2040mm x 726mm x 40mm	nr	1.00	37.63	—	67.16	104.79	41.39	—	73.88	115.27
3.01.22.03.D	2040mm x 826mm x 40mm	nr	1.00	37.63	—	59.78	97.41	41.39	—	65.76	107.15

3.01.22.04. Moulded wood fibre face; 2 light clear glazed; middleweight cellular core (c.21kg); white primed; includes hinges

3.01.22.04.A	1981mm x 762mm x 35mm	nr	0.95	35.75	—	107.69	143.44	39.32	—	118.46	157.79
3.01.22.04.B	1981mm x 838mm x 35mm	nr	0.95	35.75	—	100.08	135.83	39.32	—	110.09	149.41
3.01.22.04.C	2040mm x 726mm x 40mm	nr	1.00	37.63	—	108.69	146.32	41.39	—	119.56	160.95
3.01.22.04.D	2040mm x 826mm x 40mm	nr	1.00	37.63	—	108.69	146.32	41.39	—	119.56	160.95

3.01.22.05. Moulded wood fibre face; fire door (FD30); 6 panel; magnesium oxide board core (c.35kg); white primed; includes hinges

3.01.22.05.A	1981mm x 762mm x 44mm	nr	1.25	47.04	—	81.32	128.36	51.74	—	89.45	141.20
3.01.22.05.B	1981mm x 838mm x 44mm	nr	1.25	47.04	—	82.64	129.67	51.74	—	90.90	142.64
3.01.22.05.C	2040mm x 726mm x 44mm	nr	1.25	47.04	—	89.32	136.36	51.74	—	98.26	150.00
3.01.22.05.D	2040mm x 826mm x 44mm	nr	1.25	47.04	—	89.32	136.36	51.74	—	98.26	150.00

3.01.22.06. Moulded wood fibre face; fire door (FD30); 4 panel; magnesium oxide board core (c.35kg); white primed; includes hinges

3.01.22.06.A	1981mm x 762mm x 44mm	nr	1.25	47.04	—	81.32	128.36	51.74	—	89.45	141.20
3.01.22.06.B	1981mm x 838mm x 44mm	nr	1.25	47.04	—	82.64	129.67	51.74	—	90.90	142.64
3.01.22.06.C	2040mm x 726mm x 44mm	nr	1.25	47.04	—	89.32	136.36	51.74	—	98.26	150.00
3.01.22.06.D	2040mm x 826mm x 44mm	nr	1.25	47.04	—	89.32	136.36	51.74	—	98.26	150.00

Woodwork

Hutchins Priced Schedules 2021	Unit	Labour Hours	Labour Net £	Plant Net £	Materials Net £	Unit Rate Net £	Labour Gross £	Plant Gross £	Materials Gross £	Unit Rate Gross £
						(Gross rates include 10% profit)				
3.01. **NEW WORK**										
3.01.22. **SECOND FIXINGS INTERNAL DOORS**										
3.01.22.07. **Moulded wood fibre face; fire door (FD30); 2 panel; magnesium oxide board core (c.35kg); white primed; includes hinges**										
3.01.22.07.A 1981mm x 762mm x 44mm	nr	1.25	47.04	—	77.37	124.41	51.74	—	85.11	136.85
3.01.22.07.B 1981mm x 838mm x 44mm	nr	1.25	47.04	—	78.44	125.47	51.74	—	86.28	138.02
3.01.22.07.C 2040mm x 726mm x 44mm	nr	1.25	47.04	—	83.60	130.64	51.74	—	91.96	143.70
3.01.22.07.D 2040mm x 826mm x 44mm	nr	1.25	47.04	—	83.60	130.64	51.74	—	91.96	143.70
3.01.22.08. **Moulded wood fibre face; fire door (FD60); 6 panel; magnesium oxide board core (c.45kg); white primed; includes hinges**										
3.01.22.08.A 1981mm x 762mm x 54mm	nr	1.40	52.68	—	206.20	258.88	57.95	—	226.82	284.77
3.01.22.08.B 1981mm x 838mm x 54mm	nr	1.40	52.68	—	209.14	261.82	57.95	—	230.05	288.00
3.01.22.08.C 2040mm x 726mm x 54mm	nr	1.40	52.68	—	207.50	260.18	57.95	—	228.25	286.20
3.01.22.08.D 2040mm x 826mm x 54mm	nr	1.40	52.68	—	207.50	260.18	57.95	—	228.25	286.20
3.01.22.09. **Moulded wood fibre face; fire door (FD60); 4 panel; magnesium oxide board core (c.45kg); white primed; includes hinges**										
3.01.22.09.A 1981mm x 762mm x 54mm	nr	1.40	52.68	—	206.20	258.88	57.95	—	226.82	284.77
3.01.22.09.B 1981mm x 838mm x 54mm	nr	1.40	52.68	—	209.14	261.82	57.95	—	230.05	288.00
3.01.22.09.C 2040mm x 726mm x 54mm	nr	1.40	52.68	—	207.50	260.18	57.95	—	228.25	286.20
3.01.22.09.D 2040mm x 826mm x 54mm	nr	1.40	52.68	—	207.50	260.18	57.95	—	228.25	286.20
3.01.22.10. **Moulded wood fibre face; fire door (FD60); 2 panel; magnesium oxide board core (c.45kg); white primed; includes hinges**										
3.01.22.10.A 1981mm x 762mm x 54mm	nr	1.40	52.68	—	211.16	263.84	57.95	—	232.27	290.22
3.01.22.10.B 1981mm x 838mm x 54mm	nr	1.40	52.68	—	214.10	266.78	57.95	—	235.50	293.45
3.01.22.10.C 2040mm x 726mm x 54mm	nr	1.40	52.68	—	212.46	265.14	57.95	—	233.70	291.65
3.01.22.10.D 2040mm x 826mm x 54mm	nr	1.40	52.68	—	212.46	265.14	57.95	—	233.70	291.65
3.01.22.11. **Oregon white oak door; heavyweight with raised and fielded panels (c.34kg); 6 panel flush bead finished in clear lacquer; stain or varnish; includes hinges**										
3.01.22.11.A 1981mm x 762mm x 35mm	nr	1.24	46.66	—	124.11	170.77	51.33	—	136.52	187.85
3.01.22.11.B 1981mm x 838mm x 35mm	nr	1.24	46.66	—	124.11	170.77	51.33	—	136.52	187.85

Hutchins Priced Schedules 2021		Unit	Labour Hours	Labour Net £	Plant Net £	Materials Net £	Unit Rate Net £	Labour Gross £	Plant Gross £	Materials Gross £	Unit Rate Gross £
								(Gross rates include 10% profit)			
3.01.	**NEW WORK**										
3.01.22.	**SECOND FIXINGS INTERNAL DOORS**										
3.01.22.12.	**Oregon white oak door; heavyweight with raised and fielded panels (c.34kg); 4 panel flushed bead finished in clear lacquer; stain or varnish; includes hinges**										
3.01.22.12.A	1981mm x 762mm x 35mm	nr	1.24	46.66	—	113.44	160.10	51.33	—	124.79	176.11
3.01.22.12.B	1981mm x 838mm x 35mm	nr	1.24	46.66	—	112.53	159.19	51.33	—	123.78	175.11
3.01.22.12.C	2040mm x 726mm x 40mm	nr	1.24	46.66	—	114.48	161.14	51.33	—	125.93	177.26
3.01.22.12.D	2040mm x 826mm x 40mm	nr	1.24	46.66	—	119.95	166.61	51.33	—	131.95	183.27
3.01.22.13.	**Oregon white oak door; heavyweight with raised and fielded panels (c.34kg); 2 light clear glazed flushed bead; finished in clear lacquer; stain or varnish; includes hinges**										
3.01.22.13.A	1981mm x 762mm x 35mm	nr	1.24	46.66	—	131.27	177.93	51.33	—	144.40	195.73
3.01.22.13.B	1981mm x 838mm x 35mm	nr	1.24	46.66	—	135.58	182.24	51.33	—	149.13	200.46
3.01.22.13.C	2040mm x 726mm x 40mm	nr	1.24	46.66	—	139.48	186.14	51.33	—	153.43	204.76
3.01.22.13.D	2040mm x 826mm x 40mm	nr	1.24	46.66	—	138.43	185.09	51.33	—	152.28	203.60
3.01.22.14.	**Shaker Oregon white oak door; heavyweight (c.34kg); 1 panel; finished in clear lacquer; stain or varnish; includes hinges**										
3.01.22.14.A	1981mm x 762mm x 35mm	nr	1.24	46.66	—	94.95	141.61	51.33	—	104.45	155.77
3.01.22.14.B	1981mm x 838mm x 35mm	nr	1.24	46.66	—	97.03	143.69	51.33	—	106.73	158.06
3.01.22.14.C	2040mm x 726mm x 40mm	nr	1.24	46.66	—	120.99	167.65	51.33	—	133.09	184.42
3.01.22.14.D	2040mm x 826mm x 40mm	nr	1.24	46.66	—	119.69	166.35	51.33	—	131.66	182.99
3.01.22.15.	**Shaker Oregon white oak door; heavyweight (c.34kg); 4 panel; finished in clear lacquer; stain or varnish; includes hinges**										
3.01.22.15.A	1981mm x 762mm x 35mm	nr	1.24	46.66	—	102.76	149.42	51.33	—	113.04	164.37
3.01.22.15.B	1981mm x 838mm x 35mm	nr	1.24	46.66	—	104.07	150.73	51.33	—	114.47	165.80
3.01.22.15.C	2040mm x 726mm x 40mm	nr	1.24	46.66	—	118.00	164.66	51.33	—	129.80	181.13
3.01.22.15.D	2040mm x 826mm x 40mm	nr	1.24	46.66	—	123.33	169.99	51.33	—	135.67	186.99
3.01.22.16.	**Shaker Oregon white oak door; heavyweight (c.34kg); 4 light clear glazed; finished in clear lacquer; stain or varnish; includes hinges**										
3.01.22.16.A	1981mm x 762mm x 35mm	nr	1.24	46.66	—	108.75	155.41	51.33	—	119.62	170.95
3.01.22.16.B	1981mm x 838mm x 35mm	nr	1.24	46.66	—	111.49	158.15	51.33	—	122.64	173.97

Woodwork

Hutchins Priced Schedules 2021	Unit	Labour Hours	Labour Net £	Plant Net £	Materials Net £	Unit Rate Net £	Labour Gross £	Plant Gross £	Materials Gross £	Unit Rate Gross £
							(Gross rates include 10% profit)			

3.01. NEW WORK

3.01.22. SECOND FIXINGS INTERNAL DOORS

3.01.22.16. Shaker Oregon white oak door; heavyweight (c.34kg); 4 light clear glazed; finished in clear lacquer; stain or varnish; includes hinges

3.01.22.16.C	2040mm x 726mm x 40mm	nr	1.24	46.66	—	167.33	213.99	51.33	—	184.06	235.39
3.01.22.16.D	2040mm x 826mm x 40mm	nr	1.24	46.66	—	167.33	213.99	51.33	—	184.06	235.39

3.01.22.17. Cottage Oregon white oak door; heavyweight (c.34kg); finished in clear lacquer; stain or varnish; includes hinges

3.01.22.17.A	1981mm x 762mm x 35mm	nr	1.24	46.66	—	108.10	154.76	51.33	—	118.91	170.23
3.01.22.17.B	1981mm x 838mm x 35mm	nr	1.24	46.66	—	110.19	156.85	51.33	—	121.21	172.53
3.01.22.17.C	2040mm x 726mm x 40mm	nr	1.24	46.66	—	114.74	161.41	51.33	—	126.22	177.55
3.01.22.17.D	2040mm x 826mm x 40mm	nr	1.24	46.66	—	121.12	167.78	51.33	—	133.23	184.56

3.01.22.18. Cottage etched glazed Oregon white oak door; heavyweight (c.34kg); finished in clear lacquer; stain or varnish; includes hinges

3.01.22.18.A	1981mm x 762mm x 35mm	nr	1.24	46.66	—	127.11	173.77	51.33	—	139.82	191.15
3.01.22.18.B	1981mm x 838mm x 35mm	nr	1.24	46.66	—	133.49	180.15	51.33	—	146.84	198.16

3.01.22.19. Cottage 6 light clear glazed Oregon white oak door; heavyweight (c.34kg); finished in clear lacquer; stain or varnish; includes hinges

3.01.22.19.A	1981mm x 762mm x 35mm	nr	1.24	46.66	—	194.92	241.58	51.33	—	214.41	265.74
3.01.22.19.B	1981mm x 838mm x 35mm	nr	1.24	46.66	—	194.92	241.58	51.33	—	214.41	265.74

3.01.22.20. Flush white oak door (c.18kg); finished in clear lacquer; includes hinges

3.01.22.20.A	1981mm x 762mm x 35mm	nr	0.90	33.87	—	79.54	113.41	37.25	—	87.50	124.75
3.01.22.20.B	1981mm x 838mm x 35mm	nr	0.90	33.87	—	80.33	114.20	37.25	—	88.36	125.62
3.01.22.20.C	2040mm x 726mm x 40mm	nr	0.95	35.75	—	80.66	116.40	39.32	—	88.72	128.04
3.01.22.20.D	2040mm x 826mm x 40mm	nr	0.95	35.75	—	80.66	116.40	39.32	—	88.72	128.04

3.01.22.21. Flush white oak fire door (FD30) (c.35kg); finished in clear lacquer; includes hinges

3.01.22.21.A	1981mm x 762mm x 44mm	nr	1.25	47.04	—	90.89	137.93	51.74	—	99.98	151.72
3.01.22.21.B	1981mm x 838mm x 44mm	nr	1.25	47.04	—	90.89	137.93	51.74	—	99.98	151.72
3.01.22.21.C	2040mm x 726mm x 44mm	nr	1.25	47.04	—	99.94	146.98	51.74	—	109.93	161.67
3.01.22.21.D	2040mm x 826mm x 44mm	nr	1.25	47.04	—	99.94	146.98	51.74	—	109.93	161.67

3.01.22.22. Flush white oak fire door (FD60) (c.45kg); finished in clear lacquer; includes hinges

3.01.22.22.A	1981mm x 762mm x 54mm	nr	1.40	52.68	—	188.31	240.99	57.95	—	207.14	265.09

Hutchins Priced Schedules 2021		Unit	Labour Hours	Labour Net £	Plant Net £	Materials Net £	Unit Rate Net £	Labour Gross £	Plant Gross £	Materials Gross £	Unit Rate Gross £
									(Gross rates include 10% profit)		
3.01.	**NEW WORK**										
3.01.22.	**SECOND FIXINGS INTERNAL DOORS**										
3.01.22.22.	**Flush white oak fire door (FD60) (c.45kg); finished in clear lacquer; includes hinges**										
3.01.22.22.B	1981mm x 838mm x 54mm	nr	1.40	52.68	—	193.22	245.90	57.95	—	212.54	270.49
3.01.22.22.C	2040mm x 726mm x 54mm	nr	1.40	52.68	—	193.22	245.90	57.95	—	212.54	270.49
3.01.22.22.D	2040mm x 826mm x 54mm	nr	1.40	52.68	—	198.14	250.82	57.95	—	217.95	275.90
3.01.23.	**SECOND FIXINGS EXTERNAL DOORS**										
3.01.23.01.	**Cottage American white oak door; heavyweight (c.40kg); unfinished; includes hinges**										
3.01.23.01.A	1981mm x 838mm x 44mm	nr	1.35	50.80	—	187.97	238.77	55.88	—	206.77	262.65
3.01.23.01.B	1981mm x 914mm x 44mm	nr	1.35	50.80	—	207.33	258.13	55.88	—	228.07	283.95
3.01.23.01.C	2032mm x 813mm x 44mm	nr	1.35	50.80	—	197.95	248.75	55.88	—	217.74	273.62
3.01.23.02.	**Cottage American white oak door; heavyweight (c.40kg); 4 light clear 200mm squares; unfinished; includes hinges**										
3.01.23.02.A	1981mm x 838mm x 44mm	nr	1.35	50.80	—	222.14	272.94	55.88	—	244.35	300.23
3.01.23.02.B	1981mm x 914mm x 44mm	nr	1.35	50.80	—	236.00	286.80	55.88	—	259.60	315.48
3.01.23.02.C	2032mm x 813mm x 40mm	nr	1.35	50.80	—	222.14	272.94	55.88	—	244.35	300.23
3.01.23.03.	**American white oak door; heavyweight (c.40kg); 6 panel; unfinished; includes hinges**										
3.01.23.03.A	1981mm x 838mm x 44mm	nr	1.35	50.80	—	147.04	197.84	55.88	—	161.75	217.63
3.01.23.03.B	1981mm x 914mm x 44mm	nr	1.35	50.80	—	156.43	207.23	55.88	—	172.07	227.95
3.01.23.03.C	2032mm x 813mm x 40mm	nr	1.35	50.80	—	147.04	197.84	55.88	—	161.75	217.63
3.01.24.	**SECOND FIXINGS INTERNAL DOOR LININGS**										
3.01.24.01.	**Internal door linings; with door stop; to suit doors up to 926mm wide and 2050mm tall; unfinished**										
3.01.24.01.A	softwood; 108mm thick	nr	1.50	56.45	—	15.90	72.34	62.09	—	17.49	79.58
3.01.24.01.B	softwood; 133mm thick	nr	1.50	56.45	—	15.90	72.34	62.09	—	17.49	79.58
3.01.24.02.	**Internal fire door linings; with door stop; grooved and fitted with intumescent strip; to suit doors up to 926mm wide and 2050mm tall; unfinished**										
3.01.24.02.A	white oak; 133mm thick; FD30	nr	1.50	56.45	—	87.21	143.66	62.09	—	95.93	158.02
3.01.24.02.B	hardwood; 120mm thick; FD60	nr	1.50	56.45	—	122.65	179.10	62.09	—	134.92	197.01
3.01.24.02.C	hardwood; 133mm thick; FD60	nr	1.50	56.45	—	122.65	179.10	62.09	—	134.92	197.01

Woodwork

Hutchins Priced Schedules 2021		Unit	Labour Hours	Labour Net £	Plant Net £	Materials Net £	Unit Rate Net £	Labour Gross £	Plant Gross £	Materials Gross £	Unit Rate Gross £
								(Gross rates include 10% profit)			
3.01.	**NEW WORK**										
3.01.25.	**SECOND FIXINGS EXTERNAL DOORS FRAMES**										
3.01.25.01.	**External door frame; with door stop; to suit doors up to 914mm wide and 2032mm tall; weatherstripped; unfinished; sealed both sides**										
3.01.25.01.A	softwood 66mm thick; with no sill	nr	1.75	65.85	—	73.84	139.70	72.44	—	81.23	153.67
3.01.25.01.B	softwood 66mm thick; with sill	nr	1.75	65.85	—	93.96	159.81	72.44	—	103.36	175.79
3.01.25.01.C	hardwood 66mm thick; with no sill	nr	1.75	65.85	—	171.41	237.26	72.44	—	188.55	260.99
3.01.25.01.D	hardwood 66mm thick; with sill	nr	1.75	65.85	—	251.50	317.36	72.44	—	276.65	349.09
3.01.25.01.E	oak 66mm thick; with sill	nr	1.75	65.85	—	193.38	259.23	72.44	—	212.71	285.15
3.01.26.	**SECOND FIXINGS STANDARD WINDOWS**										
3.01.26.01.	**Stormsure standard double glazed casement windows finished in Hi-Build white paint; including frames; side hung; including trickle vent; fitted with polished chrome handles; sealed both sides**										
3.01.26.01.A	height 750mm x 630mm; 1-lite with left opener	nr	0.88	33.11	—	324.71	357.82	36.43	—	357.18	393.61
3.01.26.01.B	height 900mm x 630mm; 1-lite with left opener	nr	0.99	37.25	—	340.10	377.36	40.98	—	374.11	415.09
3.01.26.01.C	height 1050mm x 630mm; 1-lite with left opener	nr	1.15	43.27	—	356.11	399.38	47.60	—	391.72	439.32
3.01.26.01.D	height 1050mm x 915mm; 2-lite with left opener	nr	1.25	47.04	—	561.83	608.87	51.74	—	618.01	669.75
3.01.26.01.E	height 1050mm x 1200mm; 2-lite with left and right opener	nr	1.40	52.68	—	614.85	667.53	57.95	—	676.33	734.28
3.01.26.01.F	height 1200mm x 1200mm; 2-lite with left and right opener	nr	1.50	56.45	—	647.04	703.49	62.09	—	711.75	773.84
3.01.26.01.G	height 1050mm x 1770mm; 3-lite with left and right opener	nr	1.55	58.33	—	787.37	845.70	64.16	—	866.11	930.26
3.01.26.01.H	height 1200mm x 1770mm; 3-lite with left and right opener	nr	2.00	75.26	—	829.63	904.89	82.79	—	912.60	995.38
3.01.26.01.I	height 900mm x 2339mm; 4-lite with left and right opener	nr	1.05	39.51	—	907.40	946.91	43.46	—	998.14	1041.60
3.01.26.01.J	height 1050mm x 2339mm; 4-lite with left and right opener	nr	2.20	82.79	—	957.55	1040.33	91.06	—	1053.30	1144.37
3.01.26.01.K	height 1200mm x 2339mm; 4-lite with left and right opener	nr	2.25	84.67	—	1009.89	1094.56	93.13	—	1110.88	1204.02

Hutchins Priced Schedules 2021	Unit	Labour Hours	Labour Net £	Plant Net £	Materials Net £	Unit Rate Net £	Labour Gross £	Plant Gross £	Materials Gross £	Unit Rate Gross £
							(Gross rates include 10% profit)			

3.01.   NEW WORK

3.01.26.   SECOND FIXINGS STANDARD WINDOWS

3.01.26.02.   **Stormsure Energy+ high performance triple glazed casement windows finished in Hi-Build white paint; including frames; side hung; including trickle vent; fitted with polished chrome handles; sealed both sides**

Hutchins Priced Schedules 2021	Unit	Labour Hours	Labour Net £	Plant Net £	Materials Net £	Unit Rate Net £	Labour Gross £	Plant Gross £	Materials Gross £	Unit Rate Gross £
3.01.26.02.A height 750mm x 630mm; 1-lite with left opener....................	nr	0.88	33.11	—	508.52	541.64	36.43	—	559.37	595.80
3.01.26.02.B height 900mm x 630mm; 1-lite with left opener....................	nr	0.99	37.25	—	567.06	604.32	40.98	—	623.77	664.75
3.01.26.02.C height 1050mm x 630mm; 1-lite with left opener....................	nr	1.15	43.27	—	626.18	669.45	47.60	—	688.80	736.40
3.01.26.02.D height 1050mm x 915mm; 2-lite with left opener....................	nr	1.25	47.04	—	949.67	996.71	51.74	—	1044.64	1096.38
3.01.26.02.E height 1050mm x 1200mm; 2-lite with left and right opener.............	nr	1.40	52.68	—	1154.21	1206.89	57.95	—	1269.63	1327.58
3.01.26.02.F height 1200mm x 1200mm; 2-lite with left and right opener.............	nr	1.50	56.45	—	1272.84	1329.29	62.09	—	1400.13	1462.22
3.01.26.02.G height 1050mm x 1770mm; 3-lite with left and right opener.............	nr	1.55	58.33	—	1627.13	1685.45	64.16	—	1789.84	1854.00
3.01.26.02.H height 1200mm x 1770mm; 3-lite with left and right opener.............	nr	2.00	75.26	—	1801.68	1876.94	82.79	—	1981.85	2064.64
3.01.26.02.I height 900mm x 2339mm; 4-lite with left and right opener.............	nr	2.15	80.90	—	1869.90	1950.81	88.99	—	2056.89	2145.89
3.01.26.02.J height 1050mm x 2339mm; 4-lite with left and right opener.............	nr	2.20	82.79	—	2098.17	2180.96	91.06	—	2307.99	2399.06
3.01.26.02.K height 1200mm x 2339mm; 4-lite with left and right opener.............	nr	2.25	84.67	—	2328.78	2413.44	93.13	—	2561.66	2654.79

3.01.26.03.   **Stormsure oak casement double glazed windows finished in golden oak stain; including frames; side hung; including trickle vent; fitted with polished chrome handles; sealed both sides**

Hutchins Priced Schedules 2021	Unit	Labour Hours	Labour Net £	Plant Net £	Materials Net £	Unit Rate Net £	Labour Gross £	Plant Gross £	Materials Gross £	Unit Rate Gross £
3.01.26.03.A height 750mm x 630mm; 1-lite with left opener....................	nr	0.88	33.11	—	652.52	685.63	36.43	—	717.77	754.20
3.01.26.03.B height 900mm x 630mm; 1-lite with left opener....................	nr	0.99	37.25	—	694.49	731.74	40.98	—	763.94	804.92
3.01.26.03.C height 1050mm x 630mm; 1-lite with left opener....................	nr	1.15	43.27	—	737.16	780.44	47.60	—	810.88	858.48
3.01.26.03.D height 1050mm x 915mm; 2-lite with left opener....................	nr	1.25	47.04	—	1172.21	1219.24	51.74	—	1289.43	1341.17
3.01.26.03.E height 1050mm x 1200mm; 2-lite with left and right opener.............	nr	1.40	52.68	—	1302.00	1354.68	57.95	—	1432.20	1490.15

Woodwork

Hutchins Priced Schedules 2021	Unit	Labour Hours	Labour Net £	Plant Net £	Materials Net £	Unit Rate Net £	Labour Gross £	Plant Gross £	Materials Gross £	Unit Rate Gross £
								(Gross rates include 10% profit)		

3.01.   NEW WORK

3.01.26.   SECOND FIXINGS STANDARD WINDOWS

3.01.26.03.   Stormsure oak casement double glazed windows finished in golden oak stain; including frames; side hung; including trickle vent; fitted with polished chrome handles; sealed both sides

3.01.26.03.F height 1200mm x 1200mm; 2-lite with left and right opener	nr	1.50	56.45	—	1145.51	1201.96	62.09	—	1260.06	1322.15
3.01.26.03.G height 1050mm x 1770mm; 3-lite with left and right opener	nr	1.55	58.33	—	1592.36	1650.69	64.16	—	1751.60	1815.76
3.01.26.03.H height 1200mm x 1770mm; 3-lite with left and right opener	nr	2.00	75.26	—	1683.21	1758.47	82.79	—	1851.53	1934.32
3.01.26.03.I height 900mm x 2339mm; 4-lite with left and right opener	nr	2.15	80.90	—	1879.40	1960.31	88.99	—	2067.34	2156.34
3.01.26.03.J height 1050mm x 2339mm; 4-lite with left and right opener	nr	2.20	82.79	—	1992.30	2075.09	91.06	—	2191.53	2282.60
3.01.26.03.K height 1200mm x 2339mm; 4-lite with left and right opener	nr	2.25	84.67	—	2107.76	2192.43	93.13	—	2318.54	2411.67

3.01.26.04.   Velux roof windows; centre pivot; laminated Nordic red pine frame and sash; sealed unit double pre-glazing; 3mm clear float glass; exterior aluminium cladding; natural brownish-grey finish; recessed flashings for installation in tiles up to 90mm in profile; roof pitch 20 deg to 90 deg; including insulation collar

3.01.26.04.A height 780mm x 550mm	nr	6.05	227.66	—	268.76	496.42	250.43	—	295.63	546.06
3.01.26.04.B height 980mm x 550mm	nr	6.92	260.44	—	282.39	542.82	286.48	—	310.63	597.11
3.01.26.04.C height 1180mm x 660mm	nr	7.31	275.00	—	319.18	594.18	302.50	—	351.10	653.60
3.01.26.04.D height 980mm x 780mm	nr	7.29	274.13	—	310.03	584.17	301.55	—	341.04	642.58
3.01.26.04.E height 1400mm x 780mm	nr	7.81	294.04	—	370.68	664.72	323.44	—	407.75	731.19
3.01.26.04.F height 1600mm x 940mm	nr	8.48	319.06	—	444.23	763.30	350.97	—	488.66	839.63
3.01.26.04.G height 1180mm x 1140mm	nr	8.22	309.43	—	431.34	740.77	340.37	—	474.47	814.85
3.01.26.04.H height 980mm x 1340mm	nr	8.17	307.47	—	425.04	732.51	338.22	—	467.54	805.77
3.01.26.04.I height 1400mm x 1340mm	nr	9.08	341.68	—	496.23	837.91	375.85	—	545.85	921.70

3.01.26.05.   Velux roof windows; top hung; laminated Nordic red pine frame and sash; sealed unit double pre-glazing; 3mm clear float glass; exterior aluminium cladding; natural brownish-grey finish; recessed flashings for installation in tiles up to 90mm in profile; roof pitch 15 deg to 75 deg; including insulation collar

3.01.26.05.A height 780mm x 550mm	nr	6.05	227.66	—	316.00	543.66	250.43	—	347.60	598.02
3.01.26.05.B height 980mm x 550mm	nr	6.92	260.44	—	352.36	612.80	286.48	—	387.59	674.08
3.01.26.05.C height 1180mm x 660mm	nr	7.31	275.00	—	423.26	698.26	302.50	—	465.58	768.08
3.01.26.05.D height 980mm x 780mm	nr	7.29	274.13	—	416.14	690.27	301.55	—	457.75	759.30

Hutchins Priced Schedules 2021	Unit	Labour Hours	Labour Net £	Plant Net £	Materials Net £	Unit Rate Net £	Labour Gross £	Plant Gross £	Materials Gross £	Unit Rate Gross £
										(Gross rates include 10% profit)

3.01. **NEW WORK**

3.01.26. SECOND FIXINGS STANDARD WINDOWS

3.01.26.05. Velux roof windows; top hung; laminated Nordic red pine frame and sash; sealed unit double pre-glazing; 3mm clear float glass; exterior aluminium cladding; natural brownish-grey finish; recessed flashings for installation in tiles up to 90mm in profile; roof pitch 15 deg to 75 deg; including insulation collar

	Unit	Labour Hours	Labour Net £	Plant Net £	Materials Net £	Unit Rate Net £	Labour Gross £	Plant Gross £	Materials Gross £	Unit Rate Gross £
3.01.26.05.E height 1400mm x 780mm	nr	7.81	294.04	—	485.91	779.95	323.44	—	534.50	857.94
3.01.26.05.F height 1600mm x 940mm	nr	8.48	319.06	—	605.86	924.93	350.97	—	666.45	1017.42
3.01.26.05.G height 1180mm x 1140mm	nr	8.22	309.43	—	587.50	896.93	340.37	—	646.25	986.62
3.01.26.05.H height 980mm x 1340mm	nr	8.17	307.47	—	568.35	875.83	338.22	—	625.19	963.41
3.01.26.05.I height 1400mm x 1340mm	nr	9.08	341.68	—	648.68	990.36	375.85	—	713.55	1089.40

3.01.28. SECOND FIXINGS STANDARD KITCHEN UNITS AND WORKTOPS

3.01.28.01. Base units; depth 600mm; height 900mm

	Unit	Labour Hours	Labour Net £	Plant Net £	Materials Net £	Unit Rate Net £	Labour Gross £	Plant Gross £	Materials Gross £	Unit Rate Gross £
3.01.28.01.A 300mm wide; single base unit; including door and fittings	nr	1.20	45.16	—	91.07	136.22	49.67	—	100.17	149.84
3.01.28.01.B 300mm wide; single base unit with drawerline; including drawer; door and fittings	nr	1.53	57.69	—	128.00	185.68	63.46	—	140.79	204.25
3.01.28.01.C 400mm wide; single base unit; including door and fittings	nr	1.37	51.55	—	103.41	154.97	56.71	—	113.76	170.46
3.01.28.01.D 400mm wide; single base unit with drawerline; including drawer, door and fittings	nr	1.70	64.08	—	138.21	202.30	70.49	—	152.03	222.52
3.01.28.01.E 500mm wide; single base unit; including door and fittings	nr	1.54	57.95	—	113.34	171.29	63.75	—	124.67	188.41
3.01.28.01.F 500mm wide; single base unit with drawerline; including drawer, door and fittings	nr	1.87	70.48	—	148.46	218.94	77.53	—	163.31	240.83
3.01.28.01.G 500mm wide; single base unit with two pull out pan drawers; including doors and fittings	nr	2.20	82.79	—	220.71	303.50	91.06	—	242.78	333.85
3.01.28.01.H 600mm wide; single base unit; including door and fittings	nr	1.71	64.35	—	123.27	187.61	70.78	—	135.59	206.38
3.01.28.01.I 600mm wide; single base unit with drawerline; including drawer, door and fittings	nr	2.04	76.88	—	159.64	236.52	84.57	—	175.61	260.17
3.01.28.01.J 600mm wide; single base unit with two pull out pan drawers; including doors and fittings	nr	2.37	89.18	—	236.46	325.64	98.10	—	260.11	358.21
3.01.28.01.K 1000mm wide; double base unit; including doors and fittings	nr	1.98	74.51	—	167.69	242.20	81.96	—	184.46	266.42
3.01.28.01.L 1000mm wide; double base unit with drawerline; including drawer, doors and fittings	nr	2.65	99.57	—	241.30	340.87	109.53	—	265.43	374.96
3.01.28.01.M 1200mm wide; double base unit; including doors and fittings	nr	2.20	82.79	—	185.61	268.39	91.06	—	204.17	295.23

Woodwork

	Unit	Labour Hours	Labour Net £	Plant Net £	Materials Net £	Unit Rate Net £	Labour Gross £	Plant Gross £	Materials Gross £	Unit Rate Gross £
								(Gross rates include 10% profit)		
3.01.	**NEW WORK**									
3.01.28.	**SECOND FIXINGS STANDARD KITCHEN UNITS AND WORKTOPS**									
3.01.28.01.	**Base units; depth 600mm; height 900mm**									
3.01.28.01.N	1200mm wide; double base unit with drawerline; including drawer, doors and fittings	nr	2.87	107.85	—	263.88	371.72	118.63	—	290.26	408.90
3.01.28.02.	**Sink base units; depth 600mm; height 900mm**									
3.01.28.02.A	500mm wide; single base unit; including door and fittings	nr	1.54	57.95	—	113.34	171.29	63.75	—	124.67	188.41
3.01.28.02.B	500mm wide; single base unit with drawerline effect; including door and fittings	nr	1.54	57.95	—	115.00	172.95	63.75	—	126.50	190.24
3.01.28.02.C	600mm wide; single base unit; including door and fittings	nr	1.71	64.35	—	123.27	187.61	70.78	—	135.59	206.38
3.01.28.02.D	600mm wide; single base unit with drawerline effect; including door and fittings	nr	1.71	64.35	—	125.08	189.42	70.78	—	137.58	208.37
3.01.28.02.E	1000mm wide; double base unit; including doors and fittings	nr	2.00	75.26	—	167.69	242.95	82.79	—	184.46	267.25
3.01.28.02.F	1000mm wide; double base unit with drawerline effect; including doors and fittings	nr	2.00	75.26	—	169.85	245.11	82.79	—	186.83	269.62
3.01.28.02.G	1000mm wide; double base unit with working drawerline and drawerline effect; including drawer, doors and fittings	nr	2.33	87.79	—	205.07	292.86	96.57	—	225.57	322.14
3.01.28.02.H	1200mm wide; double base unit; including doors and fittings	nr	2.20	82.79	—	185.61	268.39	91.06	—	204.17	295.23
3.01.28.02.I	1200mm wide; double base unit with drawerline effect; including doors and fittings	nr	2.20	82.79	—	188.18	270.97	91.06	—	207.00	298.06
3.01.28.02.J	1200mm wide; double base unit with working drawerline and drawerline effect; including drawer, doors and fittings	nr	2.53	95.32	—	213.17	308.49	104.85	—	234.49	339.34
3.01.28.03.	**Kitchen appliance base units; depth 600mm; height 900mm**									
3.01.28.03.A	600mm wide; built under oven housing . .	nr	1.65	62.09	—	79.59	141.68	68.30	—	87.55	155.85
3.01.28.03.B	600mm wide; built under oven housing with drawer	nr	1.98	74.51	—	114.16	188.66	81.96	—	125.57	207.53
3.01.28.04.	**Wall units; depth 300mm; height 575mm**									
3.01.28.04.A	300mm wide; single wall unit; including doors and fittings	nr	1.20	45.16	—	73.09	118.25	49.67	—	80.40	130.07
3.01.28.04.B	400mm wide; single wall unit; including doors and fittings	nr	1.37	51.55	—	80.87	132.42	56.71	—	88.96	145.67

Hutchins Priced Schedules 2021	Unit	Labour Hours	Labour Net £	Plant Net £	Materials Net £	Unit Rate Net £	Labour Gross £	Plant Gross £	Materials Gross £	Unit Rate Gross £
							(Gross rates include 10% profit)			
3.01. **NEW WORK**										
3.01.28. **SECOND FIXINGS STANDARD KITCHEN UNITS AND WORKTOPS**										
3.01.28.04. **Wall units; depth 300mm; height 575mm**										
3.01.28.04.C 500mm wide; single wall unit; including doors and fittings	nr	1.54	57.95	—	90.68	148.63	63.75	—	99.75	163.50
3.01.28.04.D 600mm wide; single wall unit; including doors and fittings	nr	1.71	64.35	—	99.86	164.21	70.78	—	109.85	180.63
3.01.28.04.E 1000mm wide; double wall unit; including doors and fittings	nr	1.96	73.75	—	147.09	220.84	81.13	—	161.79	242.92
3.01.28.05. **Wall units; depth 300mm; height 720mm**										
3.01.28.05.A 300mm wide; single wall unit; including doors and fittings	nr	1.27	47.79	—	81.67	129.46	52.57	—	89.84	142.41
3.01.28.05.B 400mm wide; single wall unit; including doors and fittings	nr	1.48	55.69	—	92.55	148.24	61.26	—	101.80	163.06
3.01.28.05.C 500mm wide; single wall unit; including doors and fittings	nr	1.69	63.59	—	107.54	171.13	69.95	—	118.29	188.24
3.01.28.05.D 600mm wide; single wall unit; including doors and fittings	nr	1.90	71.50	—	117.29	188.79	78.65	—	129.02	207.66
3.01.28.05.E 1000mm wide; double wall unit; including doors and fittings	nr	2.20	82.79	—	163.52	246.30	91.06	—	179.87	270.93
3.01.28.06. **Wall units; depth 300mm; height 900mm**										
3.01.28.06.A 300mm wide; single wall unit; including doors and fittings	nr	1.34	50.42	—	103.56	153.99	55.47	—	113.92	169.38
3.01.28.06.B 400mm wide; single wall unit; including doors and fittings	nr	1.59	59.83	—	116.62	176.46	65.81	—	128.29	194.10
3.01.28.06.C 500mm wide; single wall unit; including doors and fittings	nr	1.84	69.24	—	132.51	201.75	76.16	—	145.76	221.92
3.01.28.06.D 600mm wide; single wall unit; including doors and fittings	nr	2.09	78.65	—	145.52	224.17	86.51	—	160.07	246.58
3.01.28.06.E 1000mm wide; double wall unit; including doors and fittings	nr	2.44	91.82	—	202.09	293.91	101.00	—	222.30	323.30
3.01.28.07. **Tall larder units; depth 600mm**										
3.01.28.07.A 500mm wide; 1970mm high; including split doors, shelves and fittings	nr	2.47	92.95	—	249.07	342.01	102.24	—	273.98	376.22
3.01.28.07.B 500mm wide; 2150mm high; including split doors, shelves and fittings	nr	2.67	100.47	—	295.88	396.35	110.52	—	325.47	435.99
3.01.28.07.C 600mm wide; 1970mm high; including split doors, shelves and fittings	nr	2.64	99.34	—	273.59	372.93	109.28	—	300.94	410.22
3.01.28.07.D 600mm wide; 2150mm high; including split doors, shelves and fittings	nr	2.84	106.87	—	322.46	429.32	117.56	—	354.70	472.26

Woodwork

Hutchins Priced Schedules 2021	Unit	Labour Hours	Labour Net £	Plant Net £	Materials Net £	Unit Rate Net £	Labour Gross £	Plant Gross £	Materials Gross £	Unit Rate Gross £
										(Gross rates include 10% profit)
3.01.	**NEW WORK**									
3.01.28.	**SECOND FIXINGS STANDARD KITCHEN UNITS AND WORKTOPS**									
3.01.28.08.	**Worktop; depth 600mm; cut to size; fixed to base units with adhesive**									
3.01.28.08.A	40mm thick laminate; double rounded edge (3mm radius)	m	0.96	36.12	—	36.54	72.66	39.74	—	40.19	79.93
3.01.28.08.B	40mm thick laminate; square edge	m	0.96	36.12	—	50.40	86.52	39.74	—	55.44	95.18
3.01.28.08.C	40mm thick solid beech; square edge	m	0.99	37.33	—	44.80	82.13	41.06	—	49.28	90.35
3.01.28.08.D	40mm thick solid rubberwood; square edge	m	0.99	37.33	—	57.30	94.63	41.06	—	63.03	104.09
3.01.28.08.E	40mm thick solid oak; square edge	m	0.99	37.33	—	64.56	101.89	41.06	—	71.02	112.08
3.01.28.08.F	40mm thick solid rustic oak; square edge	m	0.99	37.33	—	82.25	119.58	41.06	—	90.47	131.53
3.01.28.08.G	40mm thick solid European walnut; square edge	m	0.99	37.33	—	103.66	140.98	41.06	—	114.02	155.08
3.01.28.08.H	40mm thick solid iroko; square edge	m	0.99	37.33	—	142.40	179.73	41.06	—	156.64	197.70
3.01.28.08.I	30mm thick granite; bevelled edge	m	1.37	91.88	—	132.83	224.70	101.06	—	146.11	247.17
3.01.28.08.J	30mm thick quartz; bevelled edge	m	1.14	76.69	—	174.83	251.51	84.36	—	192.31	276.66
3.01.28.08.K	20mm thick quartz; bevelled edge	m	0.76	51.28	—	137.81	189.09	56.41	—	151.59	208.00
3.01.28.08.L	20mm thick Dekton; bevelled edge	m	0.87	58.41	—	256.73	315.13	64.25	—	282.40	346.64
3.01.28.08.M	12mm thick Dekton; bevelled edge	m	1.16	43.80	—	227.33	271.13	48.18	—	250.06	298.24
3.01.42.	**SOFTWOOD STAIRCASES COMPONENTS**									
3.01.42.01.	**Treads**									
3.01.42.01.A	25mm nosed and risers 19mm tongued	m²	5.40	203.20	—	36.11	239.31	223.52	—	39.72	263.24
3.01.42.01.B	32mm nosed and risers 25mm tongued	m²	6.00	225.78	—	40.99	266.77	248.36	—	45.08	293.44
3.01.42.01.C	extra for bullnose step	nr	2.25	84.67	—	5.86	90.52	93.13	—	6.44	99.57
3.01.42.01.D	extra for double bullnose step	nr	4.00	150.52	—	12.69	163.21	165.57	—	13.95	179.53
3.01.42.02.	**Winders**									
3.01.42.02.A	25mm cross-tongued and risers 19mm	m²	7.50	282.23	—	48.79	331.02	310.45	—	53.67	364.12
3.01.42.02.B	32mm cross-tongued and risers 25mm	m²	8.00	301.04	—	53.67	354.71	331.14	—	59.04	390.18
3.01.42.03.	**Landings**									
3.01.42.03.A	25mm cross-tongued, including bearers	m²	5.30	199.44	—	47.82	247.26	219.38	—	52.60	271.98
3.01.42.03.B	32mm cross-tongued, including bearers	m²	5.30	199.44	—	52.70	252.13	219.38	—	57.96	277.35
3.01.42.05.	**Strings**									
3.01.42.05.A	275mm x 38mm	m	0.75	28.22	—	14.64	42.86	31.04	—	16.10	47.15
3.01.42.05.B	ends of string framed to newel	nr	0.65	24.46	—	—	24.46	26.91	—	—	26.91
3.01.42.06.	**Handrails**									
3.01.42.06.A	50mm mopstick	m	0.40	15.05	—	3.35	18.40	16.56	—	3.68	20.24
3.01.42.06.B	75mm x 50mm	m	0.75	28.22	—	4.49	32.72	31.04	—	4.94	35.99
3.01.42.06.D	extra for ramp	nr	4.00	150.52	—	15.61	166.13	165.57	—	17.17	182.75

Hutchins Priced Schedules 2021	Unit	Labour Hours	Labour Net £	Plant Net £	Materials Net £	Unit Rate Net £	Labour Gross £	Plant Gross £	Materials Gross £	Unit Rate Gross £
							(Gross rates include 10% profit)			
3.01. **NEW WORK**										
3.01.42. **SOFTWOOD STAIRCASES COMPONENTS**										
3.01.42.07. **Softwood newels**										
3.01.42.07.A turned complete newel post 82mm x 1500mm	nr	1.65	62.09	—	35.44	97.53	68.30	—	38.98	107.28
3.01.42.07.B turned complete half newel post 82mm x 1500mm	nr	1.10	41.39	—	22.84	64.23	45.53	—	25.12	70.65
3.01.42.07.C stop chamfered newel post 90mm x 1500mm	nr	0.20	7.53	—	43.31	50.84	8.28	—	47.64	55.92
3.01.42.07.D stop chamfered half newel post 41mm x 1500mm	nr	0.20	7.53	—	25.20	32.73	8.28	—	27.72	36.00
3.01.42.07.E square newel cap 110mm x 20mm high	nr	1.65	62.09	—	7.25	69.33	68.30	—	7.97	76.27
3.01.42.07.F half square newel cap 28mm x 106mm x 20mm high	nr	1.10	41.39	—	5.51	46.91	45.53	—	6.06	51.60
3.01.42.07.G ball newel cap 80mm x 80mm x 75mm high	nr	0.20	7.53	—	9.45	16.98	8.28	—	10.40	18.67
3.01.42.07.H half ball newel cap 38mm x 80mm x 132mm high	nr	0.20	7.53	—	5.51	13.04	8.28	—	6.06	14.34
3.01.42.08. **Framed spandrel**										
3.01.42.08.A 38mm with plywood panelling	m²	6.50	244.60	—	31.23	275.82	269.05	—	34.35	303.40
3.01.42.09. **Balusters**										
3.01.42.09.A 38mm x 38mm	m	0.33	12.42	—	1.37	13.78	13.66	—	1.50	15.16
3.01.42.11. **Apron lining; chamfered and beaded**										
3.01.42.11.A 225mm x 19mm	m	0.45	16.93	—	6.83	23.76	18.63	—	7.51	26.14
3.01.42.11.B 225mm x 25mm	m	0.50	18.82	—	9.76	28.57	20.70	—	10.73	31.43
3.01.46. **SOFTWOOD STAIRCASES STANDARD UNITS**										
3.01.46.01. **Staircase with 25mm MDF tread with 20mm overhang rounded and 13mm pine faced risers, glued, wedged and blocked with 32mm pine wall string, including pine newel post plain square on left top and left bottom. To include pine balustrade and 59 x 59mm handrail and 56mm x 22mm baserail**										
3.01.46.01.A straight staircase; 860mm wide x 2600mm rise; 13 treads, 240mm risers	nr	18.00	677.34	—	446.38	1123.72	745.07	—	491.01	1236.09
3.01.46.01.B single winder staircase; 860mm wide x 2600mm rise; 13 treads (10 before turn), 240mm risers	nr	18.00	677.34	—	592.81	1270.15	745.07	—	652.09	1397.16

Woodwork

							(Gross rates include 10% profit)			
3.01.	**NEW WORK**									
3.01.46.	**SOFTWOOD STAIRCASES STANDARD UNITS**									
3.01.46.04.	**Landing balustrade with 50mm x 75mm hardwood handrail, 38mm x 38mm turned Regency balusters, 32mm x 140mm baluster knee rails, 2nr 32mm x 50mm stiffeners, one end joined to newel post, other end built into half newel**									
3.01.46.04.A	3m long	nr	12.00	451.56	—	310.42	761.98	496.72	—	341.46	838.18
3.01.48.	**SUNDRY ITEMS**									
3.01.48.03.	**Softwood weather mould; throated; screwed to door**									
3.01.48.03.A	50mm x 75mm	m	0.60	22.58	—	9.21	31.79	24.84	—	10.13	34.97
3.01.48.05.	**Hat and coat rails, chamfered edges, plugged to brickwork**									
3.01.48.05.A	125mm x 25mm	m	0.34	12.79	—	2.39	15.19	14.07	—	2.63	16.70
3.01.48.05.B	125mm x 25mm in short lengths	m	0.50	18.82	—	2.39	21.21	20.70	—	2.63	23.33
3.01.50.	**SUNDRY LABOURS**									
3.01.50.01.	**Extra over fixing with nails for**									
3.01.50.01.A	steel screws	m	0.04	1.51	—	0.05	1.56	1.66	—	0.06	1.72
3.01.50.01.B	steel screws; sinking; filling heads	m	0.07	2.63	—	0.05	2.69	2.90	—	0.06	2.96
3.01.50.01.C	steel screws; sinking; pellating over	m	0.19	7.15	—	0.05	7.20	7.86	—	0.06	7.92
3.01.50.01.D	brass cups and screws	m	0.12	4.52	—	0.58	5.10	4.97	—	0.64	5.61
3.01.50.02.	**Plugging blockwork**									
3.01.50.02.A	300mm centres; one way	m	0.08	3.01	—	0.20	3.21	3.31	—	0.22	3.53
3.01.50.02.B	300mm centres; both ways	m	0.14	5.27	—	0.59	5.86	5.80	—	0.65	6.44
3.01.50.03.	**Plugging brickwork**									
3.01.50.03.A	300mm centres; one way	m	0.12	4.52	—	0.20	4.71	4.97	—	0.22	5.18
3.01.50.03.B	300mm centres; both ways	m	0.24	9.03	—	0.59	9.62	9.93	—	0.65	10.58
3.01.50.04.	**Plugging concrete**									
3.01.50.04.A	300mm centres; one way	m	0.22	8.28	—	0.20	8.48	9.11	—	0.22	9.32
3.01.50.04.B	300mm centres; both ways	m	0.44	16.56	—	0.59	17.15	18.21	—	0.65	18.86
3.01.50.06.	**Holes for pipes, bars, etc.; through softwood; thickness**									
3.01.50.06.A	12mm	nr	0.05	1.88	—	—	1.88	2.07	—	—	2.07
3.01.50.06.B	25mm	nr	0.08	3.01	—	—	3.01	3.31	—	—	3.31
3.01.50.06.C	50mm	nr	0.12	4.52	—	—	4.52	4.97	—	—	4.97

Hutchins Priced Schedules 2021	Unit	Labour Hours	Labour Net £	Plant Net £	Materials Net £	Unit Rate Net £	Labour Gross £	Plant Gross £	Materials Gross £	Unit Rate Gross £
							(Gross rates include 10% profit)			
3.01. **NEW WORK**										
3.01.50. **SUNDRY LABOURS**										
3.01.50.06. **Holes for pipes, bars, etc.; through softwood; thickness**										
3.01.50.06.D 75mm	nr	0.16	6.02	—	—	6.02	6.62	—	—	6.62
3.01.50.06.E 100mm	nr	0.19	7.15	—	—	7.15	7.86	—	—	7.86
3.01.50.07. **Head or nut in softwood**										
3.01.50.07.A let in; flush	nr	0.07	2.63	—	—	2.63	2.90	—	—	2.90
3.01.50.09. **Head or nut in hardwood**										
3.01.50.09.A let in; flush	nr	0.10	3.76	—	—	3.76	4.14	—	—	4.14
3.01.50.09.B let in; pellated	nr	0.24	9.03	—	—	9.03	9.93	—	—	9.93
3.01.50.10. **Mortice**										
3.01.50.10.A for and including metal dowel	nr	0.11	4.14	—	—	4.14	4.55	—	—	4.55
3.01.50.11. **Notching and fitting**										
3.01.50.11.A timber to steel	nr	0.33	12.42	—	—	12.42	13.66	—	—	13.66
3.01.50.12. **Planing**										
3.01.50.12.A by hand	m²	0.40	15.05	—	—	15.05	16.56	—	—	16.56
3.01.50.14. **Labours on softwood**										
3.01.50.14.A chamfers	m	0.19	7.15	—	—	7.15	7.86	—	—	7.86
3.01.50.14.B rounds	m	0.25	9.41	—	—	9.41	10.35	—	—	10.35
3.01.50.14.C grooves	m	0.36	13.55	—	—	13.55	14.90	—	—	14.90
3.01.50.14.D rebates	m	0.39	14.68	—	—	14.68	16.14	—	—	16.14
3.01.50.14.E throats	m	0.39	14.68	—	—	14.68	16.14	—	—	16.14
3.01.50.14.F mouldings per 25mm girth	m	1.13	42.52	—	—	42.52	46.77	—	—	46.77
3.01.50.14.G rebate in bottom of rail or door	m	0.30	11.29	—	—	11.29	12.42	—	—	12.42
3.01.51. **INSULATION AND VENTILATION**										
3.01.51.01. **Superglass multi acoustic insulation roll**										
3.01.51.01.A 25mm; laid between joists	m²	0.15	5.64	—	1.28	6.93	6.21	—	1.41	7.62
3.01.51.01.B 25mm; fixed vertically between softwood battens	m²	0.15	5.64	—	1.28	6.93	6.21	—	1.41	7.62
3.01.51.02. **Glass fibre thermal insulating quilt laid over ceiling joists**										
3.01.51.02.A 60mm	m²	0.17	6.40	—	1.98	8.38	7.04	—	2.18	9.22
3.01.51.02.B 80mm	m²	0.17	6.40	—	2.08	8.48	7.04	—	2.29	9.32
3.01.51.02.C 100mm	m²	0.17	6.40	—	3.87	10.27	7.04	—	4.26	11.30
3.01.51.03. **Glass fibre thermal insulating quilt laid between joists**										
3.01.51.03.A 60mm	m²	0.19	7.15	—	1.98	9.13	7.86	—	2.18	10.05

Woodwork

Hutchins Priced Schedules 2021	Unit	Labour Hours	Labour Net £	Plant Net £	Materials Net £	Unit Rate Net £	Labour Gross £	Plant Gross £	Materials Gross £	Unit Rate Gross £
									(Gross rates include 10% profit)	
3.01.	**NEW WORK**									
3.01.51.	**INSULATION AND VENTILATION**									
3.01.51.03.	**Glass fibre thermal insulating quilt laid between joists**									
3.01.51.03.B	80mm	m²	0.19	7.15	—	2.08	9.23	7.86	—	2.29	10.15
3.01.51.03.C	100mm	m²	0.19	7.15	—	3.87	11.02	7.86	—	4.26	12.13
3.01.51.04.	**Vermiculite granular loose fill insulation laid between joists**									
3.01.51.04.A	50mm	m²	0.20	7.53	—	11.70	19.22	8.28	—	12.87	21.15
3.01.51.04.B	75mm	m²	0.30	11.29	—	17.58	28.87	12.42	—	19.33	31.75
3.01.51.04.C	100mm	m²	0.40	15.05	—	23.31	38.36	16.56	—	25.64	42.20
3.01.51.06.	**Expanded polystyrene insulation board**									
3.01.51.06.A	53mm	m²	0.38	14.30	—	5.25	19.55	15.73	—	5.78	21.50
3.01.51.06.B	83mm	m²	0.38	14.30	—	9.00	23.30	15.73	—	9.90	25.63
3.01.51.07.	**Bitumen impregnated insulating board**									
3.01.51.07.A	12mm	m²	0.25	9.41	—	5.90	15.31	10.35	—	6.49	16.84
3.01.51.07.B	raking cutting	m	0.36	13.55	—	—	13.55	14.90	—	—	14.90
3.01.51.07.C	curved cutting	m	0.60	22.58	—	—	22.58	24.84	—	—	24.84
3.01.51.08.	**PVCu push-in soffit ventilators, 70mm dia, including cutting hole through timber soffit board and installing discs at**									
3.01.51.08.A	140mm centres	m	1.20	45.16	—	17.49	62.65	49.67	—	19.24	68.91
3.01.51.09.	**PVCu type C slotted soffit ventilator and screwing to**									
3.01.51.09.A	back of timber fascia	m	0.33	12.42	—	1.01	13.43	13.66	—	1.11	14.77
3.01.52.	**METALWORK**									
3.01.52.01.	**Stainless steel water bars including grooves in timber**									
3.01.52.01.A	30mm x 6mm	m	0.30	11.29	—	14.79	26.08	12.42	—	16.27	28.69
3.01.52.01.B	40mm x 6mm	m	0.32	12.04	—	19.97	32.01	13.25	—	21.97	35.21
3.01.52.02.	**Black stainless steel cup square carriage bolt with hexagon nut and washer**									
3.01.52.02.A	M10 x 50mm	nr	0.12	4.52	—	1.00	5.51	4.97	—	1.10	6.06
3.01.52.02.B	M10 x 75mm	nr	0.14	5.27	—	1.12	6.39	5.80	—	1.24	7.03
3.01.52.02.C	M10 x 100mm	nr	0.14	5.27	—	1.23	6.50	5.80	—	1.35	7.15
3.01.52.02.D	M10 x 150mm	nr	0.16	6.02	—	1.49	7.52	6.62	—	1.64	8.27
3.01.52.02.E	M12 x 50mm	nr	0.14	5.27	—	1.35	6.61	5.80	—	1.48	7.28
3.01.52.02.F	M12 x 75mm	nr	0.14	5.27	—	1.55	6.82	5.80	—	1.70	7.50
3.01.52.02.G	M12 x 100mm	nr	0.15	5.64	—	1.77	7.41	6.21	—	1.95	8.16

Hutchins Priced Schedules 2021		Unit	Labour Hours	Labour Net £	Plant Net £	Materials Net £	Unit Rate Net £	Labour Gross £	Plant Gross £	Materials Gross £	Unit Rate Gross £
								(Gross rates include 10% profit)			
3.01.	**NEW WORK**										
3.01.52.	**METALWORK**										
3.01.52.02.	**Black stainless steel cup square carriage bolt with hexagon nut and washer**										
3.01.52.02.H	M12 x 150mm...............	nr	0.17	6.40	—	2.16	8.56	7.04	—	2.38	9.42
3.01.52.04.	**Galvanised mild steel joist restraint straps; twice bent; one end drilled and screwed to timber; other end built in**										
3.01.52.04.A	27.5mm x 5mm; 600mm girth bent at 100mm...................	nr	0.40	15.05	—	1.91	16.96	16.56	—	2.10	18.66
3.01.52.04.B	27.5mm x 5mm; 800mm girth bent at 100mm...................	nr	0.48	18.06	—	2.43	20.49	19.87	—	2.67	22.54
3.01.52.04.C	27.5mm x 5mm; 1000mm girth bent at 100mm...................	nr	0.52	19.57	—	3.71	23.28	21.52	—	4.08	25.61
3.01.52.06.	**Galvanised steel joist hangers built in**										
3.01.52.06.A	47mm x 100mm...............	nr	0.20	7.53	—	1.48	9.01	8.28	—	1.63	9.91
3.01.52.06.B	47mm x 125mm...............	nr	0.21	7.90	—	1.48	9.39	8.69	—	1.63	10.32
3.01.52.06.C	47mm x 150mm...............	nr	0.22	8.28	—	1.60	9.88	9.11	—	1.76	10.87
3.01.52.06.D	47mm x 175mm...............	nr	0.23	8.65	—	1.70	10.35	9.52	—	1.87	11.39
3.01.52.06.E	47mm x 200mm...............	nr	0.24	9.03	—	1.89	10.92	9.93	—	2.08	12.01
3.01.52.06.H	75mm x 100mm...............	nr	0.27	10.16	—	2.95	13.11	11.18	—	3.24	14.42
3.01.52.06.I	75mm x 175mm...............	nr	0.28	10.54	—	3.24	13.78	11.59	—	3.57	15.16
3.01.52.06.J	75mm x 200mm...............	nr	0.29	10.91	—	3.24	14.16	12.00	—	3.57	15.57
3.01.52.07.	**Galvanised steel square toothed plate timber connectors to BS EN 912:2011**										
3.01.52.07.A	M10 x 38mm; single sided.........	nr	0.03	1.13	—	0.16	1.29	1.24	—	0.17	1.42
3.01.52.07.B	M10 x 38mm; double sided.........	nr	0.03	1.13	—	0.20	1.33	1.24	—	0.22	1.46
3.01.52.07.C	M10 x 50mm; single sided.........	nr	0.04	1.51	—	0.29	1.80	1.66	—	0.32	1.98
3.01.52.07.D	M10 x 50mm; double sided.........	nr	0.04	1.51	—	0.33	1.83	1.66	—	0.36	2.01
3.01.52.07.E	M10 x 63mm; single sided.........	nr	0.05	1.88	—	0.43	2.31	2.07	—	0.47	2.54
3.01.52.07.F	M10 x 63mm; double sided.........	nr	0.05	1.88	—	0.50	2.39	2.07	—	0.55	2.62
3.01.52.07.G	M12 x 38mm; single sided.........	nr	0.03	1.13	—	0.20	1.33	1.24	—	0.22	1.46
3.01.52.07.H	M12 x 38mm; double sided.........	nr	0.03	1.13	—	0.20	1.33	1.24	—	0.22	1.46
3.01.52.07.I	M12 x 50mm; single sided.........	nr	0.04	1.51	—	0.17	1.67	1.66	—	0.18	1.84
3.01.52.07.J	M12 x 50mm; double sided.........	nr	0.04	1.51	—	0.24	1.75	1.66	—	0.27	1.92
3.01.52.07.K	M12 x 63mm; single sided.........	nr	0.05	1.88	—	0.26	2.14	2.07	—	0.29	2.36
3.01.52.07.L	M12 x 63mm; double sided.........	nr	0.05	1.88	—	0.26	2.14	2.07	—	0.29	2.36
3.01.52.07.M	M12 x 76mm; single sided.........	nr	0.06	2.26	—	0.69	2.95	2.48	—	0.76	3.25
3.01.52.07.N	M12 x 76mm; double sided.........	nr	0.06	2.26	—	0.69	2.95	2.48	—	0.76	3.25
3.01.52.07.O	M16 x 50mm; single sided.........	nr	0.05	1.88	—	0.29	2.18	2.07	—	0.32	2.39
3.01.52.07.P	M16 x 50mm; double sided.........	nr	0.05	1.88	—	0.33	2.21	2.07	—	0.36	2.43
3.01.52.07.Q	M16 x 63mm; single sided.........	nr	0.06	2.26	—	0.43	2.69	2.48	—	0.47	2.96
3.01.52.07.R	M16 x 63mm; double sided.........	nr	0.06	2.26	—	0.50	2.76	2.48	—	0.55	3.04

Woodwork

Hutchins Priced Schedules 2021	Unit	Labour Hours	Labour Net £	Plant Net £	Materials Net £	Unit Rate Net £	Labour Gross £	Plant Gross £	Materials Gross £	Unit Rate Gross £
							(Gross rates include 10% profit)			
3.01.	**NEW WORK**									
3.01.52.	**METALWORK**									
3.01.52.07.	**Galvanised steel square toothed plate timber connectors to BS EN 912:2011**									
3.01.52.07.S	M16 x 76mm; single sided	nr	0.07	2.63	—	0.95	3.58	2.90	—	1.04	3.94
3.01.52.07.T	M16 x 76mm; double sided	nr	0.07	2.63	—	0.96	3.59	2.90	—	1.05	3.95
3.01.52.07.U	M20 x 50mm; single sided	nr	0.06	2.26	—	0.32	2.57	2.48	—	0.35	2.83
3.01.52.07.V	M20 x 50mm; double sided	nr	0.06	2.26	—	0.34	2.59	2.48	—	0.37	2.85
3.01.52.07.W	M20 x 63mm; single sided	nr	0.07	2.63	—	0.45	3.09	2.90	—	0.50	3.39
3.01.52.07.X	M20 x 63mm; double sided	nr	0.07	2.63	—	0.53	3.16	2.90	—	0.58	3.48
3.01.52.07.Y	M20 x 76mm; single sided	nr	0.08	3.01	—	0.98	3.99	3.31	—	1.07	4.39
3.01.52.07.Z	M20 x 76mm; double sided	nr	0.08	3.01	—	0.99	4.00	3.31	—	1.09	4.40
3.01.54.	**IRONMONGERY**									
3.01.54.01.	**1838 pattern, pressed light steel butt hinges and labour hanging door**									
3.01.54.01.B	75mm, self coloured	pair	1.17	44.03	—	0.93	44.95	48.43	—	1.02	49.45
3.01.54.01.C	100mm, self coloured	pair	1.42	53.43	—	1.06	54.49	58.78	—	1.16	59.94
3.01.54.01.E	75mm, chrome plated	pair	1.17	44.03	—	1.47	45.50	48.43	—	1.62	50.05
3.01.54.01.F	100mm, chrome plated	pair	1.42	53.43	—	2.14	55.57	58.78	—	2.35	61.13
3.01.54.04.	**Strong pattern butts and labour hanging door**									
3.01.54.04.A	75mm zinc plated	pair	1.17	44.03	—	1.78	45.81	48.43	—	1.96	50.39
3.01.54.04.B	100mm zinc plated	pair	1.42	53.43	—	2.31	55.74	58.78	—	2.54	61.32
3.01.54.04.C	64mm brass	pair	1.05	39.51	—	15.60	55.11	43.46	—	17.16	60.63
3.01.54.04.D	76mm brass	pair	1.17	44.03	—	32.91	76.93	48.43	—	36.20	84.63
3.01.54.06.	**Rising butts and labour hanging doors**									
3.01.54.06.A	75mm steel	pair	1.45	54.56	—	2.23	56.80	60.02	—	2.46	62.48
3.01.54.06.B	100mm steel	pair	1.70	63.97	—	3.64	67.62	70.37	—	4.01	74.38
3.01.54.06.C	75mm zinc plated	pair	1.45	54.56	—	6.28	60.85	60.02	—	6.91	66.93
3.01.54.06.D	102mm zinc plated	pair	1.70	63.97	—	10.74	74.71	70.37	—	11.81	82.18
3.01.54.09.	**Steel washered brass butts and labour hanging door**									
3.01.54.09.A	75mm	pair	1.17	44.03	—	3.63	47.66	48.43	—	4.00	52.43
3.01.54.09.B	100mm	pair	1.42	53.43	—	4.22	57.65	58.78	—	4.64	63.42
3.01.54.11.	**Brass rising butts and labour hanging door**									
3.01.54.11.A	75mm x 70mm	pair	1.45	54.56	—	7.77	62.33	60.02	—	8.54	68.56
3.01.54.11.B	100mm x 75mm	pair	1.70	63.97	—	11.00	74.97	70.37	—	12.10	82.47
3.01.54.12.	**Hurlinge butts and labour hanging door**									
3.01.54.12.A	75mm chrome	pair	0.33	12.42	—	2.52	14.94	13.66	—	2.77	16.43
3.01.54.12.B	100mm chrome	pair	0.42	15.80	—	3.21	19.01	17.39	—	3.53	20.92

Hutchins Priced Schedules 2021		Unit	Labour Hours	Labour Net £	Plant Net £	Materials Net £	Unit Rate Net £	Labour Gross £	Plant Gross £	Materials Gross £	Unit Rate Gross £
								(Gross rates include 10% profit)			
3.01.	**NEW WORK**										
3.01.54.	**IRONMONGERY**										
3.01.54.12.	**Hurlinge butts and labour hanging door**										
3.01.54.12.C	75mm zinc plated	pair	0.33	12.42	—	2.47	14.89	13.66	—	2.71	16.37
3.01.54.12.D	100mm zinc plated	pair	0.42	15.80	—	3.00	18.80	17.39	—	3.30	20.68
3.01.54.15.	**Steel tee hinges and labour hanging door**										
3.01.54.15.A	305mm	pair	0.83	31.23	—	7.74	38.98	34.36	—	8.52	42.88
3.01.54.15.B	457mm	pair	0.92	34.62	—	12.78	47.40	38.08	—	14.06	52.14
3.01.54.16.	**Steel light reversible hinges and labour hanging door**										
3.01.54.16.A	350mm	pair	1.05	39.51	—	20.06	59.57	43.46	—	22.07	65.53
3.01.54.16.B	450mm	pair	1.15	43.27	—	20.75	64.03	47.60	—	22.83	70.43
3.01.54.18.	**Steel heavy reversible hinges and labour hanging door**										
3.01.54.18.A	300mm	pair	1.33	50.05	—	19.00	69.05	55.05	—	20.90	75.95
3.01.54.18.B	450mm	pair	1.50	56.45	—	22.98	79.42	62.09	—	25.27	87.36
3.01.54.18.C	600mm	pair	1.67	62.84	—	28.44	91.28	69.13	—	31.28	100.41
3.01.54.19.	**Interior straight sliding door gear; top track with wheel hangers; door guides; stops; finger pulls; steel pelmet and labour hanging**										
3.01.54.19.A	35mm to 44mm thick single softwood door	nr	3.50	131.71	—	68.07	199.78	144.88	—	74.88	219.76
3.01.54.20.	**Locks and latches**										
3.01.54.20.A	rim lock and furniture	nr	1.25	47.04	—	22.06	69.10	51.74	—	24.26	76.01
3.01.54.20.B	mortice sash lock and furniture	nr	1.45	54.56	—	18.26	72.83	60.02	—	20.09	80.11
3.01.54.20.C	mortice 5 lever dead lock	nr	1.00	37.63	—	19.88	57.51	41.39	—	21.86	63.26
3.01.54.20.D	cylinder rim night latch	nr	1.20	45.16	—	22.47	67.63	49.67	—	24.72	74.39
3.01.54.20.E	Suffolk latch	nr	1.00	37.63	—	7.06	44.69	41.39	—	7.77	49.16
3.01.54.20.F	escutcheon	pair	0.30	11.29	—	4.85	16.14	12.42	—	5.34	17.76
3.01.54.20.G	bales catch	nr	0.50	18.82	—	4.09	22.91	20.70	—	4.50	25.20
3.01.54.20.H	cupboard catch	nr	0.30	11.29	—	3.32	14.61	12.42	—	3.65	16.07
3.01.54.20.I	cupboard or drawer lock	nr	0.60	22.58	—	3.72	26.30	24.84	—	4.09	28.92
3.01.54.22.	**Door closers**										
3.01.54.22.A	overhead door closer; surface fixing	nr	1.75	65.85	—	85.68	151.53	72.44	—	94.25	166.69
3.01.54.22.B	Perko closer	nr	1.75	65.85	—	20.03	85.89	72.44	—	22.04	94.48
3.01.54.22.C	coil gate spring	nr	0.40	15.05	—	3.53	18.58	16.56	—	3.88	20.44
3.01.54.24.	**Bolts**										
3.01.54.24.A	100mm straight barrel, polished chrome .	nr	0.45	16.93	—	3.39	20.33	18.63	—	3.73	22.36

Woodwork

							(Gross rates include 10% profit)			
3.01.	**NEW WORK**									
3.01.54.	**IRONMONGERY**									
3.01.54.24.	**Bolts**									
3.01.54.24.B	150mm straight barrel, polished chrome	nr	0.55	20.70	—	4.93	25.63	22.77	—	5.42	28.19
3.01.54.24.C	150mm tower; straight	nr	0.45	16.93	—	2.92	19.85	18.63	—	3.21	21.83
3.01.54.24.D	255mm tower; straight	nr	0.55	20.70	—	4.98	25.68	22.77	—	5.48	28.25
3.01.54.24.E	450mm monkey tail, galvanised, bow handle	nr	0.55	20.70	—	12.35	33.05	22.77	—	13.58	36.35
3.01.54.24.F	200mm lever action flush, polished chrome	nr	0.85	31.99	—	17.91	49.90	35.18	—	19.71	54.89
3.01.54.24.G	indicator	nr	1.20	45.16	—	5.25	50.40	49.67	—	5.77	55.44
3.01.54.24.H	single door panic	nr	1.75	65.85	—	48.23	114.08	72.44	—	53.05	125.49
3.01.54.24.I	double door panic	nr	2.30	86.55	—	96.99	183.54	95.20	—	106.69	201.89
3.01.54.26.	**Handles and pulls**									
3.01.54.26.A	door handle	nr	0.30	11.29	—	7.05	18.34	12.42	—	7.75	20.17
3.01.54.26.B	drawer pull	nr	0.20	7.53	—	3.98	11.50	8.28	—	4.37	12.65
3.01.54.26.C	cupboard knob	nr	0.20	7.53	—	1.67	9.20	8.28	—	1.84	10.12
3.01.54.28.	**Plates**									
3.01.54.28.A	200mm door push plate	nr	0.40	15.05	—	5.48	20.53	16.56	—	6.03	22.59
3.01.54.28.B	letter plate and opening through door	nr	1.75	65.85	—	14.13	79.98	72.44	—	15.54	87.98
3.01.54.30.	**Window fittings**									
3.01.54.30.A	casement stay; 250mm; with two pins	nr	0.40	15.05	—	9.71	24.76	16.56	—	10.68	27.24
3.01.54.30.B	casement fastener; wedge pattern	nr	0.50	18.82	—	4.15	22.96	20.70	—	4.56	25.26
3.01.54.30.C	casement fastener; locking with keys	nr	0.55	20.70	—	6.81	27.51	22.77	—	7.50	30.26
3.01.54.30.D	sliding sash fastener	nr	1.20	45.16	—	3.06	48.21	49.67	—	3.36	53.03
3.01.54.30.E	sash lift	nr	0.30	11.29	—	0.99	12.28	12.42	—	1.09	13.50
3.01.54.30.F	quadrant stay	nr	0.40	15.05	—	10.40	25.45	16.56	—	11.43	27.99
3.01.54.32.	**Sundry items**									
3.01.54.32.A	hat and coat hooks	nr	0.21	7.90	—	2.63	10.53	8.69	—	2.89	11.58
3.01.54.32.B	cabin hook and eye	nr	0.30	11.29	—	3.50	14.79	12.42	—	3.85	16.26
3.01.54.32.C	padlock hasp and staple	nr	0.30	11.29	—	4.36	15.65	12.42	—	4.79	17.21
3.01.54.32.D	swivel locking bar	nr	0.40	15.05	—	9.90	24.95	16.56	—	10.89	27.45
3.01.54.32.E	rubber door stop	nr	0.20	7.53	—	0.07	7.60	8.28	—	0.08	8.36
3.01.54.32.F	door buffer	nr	0.30	11.29	—	1.58	12.86	12.42	—	1.73	14.15
3.01.54.32.G	shelf bracket	nr	0.40	15.05	—	2.32	17.37	16.56	—	2.55	19.11
3.01.54.32.H	security door chain	nr	0.30	11.29	—	3.76	15.05	12.42	—	4.13	16.55
3.01.54.32.I	numbers; 75mm high	nr	0.20	7.53	—	3.50	11.02	8.28	—	3.85	12.12

Hutchins Priced Schedules 2021	Unit	Labour Hours	Labour Net £	Plant Net £	Materials Net £	Unit Rate Net £	Labour Gross £	Plant Gross £	Materials Gross £	Unit Rate Gross £
							(Gross rates include 10% profit)			

3.02. REPAIRS AND ALTERATIONS

3.02.01. REMOVE TIMBERS

3.02.01.01. Roof timbers

| 3.02.01.01.A | complete including rafters, purlins, ceiling joists, plates and the like (measured flat on plan) . . . . | m² | 0.37 | 10.94 | — | — | 10.94 | 12.04 | — | — | 12.04 |

3.02.01.02. Floor construction

3.02.01.02.A	joists; softwood; at ground level	m²	0.28	8.28	—	—	8.28	9.11	—	—	9.11
3.02.01.02.B	joists; softwood; at first floor level	m²	0.55	16.27	—	—	16.27	17.90	—	—	17.90
3.02.01.02.C	joists; softwood; at roof level	m²	0.77	22.78	—	—	22.78	25.05	—	—	25.05
3.02.01.02.D	individual floor or roof members	m	0.30	8.87	—	—	8.87	9.76	—	—	9.76
3.02.01.02.E	extra for cutting off end flush with wall . .	nr	0.50	14.79	—	—	14.79	16.27	—	—	16.27
3.02.01.02.F	decayed or infected floor plates	m	0.40	11.83	—	—	11.83	13.02	—	—	13.02
3.02.01.02.G	tilting fillet or roll	m	0.17	5.03	—	—	5.03	5.53	—	—	5.53
3.02.01.02.H	fascia or barge board	m	0.65	19.23	—	—	19.23	21.15	—	—	21.15

3.02.01.03. Boarding and flooring; softwood; including withdrawing nails; at

3.02.01.03.A	ground floor	m²	0.42	12.42	—	—	12.42	13.67	—	—	13.67
3.02.01.03.B	first floor	m²	0.68	20.11	—	—	20.11	22.13	—	—	22.13
3.02.01.03.C	roof; softwood	m²	0.80	23.66	—	—	23.66	26.03	—	—	26.03
3.02.01.03.D	gutter; softwood	m²	0.88	26.03	—	—	26.03	28.63	—	—	28.63
3.02.01.03.E	ground level; chipboard	m²	0.17	5.03	—	—	5.03	5.53	—	—	5.53
3.02.01.03.F	first floor level; chipboard	m²	0.42	12.42	—	—	12.42	13.67	—	—	13.67
3.02.01.03.G	ground level; plywood	m²	0.25	7.40	—	—	7.40	8.13	—	—	8.13
3.02.01.03.H	first floor level; plywood	m²	0.48	14.20	—	—	14.20	15.62	—	—	15.62

3.02.01.04. Stud partition; softwood; including finishings both sides

| 3.02.01.04.A | solid . . . . | m² | 0.50 | 14.79 | — | — | 14.79 | 16.27 | — | — | 16.27 |
| 3.02.01.04.B | glazed, including removal of glass . . . . . | m² | 0.67 | 19.82 | — | — | 19.82 | 21.80 | — | — | 21.80 |

3.02.01.05. Wall linings; including battening behind

| 3.02.01.05.A | plain sheeting . . . . | m² | 0.33 | 9.76 | — | — | 9.76 | 10.74 | — | — | 10.74 |
| 3.02.01.05.B | matchboarding . . . . | m² | 0.45 | 13.31 | — | — | 13.31 | 14.64 | — | — | 14.64 |

3.02.01.06. Ceiling linings; including battening behind

| 3.02.01.06.A | plain sheeting . . . . | m² | 0.50 | 14.79 | — | — | 14.79 | 16.27 | — | — | 16.27 |
| 3.02.01.06.B | matchboarding . . . . | m² | 0.67 | 19.82 | — | — | 19.82 | 21.80 | — | — | 21.80 |

3.02.01.07. Skirtings, etc.

| 3.02.01.07.A | skirtings, picture rails, dado rails architraves and the like . . . . | m | 0.12 | 3.55 | — | — | 3.55 | 3.90 | — | — | 3.90 |

Woodwork

	Unit	Labour Hours	Labour Net £	Plant Net £	Materials Net £	Unit Rate Net £	Labour Gross £	Plant Gross £	Materials Gross £	Unit Rate Gross £
3.02.	**REPAIRS AND ALTERATIONS**										
3.02.01.	**REMOVE TIMBERS**										
3.02.01.08.	**Shelves, etc.**										
3.02.01.08.A	shelves, window boards and the like . . .	m	0.35	10.35	—	—	10.35	11.39	—	—	11.39
3.02.01.09.	**Doors**										
3.02.01.09.A	single	nr	0.45	13.31	—	—	13.31	14.64	—	—	14.64
3.02.01.09.B	single with frame or lining	nr	0.88	26.03	—	—	26.03	28.63	—	—	28.63
3.02.01.09.C	pair	nr	0.77	22.78	—	—	22.78	25.05	—	—	25.05
3.02.01.09.D	pair with frame or lining	nr	1.32	39.05	—	—	39.05	42.95	—	—	42.95
3.02.01.09.E	extra for taking out spring box	nr	0.83	24.55	—	—	24.55	27.01	—	—	27.01
3.02.01.10.	**Windows**										
3.02.01.10.A	casement; with frame	nr	1.32	39.05	—	—	39.05	42.95	—	—	42.95
3.02.01.10.B	double hung sash; with frame	nr	1.77	52.36	—	—	52.36	57.59	—	—	57.59
3.02.01.10.C	French with frame	pair	4.40	130.15	—	—	130.15	143.17	—	—	143.17
3.02.01.11.	**Staircase; balustrade**										
3.02.01.11.A	single straight flight	nr	3.85	113.88	—	—	113.88	125.27	—	—	125.27
3.02.01.11.B	dogleg flight	nr	5.50	162.69	—	—	162.69	178.96	—	—	178.96
3.02.01.11.C	handrail and brackets	m	0.12	3.55	—	—	3.55	3.90	—	—	3.90
3.02.01.12.	**Bath panels**										
3.02.01.12.A	including frame	nr	0.45	13.31	—	—	13.31	14.64	—	—	14.64
3.02.01.14.	**Kitchen fittings**										
3.02.01.14.A	wall units	nr	0.50	14.79	—	—	14.79	16.27	—	—	16.27
3.02.01.14.B	floor units	nr	0.33	9.76	—	—	9.76	10.74	—	—	10.74
3.02.01.14.C	larder units	nr	0.45	13.31	—	—	13.31	14.64	—	—	14.64
3.02.01.14.D	built-in cupboards	nr	1.55	45.85	—	—	45.85	50.43	—	—	50.43
3.02.01.15.	**Casings**										
3.02.01.15.A	for pipes	m	0.33	9.76	—	—	9.76	10.74	—	—	10.74
3.02.02.	**ERECT TEMPORARY HOARDING**										
3.02.02.02.	**Enclose frontage to site with chestnut fencing with posts at 1.80m intervals and dismantle on completion**										
3.02.02.02.A	1.20m high	m	0.14	9.41	—	9.27	18.68	10.35	—	10.20	20.55
3.02.04.	**REMOVE DEFECTIVE AND RENEW**										
3.02.04.01.	**Take up defective gutter boards and bearers and supply and fix**										
3.02.04.01.A	new	m²	1.85	124.14	—	23.57	147.71	136.55	—	25.93	162.48

Woodwork

Hutchins Priced Schedules 2021	Unit	Labour Hours	Labour Net £	Plant Net £	Materials Net £	Unit Rate Net £	Labour Gross £	Plant Gross £	Materials Gross £	Unit Rate Gross £
								(Gross rates include 10% profit)		
3.02.	**REPAIRS AND ALTERATIONS**									
3.02.04.	**REMOVE DEFECTIVE AND RENEW**									
3.02.04.02.	**Take off defective rounded wood rolls to flats and supply and fix**									
3.02.04.02.A	new......................	m	0.11	7.39	—	1.71	9.10	8.13	—	1.88	10.02
3.02.04.03.	**Renew roof timbers**									
3.02.04.03.A	47mm x 100mm................	m	0.08	5.04	—	3.50	8.54	5.54	—	3.85	9.39
3.02.04.03.B	47mm x 125mm................	m	0.10	6.72	—	4.23	10.95	7.39	—	4.65	12.05
3.02.04.03.C	47mm x 150mm................	m	0.13	8.40	—	5.07	13.47	9.24	—	5.58	14.82
3.02.04.04.	**Take down defective hips and ridges and supply and fix new**									
3.02.04.04.A	175mm x 31mm................	m	0.20	13.44	—	4.07	17.52	14.79	—	4.48	19.27
3.02.04.05.	**Take down defective fascia and supply and fix new**									
3.02.04.05.A	150mm x 25mm................	m	0.20	13.44	—	2.76	16.20	14.79	—	3.04	17.82
3.02.04.06.	**Take down defective soffit and bearers and supply and fix new**									
3.02.04.06.A	225mm x 19mm................	m	0.30	20.16	—	3.03	23.20	22.18	—	3.34	25.52
3.02.04.07.	**Take off front gate; remove defective timber posts; grub up concrete; supply new post approximately 1.50m long set in new concrete and rehang gate**									
3.02.04.07.A	150mm x 150mm treated fir post......	nr	1.50	100.68	—	35.92	136.60	110.75	—	39.51	150.26
3.02.04.07.B	150mm x 150mm oak post.........	nr	1.62	109.08	—	67.88	176.96	119.99	—	74.67	194.66
3.02.04.08.	**Excavate for and bolt to wood gate post**									
3.02.04.08.A	oak spur set in concrete...........	nr	1.00	67.14	—	35.73	102.87	73.86	—	39.30	113.16
3.02.04.09.	**Take down and remove all temporary weatherproofing together with all associated timber work to windows and doors; make good all existing joinery work including withdrawing all nails**									
3.02.04.09.A	polythene sheet, hardboard, chipboard and the like..................	m²	0.60	40.26	—	0.67	40.93	44.28	—	0.74	45.02
3.02.06.	**TEMPORARY SCREENS**									
3.02.06.01.	**Temporary screen comprising**									
3.02.06.01.A	100mm x 50mm framing; lined both sides with building paper.............	m²	0.13	8.74	—	5.91	14.65	9.61	—	6.51	16.12
3.02.06.01.B	100mm x 50mm framing; lined one side with 19mm matchboard...........	m²	0.30	20.16	—	8.60	28.77	22.18	—	9.46	31.64

Woodwork

Hutchins Priced Schedules 2021	Unit	Labour Hours	Labour Net £	Plant Net £	Materials Net £	Unit Rate Net £	Labour Gross £	Plant Gross £	Materials Gross £	Unit Rate Gross £
							(Gross rates include 10% profit)			
3.02.	**REPAIRS AND ALTERATIONS**									
3.02.06.	**TEMPORARY SCREENS**									
3.02.06.01.	**Temporary screen comprising**									
3.02.06.01.C	47mm x 50mm framing; lined one side with hardboard	m²	0.23	15.12	—	3.52	18.64	16.63	—	3.87	20.51
3.02.06.02.	**Strut up ceiling and remove struts on completion**									
3.02.06.02.A	floor to ceiling average 2.60m	m	0.55	36.90	—	1.81	38.71	40.59	—	2.00	42.58
3.02.06.03.	**Strut and support window openings; area of window**									
3.02.06.03.A	1.00m2	nr	0.23	15.12	—	1.96	17.08	16.63	—	2.16	18.79
3.02.06.03.B	1.50m2	nr	0.25	16.80	—	2.93	19.74	18.48	—	3.23	21.71
3.02.06.03.C	2.00m2	nr	0.28	18.48	—	3.92	22.41	20.33	—	4.32	24.65
3.02.08.	**REPAIRS TO FLOORS**									
3.02.08.01.	**Remove all grease and dirt from existing flooring; remove all projecting lino nails or tacks; punch down all floor brads; resecure any loose boards; plane off and leave smooth**									
3.02.08.01.A	generally	m²	0.43	28.56	—	—	28.56	31.42	—	—	31.42
3.02.08.01.B	in small areas; less than 1.00m2	m²	0.55	36.90	—	—	36.90	40.59	—	—	40.59
3.02.08.01.C	take up loose floor blocks and relay in Lecol 7500 or the like; single block	nr	0.18	11.76	—	0.79	12.55	12.94	—	0.87	13.80
3.02.08.01.D	take up loose floor blocks and relay in Lecol 7500 or the like; in patches up to six blocks	nr	0.11	7.39	—	0.79	8.18	8.13	—	0.87	9.00
3.02.08.02.	**Smooth hardwood floor with**									
3.02.08.02.A	electric sanding machine	m²	0.50	33.54	0.72	—	34.26	36.89	0.79	—	37.68
3.02.08.03.	**Take off existing skirting and refix**									
3.02.08.03.A	replug grounds	m	0.18	12.10	—	—	12.10	13.31	—	—	13.31
3.02.08.04.	**Take off existing softwood skirting and supply and fix new**									
3.02.08.04.A	25mm x 150mm	m	0.25	16.80	—	4.44	21.24	18.48	—	4.89	23.37
3.02.08.06.	**Take up existing shrunk or worn flooring, any thickness; draw all nails, relay, cramp up, make up width or length with extra boarding of same thickness and clean off on completion; areas over 0.50m2**									
3.02.08.06.A	plain edge	m²	0.45	30.18	—	1.12	31.29	33.20	—	1.23	34.42
3.02.08.06.B	tongued and grooved	m²	0.55	36.90	—	1.14	38.04	40.59	—	1.26	41.85

Hutchins Priced Schedules 2021	Unit	Labour Hours	Labour Net £	Plant Net £	Materials Net £	Unit Rate Net £	Labour Gross £	Plant Gross £	Materials Gross £	Unit Rate Gross £
								(Gross rates include 10% profit)		

3.02. **REPAIRS AND ALTERATIONS**

3.02.08. **REPAIRS TO FLOORS**

3.02.08.07. **Remove damaged 25mm softwood floor boards; clean joists and renew**

	Unit	Labour Hours	Labour Net £	Plant Net £	Materials Net £	Unit Rate Net £	Labour Gross £	Plant Gross £	Materials Gross £	Unit Rate Gross £
3.02.08.07.A plain edge	m²	0.43	28.56	—	8.82	37.38	31.42	—	9.70	41.12
3.02.08.07.B tongued and grooved	m²	0.50	33.54	—	9.03	42.57	36.89	—	9.93	46.82
3.02.08.07.C plain edge; in small detached areas not exceeding 1.00m2	m²	1.10	73.80	—	8.82	82.62	81.18	—	9.70	90.88
3.02.08.07.D tongued and grooved in small detached areas not exceeding 1.00m2	m²	1.35	90.60	—	9.03	99.63	99.66	—	9.93	109.59
3.02.08.07.E plain edge exceeding 1.00m2, not exceeding 2.50m2	m²	1.00	67.14	—	8.82	75.96	73.86	—	9.70	83.56
3.02.08.07.F tongued and grooved exceeding 1.00m2, not exceeding 2.50m2	m²	1.25	83.88	—	9.03	92.91	92.27	—	9.93	102.20

3.02.08.09. **Renewing softwood joists and flooring including treating joists and underside of boards with creocote or other preservative**

	Unit	Labour Hours	Labour Net £	Plant Net £	Materials Net £	Unit Rate Net £	Labour Gross £	Plant Gross £	Materials Gross £	Unit Rate Gross £
3.02.08.09.A 100mm x 50mm floor joists; 25mm plain edge flooring	m²	1.00	37.63	—	34.37	72.00	41.39	—	37.80	79.20

3.02.08.10. **Oak strip flooring pinned and glued to existing softwood floor; clean off and wax polish**

	Unit	Labour Hours	Labour Net £	Plant Net £	Materials Net £	Unit Rate Net £	Labour Gross £	Plant Gross £	Materials Gross £	Unit Rate Gross £
3.02.08.10.A 14mm (PC sum £35 per m2)	m²	0.90	33.87	—	36.75	70.62	37.25	—	40.43	77.68

3.02.10. **REPAIRS TO DOOR FRAMES, LININGS, ETC.**

3.02.10.01. **Take down door**

	Unit	Labour Hours	Labour Net £	Plant Net £	Materials Net £	Unit Rate Net £	Labour Gross £	Plant Gross £	Materials Gross £	Unit Rate Gross £
3.02.10.01.A cut 13mm off bottom edge and rehang	nr	1.60	60.21	—	—	60.21	66.23	—	—	66.23

3.02.10.02. **Cut down architraves (one side)**

	Unit	Labour Hours	Labour Net £	Plant Net £	Materials Net £	Unit Rate Net £	Labour Gross £	Plant Gross £	Materials Gross £	Unit Rate Gross £
3.02.10.02.A reduce length by 13mm and refix	set	0.55	20.70	—	—	20.70	22.77	—	—	22.77
3.02.10.02.B reduce length by 13mm without removal	set	0.50	18.82	—	—	18.82	20.70	—	—	20.70

3.02.10.03. **Take off skirting**

	Unit	Labour Hours	Labour Net £	Plant Net £	Materials Net £	Unit Rate Net £	Labour Gross £	Plant Gross £	Materials Gross £	Unit Rate Gross £
3.02.10.03.A refix at higher level	m	0.30	11.29	—	—	11.29	12.42	—	—	12.42

3.02.10.04. **Hardwood border to hearth**

	Unit	Labour Hours	Labour Net £	Plant Net £	Materials Net £	Unit Rate Net £	Labour Gross £	Plant Gross £	Materials Gross £	Unit Rate Gross £
3.02.10.04.A mitred	nr	0.60	22.58	—	18.86	41.44	24.84	—	20.74	45.58

3.02.10.06. **Take down door; take out lining or frame; realign and refix. Refix existing architraves; make good work disturbed**

	Unit	Labour Hours	Labour Net £	Plant Net £	Materials Net £	Unit Rate Net £	Labour Gross £	Plant Gross £	Materials Gross £	Unit Rate Gross £
3.02.10.06.A ease, adjust and rehang door	nr	5.00	188.15	—	—	188.15	206.97	—	—	206.97

Woodwork

Hutchins Priced Schedules 2021	Unit	Labour Hours	Labour Net £	Plant Net £	Materials Net £	Unit Rate Net £	Labour Gross £	Plant Gross £	Materials Gross £	Unit Rate Gross £
										(Gross rates include 10% profit)
3.02.	**REPAIRS AND ALTERATIONS**									
3.02.10.	**REPAIRS TO DOOR FRAMES, LININGS, ETC.**									
3.02.10.10.	**Take down, ease and rehang**									
3.02.10.10.A	door	nr	1.80	67.73	—	—	67.73	74.51	—	—	74.51
3.02.10.10.B	door on new butt hinges; remove lock and furniture and supply and fit new rim lock and furniture	nr	3.00	112.89	—	25.24	138.13	124.18	—	27.76	151.94
3.02.10.10.C	door on new butt hinges; remove lock and furniture and supply and fit new mortice lock and furniture	nr	3.50	131.71	—	21.44	153.15	144.88	—	23.59	168.46
3.02.10.10.D	door; take apart and fit new panel or rail .	nr	4.80	180.62	—	4.13	184.75	198.69	—	4.54	203.22
3.02.10.10.E	renew weatherboard to external softwood door	nr	1.00	37.63	—	8.09	45.72	41.39	—	8.89	50.29
3.02.10.10.F	casement sash	nr	1.20	45.16	—	—	45.16	49.67	—	—	49.67
3.02.10.10.G	defective staff and parting beads to double hung sash window and renew . . .	nr	0.80	30.10	—	1.13	31.24	33.11	—	1.24	34.36
3.02.10.10.H	double hung sashes; including new cords	nr	1.35	50.80	—	2.80	53.60	55.88	—	3.08	58.96
3.02.10.10.I	cut out defective glazing bars to skylights, windows, doors or greenhouses and renew	m	0.90	33.87	—	2.08	35.95	37.25	—	2.29	39.55
3.02.12.	**REPAIRS TO STAIRS AND HANDRAILS**									
3.02.12.01.	**Strengthening handrail and balusters**									
3.02.12.01.A	including renewing defective balusters . .	m	1.25	47.04	—	2.95	49.99	51.74	—	3.25	54.99
3.02.12.02.	**Cutting out defective or worn portion of tread**									
3.02.12.02.A	piecing in new	nr	0.80	30.10	—	3.90	34.01	33.11	—	4.29	37.41
3.02.13.	**TAKE OFF AND REPLACE IRONMONGERY**									
3.02.13.01.	**Take off and replace ironmongery fixed to softwood**									
3.02.13.01.A	75mm strong pattern zinc plated butts . .	pair	1.33	50.05	—	1.78	51.83	55.05	—	1.96	57.01
3.02.13.01.B	100mm strong pattern zinc plated butts . .	pair	1.58	59.46	—	2.31	61.76	65.40	—	2.54	67.94
3.02.13.01.C	75mm steel washered brass butts	pair	1.33	50.05	—	3.63	53.68	55.05	—	4.00	59.05
3.02.13.01.D	100mm steel washered brass butts	pair	1.58	59.46	—	4.22	63.67	65.40	—	4.64	70.04
3.02.13.01.E	75mm brass rising butts	pair	1.61	60.58	—	7.77	68.35	66.64	—	8.54	75.19
3.02.13.01.F	100mm brass rising butts	pair	1.86	69.99	—	11.00	80.99	76.99	—	12.10	89.09
3.02.13.01.G	300mm steel tee hinges	pair	1.13	42.52	—	19.00	61.52	46.77	—	20.90	67.68
3.02.13.01.H	450mm steel tee hinges	pair	1.22	45.91	—	22.98	68.89	50.50	—	25.27	75.77
3.02.13.01.I	rim lock and furniture	nr	1.58	59.46	—	22.06	81.51	65.40	—	24.26	89.67
3.02.13.01.J	mortice lock and furniture	nr	1.84	69.24	—	18.26	87.50	76.16	—	20.09	96.25
3.02.13.01.K	Suffolk latch	nr	1.33	50.05	—	7.06	57.11	55.05	—	7.77	62.82

Hutchins Priced Schedules 2021		Unit	Labour Hours	Labour Net £	Plant Net £	Materials Net £	Unit Rate Net £	Labour Gross £	Plant Gross £	Materials Gross £	Unit Rate Gross £
							(Gross rates include 10% profit)				
3.02.	**REPAIRS AND ALTERATIONS**										
3.02.13.	**TAKE OFF AND REPLACE IRONMONGERY**										
3.02.13.01.	**Take off and replace ironmongery fixed to softwood**										
3.02.13.01.L	100mm straight barrel bolt; polished chrome	nr	0.68	25.59	—	3.39	28.98	28.15	—	3.73	31.88
3.02.13.01.M	150mm straight barrel bolt; polished chrome	nr	0.78	29.35	—	4.93	34.28	32.29	—	5.42	37.71
3.02.13.01.N	250mm casement stay with two pins . . .	nr	0.62	23.33	—	9.71	33.04	25.66	—	10.68	36.34
3.02.13.01.O	casement fastener; wedge pattern	nr	0.72	27.09	—	4.19	31.28	29.80	—	4.61	34.41
3.02.13.01.P	sliding sash fastener	nr	1.47	55.32	—	10.43	65.75	60.85	—	11.47	72.32
3.02.13.01.Q	sash lift	nr	0.47	17.69	—	1.00	18.68	19.45	—	1.10	20.55
3.02.14.	**SHORING**										
3.02.14.01.	**Erect temporary dead shoring to form opening using three pairs 150mm x 150mm uprights and three 225mm x 150mm needles, braces and 225mm base plates; hole brickwork for needles; all cartage; make good and remove on completion**										
3.02.14.01.A	volume of timber 0.90m3	item	34.95	2348.99	—	168.01	2517.00	2583.89	—	184.81	2768.70
3.02.14.01.B	add or deduct for every 0.30m3 more or less than 0.90m3	item	1.00	67.14	—	5.64	72.78	73.86	—	6.20	80.06
3.02.14.02.	**Erect temporary raking shoring including rakers, wall plates, needles, holing brickwork; all cartage; make good on completion**										
3.02.14.02.A	volume of timber 0.30m3	item	15.98	1073.81	—	56.00	1129.82	1181.20	—	61.60	1242.80
3.02.14.02.B	add or deduct for every 0.03m3 more or less than 0.30m3	item	1.50	100.68	—	5.64	106.32	110.75	—	6.20	116.95
3.02.14.04.	**Erect temporary flying shoring including horizontal shores, struts, wall plates, posts, needles, holing brickwork; all cartage; make good on completion**										
3.02.14.04.A	volume of timber 0.60m3	item	39.94	2684.57	—	112.01	2796.57	2953.03	—	123.21	3076.23
3.02.14.04.B	add or deduct for every 0.03m3 more or less than 0.60m3	item	2.00	134.22	—	5.64	139.85	147.64	—	6.20	153.84
3.02.14.06.	**Erect permanent raking shoring including horizontal shores, struts, wall plates, posts, needles, holing brickwork; all cartage; left in position for an indefinite period**										
3.02.14.06.A	volume of timber 0.30m3	item	9.99	671.16	—	121.23	792.39	738.27	—	133.35	871.63
3.02.14.06.B	add or deduct for each 0.30m3 more or less than 0.30m3	item	1.00	67.14	—	12.20	79.34	73.86	—	13.42	87.28

Woodwork

Hutchins Priced Schedules 2021	Unit	Labour Hours	Labour Net £	Plant Net £	Materials Net £	Unit Rate Net £	Labour Gross £	Plant Gross £	Materials Gross £	Unit Rate Gross £
							(Gross rates include 10% profit)			
3.02. **REPAIRS AND ALTERATIONS**										
3.02.14. **SHORING**										
3.02.14.07. **Erect permanent flying shoring, including horizontal shores, struts, wall plates, posts, needles, holing brickwork; all cartage; left in position for an indefinite period**										
3.02.14.07.A volume of timber 0.60m3	item	26.96	1812.05	—	242.46	2054.51	1993.25	—	266.71	2259.96
3.02.14.07.B add or deduct for every 0.03m3 more or less than 0.60m3	item	1.35	90.60	—	12.20	102.80	99.66	—	13.42	113.08

Hutchins Priced Schedules 2021	Unit	Labour Hours	Labour Net £	Plant Net £	Materials Net £	Unit Rate Net £	Labour Gross £	Plant Gross £	Materials Gross £	Unit Rate Gross £
						(Gross rates include 10% profit)				

4.01.   NEW WORK

4.01.01.   CARLITE PLASTER

4.01.01.01.   Plaster; 8mm bonding; 2mm finish; steel trowelled; internal; 10mm work; concrete or plasterboard base

	Unit	Labour Hours	Labour Net £	Plant Net £	Materials Net £	Unit Rate Net £	Labour Gross £	Plant Gross £	Materials Gross £	Unit Rate Gross £
4.01.01.01.A over 300mm wide to walls	m²	0.26	17.54	—	3.19	20.73	19.30	—	3.51	22.80
4.01.01.01.B not exceeding 300mm wide to walls	m²	0.39	26.35	—	3.19	29.54	28.98	—	3.51	32.49
4.01.01.01.C over 300mm wide to ceilings	m²	0.33	21.91	—	3.19	25.10	24.10	—	3.51	27.61
4.01.01.01.D not exceeding 300mm wide to ceilings . .	m²	0.49	32.73	—	3.19	35.92	36.00	—	3.51	39.51

4.01.01.02.   Plaster; 11mm undercoat; 2mm finish; steel trowelled; internal; 13mm work; to brick or block base

	Unit	Labour Hours	Labour Net £	Plant Net £	Materials Net £	Unit Rate Net £	Labour Gross £	Plant Gross £	Materials Gross £	Unit Rate Gross £
4.01.01.02.A over 300mm wide to walls	m²	0.27	17.88	—	3.19	21.07	19.67	—	3.51	23.18
4.01.01.02.B not exceeding 300mm wide to ceilings . .	m²	0.40	26.68	—	3.19	29.88	29.35	—	3.51	32.86

4.01.01.04.   Plaster; 11mm metal lathing undercoat; 2mm finish; steel trowelled; internal; 13mm work; metal lathing base

	Unit	Labour Hours	Labour Net £	Plant Net £	Materials Net £	Unit Rate Net £	Labour Gross £	Plant Gross £	Materials Gross £	Unit Rate Gross £
4.01.01.04.A over 300mm wide to walls	m²	0.27	17.88	—	6.70	24.57	19.67	—	7.36	27.03
4.01.01.04.B not exceeding 300mm wide to walls	m²	0.40	26.68	—	6.70	33.38	29.35	—	7.36	36.72
4.01.01.04.C over 300mm wide to ceilings	m²	0.34	22.99	—	6.70	29.68	25.28	—	7.36	32.65
4.01.01.04.D not exceeding 300mm wide to ceilings . .	m²	0.51	34.41	—	6.70	41.11	37.85	—	7.36	45.22

4.01.03.   THISTLE PLASTER

4.01.03.01.   Plaster; 11mm renovating; 2mm finish; steel trowelled; internal; 13mm work; to existing concrete, brick or block base

	Unit	Labour Hours	Labour Net £	Plant Net £	Materials Net £	Unit Rate Net £	Labour Gross £	Plant Gross £	Materials Gross £	Unit Rate Gross £
4.01.03.01.A over 300mm wide to walls	m²	0.27	17.88	—	6.03	23.91	19.67	—	6.63	26.30
4.01.03.01.B not exceeding 300mm wide to walls	m²	0.40	26.68	—	6.03	32.71	29.35	—	6.63	35.98

4.01.03.02.   Plaster; Universal one coat; steel trowelled; internal; 10mm work; to concrete base

	Unit	Labour Hours	Labour Net £	Plant Net £	Materials Net £	Unit Rate Net £	Labour Gross £	Plant Gross £	Materials Gross £	Unit Rate Gross £
4.01.03.02.A over 300mm wide to walls	m²	0.20	13.51	—	5.28	18.79	14.86	—	5.81	20.67
4.01.03.02.B not exceeding 300mm wide to walls	m²	0.30	20.23	—	5.28	25.51	22.25	—	5.81	28.06
4.01.03.02.C over 300mm wide to ceilings	m²	0.26	17.54	—	5.28	22.82	19.30	—	5.81	25.10
4.01.03.02.D not exceeding 300mm wide to ceilings . .	m²	0.39	26.01	—	5.28	31.29	28.61	—	5.81	34.42

Floor, Wall and Ceiling Finishes

Hutchins Priced Schedules 2021		Unit	Labour Hours	Labour Net £	Plant Net £	Materials Net £	Unit Rate Net £	Labour Gross £	Plant Gross £	Materials Gross £	Unit Rate Gross £
									(Gross rates include 10% profit)		
4.01.	**NEW WORK**										
4.01.03.	**THISTLE PLASTER**										
4.01.03.03.	**Plaster; Universal one coat; steel trowelled; internal; 13mm work; to brick or block base**										
4.01.03.03.A	over 300mm wide to walls	m²	0.21	14.18	—	7.26	21.44	15.60	—	7.99	23.59
4.01.03.03.B	not exceeding 300mm wide to walls	m²	0.31	20.90	—	7.26	28.16	22.99	—	7.99	30.98
4.01.03.04.	**Plaster; Universal one coat; steel trowelled; internal; 5mm work; to plasterboard base**										
4.01.03.04.A	over 300mm wide to walls	m²	0.19	12.50	—	2.64	15.14	13.75	—	2.90	16.66
4.01.03.04.B	not exceeding 300mm wide to walls	m²	0.29	19.22	—	2.64	21.86	21.14	—	2.90	24.05
4.01.03.04.C	over 300mm wide to ceilings	m²	0.25	16.53	—	2.64	19.17	18.19	—	2.90	21.09
4.01.03.04.D	not exceeding 300mm wide to ceilings . .	m²	0.37	25.00	—	2.64	27.64	27.50	—	2.90	30.41
4.01.03.05.	**Plaster; Thistle; 3mm one coat board finish; steel trowelled; internal; to plasterboard base**										
4.01.03.05.A	over 300mm wide to walls	m²	0.19	12.50	—	0.87	13.37	13.75	—	0.96	14.71
4.01.03.05.B	not exceeding 300mm wide to walls	m²	0.29	19.22	—	0.87	20.09	21.14	—	0.96	22.10
4.01.03.05.C	over 300mm wide to ceilings	m²	0.25	16.53	—	0.87	17.40	18.19	—	0.96	19.14
4.01.03.05.D	not exceeding 300mm wide to ceilings . .	m²	0.37	25.00	—	0.87	25.87	27.50	—	0.96	28.46
4.01.03.07.	**Plaster; Thistle; 10mm cement and sand (1:3); 3mm finish; 13mm work to concrete, brick or block base**										
4.01.03.07.A	over 300mm wide to walls	m²	0.28	18.55	—	2.29	20.84	20.40	—	2.52	22.93
4.01.03.07.B	not exceeding 300mm wide to walls	m²	0.42	28.03	—	2.29	30.32	30.83	—	2.52	33.35
4.01.03.07.C	over 300mm wide to ceilings	m²	0.36	24.33	—	2.29	26.62	26.76	—	2.52	29.29
4.01.03.07.D	not exceeding 300mm wide to ceilings . .	m²	0.54	36.09	—	2.29	38.39	39.70	—	2.52	42.22
4.01.05.	**LABOURS ON PLASTERING**										
4.01.05.01.	**Rounded internal angle**										
4.01.05.01.A	not exceeding 10mm radius	m	0.03	2.02	—	—	2.02	2.22	—	—	2.22
4.01.05.01.B	over 10mm radius	m	0.04	2.69	—	—	2.69	2.96	—	—	2.96
4.01.05.02.	**Rounded external angle**										
4.01.05.02.A	not exceeding 10mm radius	m	0.04	2.35	—	—	2.35	2.59	—	—	2.59
4.01.05.02.B	over 10mm radius	m	0.05	3.36	—	—	3.36	3.70	—	—	3.70
4.01.05.03.	**Make good plaster around pipes, angles and the like**										
4.01.05.03.A	not exceeding 300mm girth	nr	0.04	2.35	—	—	2.35	2.59	—	—	2.59
4.01.05.03.B	over 300mm girth	nr	0.04	2.69	—	—	2.69	2.96	—	—	2.96

Hutchins Priced Schedules 2021	Unit	Labour Hours	Labour Net £	Plant Net £	Materials Net £	Unit Rate Net £	Labour Gross £	Plant Gross £	Materials Gross £	Unit Rate Gross £
							(Gross rates include 10% profit)			

4.01.   **NEW WORK**

4.01.06.   **PLASTER BEADS AND THE LIKE**

4.01.06.01.   Catnic galvanised steel beads; fixed with plaster dabs

4.01.06.01.A standard angle bead	m	0.07	4.70	—	0.48	5.19	5.18	—	0.53	5.71
4.01.06.01.B Supasave angle bead	m	0.07	4.70	—	0.41	5.11	5.18	—	0.45	5.63
4.01.06.01.C micro mesh angle bead	m	0.07	4.70	—	0.43	5.14	5.18	—	0.47	5.65
4.01.06.01.D dry wall angle bead	m	0.07	4.70	—	0.40	5.10	5.18	—	0.44	5.61
4.01.06.01.E dry wall stop bead; 3mm	m	0.07	4.70	—	0.61	5.31	5.18	—	0.67	5.85
4.01.06.01.F dry wall stop bead; 6mm	m	0.07	4.70	—	0.53	5.23	5.18	—	0.58	5.75
4.01.06.01.G renderstop	m	0.07	4.70	—	1.26	5.96	5.18	—	1.39	6.56
4.01.06.01.H plaster stop; 10mm	m	0.07	4.70	—	1.58	6.28	5.18	—	1.73	6.91
4.01.06.01.I plaster stop; 13mm	m	0.07	4.70	—	1.58	6.28	5.18	—	1.73	6.91
4.01.06.01.J plaster stop; 16mm	m	0.07	4.70	—	1.71	6.42	5.18	—	1.88	7.06
4.01.06.01.K plaster stop; 19mm	m	0.07	4.70	—	2.78	7.49	5.18	—	3.06	8.24
4.01.06.01.L plasterboard edging bead; 10mm	m	0.07	4.70	—	0.80	5.50	5.18	—	0.88	6.05
4.01.06.01.M plasterboard edging bead; 12.5mm	m	0.07	4.70	—	0.80	5.50	5.18	—	0.88	6.05
4.01.06.01.N architrave bead; 10mm	m	0.07	4.70	—	2.49	7.19	5.18	—	2.74	7.91
4.01.06.01.O movement bead; 12mm	m	0.07	4.70	—	5.88	10.58	5.18	—	6.47	11.64

4.01.08.   **PORTLAND CEMENT FINISHES**

4.01.08.01.   Render; cement and sand (1:3); 6mm work; dubbing out; internal; to walls; brick or block base

4.01.08.01.A over 300mm wide	m²	0.15	9.81	—	0.71	10.52	10.79	—	0.78	11.58
4.01.08.01.B not exceeding 300mm wide	m²	0.22	14.52	—	0.71	15.23	15.97	—	0.78	16.75

4.01.08.02.   Render; cement and sand (1:3); 13mm work dubbing out; internal; to walls; brick or block base

4.01.08.02.A over 300mm wide	m²	0.21	14.18	—	1.78	15.96	15.60	—	1.96	17.56
4.01.08.02.B not exceeding 300mm wide	m²	0.31	20.90	—	1.78	22.68	22.99	—	1.96	24.95

4.01.08.03.   Render; cement and sand (1:3); 19mm work dubbing out; internal; to walls; brick or block base

4.01.08.03.A over 300mm wide	m²	0.27	17.88	—	2.49	20.37	19.67	—	2.74	22.41
4.01.08.03.B not exceeding 300mm wide	m²	0.39	26.35	—	2.49	28.84	28.98	—	2.74	31.72

4.01.08.04.   Render; cement and sand (1:3); 25mm work dubbing out; internal; to walls; brick or block base

4.01.08.04.A over 300mm wide	m²	0.30	20.23	—	3.32	23.55	22.25	—	3.65	25.91
4.01.08.04.B not exceeding 300mm wide	m²	0.42	28.36	—	3.32	31.68	31.20	—	3.65	34.85

Floor, Wall and Ceiling Finishes

Hutchins Priced Schedules 2021	Unit	Labour Hours	Labour Net £	Plant Net £	Materials Net £	Unit Rate Net £	Labour Gross £	Plant Gross £	Materials Gross £	Unit Rate Gross £
							(Gross rates include 10% profit)			
4.01. **NEW WORK**										
4.01.08. PORTLAND CEMENT FINISHES										
4.01.08.06. **Render; cement and sand (1:3); 13mm one coat work to walls; wood floated; plain face; internal; to brick or block base**										
4.01.08.06.A over 300mm wide	m²	0.21	14.18	—	1.78	15.96	15.60	—	1.96	17.56
4.01.08.06.B not exceeding 300mm wide	m²	0.31	20.90	—	1.78	22.68	22.99	—	1.96	24.95
4.01.08.07. **Render; cement and sand (1:3); 20mm two coat work to walls; wood floated; plain face; internal; to brick or block base**										
4.01.08.07.A over 300mm wide	m²	0.30	20.23	—	2.61	22.84	22.25	—	2.87	25.12
4.01.08.07.B not exceeding 300mm wide	m²	0.45	30.38	—	2.61	32.99	33.42	—	2.87	36.29
4.01.08.07.C kerb to gulley	nr	0.43	28.70	—	1.54	30.24	31.57	—	1.70	33.27
4.01.08.07.D plinth at back of gulley	nr	0.20	13.51	—	0.95	14.46	14.86	—	1.04	15.90
4.01.08.08. **Render; cement and sand (1:3); rough cast face; wood floated; external; 20mm two coat work to walls; brick or block base**										
4.01.08.08.A over 300mm wide	m²	0.45	30.38	—	2.61	32.99	33.42	—	2.87	36.29
4.01.08.08.B not exceeding 300mm wide	m²	0.60	40.53	—	2.61	43.14	44.58	—	2.87	47.45
4.01.08.09. **Render; cement and sand (1:3); pebble dash finish; wood floated; external; 20mm two coat work to walls; brick or block base**										
4.01.08.09.A over 300mm wide	m²	0.48	32.40	—	2.43	34.83	35.63	—	2.68	38.31
4.01.08.09.B not exceeding 300mm wide	m²	0.63	42.54	—	2.43	44.98	46.80	—	2.68	49.47
4.01.08.11. **Render; cement-lime-sand (1:1:6); wood floated; plain face; external; 13mm one coat work to walls; brick or block base**										
4.01.08.11.A over 300mm wide	m²	0.21	14.18	—	1.93	16.11	15.60	—	2.12	17.72
4.01.08.11.B not exceeding 300mm wide	m²	0.31	20.90	—	1.93	22.83	22.99	—	2.12	25.12
4.01.08.12. **Render; cement-lime-sand (1:1:6); wood floated; plain face; external; 20mm two coat work to walls; brick or block base**										
4.01.08.12.A over 300mm wide	m²	0.30	20.23	—	2.83	23.06	22.25	—	3.11	25.37
4.01.08.12.B not exceeding 300mm wide	m²	0.45	30.38	—	2.83	33.21	33.42	—	3.11	36.53

Hutchins Priced Schedules 2021	Unit	Labour Hours	Labour Net £	Plant Net £	Materials Net £	Unit Rate Net £	Labour Gross £	Plant Gross £	Materials Gross £	Unit Rate Gross £
							(Gross rates include 10% profit)			

4.01.   NEW WORK

4.01.08.   PORTLAND CEMENT FINISHES

4.01.08.13.   Render; cement-lime-sand (1:1:6); wood floated; rough cast face; external; 20mm two coat work to walls; brick or block base

4.01.08.13.A over 300mm wide	m²	0.45	30.38	—	2.83	33.21	33.42	—	3.11	36.53
4.01.08.13.B not exceeding 300mm wide	m²	0.60	40.53	—	2.83	43.36	44.58	—	3.11	47.70

4.01.08.14.   Render; cement-lime-sand (1:1:6); wood floated; pebbledash finish; external; 20mm two coat work to walls; brick or block base

4.01.08.14.A over 300mm wide	m²	0.48	32.40	—	3.25	35.64	35.63	—	3.57	39.21
4.01.08.14.B not exceeding 300mm wide	m²	0.63	42.54	—	3.25	45.79	46.80	—	3.57	50.37

4.01.08.15.   Render; cement-lime-sand (1:1:6); Snowcem; Weber Cullamix Tyrolean; external; 16mm one coat work to walls; brick or block base

4.01.08.15.A over 300mm wide	m²	0.39	26.35	—	7.44	33.79	28.98	—	8.18	37.16
4.01.08.15.B not exceeding 300mm wide	m²	0.58	39.18	—	7.44	46.62	43.10	—	8.18	51.28

4.01.09.   LABOURS ON RENDERING

4.01.09.01.   Rounded internal angle

4.01.09.01.A not exceeding 10mm radius	m	0.04	2.69	—	—	2.69	2.96	—	—	2.96
4.01.09.01.B over 10mm radius	m	0.06	3.70	—	—	3.70	4.07	—	—	4.07

4.01.09.02.   Rounded external angle

4.01.09.02.A not exceeding 10mm radius	m	0.05	3.36	—	—	3.36	3.70	—	—	3.70
4.01.09.02.B over 10mm radius	m	0.06	4.03	—	—	4.03	4.44	—	—	4.44

4.01.09.03.   Make good rendering around pipes, angles and the like

4.01.09.03.A not exceeding 300mm girth	nr	0.04	2.35	—	—	2.35	2.59	—	—	2.59
4.01.09.03.B over 300mm girth	nr	0.04	2.69	—	—	2.69	2.96	—	—	2.96

4.01.10.   RENDER BEADS AND THE LIKE

4.01.10.01.   Expamet beads for external use; stainless steel

4.01.10.01.A angle bead	m	0.07	4.70	—	4.46	9.17	5.18	—	4.91	10.08
4.01.10.01.B movement bead	m	0.07	4.70	—	8.31	13.01	5.18	—	9.14	14.31
4.01.10.01.C stop bead; 10mm	m	0.07	4.70	—	4.42	9.13	5.18	—	4.86	10.04
4.01.10.01.D stop bead; 13mm	m	0.07	4.70	—	4.52	9.22	5.18	—	4.97	10.14
4.01.10.01.E stop bead; 16mm	m	0.07	4.70	—	4.65	9.36	5.18	—	5.12	10.29
4.01.10.01.F stop bead; 19mm	m	0.07	4.70	—	4.65	9.36	5.18	—	5.12	10.29
4.01.10.01.G external render stop	m	0.07	4.70	—	1.40	6.10	5.18	—	1.54	6.71

Floor, Wall and Ceiling Finishes

Hutchins Priced Schedules 2021	Unit	Labour Hours	Labour Net £	Plant Net £	Materials Net £	Unit Rate Net £	Labour Gross £	Plant Gross £	Materials Gross £	Unit Rate Gross £
								(Gross rates include 10% profit)		
4.01.	**NEW WORK**									
4.01.11.	**CEMENT SCREEDS**									
4.01.11.01.	**Screed; cement and sand (1:3); steel trowelled smooth; floors; level and to falls**									
4.01.11.01.A	25mm; over 300mm wide	m²	0.45	30.38	—	3.56	33.94	33.42	—	3.92	37.33
4.01.11.01.B	25mm; not exceeding 300mm wide	m²	0.90	60.76	—	3.56	64.32	66.83	—	3.92	70.75
4.01.11.01.C	32mm; over 300mm wide	m²	0.50	33.74	—	4.39	38.13	37.11	—	4.83	41.94
4.01.11.01.D	32mm; not exceeding 300mm wide	m²	1.00	67.48	—	4.39	71.87	74.23	—	4.83	79.06
4.01.11.01.E	38mm; over 300mm wide	m²	0.55	37.10	—	5.34	42.44	40.81	—	5.87	46.68
4.01.11.01.F	38mm; not exceeding 300mm wide	m²	1.11	74.27	—	5.34	79.61	81.69	—	5.87	87.57
4.01.11.01.G	50mm; over 300mm wide	m²	0.65	43.89	—	6.88	50.77	48.28	—	7.57	55.85
4.01.11.01.H	50mm; not exceeding 300mm wide	m²	1.31	87.78	—	6.88	94.66	96.55	—	7.57	104.12
4.01.11.02.	**Skirtings; cement and sand (1:3); 150mm high; fair edge; ends and the like**									
4.01.11.02.A	straight	m	0.28	18.55	—	0.47	19.02	20.40	—	0.52	20.93
4.01.11.02.B	curved	m	0.43	28.70	—	0.47	29.17	31.57	—	0.52	32.09
4.01.11.03.	**Make good paving to**									
4.01.11.03.A	floor channel	m	0.13	8.47	—	—	8.47	9.32	—	—	9.32
4.01.11.03.B	rainwater, soil and ventilation pipes	nr	0.20	13.51	—	—	13.51	14.86	—	—	14.86
4.01.11.03.C	yard gulley	nr	0.25	16.87	—	—	16.87	18.56	—	—	18.56
4.01.11.03.D	manhole cover frame	nr	0.50	33.74	—	—	33.74	37.11	—	—	37.11
4.01.12.	**GRANOLITHIC; CEMENT AND GRANITE CHIPPINGS (2:7)**									
4.01.12.01.	**Granolithic; steel trowelled smooth floors; level and to falls**									
4.01.12.01.A	25mm; over 300mm wide	m²	0.23	15.53	—	3.87	19.40	17.08	—	4.26	21.34
4.01.12.01.B	25mm; not exceeding 300mm wide	m²	0.34	22.99	—	3.87	26.86	25.28	—	4.26	29.54
4.01.12.01.C	32mm; over 300mm wide	m²	0.25	16.87	—	4.77	21.64	18.56	—	5.25	23.81
4.01.12.01.D	32mm; not exceeding 300mm wide	m²	0.38	25.67	—	4.77	30.45	28.24	—	5.25	33.49
4.01.12.01.E	38mm; over 300mm wide	m²	0.27	18.21	—	5.81	24.02	20.04	—	6.39	26.42
4.01.12.01.F	38mm; not exceeding 300mm wide	m²	0.41	27.69	—	5.81	33.50	30.46	—	6.39	36.85
4.01.12.01.G	50mm; over 300mm wide	m²	0.29	19.56	—	7.48	27.04	21.51	—	8.23	29.74
4.01.12.01.H	50mm; not exceeding 300mm wide	m²	0.43	29.03	—	7.48	36.52	31.94	—	8.23	40.17
4.01.12.02.	**Lining to channels; to falls; rounded arrises; coved junction; ends, angles and the like; 150mm girth**									
4.01.12.02.A	25mm	m	0.26	17.54	—	0.52	18.06	19.30	—	0.57	19.86
4.01.12.02.B	32mm	m	0.28	18.55	—	0.65	19.20	20.40	—	0.71	21.11

Hutchins Priced Schedules 2021	Unit	Labour Hours	Labour Net £	Plant Net £	Materials Net £	Unit Rate Net £	Labour Gross £	Plant Gross £	Materials Gross £	Unit Rate Gross £
						(Gross rates include 10% profit)				
4.01.	**NEW WORK**									
4.01.12.	**GRANOLITHIC; CEMENT AND GRANITE CHIPPINGS (2:7)**									
4.01.12.02.	Lining to channels; to falls; rounded arrises; coved junction; ends, angles and the like; 150mm girth									
4.01.12.02.C	38mm......................	m	0.46	30.71	—	0.77	31.49	33.79	—	0.85	34.64
4.01.12.02.D	50mm......................	m	0.48	32.40	—	1.03	33.43	35.63	—	1.14	36.77
4.01.12.03.	Treads; rounded nosing; ends, angles and the like; 275mm wide									
4.01.12.03.A	25mm......................	m	0.21	13.85	—	0.90	14.75	15.23	—	0.99	16.22
4.01.12.03.B	32mm......................	m	0.24	15.86	—	1.29	17.15	17.45	—	1.42	18.87
4.01.12.03.C	38mm......................	m	0.27	17.88	—	1.55	19.43	19.67	—	1.70	21.37
4.01.12.03.D	50mm......................	m	0.30	19.89	—	1.94	21.83	21.88	—	2.13	24.01
4.01.12.04.	Risers; coved junction to tread; undercut; 150mm high									
4.01.12.04.A	13mm......................	m	0.16	10.82	—	0.26	11.08	11.90	—	0.28	12.19
4.01.12.04.B	19mm......................	m	0.19	12.50	—	0.39	12.89	13.75	—	0.43	14.18
4.01.12.05.	Strings or aprons; rounded top edge; ends, angles and the like; 275mm wide									
4.01.12.05.A	13mm......................	m	0.30	19.89	—	0.52	20.41	21.88	—	0.57	22.45
4.01.12.05.B	19mm......................	m	0.39	26.35	—	0.77	27.12	28.98	—	0.85	29.83
4.01.12.06.	Skirtings; rounded top edge; coved junction to paving; ends, angles and the like; 150mm high									
4.01.12.06.A	13mm......................	m	0.26	17.54	—	0.26	17.80	19.30	—	0.28	19.58
4.01.12.06.B	19mm......................	m	0.36	24.33	—	0.39	24.72	26.76	—	0.43	27.19
4.01.12.07.	Carborundum surface dressing									
4.01.12.07.A	1kg/m2......................	m²	0.06	4.03	—	3.31	7.34	4.44	—	3.64	8.07
4.01.12.08.	Fair joint to flush edge									
4.01.12.08.A	of existing finishes..............	m	0.07	4.70	—	—	4.70	5.18	—	—	5.18
4.01.12.09.	Making good around pipes and the like; not exceeding 300mm girth									
4.01.12.09.A	25mm......................	nr	0.06	4.03	—	—	4.03	4.44	—	—	4.44
4.01.12.09.B	32mm......................	nr	0.06	4.03	—	—	4.03	4.44	—	—	4.44
4.01.12.09.C	38mm......................	nr	0.07	4.70	—	—	4.70	5.18	—	—	5.18
4.01.12.09.D	50mm......................	nr	0.07	4.70	—	—	4.70	5.18	—	—	5.18
4.01.14.	**METAL LATHING**									
4.01.14.01.	Expamet; 9mm galvanised expanded metal lathing; BB263; 0.50mm thick; to walls									
4.01.14.01.A	over 300mm wide; fixed to softwood with galvanised nails..............	m²	0.10	6.72	—	4.87	11.59	7.39	—	5.36	12.75

Floor, Wall and Ceiling Finishes

Hutchins Priced Schedules 2021		Unit	Labour Hours	Labour Net £	Plant Net £	Materials Net £	Unit Rate Net £	Labour Gross £	Plant Gross £	Materials Gross £	Unit Rate Gross £
										(Gross rates include 10% profit)	
4.01.	**NEW WORK**										
4.01.14.	**METAL LATHING**										
4.01.14.01.	**Expamet; 9mm galvanised expanded metal lathing; BB263; 0.50mm thick; to walls**										
4.01.14.01.B	not exceeding 300mm wide; fixed to softwood with galvanised nails	m²	0.16	10.48	—	4.87	15.36	11.53	—	5.36	16.89
4.01.14.01.C	over 300mm wide; fixed to steel with tying wire	m²	0.13	8.47	—	4.80	13.27	9.32	—	5.28	14.59
4.01.14.01.D	not exceeding 300mm wide; fixed to steel with tying wire	m²	0.18	12.17	—	4.80	16.96	13.38	—	5.28	18.66
4.01.14.02.	**Expamet; 9mm galvanised expanded metal lathing; BB263; 0.50mm thick; to ceilings**										
4.01.14.02.A	over 300mm wide; fixed to softwood with galvanised nails	m²	0.12	8.13	—	4.87	13.00	8.95	—	5.36	14.30
4.01.14.02.B	not exceeding 300mm wide; fixed to softwood with galvanised nails	m²	0.18	12.17	—	4.87	17.04	13.38	—	5.36	18.74
4.01.14.02.C	over 300mm wide; fixed to steel with tying wire	m²	0.15	9.81	—	4.80	14.61	10.79	—	5.28	16.07
4.01.14.02.D	not exceeding 300mm wide; fixed to steel with tying wire	m²	0.21	13.85	—	4.80	18.64	15.23	—	5.28	20.51
4.01.14.03.	**Expamet; 9mm galvanised expanded metal lathing; BB263; 0.50mm thick; to beams, sides, soffits and tops**										
4.01.14.03.A	over 300mm wide; fixed to steel with tying wire	m²	0.18	11.83	—	4.80	16.63	13.01	—	5.28	18.29
4.01.14.03.B	not exceeding 300mm wide; fixed to steel with tying wire	m²	0.25	16.87	—	4.80	21.67	18.56	—	5.28	23.84
4.01.14.03.C	raking cutting	m	0.06	4.03	—	—	4.03	4.44	—	—	4.44
4.01.14.03.D	curved cutting	m	0.09	6.05	—	—	6.05	6.65	—	—	6.65
4.01.14.04.	**Expamet; 9mm galvanised expanded metal lathing; BB264; 0.725mm thick; to walls**										
4.01.14.04.A	over 300mm wide; fixed to softwood with galvanised nails	m²	0.12	7.80	—	5.48	13.28	8.58	—	6.03	14.60
4.01.14.04.B	not exceeding 300mm wide; fixed to softwood with galvanised nails	m²	0.17	11.49	—	5.48	16.97	12.64	—	6.03	18.67
4.01.14.04.C	over 300mm wide; fixed to steel with tying wire	m²	0.14	9.48	—	5.41	14.88	10.42	—	5.95	16.37
4.01.14.04.D	not exceeding 300mm wide; fixed to steel with tying wire	m²	0.20	13.17	—	5.41	18.58	14.49	—	5.95	20.44
4.01.14.05.	**Expamet; 9mm galvanised expanded metal lathing; BB264; 0.725mm thick; to ceilings**										
4.01.14.05.A	over 300mm wide; fixed to softwood with galvanised nails	m²	0.13	8.80	—	5.48	14.29	9.68	—	6.03	15.71

Hutchins Priced Schedules 2021	Unit	Labour Hours	Labour Net £	Plant Net £	Materials Net £	Unit Rate Net £	Labour Gross £	Plant Gross £	Materials Gross £	Unit Rate Gross £
								(Gross rates include 10% profit)		

4.01. NEW WORK

4.01.14. METAL LATHING

4.01.14.05. Expamet; 9mm galvanised expanded metal lathing; BB264; 0.725mm thick; to ceilings

	Unit	Labour Hours	Labour Net £	Plant Net £	Materials Net £	Unit Rate Net £	Labour Gross £	Plant Gross £	Materials Gross £	Unit Rate Gross £
4.01.14.05.B not exceeding 300mm wide; fixed to softwood with galvanised nails	m²	0.19	12.84	—	5.48	18.32	14.12	—	6.03	20.15
4.01.14.05.C over 300mm wide; fixed to steel with tying wire	m²	0.16	10.48	—	5.41	15.89	11.53	—	5.95	17.48
4.01.14.05.D not exceeding 300mm wide; fixed to steel with tying wire	m²	0.22	14.52	—	5.41	19.92	15.97	—	5.95	21.92
4.01.14.06. Expamet; 9mm galvanised expanded metal lathing; BB264; 0.725mm thick; to beams, sides, soffits and tops										
4.01.14.06.A over 300mm wide; fixed to steel with tying wire	m²	0.19	12.50	—	5.41	17.91	13.75	—	5.95	19.70
4.01.14.06.B not exceeding 300mm wide; fixed to steel with tying wire	m²	0.27	17.88	—	5.41	23.29	19.67	—	5.95	25.61
4.01.14.06.C raking cutting	m	0.07	4.37	—	—	4.37	4.81	—	—	4.81
4.01.14.06.D curved cutting	m	0.10	6.38	—	—	6.38	7.02	—	—	7.02
4.01.16. DRY LININGS AND PARTITIONS										
4.01.16.01. Baseboard; 9.5mm work; fixed to timber base with galvanised nails; taped butt joints										
4.01.16.01.A over 300mm wide; to walls	m²	0.16	10.48	—	3.66	14.15	11.53	—	4.03	15.56
4.01.16.01.B not exceeding 300mm wide; to walls . . .	m²	0.24	16.20	—	3.66	19.86	17.82	—	4.03	21.85
4.01.16.01.C over 300mm wide; to ceilings	m²	0.19	12.50	—	3.66	16.17	13.75	—	4.03	17.78
4.01.16.01.D not exceeding 300mm wide; to ceilings . .	m²	0.28	18.55	—	3.66	22.21	20.40	—	4.03	24.44
4.01.16.03. Gyproc square edge wallboard; 9.5mm work; fixed to timber base with galvanised nails; taped butt joints										
4.01.16.03.A over 300mm wide; to walls	m²	0.13	8.80	—	2.13	10.94	9.68	—	2.34	12.03
4.01.16.03.B not exceeding 300mm wide; to walls . . .	m²	0.20	13.51	—	2.13	15.64	14.86	—	2.34	17.20
4.01.16.03.C over 300mm wide; to ceilings	m²	0.16	10.48	—	2.13	12.62	11.53	—	2.34	13.88
4.01.16.03.D not exceeding 300mm wide; to ceilings . .	m²	0.24	15.86	—	2.13	17.99	17.45	—	2.34	19.79
4.01.16.04. Gyproc square edge wallboard; 12.5mm work; fixed to timber base with galvanised nails; taped butt joints										
4.01.16.04.A over 300mm wide; to walls	m²	0.16	10.48	—	2.23	12.71	11.53	—	2.45	13.98
4.01.16.04.B not exceeding 300mm wide; to walls . . .	m²	0.24	16.20	—	2.23	18.42	17.82	—	2.45	20.27
4.01.16.04.C over 300mm wide; to ceilings	m²	0.19	12.50	—	2.23	14.73	13.75	—	2.45	16.20
4.01.16.04.D not exceeding 300mm wide; to ceilings . .	m²	0.28	18.89	—	2.23	21.11	20.77	—	2.45	23.22

Floor, Wall and Ceiling Finishes

Hutchins Priced Schedules 2021		Unit	Labour Hours	Labour Net £	Plant Net £	Materials Net £	Unit Rate Net £	Labour Gross £	Plant Gross £	Materials Gross £	Unit Rate Gross £
								(Gross rates include 10% profit)			
4.01.	**NEW WORK**										
4.01.16.	**DRY LININGS AND PARTITIONS**										
4.01.16.06.	**Gyproc taper edge thermal board; 22mm work; fixed to timber base with galvanised nails; taped butt joints**										
4.01.16.06.A	over 300mm wide; to walls	m²	0.19	12.50	—	7.94	20.44	13.75	—	8.74	22.49
4.01.16.06.B	not exceeding 300mm wide; to walls . . .	m²	0.27	18.21	—	7.94	26.16	20.04	—	8.74	28.77
4.01.16.06.C	over 300mm wide; to ceilings	m²	0.22	14.52	—	7.94	22.46	15.97	—	8.74	24.70
4.01.16.06.D	not exceeding 300mm wide; to ceilings . .	m²	0.31	20.57	—	7.94	28.51	22.62	—	8.74	31.36
4.01.16.08.	**Gyproc taper edge thermal board; 40mm work; fixed to timber base with galvanised nails; taped butt joints**										
4.01.16.08.A	over 300mm wide; to walls	m²	0.22	14.52	—	10.30	24.82	15.97	—	11.33	27.30
4.01.16.08.B	not exceeding 300mm wide; to walls . . .	m²	0.30	20.23	—	10.30	30.53	22.25	—	11.33	33.58
4.01.16.08.C	over 300mm wide; to ceilings	m²	0.25	16.53	—	10.30	26.83	18.19	—	11.33	29.52
4.01.16.08.D	not exceeding 300mm wide; to ceilings . .	m²	0.34	22.58	—	10.30	32.88	24.84	—	11.33	36.17
4.01.16.10.	**Gyproc taper edge vapour check thermal board; 27mm work; fixed to timber base with galvanised nails; taped butt joints**										
4.01.16.10.A	over 300mm wide; to walls	m²	0.19	12.50	—	12.55	25.05	13.75	—	13.81	27.56
4.01.16.10.B	not exceeding 300mm wide; to walls . . .	m²	0.27	18.21	—	12.55	30.76	20.04	—	13.81	33.84
4.01.16.10.C	over 300mm wide; to ceilings	m²	0.22	14.52	—	12.55	27.07	15.97	—	13.81	29.77
4.01.16.10.D	not exceeding 300mm wide; to ceilings . .	m²	0.31	20.57	—	12.55	33.12	22.62	—	13.81	36.43
4.01.16.12.	**Gyproc taper edge vapour check thermal board; 40mm work; fixed to timber base with galvanised nails; taped butt joints**										
4.01.16.12.A	over 300mm wide; to walls	m²	0.22	14.52	—	15.79	30.31	15.97	—	17.37	33.34
4.01.16.12.B	not exceeding 300mm wide; to walls . . .	m²	0.30	20.23	—	15.79	36.02	22.25	—	17.37	39.62
4.01.16.12.C	over 300mm wide; to ceilings	m²	0.25	16.53	—	15.79	32.32	18.19	—	17.37	35.56
4.01.16.12.D	not exceeding 300mm wide; to ceilings . .	m²	0.34	22.58	—	15.79	38.37	24.84	—	17.37	42.21
4.01.16.16.	**Extra for taped joints**										
4.01.16.16.A	filled; one coat Gyproc Drywall Sealer to plasterboard; (both sides measured) . . .	m²	0.09	6.05	—	1.16	7.21	6.65	—	1.28	7.93
4.01.16.20.	**Perimeter fixing; battens for 57mm or 63mm partitions**										
4.01.16.20.A	38mm x 19mm	m	0.07	4.37	—	0.35	4.72	4.81	—	0.38	5.19
4.01.16.20.B	38mm x 19mm; plugged to concrete of brickwork	m	0.23	15.53	—	0.58	16.11	17.08	—	0.64	17.72
4.01.16.24.	**Angle junction fixings; battens for 57mm partitions**										
4.01.16.24.A	37mm x 19mm and 37mm x 37mm	m	0.22	14.52	—	1.61	16.12	15.97	—	1.77	17.74

Hutchins Priced Schedules 2021	Unit	Labour Hours	Labour Net £	Plant Net £	Materials Net £	Unit Rate Net £	Labour Gross £	Plant Gross £	Materials Gross £	Unit Rate Gross £
						(Gross rates include 10% profit)				
4.01. **NEW WORK**										
4.01.16. **DRY LININGS AND PARTITIONS**										
4.01.16.26. **Angle junction fixings; battens for 63mm partitions**										
4.01.16.26.A 37mm x 19mm and 37mm x 37mm	m	0.23	15.19	—	1.61	16.80	16.71	—	1.77	18.48
4.01.16.30. **Tee junction fixings; battens for 57mm partitions**										
4.01.16.30.A 37mm x 19mm and 37mm x 37mm	m	0.17	11.49	—	1.61	13.10	12.64	—	1.77	14.41
4.01.16.32. **Tee junction fixings; battens for 63mm partitions**										
4.01.16.32.A 37mm x 19mm and 37mm x 37mm	m	0.18	11.83	—	1.61	13.44	13.01	—	1.77	14.78
4.01.16.36. **Gyproc metal stud partition; 75mm; comprising 50mm metal stud framing, clad both sides with one layer of 12.5mm Gyproc wallboard**										
4.01.16.36.A over 300mm wide	m²	0.65	43.89	—	5.62	49.50	48.28	—	6.18	54.45
4.01.16.36.B not exceeding 300mm wide	m²	1.63	109.69	—	8.48	118.17	120.66	—	9.33	129.99
4.01.16.38. **Gyproc metal stud partition; 95mm; comprising 70mm metal stud framing clad both sides with one layer of 12.5mm Gyproc wallboard**										
4.01.16.38.A over 300mm wide	m²	0.65	43.89	—	6.43	50.31	48.28	—	7.07	55.35
4.01.16.38.B not exceeding 300mm wide	m²	1.63	109.69	—	10.08	119.77	120.66	—	11.09	131.74
4.01.16.39. **Tape and fill joints**										
4.01.16.39.A with mechanical jointer	m	0.13	8.47	—	0.29	8.76	9.32	—	0.32	9.63
4.01.16.43. **Gyproc DriLyner system; Gyproc square edge wallboard; fixed with adhesive dab to masonry walls; taped butt joints; 9mm work**										
4.01.16.43.A over 300mm wide	m²	0.21	13.85	—	2.34	16.19	15.23	—	2.57	17.80
4.01.16.43.B not exceeding 300mm wide	m²	0.32	21.24	—	2.34	23.58	23.36	—	2.57	25.94
4.01.16.45. **Gyproc DriLyner system; Gyproc square edge wallboard; fixed with adhesive dab to masonry walls; taped butt joints; 12.5mm work**										
4.01.16.45.A over 300mm wide	m²	0.25	16.53	—	2.34	18.87	18.19	—	2.57	20.76
4.01.16.45.B not exceeding 300mm wide	m²	0.38	25.67	—	2.34	28.01	28.24	—	2.57	30.82

Floor, Wall and Ceiling Finishes

Hutchins Priced Schedules 2021		Unit	Labour Hours	Labour Net £	Plant Net £	Materials Net £	Unit Rate Net £	Labour Gross £	Plant Gross £	Materials Gross £	Unit Rate Gross £
								(Gross rates include 10% profit)			
4.01.	**NEW WORK**										
4.01.16.	**DRY LININGS AND PARTITIONS**										
4.01.16.47.	Gyproc DriLyner system; taper edge thermal board; fixed with adhesive dabs to masonry walls; taped butt joints; 22mm work										
4.01.16.47.A	over 300mm wide	m²	0.29	19.56	—	8.09	27.65	21.51	—	8.90	30.42
4.01.16.47.B	not exceeding 300mm wide	m²	0.43	28.70	—	8.09	36.79	31.57	—	8.90	40.47
4.01.16.49.	Gyproc DriLyner system; taper edge wallboard; fixed with adhesive dabs to masonry walls; taped butt joints; 40mm work										
4.01.16.49.A	over 300mm wide	m²	0.34	22.99	—	10.41	33.40	25.28	—	11.46	36.74
4.01.16.49.B	not exceeding 300mm wide	m²	0.48	32.06	—	10.41	42.47	35.27	—	11.46	46.72
4.01.18.	**FIBROUS PLASTER**										
4.01.18.02.	Ventilator; louvred; fixing to plastered wall										
4.01.18.02.A	225mm x 75mm	nr	0.08	5.04	—	8.38	13.43	5.54	—	9.22	14.77
4.01.18.02.B	225mm x 150mm	nr	0.10	6.72	—	9.44	16.17	7.39	—	10.39	17.78
4.01.18.02.C	225mm x 225mm	nr	0.13	8.47	—	10.50	18.97	9.32	—	11.56	20.87
4.01.18.04.	Ventilator; louvred; perforated zinc flyscreen; fixing to plastered wall										
4.01.18.04.A	225mm x 75mm	nr	0.10	6.72	—	8.38	15.11	7.39	—	9.22	16.62
4.01.18.04.B	225mm x 150mm	nr	0.13	8.47	—	9.44	17.91	9.32	—	10.39	19.70
4.01.18.04.C	225mm x 225mm	nr	0.15	10.15	—	10.50	20.65	11.16	—	11.56	22.72
4.01.18.06.	Gyproc plaster core cornice cove; with mitres and ends; fixed with adhesive										
4.01.18.06.A	100mm girth	m	0.20	13.51	—	2.38	15.88	14.86	—	2.61	17.47
4.01.18.06.B	127mm girth	m	0.22	14.85	—	2.56	17.42	16.34	—	2.82	19.16
4.01.20.	**BEDS AND BACKINGS PORTLAND CEMENT**										
4.01.20.01.	Cement and sand (1:3); floated bed; laid level and to falls on concrete										
4.01.20.01.A	19mm; over 300mm wide	m²	0.30	20.23	—	2.76	22.99	22.25	—	3.04	25.29
4.01.20.01.B	19mm; not exceeding 300mm wide	m²	0.60	40.53	—	2.76	43.29	44.58	—	3.04	47.62
4.01.20.01.C	25mm; over 300mm wide	m²	0.35	23.66	—	3.76	27.42	26.02	—	4.14	30.16
4.01.20.01.D	25mm; not exceeding 300mm wide	m²	0.70	47.25	—	3.76	51.01	51.97	—	4.14	56.11
4.01.20.01.E	32mm; over 300mm wide	m²	0.40	27.02	—	4.64	31.66	29.72	—	5.11	34.83
4.01.20.01.F	32mm; not exceeding 300mm wide	m²	0.80	54.04	—	4.64	58.68	59.44	—	5.11	64.55
4.01.20.01.G	38mm; over 300mm wide	m²	0.45	30.38	—	5.65	36.03	33.42	—	6.21	39.63
4.01.20.01.H	38mm; not exceeding 300mm wide	m²	0.90	60.76	—	5.65	66.40	66.83	—	6.21	73.04
4.01.20.01.I	50mm; over 300mm wide	m²	0.55	37.10	—	7.28	44.38	40.81	—	8.01	48.82
4.01.20.01.J	50mm; not exceeding 300mm wide	m²	1.11	74.27	—	7.28	81.54	81.69	—	8.01	89.70

Hutchins Priced Schedules 2021	Unit	Labour Hours	Labour Net £	Plant Net £	Materials Net £	Unit Rate Net £	Labour Gross £	Plant Gross £	Materials Gross £	Unit Rate Gross £
							(Gross rates include 10% profit)			
4.01. **NEW WORK**										
4.01.20. **BEDS AND BACKINGS PORTLAND CEMENT**										
4.01.20.02. **Cement and sand (1:3) trowelled paving; laid level and to falls on concrete**										
4.01.20.02.A 19mm; over 300mm wide	m²	0.40	27.02	—	2.76	29.78	29.72	—	3.04	32.76
4.01.20.02.B 19mm; not exceeding 300mm wide	m²	0.80	54.04	—	2.76	56.80	59.44	—	3.04	62.48
4.01.20.02.C 25mm; over 300mm wide	m²	0.45	30.38	—	3.76	34.14	33.42	—	4.14	37.56
4.01.20.02.D 25mm; not exceeding 300mm wide	m²	0.90	60.76	—	3.76	64.52	66.83	—	4.14	70.97
4.01.20.02.E 32mm; over 300mm wide	m²	0.50	33.74	—	4.64	38.38	37.11	—	5.11	42.22
4.01.20.02.F 32mm; not exceeding 300mm wide	m²	1.00	67.48	—	4.64	72.12	74.23	—	5.11	79.33
4.01.20.02.G 38mm; over 300mm wide	m²	0.55	37.10	—	5.65	42.75	40.81	—	6.21	47.02
4.01.20.02.H 38mm; not exceeding 300mm wide	m²	1.11	74.27	—	5.65	79.91	81.69	—	6.21	87.91
4.01.20.02.I 50mm; over 300mm wide	m²	0.65	43.89	—	7.28	51.17	48.28	—	8.01	56.28
4.01.20.02.J 50mm; not exceeding 300mm wide	m²	1.31	87.78	—	7.28	95.05	96.55	—	8.01	104.56
4.01.20.06. **Cement and sand (1:3); screeded backings; to walls**										
4.01.20.06.A 13mm work over 300mm wide	m²	0.27	18.21	—	1.78	19.99	20.04	—	1.96	21.99
4.01.20.06.B 13mm work not exceeding 300mm wide .	m²	0.54	36.43	—	1.78	38.21	40.07	—	1.96	42.03
4.01.20.08. **Cement and sand (1:3); trowelled backings; to walls**										
4.01.20.08.A 13mm work over 300mm wide	m²	0.37	25.00	—	1.88	26.88	27.50	—	2.07	29.57
4.01.20.08.B 13mm work not exceeding 300mm wide .	m²	0.74	49.94	—	1.88	51.82	54.93	—	2.07	57.00
4.01.22. **FLOOR LEVELLING COMPOUND**										
4.01.22.01. **Floor levelling compound; screeded bed; laid level and to falls on existing concrete**										
4.01.22.01.A 3mm; over 300mm wide	m²	0.10	6.72	—	3.54	10.26	7.39	—	3.90	11.29
4.01.22.01.B 3mm; not exceeding 300mm wide	m²	0.20	13.51	—	3.54	17.05	14.86	—	3.90	18.76
4.01.22.01.C 5mm; over 300mm wide	m²	0.13	8.80	—	5.98	14.79	9.68	—	6.58	16.26
4.01.22.01.D 5mm; not exceeding 300mm wide	m²	0.26	17.54	—	5.98	23.52	19.30	—	6.58	25.88
4.01.24. **INSULATING SCREEDS**										
4.01.24.01. **Lightweight concrete screed; cement and vermiculite aggregate (1:8) on 13mm cement and sand (1:4) screeded bed; laid level and to falls**										
4.01.24.01.A 25mm; over 300mm wide	m²	0.13	8.47	—	5.99	14.46	9.32	—	6.59	15.90
4.01.24.01.B 25mm; not exceeding 300mm wide	m²	0.25	16.87	—	5.99	22.86	18.56	—	6.59	25.14
4.01.24.01.C 38mm; over 300mm wide	m²	0.15	10.15	—	10.93	21.08	11.16	—	12.03	23.19
4.01.24.01.D 38mm; not exceeding 300mm wide	m²	0.30	20.23	—	10.93	31.16	22.25	—	12.03	34.28

Floor, Wall and Ceiling Finishes

Hutchins Priced Schedules 2021	Unit	Labour Hours	Labour Net £	Plant Net £	Materials Net £	Unit Rate Net £	Labour Gross £	Plant Gross £	Materials Gross £	Unit Rate Gross £
								(Gross rates include 10% profit)		

4.01.   **NEW WORK**

4.01.24.   **INSULATING SCREEDS**

4.01.24.01.   Lightweight concrete screed; cement and vermiculite aggregate (1:8) on 13mm cement and sand (1:4) screeded bed; laid level and to falls

4.01.24.01.E	50mm; over 300mm wide	m²	0.18	11.83	—	14.95	26.78	13.01	—	16.44	29.46
4.01.24.01.F	50mm; not exceeding 300mm wide	m²	0.35	23.66	—	14.95	38.61	26.02	—	16.44	42.47
4.01.24.01.G	75mm; over 300mm wide	m²	0.23	15.19	—	23.91	39.10	16.71	—	26.30	43.01
4.01.24.01.H	75mm; not exceeding 300mm wide	m²	0.45	30.38	—	23.91	54.29	33.42	—	26.30	59.72

4.01.26.   **SCREED REINFORCEMENT**

4.01.26.01.   Galvanised wire netting reinforcement; 150mm laps; placed in floors

4.01.26.01.A	19mm mesh	m²	0.06	4.03	—	7.29	11.32	4.44	—	8.02	12.45
4.01.26.01.B	25mm mesh	m²	0.06	4.03	—	4.54	8.57	4.44	—	4.99	9.43
4.01.26.01.C	50mm mesh	m²	0.06	4.03	—	2.59	6.63	4.44	—	2.85	7.29

4.01.28.   **GLAZED CERAMIC WALL TILING**

4.01.28.01.   Glazed ceramic wall tiles; fixed with adhesive; butt joints; straight both ways; flush pointing with white grout; to plastered backings; to walls

4.01.28.01.A	100mm x 100mm x 6.5mm gloss tile; over 300mm wide	m²	0.93	62.51	—	18.57	81.08	68.76	—	20.43	89.18
4.01.28.01.B	100mm x 100mm x 6.5mm gloss tile; not exceeding 300mm wide	m²	1.80	120.98	—	18.57	139.55	133.08	—	20.43	153.50
4.01.28.01.C	200mm x 100mm x 8mm metro tile; over 300mm wide	m²	0.87	58.74	—	18.42	77.17	64.62	—	20.27	84.88
4.01.28.01.D	200mm x 100mm x 8mm metro tile; not exceeding 300mm wide	m²	1.75	117.48	—	18.42	135.91	129.23	—	20.27	149.50
4.01.28.01.E	297mm x 97mm x 9mm diamante tile; over 300mm wide	m²	0.81	54.71	—	38.55	93.26	60.18	—	42.41	102.59
4.01.28.01.F	297mm x 97mm x 9mm diamante tile; not exceeding 300mm wide	m²	1.63	109.35	—	38.55	147.90	120.29	—	42.41	162.69
4.01.28.01.G	300mm x 100mm x 7mm linear tile; over 300mm wide	m²	0.81	54.71	—	23.38	78.09	60.18	—	25.72	85.90
4.01.28.01.H	300mm x 100mm x 7mm linear tile; not exceeding 300mm wide	m²	1.63	109.35	—	23.38	132.73	120.29	—	25.72	146.00
4.01.28.01.I	450mm x 450mm x 10.5mm Stamford tile; over 300mm wide	m²	0.70	47.05	—	49.00	96.05	51.75	—	53.90	105.65
4.01.28.01.J	450mm x 450mm x 10.5mm Stamford tile; not exceeding 300mm wide	m²	1.40	94.09	—	49.00	143.10	103.50	—	53.90	157.41
4.01.28.01.K	600mm x 300mm x 13mm gloss tile; over 300mm wide	m²	0.70	47.05	—	22.10	69.15	51.75	—	24.31	76.06
4.01.28.01.L	600mm x 300mm x 13mm gloss tile; not exceeding 300mm wide	m²	1.40	94.09	—	22.10	116.19	103.50	—	24.31	127.81
4.01.28.01.M	straight cutting	m	0.12	8.13	—	—	8.13	8.95	—	—	8.95
4.01.28.01.N	raking cutting	m	0.11	7.46	—	—	7.46	8.21	—	—	8.21

Hutchins Priced Schedules 2021	Unit	Labour Hours	Labour Net £	Plant Net £	Materials Net £	Unit Rate Net £	Labour Gross £	Plant Gross £	Materials Gross £	Unit Rate Gross £
							(Gross rates include 10% profit)			

4.01. **NEW WORK**

4.01.28. **GLAZED CERAMIC WALL TILING**

4.01.28.01. Glazed ceramic wall tiles; fixed with adhesive; butt joints; straight both ways; flush pointing with white grout; to plastered backings; to walls

4.01.28.01.O curved cutting	m	0.18	11.83	—	—	11.83	13.01	—	—	13.01
4.01.28.01.P cut and fit tiling around small pipe	nr	0.11	7.46	—	—	7.46	8.21	—	—	8.21
4.01.28.01.Q cut and fit tiling around large pipe	nr	0.18	11.83	—	—	11.83	13.01	—	—	13.01

4.01.32. **CLAY FLOOR TILING**

4.01.32.03. Heather brown quarry tiles; bedded and jointed in cement mortar (1:3); butt joints; flush pointed with grout; to cement and sand backing

4.01.32.03.A 200mm x 200mm x 14mm; to floors over 300mm wide	m²	0.60	40.53	—	36.23	76.75	44.58	—	39.85	84.43
4.01.32.03.B 200mm x 200mm x 14mm; to floors not exceeding 300mm wide	m²	1.21	80.99	—	36.23	117.21	89.09	—	39.85	128.93
4.01.32.03.C 150mm x 150mm x 12mm; to floors; over 300mm wide	m²	0.78	52.63	—	36.77	89.40	57.89	—	40.45	98.34
4.01.32.03.D 150mm x 150mm x 12mm; to floors; not exceeding 300mm wide	m²	1.57	105.32	—	36.77	142.09	115.85	—	40.45	156.30

4.01.32.05. Red quarry tiles; bedded and jointed in cement mortar (1:3); butt joints; flush pointed with grout; to cement and sand backing

4.01.32.05.A 200mm x 200mm x 14mm; to floors over 300mm wide	m²	0.60	40.53	—	36.23	76.75	44.58	—	39.85	84.43
4.01.32.05.B 200mm x 200mm x 14mm; to floors not exceeding 300mm wide	m²	1.21	80.99	—	36.23	117.21	89.09	—	39.85	128.93
4.01.32.05.C 150mm x 150mm x 12mm; to floors; over 300mm wide	m²	0.78	52.63	—	36.77	89.40	57.89	—	40.45	98.34
4.01.32.05.D 150mm x 150mm x 12mm; to floors; not exceeding 300mm wide	m²	1.57	105.32	—	36.77	142.09	115.85	—	40.45	156.30

4.01.32.06. Coved quarry tile skirtings; rounded top edge

4.01.32.06.A 200mm x 110mm x 12mm	m	0.27	18.21	—	20.16	38.38	20.04	—	22.18	42.21
4.01.32.06.C 150mm x 110mm x 12mm	m	0.24	16.20	—	18.17	34.37	17.82	—	19.98	37.80

4.01.32.10. Labours on tiling

4.01.32.10.A straight cutting	m	0.03	2.02	—	—	2.02	2.22	—	—	2.22
4.01.32.10.B raking cutting	m	0.08	5.38	—	—	5.38	5.91	—	—	5.91
4.01.32.10.C curved cutting	m	0.16	10.82	—	—	10.82	11.90	—	—	11.90
4.01.32.10.D cut and fit tiling around small pipe	nr	0.25	16.87	—	—	16.87	18.56	—	—	18.56
4.01.32.10.E cut and fit tiling around large pipe	nr	0.35	23.66	—	—	23.66	26.02	—	—	26.02
4.01.32.10.F cut and fit tiling around pedestal of WC or lavatory basin	nr	0.50	33.74	—	—	33.74	37.11	—	—	37.11

Floor, Wall and Ceiling Finishes

Hutchins Priced Schedules 2021	Unit	Labour Hours	Labour Net £	Plant Net £	Materials Net £	Unit Rate Net £	Labour Gross £	Plant Gross £	Materials Gross £	Unit Rate Gross £
						(Gross rates include 10% profit)				
4.01.	**NEW WORK**									
4.01.34.	**HARDWOOD FLOORING**									
4.01.34.01.	**Supply and lay nominal tongued and grooved and end matched flooring; secret nailed to joists or battens provided by contractor; filled, sanded and sealed by either wax finish or lacquer and then polished**									
4.01.34.01.A	Canadian prime maple	m²	1.48	55.69	0.57	71.49	127.75	61.26	0.63	78.64	140.53
4.01.34.01.C	American white oak prime	m²	1.48	55.69	0.57	60.82	117.09	61.26	0.63	66.90	128.79
4.01.34.01.D	European prime oak	m²	1.48	55.69	0.57	88.04	144.30	61.26	0.63	96.84	158.73
4.01.34.02.	**Supply and lay 70mm x 230mm solid tongued and grooved hardwood blocks in mastic composition on cement floated level concrete surfacing provided by contractor; wax polishing or sealing on completion; herring-bone pattern with two-block border**									
4.01.34.02.A	merbau prime	m²	1.48	55.69	—	66.30	121.99	61.26	—	72.93	134.19
4.01.34.02.B	Canadian maple	m²	1.48	55.69	—	58.37	114.07	61.26	—	64.21	125.47
4.01.34.02.C	white oak prime	m²	1.48	55.69	—	62.10	117.79	61.26	—	68.31	129.57
4.01.34.02.E	European oak prime	m²	1.48	55.69	—	55.80	111.49	61.26	—	61.38	122.64
4.01.34.03.	**Supply and lay 8mm parquet mosaic panels; basket pattern; sanded off; preparing and wax polishing or sealing on completion**									
4.01.34.03.A	natural oak	m²	0.85	31.99	—	45.30	77.29	35.18	—	49.83	85.02
4.01.34.03.B	prime oak	m²	0.85	31.99	—	47.49	79.47	35.18	—	52.23	87.42
4.01.34.03.C	Mahogany	m²	0.85	31.99	—	55.80	87.79	35.18	—	61.38	96.57
4.01.35.	**SAFETY FLOORING**									
4.01.35.01.	**Slip-resistant floor coverings; including welded seams and fixing with adhesive**									
4.01.35.01.A	Safetred Universal; 2.00mm	m²	0.55	20.70	—	12.87	33.57	22.77	—	14.16	36.93
4.01.35.01.B	Safetred Universal Plus; 2.50mm	m²	0.55	20.70	—	22.85	43.55	22.77	—	25.13	47.90
4.01.35.01.C	Safetred Spectrum; 2.00mm	m²	0.55	20.70	—	13.39	34.09	22.77	—	14.73	37.49
4.01.35.01.D	Safetred ION contrast; 2.00mm	m²	0.55	20.70	—	15.48	36.18	22.77	—	17.03	39.79
4.01.35.01.E	Safetred ION linen; 2.00mm	m²	0.55	20.70	—	15.51	36.21	22.77	—	17.06	39.83
4.01.35.01.F	Safetred Wood; 2.00mm	m²	0.55	20.70	—	16.75	37.44	22.77	—	18.42	41.19
4.01.35.01.G	Safetred Natural; 2.00mm	m²	0.55	20.70	—	16.55	37.25	22.77	—	18.20	40.97

Floor, Wall and Ceiling Finishes

Hutchins Priced Schedules 2021	Unit	Labour Hours	Labour Net £	Plant Net £	Materials Net £	Unit Rate Net £	Labour Gross £	Plant Gross £	Materials Gross £	Unit Rate Gross £
								(Gross rates include 10% profit)		
4.01.	**NEW WORK**									
4.01.36.	**RESILIENT FLOORING COVERS**									
4.01.36.01.	**Vinyl click flooring with underlay**									
4.01.36.01.A	Novocare Ascot luxury..............	m²	0.50	18.82	—	22.84	41.65	20.70	—	25.12	45.82
4.01.36.01.B	Novocare York luxury..............	m²	0.50	18.82	—	18.90	37.72	20.70	—	20.79	41.49
4.01.36.01.C	GoodHome parquet effect..........	m²	0.50	18.82	—	20.48	39.29	20.70	—	22.52	43.22
4.01.36.01.D	GoodHome tile effect..............	m²	0.50	18.82	—	18.11	36.93	20.70	—	19.92	40.62
4.01.36.01.E	GoodHome wood effect............	m²	0.50	18.82	—	20.48	39.29	20.70	—	22.52	43.22
4.01.36.01.F	GoodHome mosaic effect..........	m²	0.50	18.82	—	20.48	39.29	20.70	—	22.52	43.22
4.01.36.02.	**Vinyl flooring tiles; peel and stick; 305mm x 305mm**									
4.01.36.02.A	slate effect; 1.10mm thick..........	m²	0.25	9.41	—	7.72	17.13	10.35	—	8.49	18.84
4.01.36.02.B	Ilkley slate effect; 1.10mm thick......	m²	0.25	9.41	—	9.85	19.26	10.35	—	10.83	21.18
4.01.36.02.C	parquet effect; 1.10mm thick........	m²	0.25	9.41	—	8.43	17.84	10.35	—	9.27	19.62
4.01.36.02.D	black; 2.5mm thick..............	m²	0.25	9.41	—	14.32	23.73	10.35	—	15.75	26.10
4.01.36.02.E	grey; 2.5mm thick...............	m²	0.25	9.41	—	21.46	30.87	10.35	—	23.61	33.96
4.01.36.12.	**Gradus PVCu skirting**									
4.01.36.12.A	70mm......................	m	0.15	5.64	—	2.83	8.47	6.21	—	3.11	9.32
4.01.36.12.B	100mm.....................	m	0.15	5.64	—	4.93	10.57	6.21	—	5.42	11.63
4.01.36.12.C	100mm, set in.................	m	0.23	8.65	—	7.16	15.82	9.52	—	7.88	17.40
4.01.38.	**CARPETING**									
4.01.38.01.	**Take up loose carpet; set aside**									
4.01.38.01.A	loose lay....................	m²	0.02	1.01	—	—	1.01	1.11	—	—	1.11
4.01.38.01.B	gripper battens................	m²	0.02	1.34	—	—	1.34	1.48	—	—	1.48
4.01.38.01.C	stuck around edge..............	m²	0.03	1.68	—	—	1.68	1.85	—	—	1.85
4.01.38.01.D	stuck direct..................	m²	0.05	3.36	—	—	3.36	3.70	—	—	3.70
4.01.38.02.	**Lay only carpet**									
4.01.38.02.A	loose lay....................	m²	0.02	1.01	—	—	1.01	1.11	—	—	1.11
4.01.38.02.B	gripper battens................	m²	0.03	1.68	—	0.50	2.18	1.85	—	0.55	2.40
4.01.38.02.C	stuck around edge..............	m²	0.04	2.35	—	11.47	13.82	2.59	—	12.61	15.20
4.01.38.02.D	stuck direct..................	m²	0.07	4.37	—	11.47	15.83	4.81	—	12.61	17.42
4.01.38.04.	**Fitted carpeting; domestic grade (PC sum £20 per m2); to floors over 300mm wide**									
4.01.38.04.A	loose lay....................	m²	0.15	10.15	—	21.00	31.15	11.16	—	23.10	34.26
4.01.38.04.B	gripper battens................	m²	0.18	11.83	—	21.50	33.33	13.01	—	23.65	36.67
4.01.38.04.C	stuck around edge..............	m²	0.20	13.51	—	32.47	45.98	14.86	—	35.71	50.57
4.01.38.04.D	stuck direct..................	m²	0.30	20.23	—	32.47	52.70	22.25	—	35.71	57.97

Floor, Wall and Ceiling Finishes

Hutchins Priced Schedules 2021	Unit	Labour Hours	Labour Net £	Plant Net £	Materials Net £	Unit Rate Net £	Labour Gross £	Plant Gross £	Materials Gross £	Unit Rate Gross £
									(Gross rates include 10% profit)	
4.01.	**NEW WORK**									
4.01.38.	CARPETING									
4.01.38.05.	**Fitted carpeting; domestic grade (PC sum £20 per m2); to floors not exceeding 300mm wide**									
4.01.38.05.A	loose lay	m²	0.18	11.83	—	21.00	32.83	13.01	—	23.10	36.11
4.01.38.05.B	gripper battens	m²	0.20	13.51	—	21.50	35.01	14.86	—	23.65	38.51
4.01.38.05.C	stuck around edge	m²	0.23	15.19	—	32.47	47.66	16.71	—	35.71	52.42
4.01.38.05.D	stuck direct	m²	0.33	21.91	—	32.47	54.38	24.10	—	35.71	59.81
4.01.38.06.	**Fitted carpeting; domestic grade (PC sum £20 per m2); to treads and risers over 300mm wide**									
4.01.38.06.A	gripper battens	m²	0.70	47.25	—	21.50	68.75	51.97	—	23.65	75.63
4.01.38.08.	**11mm PU foam underlay to carpeting; to floors**									
4.01.38.08.A	over 300mm wide	m²	0.09	6.05	—	3.31	9.36	6.65	—	3.64	10.29
4.01.38.08.B	not exceeding 300mm wide	m²	0.10	6.72	—	3.31	10.03	7.39	—	3.64	11.03
4.01.38.09.	**Tackless wood gripper; fixing carpet**									
4.01.38.09.A	to perimeter of floor	m	0.05	3.36	—	0.50	3.86	3.70	—	0.55	4.25
4.01.38.10.	**Cutting carpet**									
4.01.38.10.A	raking cutting	m	0.05	3.36	—	—	3.36	3.70	—	—	3.70
4.01.38.10.B	curved cutting	m	0.08	5.04	—	—	5.04	5.54	—	—	5.54
4.01.38.12.	**Aluminium cover strip**									
4.01.38.12.A	to openings	m	0.08	5.04	—	4.18	9.22	5.54	—	4.60	10.14
4.02.	**REPAIRS AND ALTERATIONS**									
4.02.01.	REMOVE SURFACE FINISHES									
4.02.01.01.	**Floors**									
4.02.01.01.A	linoleum sheeting	m²	0.12	3.55	—	—	3.55	3.90	—	—	3.90
4.02.01.01.B	carpet and underlay	m²	0.13	3.85	—	—	3.85	4.23	—	—	4.23
4.02.01.01.C	screed	m²	0.50	14.79	—	—	14.79	16.27	—	—	16.27
4.02.01.01.D	granolithic and screed	m²	0.67	19.82	—	—	19.82	21.80	—	—	21.80
4.02.01.01.E	terrazzo or ceramic tiles; screed	m²	1.10	32.54	—	—	32.54	35.79	—	—	35.79
4.02.01.02.	**Walls**									
4.02.01.02.A	plasterboard	m²	0.45	13.31	—	—	13.31	14.64	—	—	14.64
4.02.01.02.B	plaster	m²	0.22	6.51	—	—	6.51	7.16	—	—	7.16
4.02.01.02.C	cement rendering; pebbledashing	m²	0.45	13.31	—	—	13.31	14.64	—	—	14.64
4.02.01.02.D	tiling and screed	m²	0.55	16.27	—	—	16.27	17.90	—	—	17.90

Hutchins Priced Schedules 2021	Unit	Labour Hours	Labour Net £	Plant Net £	Materials Net £	Unit Rate Net £	Labour Gross £	Plant Gross £	Materials Gross £	Unit Rate Gross £
							(Gross rates include 10% profit)			
4.02.	**REPAIRS AND ALTERATIONS**									
4.02.01.	**REMOVE SURFACE FINISHES**									
4.02.01.03.	**Ceilings**									
4.02.01.03.A	plasterboard and skim including withdrawing nails.............	m²	0.33	9.76	—	—	9.76	10.74	—	—	10.74
4.02.01.03.B	wood lath and plaster including withdrawing nails.............	m²	0.55	16.27	—	—	16.27	17.90	—	—	17.90
4.02.01.03.C	suspended.................	m²	0.83	24.55	—	—	24.55	27.01	—	—	27.01
4.02.01.03.D	plaster moulded cornice; per 25mm girth .	m	0.17	5.03	—	—	5.03	5.53	—	—	5.53
4.02.02.	**PREPARE SURFACES**									
4.02.02.01.	**Prepare surface to be sound and clean, apply Unibond universal PVA adhesive and sealer to receive plaster or cement rendering**									
4.02.02.01.A	walls; existing cement and sand base over 300mm wide.............	m²	0.15	10.15	—	0.36	10.51	11.16	—	0.39	11.56
4.02.02.01.B	walls; existing glazed tile base over 300mm wide.................	m²	0.12	8.13	—	0.24	8.37	8.95	—	0.26	9.21
4.02.02.01.C	walls; existing painted base over 300mm wide....................	m²	0.14	9.48	—	0.28	9.76	10.42	—	0.31	10.73
4.02.02.01.D	walls; existing concrete base over 300mm wide..................	m²	0.15	10.15	—	0.30	10.45	11.16	—	0.33	11.49
4.02.02.01.E	ceilings; existing cement and sand base over 300mm wide.............	m²	0.19	12.50	—	0.36	12.86	13.75	—	0.39	14.15
4.02.02.01.F	ceilings; existing painted base over 300mm wide..................	m²	0.17	11.49	—	0.28	11.77	12.64	—	0.31	12.95
4.02.02.01.G	ceilings; existing concrete base over 300mm wide.................	m²	0.19	12.50	—	0.30	12.80	13.75	—	0.33	14.08
4.02.02.02.	**Hack down defective ceiling plaster and laths**									
4.02.02.02.A	clean out old nails ready for new plaster .	m²	0.43	28.70	—	—	28.70	31.57	—	—	31.57
4.02.02.04.	**Take down temporary boarded linings and clean joists**									
4.02.02.04.A	to ceilings..................	m²	0.40	11.83	—	—	11.83	13.02	—	—	13.02
4.02.04.	**CEILINGS**									
4.02.04.01.	**Expanded metal lathing and 13mm Carlite plaster to**									
4.02.04.01.A	ceiling joists................	m²	0.85	57.40	—	12.74	70.14	63.14	—	14.01	77.15
4.02.04.03.	**Baseboard, scrim and 3mm Thistle finish**									
4.02.04.03.A	to ceilings..................	m²	0.50	33.74	—	6.56	40.30	37.11	—	7.22	44.33

Floor, Wall and Ceiling Finishes

Hutchins Priced Schedules 2021	Unit	Labour Hours	Labour Net £	Plant Net £	Materials Net £	Unit Rate Net £	Labour Gross £	Plant Gross £	Materials Gross £	Unit Rate Gross £
							(Gross rates include 10% profit)			
4.02. **REPAIRS AND ALTERATIONS**										
4.02.04. **CEILINGS**										
4.02.04.04. **Hack down defective ceiling plaster and fix baseboard, scrim and 3mm Thistle finish including jointing to existing**										
4.02.04.04.A area not exceeding 1.00m2	m²	1.00	67.48	—	6.56	74.04	74.23	—	7.22	81.44
4.02.04.04.B area 1.00m2 - 4.00m2	m²	0.70	47.25	—	6.56	53.81	51.97	—	7.22	59.19
4.02.06. **PLASTERWORK TO WALLS**										
4.02.06.01. **Make good at intersection of wall and ceiling plaster after replastering**										
4.02.06.01.A wall or ceiling	m	0.28	18.55	—	0.56	19.11	20.40	—	0.62	21.02
4.02.06.02. **Make good cracks in**										
4.02.06.02.A ceiling plaster	m	0.23	15.19	—	0.56	15.75	16.71	—	0.62	17.32
4.02.06.03. **Hack brick, stone or concrete walls to form key for**										
4.02.06.03.A plaster	m²	0.75	22.19	—	—	22.19	24.40	—	—	24.40
4.02.06.04. **Hack off wall plaster and rake out brick joints to form key for**										
4.02.06.04.A new plaster	m²	0.75	22.19	—	—	22.19	24.40	—	—	24.40
4.02.06.05. **Rake out joints of brickwork to**										
4.02.06.05.A form key	m²	0.40	11.83	—	—	11.83	13.02	—	—	13.02
4.02.06.06. **Dub out uneven walls to receive**										
4.02.06.06.A new plaster	m²	0.20	13.51	—	2.33	15.84	14.86	—	2.56	17.42
4.02.06.07. **13mm Thistle plaster on**										
4.02.06.07.A brick or breeze walls	m²	0.43	28.70	—	7.26	35.96	31.57	—	7.99	39.55
4.02.06.10. **Hack down defective wall plaster in small quantities; apply 13mm Thistle plaster including jointing to existing**										
4.02.06.10.A area not exceeding 1.00m2	m²	0.88	59.08	—	7.26	66.34	64.99	—	7.99	72.97
4.02.06.10.B area 1.00m2 - 4.00m2	m²	0.45	30.38	—	7.26	37.64	33.42	—	7.99	41.40
4.02.06.11. **13mm Thistle plaster to brick walls including**										
4.02.06.11.A jointing new to old and a small amount of dubbing out	m²	0.55	37.10	—	6.03	43.13	40.81	—	6.63	47.44
4.02.06.12. **Make good cracks in plaster**										
4.02.06.12.A walls	m	0.23	15.19	—	0.52	15.71	16.71	—	0.57	17.28
4.02.06.12.B moulded cornice, per 25mm girth of cornice	m	0.13	8.47	—	0.52	8.99	9.32	—	0.57	9.88

Hutchins Priced Schedules 2021	Unit	Labour Hours	Labour Net £	Plant Net £	Materials Net £	Unit Rate Net £	Labour Gross £	Plant Gross £	Materials Gross £	Unit Rate Gross £
							(Gross rates include 10% profit)			
4.02.	**REPAIRS AND ALTERATIONS**									
4.02.06.	**PLASTERWORK TO WALLS**									
4.02.06.12.	**Make good cracks in plaster**									
4.02.06.12.C	around door and window frames and repoint	m	0.20	13.51	—	0.52	14.03	14.86	—	0.57	15.43
4.02.06.13.	**Hack down and re-run plaster cornices per 25mm girth of cornice**									
4.02.06.13.A	coved	m	0.13	8.47	—	0.52	8.99	9.32	—	0.57	9.88
4.02.06.13.B	moulded	m	0.17	11.16	—	0.52	11.67	12.27	—	0.57	12.84
4.02.06.14.	**Make good plaster around pipes**									
4.02.06.14.A	small pipes	nr	0.15	10.15	—	0.52	10.67	11.16	—	0.57	11.73
4.02.06.14.B	large pipes	nr	0.17	11.16	—	0.52	11.67	12.27	—	0.57	12.84
4.02.08.	**CERAMIC TILING**									
4.02.08.01.	**Glazed ceramic wall tiles; fixed with adhesive, pointed in white cement grout in small quantities in repairs**									
4.02.08.01.A	100mm x 100mm x 6.5mm	m²	2.35	157.94	—	17.68	175.63	173.74	—	19.45	193.19
4.02.08.01.B	150mm x 150mm x 5mm	m²	1.63	109.35	—	6.46	115.81	120.29	—	7.10	127.39
4.02.08.04.	**Hack off glazed tiles to wall in**									
4.02.08.04.A	detached areas 2.00m2 - 5.00m2	m²	1.00	29.58	—	—	29.58	32.54	—	—	32.54
4.02.08.04.B	patches 0.50m2 - 2.00m2	m²	2.50	73.95	—	—	73.95	81.35	—	—	81.35
4.02.08.04.C	single tiles in patches up to 0.50m2	nr	0.30	8.87	—	—	8.87	9.76	—	—	9.76
4.02.08.05.	**Take out broken angle beads horizontal or vertical and renew**									
4.02.08.05.A	150mm x 25mm	nr	0.09	6.05	—	0.55	6.60	6.65	—	0.60	7.26
4.02.08.05.B	150mm x 50mm moulded cappings	nr	0.10	6.72	—	0.56	7.28	7.39	—	0.61	8.01
4.02.08.06.	**Holes through wall tiling; any colour; make good**									
4.02.08.06.A	for small pipes	nr	0.28	8.28	—	—	8.28	9.11	—	—	9.11
4.02.08.06.B	for large pipes	nr	0.38	11.24	—	—	11.24	12.36	—	—	12.36
4.02.08.07.	**Strip loose tiles and clean**									
4.02.08.07.A	for reuse	m²	0.40	11.83	—	—	11.83	13.02	—	—	13.02
4.02.08.08.	**Refix only salvaged tiles**									
4.02.08.08.A	with adhesive	m²	1.13	75.95	—	1.23	77.18	83.54	—	1.36	84.90
4.02.09.	**RENDERING**									
4.02.09.01.	**Hack off defective rendering to concreted areas; grout and render in cement mortar**									
4.02.09.01.A	19mm thick	m²	0.50	33.74	—	2.66	36.40	37.11	—	2.93	40.04

Floor, Wall and Ceiling Finishes

Hutchins Priced Schedules 2021	Unit	Labour Hours	Labour Net £	Plant Net £	Materials Net £	Unit Rate Net £	Labour Gross £	Plant Gross £	Materials Gross £	Unit Rate Gross £
							(Gross rates include 10% profit)			
4.02.	**REPAIRS AND ALTERATIONS**									
4.02.09.	**RENDERING**									
4.02.09.02.	**Hack off and renew cement rendered plinth**									
4.02.09.02.A	225mm high including joints new to old . .	m	0.15	10.15	—	0.50	10.65	11.16	—	0.55	11.72
4.02.09.03.	**Cut out cracks in rough cast rendering**									
4.02.09.03.A	make good to existing	m	0.40	27.02	—	0.67	27.69	29.72	—	0.74	30.46
4.02.09.03.B	make good to existing to match adjacent work around reset window and door frames	m	0.24	16.20	—	0.22	16.42	17.82	—	0.25	18.06
4.02.09.05.	**Hack off defective rendering; prepare for and re-render in cement and sand (1:3); plain face**									
4.02.09.05.A	to walls	m²	0.50	33.74	—	2.77	36.51	37.11	—	3.05	40.16
4.02.09.07.	**Hack off broken cement rendering and renew rendering to**									
4.02.09.07.A	three-sided kerb to gulley	nr	0.45	30.38	—	2.49	32.87	33.42	—	2.74	36.16
4.02.09.08.	**Hack off all loose rendering, hack back brick (or concrete) to form key and re-render with cement and sand (1:3); plain face including reproducing all profiles and ruled joints**									
4.02.09.08.A	to match existing	m²	1.00	67.48	—	2.77	70.25	74.23	—	3.05	77.28
4.02.09.08.B	to match existing in patches not exceeding 1.00m2	nr	1.26	84.35	—	2.77	87.12	92.78	—	3.05	95.83

Hutchins Priced Schedules 2021	Unit	Labour Hours	Labour Net £	Plant Net £	Materials Net £	Unit Rate Net £	Labour Gross £	Plant Gross £	Materials Gross £	Unit Rate Gross £
								(Gross rates include 10% profit)		
5.01.	**NEW WORK**									
5.01.20.	**GUTTERWORK AND FITTINGS**									
5.01.20.01.	**PVCu FloPlast Miniflo system; fixing with standard brackets**									
5.01.20.01.A	76mm	m	0.15	14.37	—	3.45	17.82	15.81	—	3.80	19.60
5.01.20.01.B	extra for stopend outlet	nr	0.09	8.66	—	4.22	12.88	9.53	—	4.64	14.17
5.01.20.01.C	extra for stopend; external	nr	0.08	7.52	—	1.86	9.37	8.27	—	2.04	10.31
5.01.20.01.D	extra for running outlet	nr	0.15	14.37	—	4.58	18.95	15.81	—	5.04	20.84
5.01.20.01.E	extra for angle	nr	0.15	14.37	—	3.82	18.19	15.81	—	4.20	20.01
5.01.20.02.	**PVCu Osma RoundLine System; fixing with standard brackets**									
5.01.20.02.A	112mm	m	0.16	15.51	—	10.58	26.09	17.06	—	11.63	28.70
5.01.20.02.B	extra for Swivelock running outlet with fitted offset bend	nr	0.16	15.51	—	17.92	33.44	17.06	—	19.72	36.78
5.01.20.02.C	extra for stopend; external	nr	0.09	8.66	—	7.36	16.02	9.53	—	8.09	17.62
5.01.20.02.D	extra for angle	nr	0.16	15.51	—	13.70	29.21	17.06	—	15.06	32.13
5.01.20.02.F	extra for connector to cast iron half round gutter	nr	0.18	16.65	—	13.05	29.70	18.32	—	14.35	32.67
5.01.20.02.G	extra for connector to cast iron ogee gutter	nr	0.18	16.65	—	22.34	39.00	18.32	—	24.58	42.90
5.01.20.03.	**PVCu Osma SuperLine System; fixing with standard brackets**									
5.01.20.03.A	125mm	m	0.16	15.51	—	16.30	31.82	17.06	—	17.93	35.00
5.01.20.03.B	extra for stopend; external	nr	0.09	8.66	—	12.08	20.74	9.53	—	13.29	22.82
5.01.20.03.D	extra for running outlet	nr	0.16	15.51	—	20.47	35.98	17.06	—	22.52	39.58
5.01.20.03.E	extra for angle	nr	0.16	15.51	—	22.42	37.93	17.06	—	24.66	41.72
5.01.20.04.	**PVCu Osma RoofLine System; fixing with standard brackets**									
5.01.20.04.A	150mm	m	0.19	17.89	—	35.14	53.03	19.68	—	38.65	58.33
5.01.20.04.B	extra for stopend; external	nr	0.10	9.80	—	20.17	29.97	10.78	—	22.19	32.97
5.01.20.04.C	extra for running outlet	nr	0.18	16.65	—	40.65	57.31	18.32	—	44.72	63.04
5.01.20.04.D	extra for angle	nr	0.18	16.65	—	67.47	84.12	18.32	—	74.22	92.54
5.01.20.05.	**PVCu Osma SquareLine System; fixing with standard brackets**									
5.01.20.05.A	100mm	m	0.16	15.51	—	11.75	27.26	17.06	—	12.92	29.99
5.01.20.05.B	extra for stopend; external	nr	0.09	8.66	—	4.81	13.47	9.53	—	5.29	14.82
5.01.20.05.C	extra for running outlet	nr	0.16	15.51	—	15.78	31.30	17.06	—	17.36	34.43

Plumbing and Heating

Hutchins Priced Schedules 2021		Unit	Labour Hours	Labour Net £	Plant Net £	Materials Net £	Unit Rate Net £	Labour Gross £	Plant Gross £	Materials Gross £	Unit Rate Gross £
								(Gross rates include 10% profit)			
5.01.	**NEW WORK**										
5.01.20.	**GUTTERWORK AND FITTINGS**										
5.01.20.05.	**PVCu Osma SquareLine System; fixing with standard brackets**										
5.01.20.05.D	extra for angle	nr	0.16	15.51	—	15.41	30.93	17.06	—	16.95	34.02
5.01.20.06.	**Cast iron; half round; fixing with standard brackets**										
5.01.20.06.A	100mm	m	0.22	20.75	—	22.88	43.62	22.82	—	25.17	47.99
5.01.20.06.B	extra for stopend	nr	0.12	10.94	—	6.07	17.02	12.04	—	6.68	18.72
5.01.20.06.C	extra for stopend outlet	nr	0.13	12.09	—	14.62	26.70	13.30	—	16.08	29.37
5.01.20.06.D	extra for running outlet	nr	0.22	20.75	—	18.88	39.63	22.82	—	20.77	43.59
5.01.20.06.E	extra for angle	nr	0.22	20.75	—	18.88	39.63	22.82	—	20.77	43.59
5.01.20.07.	**Cast iron; half round; fixing with standard brackets**										
5.01.20.07.A	115mm	m	0.22	20.75	—	23.28	44.03	22.82	—	25.61	48.43
5.01.20.07.B	extra for stopend	nr	0.12	10.94	—	7.54	18.48	12.04	—	8.29	20.33
5.01.20.07.C	extra for stopend outlet	nr	0.13	12.09	—	19.49	31.58	13.30	—	21.44	34.74
5.01.20.07.D	extra for running outlet	nr	0.22	20.75	—	19.49	40.24	22.82	—	21.44	44.26
5.01.20.07.E	extra for angle	nr	0.22	20.75	—	19.49	40.24	22.82	—	21.44	44.26
5.01.20.08.	**Cast iron; half round; fixing with standard brackets**										
5.01.20.08.A	125mm	m	0.25	23.60	—	26.88	50.48	25.96	—	29.57	55.53
5.01.20.08.B	extra for stopend	nr	0.15	13.80	—	13.03	26.83	15.18	—	14.33	29.51
5.01.20.08.C	extra for stopend outlet	nr	0.16	14.94	—	22.44	37.38	16.44	—	24.68	41.12
5.01.20.08.D	extra for running outlet	nr	0.25	23.60	—	22.44	46.04	25.96	—	24.68	50.65
5.01.20.08.E	extra for angle	nr	0.25	23.60	—	22.44	46.04	25.96	—	24.68	50.65
5.01.20.09.	**Cast iron; half round; fixing with standard brackets**										
5.01.20.09.A	150mm	m	0.28	26.46	—	39.72	66.18	29.10	—	43.69	72.79
5.01.20.09.B	extra for stopend	nr	0.18	16.65	—	17.19	33.85	18.32	—	18.91	37.23
5.01.20.09.C	extra for stopend outlet	nr	0.19	17.89	—	41.54	59.43	19.68	—	45.70	65.38
5.01.20.09.D	extra for running outlet	nr	0.28	26.46	—	41.54	68.00	29.10	—	45.70	74.80
5.01.20.09.E	extra for angle	nr	0.28	26.46	—	41.54	68.00	29.10	—	45.70	74.80
5.01.20.12.	**Cast iron; ogee; fixing with standard brackets**										
5.01.20.12.A	100mm	m	0.23	21.89	—	25.07	46.96	24.08	—	27.58	51.66
5.01.20.12.B	extra for stopend	nr	0.13	12.09	—	9.82	21.90	13.30	—	10.80	24.09
5.01.20.12.C	extra for stopend outlet	nr	0.14	13.23	—	23.11	36.34	14.55	—	25.42	39.97
5.01.20.12.D	extra for running outlet	nr	0.23	21.89	—	23.11	45.00	24.08	—	25.42	49.50
5.01.20.12.E	extra for angle	nr	0.23	21.89	—	23.11	45.00	24.08	—	25.42	49.50

Plumbing and Heating

Hutchins Priced Schedules 2021	Unit	Labour Hours	Labour Net £	Plant Net £	Materials Net £	Unit Rate Net £	Labour Gross £	Plant Gross £	Materials Gross £	Unit Rate Gross £
							(Gross rates include 10% profit)			

5.01. NEW WORK

5.01.20. GUTTERWORK AND FITTINGS

5.01.20.13. Cast iron; ogee; fixing with standard brackets

5.01.20.13.A	115mm......................	m	0.23	21.89	—	27.07	48.96	24.08	—	29.77	53.85
5.01.20.13.B	extra for stopend...............	nr	0.13	12.09	—	10.32	22.41	13.30	—	11.36	24.65
5.01.20.13.C	extra for stopend outlet...........	nr	0.14	13.23	—	19.95	33.18	14.55	—	21.94	36.50
5.01.20.13.D	extra for running outlet............	nr	0.23	21.89	—	20.48	42.37	24.08	—	22.53	46.61
5.01.20.13.E	extra for angle.................	nr	0.23	21.89	—	20.48	42.37	24.08	—	22.53	46.61

5.01.20.14. Cast iron; ogee; fixing with standard brackets

5.01.20.14.A	125mm......................	m	0.26	24.74	—	29.11	53.86	27.22	—	32.02	59.24
5.01.20.14.B	extra for stopend...............	nr	0.16	14.94	—	14.35	29.29	16.44	—	15.79	32.22
5.01.20.14.C	extra for stopend outlet...........	nr	0.17	16.08	—	27.41	43.50	17.69	—	30.15	47.84
5.01.20.14.D	extra for running outlet............	nr	0.26	24.74	—	29.99	54.73	27.22	—	32.99	60.20
5.01.20.14.E	extra for angle.................	nr	0.26	24.74	—	29.99	54.73	27.22	—	32.99	60.20

5.01.22. RAINWATER PIPEWORK

5.01.22.01. PVCu FloPlast Miniflo System; fixing with standard brackets

5.01.22.01.A	50mm.......................	m	0.13	12.09	—	4.19	16.27	13.30	—	4.61	17.90
5.01.22.01.B	extra for offset bend.............	nr	0.09	8.66	—	3.29	11.95	9.53	—	3.61	13.14
5.01.22.01.C	extra for bend.................	nr	0.09	8.66	—	7.32	15.99	9.53	—	8.06	17.58
5.01.22.01.G	connection to back inlet gulley; cement mortar (1:3).................	nr	0.09	8.66	—	0.36	9.02	9.53	—	0.39	9.92

5.01.22.02. PVCu; Osma RoundLine System; fixing with standard brackets

5.01.22.02.A	68mm.......................	m	0.16	14.94	—	11.19	26.13	16.44	—	12.31	28.75
5.01.22.02.B	extra for offset bend.............	nr	0.10	9.23	—	6.78	16.02	10.15	—	7.46	17.62
5.01.22.02.D	extra for shoe.................	nr	0.19	17.89	—	7.35	25.24	19.68	—	8.08	27.76
5.01.22.02.E	extra for branch.................	nr	0.13	12.09	—	22.08	34.17	13.30	—	24.29	37.58
5.01.22.02.F	extra for access pipe; bolted access door	nr	0.13	12.09	—	39.87	51.95	13.30	—	43.85	57.15
5.01.22.02.G	hopper head.................	nr	0.28	26.46	—	29.65	56.11	29.10	—	32.61	61.72
5.01.22.02.H	connection to back inlet gulley; cement mortar (1:3).................	nr	0.10	9.23	—	0.36	9.59	10.15	—	0.39	10.55
5.01.22.02.I	connector to 63mm cast iron........	nr	0.12	10.94	—	13.37	24.31	12.04	—	14.70	26.74
5.01.22.02.K	adaptor to 110mm PVCu drain.......	nr	0.07	6.38	—	14.14	20.52	7.01	—	15.55	22.57

5.01.22.03. PVCu; Osma SquareLine System; fixing with standard brackets

5.01.22.03.A	61mm.......................	m	0.16	14.94	—	10.89	25.83	16.44	—	11.98	28.41
5.01.22.03.B	extra for offset bend.............	nr	0.10	9.23	—	7.24	16.47	10.15	—	7.96	18.12
5.01.22.03.D	extra for shoe.................	nr	0.19	17.89	—	10.73	28.62	19.68	—	11.80	31.48
5.01.22.03.E	extra for branch.................	nr	0.13	12.09	—	29.69	41.78	13.30	—	32.66	45.95

Plumbing and Heating

Hutchins Priced Schedules 2021		Unit	Labour Hours	Labour Net £	Plant Net £	Materials Net £	Unit Rate Net £	Labour Gross £	Plant Gross £	Materials Gross £	Unit Rate Gross £
							(Gross rates include 10% profit)				
5.01.	**NEW WORK**										
5.01.22.	**RAINWATER PIPEWORK**										
5.01.22.03.	**PVCu; Osma SquareLine System; fixing with standard brackets**										
5.01.22.03.F	adaptor; square to round	nr	0.07	6.38	—	10.37	16.74	7.01	—	11.40	18.42
5.01.22.03.G	hopper head	nr	0.25	23.60	—	37.46	61.06	25.96	—	41.21	67.17
5.01.22.03.H	connection to back inlet gulley; cement mortar (1:3)	nr	0.10	9.23	—	0.36	9.59	10.15	—	0.39	10.55
5.01.22.03.J	adaptor to 110mm PVCu drain	nr	0.13	12.09	—	14.14	26.23	13.30	—	15.55	28.85
5.01.22.03.K	access pipe	nr	0.13	12.09	—	40.76	52.84	13.30	—	44.83	58.13
5.01.22.05.	**Cast iron; round; fixing with standard brackets**										
5.01.22.05.A	65mm	m	0.18	16.65	—	46.71	63.37	18.32	—	51.38	69.70
5.01.22.05.B	extra for bend	nr	0.12	10.94	—	20.41	31.35	12.04	—	22.45	34.48
5.01.22.05.C	extra for branch	nr	0.15	13.80	—	44.65	58.45	15.18	—	49.11	64.29
5.01.22.05.F	extra for offset; 150mm projection	nr	0.15	13.80	—	31.24	45.04	15.18	—	34.36	49.54
5.01.22.05.G	extra for offset; 230mm projection	nr	0.15	13.80	—	36.32	50.12	15.18	—	39.95	55.13
5.01.22.05.H	extra for shoe	nr	0.21	19.61	—	27.57	47.18	21.57	—	30.33	51.89
5.01.22.05.J	hopper head; flat back	nr	0.25	23.60	—	34.50	58.11	25.96	—	37.95	63.92
5.01.22.05.L	connection to back inlet gulley; cement mortar (1:3)	nr	0.10	9.23	—	0.36	9.59	10.15	—	0.39	10.55
5.01.22.06.	**Cast iron; round; fixing with standard brackets**										
5.01.22.06.A	75mm	m	0.22	20.75	—	46.71	67.46	22.82	—	51.38	74.20
5.01.22.06.B	extra for bend	nr	0.16	14.94	—	30.45	45.40	16.44	—	33.50	49.94
5.01.22.06.C	extra for branch	nr	0.19	17.89	—	49.16	67.05	19.68	—	54.08	73.76
5.01.22.06.F	extra for offset; 150mm projection	nr	0.19	17.89	—	31.24	49.13	19.68	—	34.36	54.04
5.01.22.06.G	extra for offset; 230mm projection	nr	0.19	17.89	—	44.95	62.85	19.68	—	49.45	69.13
5.01.22.06.H	extra for shoe	nr	0.25	23.60	—	29.84	53.44	25.96	—	32.82	58.79
5.01.22.06.J	hopper head; flat back	nr	0.25	23.60	—	34.50	58.11	25.96	—	37.95	63.92
5.01.22.06.L	connection to back inlet gulley; cement mortar (1:3)	nr	0.11	10.37	—	0.36	10.73	11.41	—	0.39	11.80
5.01.22.07.	**Cast iron; round; fixing with standard brackets**										
5.01.22.07.A	100mm	m	0.25	23.60	—	77.85	101.45	25.96	—	85.63	111.59
5.01.22.07.B	extra for bend	nr	0.19	17.89	—	47.90	65.79	19.68	—	52.69	72.37
5.01.22.07.C	extra for branch	nr	0.22	20.75	—	58.37	79.12	22.82	—	64.21	87.03
5.01.22.07.F	extra for offset; 150mm projection	nr	0.22	20.75	—	82.19	102.94	22.82	—	90.41	113.23
5.01.22.07.G	extra for offset; 230mm projection	nr	0.22	20.75	—	105.39	126.13	22.82	—	115.92	138.75
5.01.22.07.H	extra for shoe	nr	0.28	26.46	—	63.83	90.29	29.10	—	70.22	99.32
5.01.22.07.J	hopper head; flat back	nr	0.25	23.60	—	70.74	94.35	25.96	—	77.82	103.78
5.01.22.07.L	connection to back inlet gulley; cement mortar (1:3)	nr	0.13	12.09	—	0.59	12.68	13.30	—	0.65	13.95

Plumbing and Heating

|---|---|---|---|---|---|---|---|---|---|---|
| | | | | | | | | | | (Gross rates include 10% profit) |

5.01. **NEW WORK**

5.01.24. **RAINWATER ANCILLARIES**

5.01.24.02. **Balloon guards; galvanised wire**

5.01.24.02.A	2.5in	nr	0.09	8.09	—	2.95	11.04	8.90	—	3.24	12.14
5.01.24.02.B	3in	nr	0.10	9.23	—	3.52	12.75	10.15	—	3.87	14.03
5.01.24.02.C	4in	nr	0.11	10.37	—	4.05	14.42	11.41	—	4.45	15.87
5.01.24.02.D	6in	nr	0.13	12.09	—	6.69	18.78	13.30	—	7.36	20.65
5.01.24.02.E	8in	nr	0.15	13.80	—	12.85	26.65	15.18	—	14.13	29.31
5.01.24.02.F	9in	nr	0.17	15.70	—	13.70	29.41	17.27	—	15.07	32.35

5.01.26. **SOIL AND VENT PIPEWORK**

5.01.26.01. **PVCu OsmaSoil ring seal; fixing with standard brackets**

5.01.26.01.A	82mm	m	0.21	19.61	—	27.51	47.11	21.57	—	30.26	51.82
5.01.26.01.B	extra for bend	nr	0.19	17.89	—	28.55	46.44	19.68	—	31.40	51.08
5.01.26.01.C	extra for offset bend	nr	0.19	17.89	—	29.74	47.64	19.68	—	32.72	52.40
5.01.26.01.D	extra for branch	nr	0.22	20.75	—	50.80	71.54	22.82	—	55.87	78.70
5.01.26.01.E	extra for bossed pipe	nr	0.16	14.94	—	27.24	42.18	16.44	—	29.97	46.40
5.01.26.01.F	extra for access pipe; bolted access door	nr	0.22	20.75	—	56.08	76.83	22.82	—	61.69	84.51

5.01.26.02. **PVCu OsmaSoil ring seal; fixing with standard brackets**

5.01.26.02.A	110mm	m	0.23	21.89	—	23.25	45.14	24.08	—	25.57	49.65
5.01.26.02.B	extra for bend	nr	0.21	20.18	—	33.23	53.41	22.19	—	36.55	58.75
5.01.26.02.C	extra for offset bend	nr	0.21	20.18	—	31.03	51.20	22.19	—	34.13	56.32
5.01.26.02.D	extra for branch	nr	0.24	23.03	—	46.92	69.95	25.33	—	51.61	76.94
5.01.26.02.E	extra for bossed pipe	nr	0.21	20.18	—	26.99	47.16	22.19	—	29.68	51.88
5.01.26.02.F	extra for access pipe; bossed; screwed	nr	0.21	20.18	—	56.28	76.45	22.19	—	61.90	84.10
5.01.26.02.G	extra for WC connector; 14 deg	nr	0.18	17.23	—	42.69	59.91	18.95	—	46.95	65.90
5.01.26.02.H	extra for WC connector; 90 deg	nr	0.18	17.23	—	29.84	47.06	18.95	—	32.82	51.77

5.01.26.03. **Cast iron; Timesaver System; flexible joints; fixing with standard brackets**

5.01.26.03.A	50mm	m	0.25	23.60	—	40.93	64.54	25.96	—	45.03	70.99
5.01.26.03.B	extra for bend	nr	0.19	17.89	—	27.83	45.72	19.68	—	30.61	50.29
5.01.26.03.C	extra for access bend	nr	0.19	17.89	—	68.90	86.79	19.68	—	75.79	95.47
5.01.26.03.D	extra for branch; single	nr	0.22	20.75	—	41.79	62.53	22.82	—	45.96	68.79
5.01.26.03.F	extra for access pipe; oval door	nr	0.19	17.89	—	47.68	65.57	19.68	—	52.45	72.13
5.01.26.03.M	connection to stoneware drain; cement mortar (1:3) joint	nr	0.10	9.23	—	0.36	9.59	10.15	—	0.39	10.55

Plumbing and Heating

Hutchins Priced Schedules 2021	Unit	Labour Hours	Labour Net £	Plant Net £	Materials Net £	Unit Rate Net £	Labour Gross £	Plant Gross £	Materials Gross £	Unit Rate Gross £
							(Gross rates include 10% profit)			
5.01.	**NEW WORK**									
5.01.26.	**SOIL AND VENT PIPEWORK**									
5.01.26.04.	**Cast iron; Timesaver System; flexible joints; fixing with standard brackets**									
5.01.26.04.A	75mm	m	0.28	26.46	—	45.41	71.86	29.10	—	49.95	79.05
5.01.26.04.B	extra for bend	nr	0.22	20.75	—	31.45	52.20	22.82	—	34.60	57.42
5.01.26.04.C	extra for access bend	nr	0.22	20.75	—	75.08	95.83	22.82	—	82.59	105.41
5.01.26.04.D	extra for branch; single	nr	0.25	23.60	—	47.70	71.30	25.96	—	52.47	78.43
5.01.26.04.E	extra for branch; double	nr	0.28	26.46	—	70.67	97.13	29.10	—	77.74	106.84
5.01.26.04.F	extra for access pipe; round door	nr	0.22	20.75	—	67.13	87.88	22.82	—	73.84	96.66
5.01.26.04.H	extra for offset; 75mm projection	nr	0.25	23.60	—	41.08	64.68	25.96	—	45.18	71.14
5.01.26.04.I	extra for offset; 150mm projection	nr	0.28	26.46	—	53.00	79.46	29.10	—	58.30	87.40
5.01.26.04.N	connection to stoneware drain; cement mortar (1:3) joint	nr	0.11	10.37	—	0.47	10.85	11.41	—	0.52	11.93
5.01.26.05.	**Cast iron; Timesaver System; flexible joints; fixing with standard brackets**									
5.01.26.05.A	100mm	m	0.34	32.26	—	52.76	85.02	35.49	—	58.04	93.52
5.01.26.05.B	extra for bend	nr	0.28	26.46	—	38.51	64.97	29.10	—	42.36	71.46
5.01.26.05.C	extra for access bend	nr	0.28	26.46	—	82.15	108.61	29.10	—	90.37	119.47
5.01.26.05.D	extra for branch; single	nr	0.31	29.31	—	60.07	89.38	32.24	—	66.08	98.32
5.01.26.05.E	extra for branch; double	nr	0.34	32.26	—	74.20	106.46	35.49	—	81.62	117.11
5.01.26.05.F	extra for access pipe; round door	nr	0.28	26.46	—	151.93	178.39	29.10	—	167.12	196.23
5.01.26.05.H	extra for offset; 150mm projection	nr	0.31	29.31	—	57.42	86.73	32.24	—	63.16	95.41
5.01.26.05.I	extra for offset; 300mm projection	nr	0.34	32.26	—	62.72	94.98	35.49	—	68.99	104.48
5.01.26.05.K	WC connecting pipe	nr	0.31	29.31	—	52.06	81.37	32.24	—	57.26	89.51
5.01.26.05.L	roof connector; for asphalt	nr	0.34	32.26	—	67.13	99.39	35.49	—	73.84	109.33
5.01.26.05.N	roof outlet; circular; flat grate	nr	0.37	35.12	—	109.45	144.56	38.63	—	120.39	159.02
5.01.26.05.O	connection to stoneware drain; cement mortar (1:3) joint	nr	0.13	12.09	—	0.59	12.68	13.30	—	0.65	13.95
5.01.28.	**WASTE PIPEWORK**									
5.01.28.01.	**OsmaWeld ABS Solvent Weld Waste system; solvent welded joints; fixing with clips or brackets**									
5.01.28.01.A	32mm	m	0.15	14.37	—	5.26	19.63	15.81	—	5.79	21.60
5.01.28.01.C	bend	nr	0.14	13.23	—	3.25	16.48	14.55	—	3.58	18.13
5.01.28.01.F	tee	nr	0.16	14.94	—	3.86	18.80	16.44	—	4.24	20.68
5.01.28.01.H	straight tank connector	nr	0.15	14.37	—	7.71	22.08	15.81	—	8.48	24.28
5.01.28.01.I	bottle trap; 38mm seal	nr	0.16	14.94	—	12.95	27.89	16.44	—	14.25	30.68
5.01.28.01.J	bottle trap; 75mm seal	nr	0.16	14.94	—	16.12	31.06	16.44	—	17.73	34.17
5.01.28.01.K	tubular P trap; 76mm seal	nr	0.17	16.08	—	12.35	28.43	17.69	—	13.58	31.28

Plumbing and Heating

Hutchins Priced Schedules 2021		Unit	Labour Hours	Labour Net £	Plant Net £	Materials Net £	Unit Rate Net £	Labour Gross £	Plant Gross £	Materials Gross £	Unit Rate Gross £
							(Gross rates include 10% profit)				
5.01.	**NEW WORK**										
5.01.28.	**WASTE PIPEWORK**										
5.01.28.01.	**OsmaWeld ABS Solvent Weld Waste system; solvent welded joints; fixing with clips or brackets**										
5.01.28.01.L	tubular S trap; 76mm seal	nr	0.17	16.08	—	15.12	31.20	17.69	—	16.63	34.32
5.01.28.02.	**OsmaWeld ABS Solvent Weld Waste system; solvent welded joints; fixing with clips or brackets**										
5.01.28.02.A	40mm	m	0.17	16.08	—	6.14	22.22	17.69	—	6.75	24.44
5.01.28.02.C	bend	nr	0.15	14.37	—	3.89	18.26	15.81	—	4.28	20.09
5.01.28.02.F	tee	nr	0.18	16.65	—	4.07	20.73	18.32	—	4.48	22.80
5.01.28.02.H	straight tank connector	nr	0.16	15.51	—	8.41	23.92	17.06	—	9.25	26.31
5.01.28.02.I	bottle trap; 38mm seal	nr	0.19	17.89	—	17.20	35.10	19.68	—	18.92	38.61
5.01.28.02.J	bottle trap; 75mm seal	nr	0.19	17.89	—	19.75	37.64	19.68	—	21.72	41.40
5.01.28.02.K	tubular P trap; 76mm seal	nr	0.20	19.03	—	14.06	33.09	20.94	—	15.46	36.40
5.01.28.02.L	tubular S trap; 76mm seal	nr	0.20	19.03	—	22.64	41.68	20.94	—	24.91	45.84
5.01.28.02.N	bath trap; 76mm seal	nr	0.21	19.61	—	20.84	40.44	21.57	—	22.92	44.49
5.01.28.02.P	washing machine half trap; 75mm seal	nr	0.25	23.60	—	24.90	48.50	25.96	—	27.39	53.35
5.01.28.03.	**OsmaWeld ABS Solvent Weld Waste system; solvent welded joints; fixing with clips or brackets**										
5.01.28.03.A	50mm	m	0.19	17.89	—	7.39	25.28	19.68	—	8.13	27.81
5.01.28.03.B	bend	nr	0.16	15.51	—	6.07	21.59	17.06	—	6.68	23.75
5.01.28.03.E	tee	nr	0.19	17.89	—	8.37	26.27	19.68	—	9.21	28.89
5.01.28.04.	**Osma Waste System; PVC-C Solvent Weld; fixing with clips or brackets**										
5.01.28.04.A	32mm	m	0.13	12.66	—	6.73	19.39	13.92	—	7.40	21.33
5.01.28.04.B	bend	nr	0.12	11.52	—	4.99	16.50	12.67	—	5.49	18.15
5.01.28.04.C	tee	nr	0.14	13.23	—	7.97	21.20	14.55	—	8.77	23.32
5.01.28.05.	**Osma Waste System; PVC-C Solvent Weld; fixing with clips or brackets**										
5.01.28.05.A	40mm	m	0.15	14.56	—	7.28	21.84	16.02	—	8.00	24.02
5.01.28.05.B	bend	nr	0.13	12.47	—	5.87	18.34	13.71	—	6.46	20.17
5.01.28.05.C	tee	nr	0.16	14.94	—	9.08	24.02	16.44	—	9.99	26.43
5.01.28.06.	**Osma Waste System; PVC-C Solvent Weld; fixing with clips or brackets**										
5.01.28.06.A	50mm	m	0.17	16.08	—	11.59	27.68	17.69	—	12.75	30.44
5.01.28.06.B	bend	nr	0.15	14.37	—	10.48	24.85	15.81	—	11.53	27.33
5.01.28.06.C	tee	nr	0.17	16.08	—	14.14	30.23	17.69	—	15.56	33.25
5.01.28.07.	**Osma Waste System; Push-fit; fixing with clips or brackets**										
5.01.28.07.A	32mm	m	0.12	11.71	—	3.35	15.06	12.88	—	3.68	16.56

Plumbing and Heating

Hutchins Priced Schedules 2021		Unit	Labour Hours	Labour Net £	Plant Net £	Materials Net £	Unit Rate Net £	Labour Gross £	Plant Gross £	Materials Gross £	Unit Rate Gross £
									(Gross rates include 10% profit)		
5.01.	**NEW WORK**										
5.01.28.	**WASTE PIPEWORK**										
5.01.28.07.	Osma Waste System; Push-fit; fixing with clips or brackets										
5.01.28.07.B	bend	nr	0.11	10.56	—	2.30	12.86	11.62	—	2.53	14.15
5.01.28.07.C	tee	nr	0.13	12.28	—	2.65	14.92	13.50	—	2.91	16.42
5.01.28.08.	Osma Waste System; Push-fit; fixing with clips or brackets										
5.01.28.08.A	40mm	m	0.14	13.61	—	3.76	17.37	14.97	—	4.13	19.11
5.01.28.08.B	bend	nr	0.12	11.52	—	2.31	13.83	12.67	—	2.54	15.21
5.01.28.08.C	tee	nr	0.15	13.99	—	2.31	16.30	15.39	—	2.54	17.93
5.01.28.09.	Osma Waste System; Push-fit; fixing with clips or brackets										
5.01.28.09.A	50mm	m	0.16	15.13	—	6.60	21.74	16.65	—	7.26	23.91
5.01.28.09.B	bend	nr	0.14	13.42	—	4.48	17.90	14.76	—	4.93	19.69
5.01.28.09.C	tee	nr	0.16	15.13	—	4.39	19.52	16.65	—	4.83	21.47
5.01.28.10.	Osma Waste System; waste connectors										
5.01.28.10.A	32mm bottle trap; 38mm seal	nr	0.16	15.13	—	12.83	27.96	16.65	—	14.11	30.76
5.01.28.10.B	32mm bottle trap; 75mm seal	nr	0.16	15.13	—	15.97	31.10	16.65	—	17.57	34.21
5.01.28.10.C	32mm tubular 'P' trap; 75mm seal	nr	0.17	16.08	—	12.23	28.32	17.69	—	13.46	31.15
5.01.28.10.D	32mm tubular 'S' trap; 75mm seal	nr	0.17	16.08	—	14.97	31.06	17.69	—	16.47	34.16
5.01.28.10.E	40mm bottle trap; 38mm seal	nr	0.19	17.99	—	17.04	35.03	19.79	—	18.75	38.53
5.01.28.10.F	40mm bottle trap; 75mm seal	nr	0.19	17.99	—	19.56	37.55	19.79	—	21.52	41.30
5.01.28.10.G	40mm tubular 'P' trap; 75mm seal	nr	0.20	19.03	—	13.92	32.96	20.94	—	15.32	36.25
5.01.28.10.H	40mm tubular 'S' trap; 75mm seal	nr	0.20	19.03	—	22.43	41.46	20.94	—	24.67	45.61
5.01.28.10.I	50mm tubular 'P' trap; 75mm seal	nr	0.22	20.94	—	36.19	57.13	23.03	—	39.81	62.84
5.01.30.	**OVERFLOW PIPEWORK**										
5.01.30.01.	Osma Overflow System; solvent welded joints; fixing with clips or brackets										
5.01.30.01.A	21.5mm	m	0.12	11.52	—	3.43	14.94	12.67	—	3.77	16.44
5.01.30.01.B	bend	nr	0.11	10.37	—	2.64	13.01	11.41	—	2.90	14.31
5.01.30.01.C	tee	nr	0.13	12.09	—	3.25	15.34	13.30	—	3.58	16.87
5.01.30.01.D	tank connector; straight	nr	0.13	12.09	—	3.95	16.04	13.30	—	4.35	17.64
5.01.30.01.E	tank connector; bent	nr	0.13	12.09	—	4.42	16.51	13.30	—	4.86	18.16
5.01.30.02.	Osma Overflow System; push-fit joints; fixing with clips or brackets										
5.01.30.02.A	21.5mm	m	0.11	10.56	—	3.22	13.78	11.62	—	3.54	15.16
5.01.30.02.B	bend	nr	0.10	9.42	—	2.46	11.88	10.36	—	2.71	13.07
5.01.30.02.C	tee	nr	0.12	11.13	—	3.53	14.66	12.25	—	3.88	16.13
5.01.30.02.D	tank connector; straight	nr	0.13	12.09	—	3.95	16.04	13.30	—	4.35	17.64
5.01.30.02.E	tank connector; bent	nr	0.12	11.13	—	4.61	15.75	12.25	—	5.07	17.32

Hutchins Priced Schedules 2021	Unit	Labour Hours	Labour Net £	Plant Net £	Materials Net £	Unit Rate Net £	Labour Gross £	Plant Gross £	Materials Gross £	Unit Rate Gross £
							(Gross rates include 10% profit)			

5.01.   **NEW WORK**

5.01.32.   **WATER MAINS**

5.01.32.01.   **Form stopcock pit with 100mm concrete base and with one length 100mm clayware pipe set vertically; surround with concrete; Hepworth cast iron stopcock box with flanged hinged lid**

	Unit	Labour Hours	Labour Net £	Plant Net £	Materials Net £	Unit Rate Net £	Labour Gross £	Plant Gross £	Materials Gross £	Unit Rate Gross £
5.01.32.01.A 650mm deep	nr	1.47	76.85	—	53.92	130.77	84.53	—	59.31	143.85
5.01.32.01.B 650mm deep with half class B engineering brick sides (PC sum £500 per 1000)	nr	2.20	115.22	—	63.99	179.21	126.74	—	70.39	197.13

5.01.32.02.   **Stopcock, jointing to MDPE pipe**

	Unit	Labour Hours	Labour Net £	Plant Net £	Materials Net £	Unit Rate Net £	Labour Gross £	Plant Gross £	Materials Gross £	Unit Rate Gross £
5.01.32.02.A 20mm	nr	0.14	13.23	—	18.25	31.48	14.55	—	20.08	34.63
5.01.32.02.B 25mm	nr	0.16	14.94	—	19.66	34.60	16.44	—	21.63	38.07

5.01.32.03.   **Excavate trench 750mm deep and supply and lay blue MDPE pressure water service pipe with push fit joints; refill and consolidate trench; pipe dia**

	Unit	Labour Hours	Labour Net £	Plant Net £	Materials Net £	Unit Rate Net £	Labour Gross £	Plant Gross £	Materials Gross £	Unit Rate Gross £
5.01.32.03.A 20mm	m	2.26	118.36	—	0.65	119.01	130.20	—	0.71	130.91
5.01.32.03.B 25mm	m	2.27	119.10	—	0.71	119.81	131.01	—	0.78	131.79

5.01.34.   **SERVICE PIPEWORK COPPER**

5.01.34.01.   **Pipes; half hard and annealed (formerly Table Y); laid in trench**

	Unit	Labour Hours	Labour Net £	Plant Net £	Materials Net £	Unit Rate Net £	Labour Gross £	Plant Gross £	Materials Gross £	Unit Rate Gross £
5.01.34.01.A 15mm	m	0.09	8.66	—	7.29	15.95	9.53	—	8.02	17.55
5.01.34.01.B 22mm	m	0.10	9.23	—	11.36	20.59	10.15	—	12.49	22.65
5.01.34.01.C 28mm	m	0.10	9.80	—	14.02	23.82	10.78	—	15.42	26.20

5.01.34.02.   **Pipes; half hard (formerly Table X); fixing with pipe clips plugged and screwed to walls**

	Unit	Labour Hours	Labour Net £	Plant Net £	Materials Net £	Unit Rate Net £	Labour Gross £	Plant Gross £	Materials Gross £	Unit Rate Gross £
5.01.34.02.A 15mm	m	0.18	17.23	—	3.76	20.98	18.95	—	4.13	23.08
5.01.34.02.B 15mm in short lengths	m	0.19	18.46	—	3.76	22.22	20.31	—	4.13	24.44
5.01.34.02.C 22mm	m	0.19	17.89	—	6.23	24.12	19.68	—	6.85	26.53
5.01.34.02.D 22mm in short lengths	m	0.20	19.03	—	6.23	25.26	20.94	—	6.85	27.79
5.01.34.02.E 28mm	m	0.21	20.18	—	8.04	28.22	22.19	—	8.85	31.04

5.01.34.03.   **Made bends**

	Unit	Labour Hours	Labour Net £	Plant Net £	Materials Net £	Unit Rate Net £	Labour Gross £	Plant Gross £	Materials Gross £	Unit Rate Gross £
5.01.34.03.A 15mm	nr	0.11	10.37	—	—	10.37	11.41	—	—	11.41
5.01.34.03.B 22mm	nr	0.15	13.80	—	—	13.80	15.18	—	—	15.18
5.01.34.03.C 28mm	nr	0.18	17.23	—	—	17.23	18.95	—	—	18.95

Plumbing and Heating

Hutchins Priced Schedules 2021	Unit	Labour Hours	Labour Net £	Plant Net £	Materials Net £	Unit Rate Net £	Labour Gross £	Plant Gross £	Materials Gross £	Unit Rate Gross £
							(Gross rates include 10% profit)			
5.01.	**NEW WORK**									
5.01.34.	**SERVICE PIPEWORK COPPER**									
5.01.34.04.	**Made offsets**									
5.01.34.04.A	15mm.....................	nr	0.29	27.60	—	—	27.60	30.36	—	—	30.36
5.01.34.04.B	22mm.....................	nr	0.33	31.12	—	—	31.12	34.23	—	—	34.23
5.01.34.04.C	28mm.....................	nr	0.36	34.55	—	—	34.55	38.00	—	—	38.00
5.01.34.05.	**Pipes; hard drawn (formerly Table Z); fixing with pipe clips plugged and screwed to walls**									
5.01.34.05.A	15mm.....................	m	0.18	17.23	—	3.06	20.29	18.95	—	3.37	22.32
5.01.34.05.B	15mm in short lengths............	m	0.19	18.46	—	3.09	21.56	20.31	—	3.40	23.71
5.01.34.05.C	22mm.....................	m	0.19	17.89	—	5.62	23.51	19.68	—	6.18	25.86
5.01.34.05.D	22mm in short lengths............	m	0.20	19.03	—	5.62	24.65	20.94	—	6.18	27.12
5.01.34.05.E	28mm.....................	m	0.21	20.18	—	7.57	27.75	22.19	—	8.33	30.52
5.01.34.06.	**Capillary fittings; straight union couplings**									
5.01.34.06.A	15mm.....................	nr	0.12	11.52	—	0.20	11.72	12.67	—	0.22	12.89
5.01.34.06.B	22mm.....................	nr	0.16	14.94	—	0.49	15.43	16.44	—	0.54	16.97
5.01.34.06.C	28mm.....................	nr	0.21	19.61	—	1.39	20.99	21.57	—	1.53	23.09
5.01.34.08.	**Capillary fittings; elbows**									
5.01.34.08.A	15mm.....................	nr	0.12	11.52	—	0.70	12.22	12.67	—	0.77	13.44
5.01.34.08.B	22mm.....................	nr	0.16	14.94	—	0.80	15.74	16.44	—	0.87	17.31
5.01.34.08.C	28mm.....................	nr	0.21	19.61	—	1.99	21.60	21.57	—	2.19	23.76
5.01.34.09.	**Capillary fittings; equal tees**									
5.01.34.09.A	15mm.....................	nr	0.18	17.23	—	0.70	17.93	18.95	—	0.77	19.72
5.01.34.09.B	22mm.....................	nr	0.24	23.03	—	1.99	25.02	25.33	—	2.19	27.53
5.01.34.09.C	28mm.....................	nr	0.30	28.17	—	9.97	38.14	30.99	—	10.97	41.96
5.01.34.10.	**Capillary fittings; straight tap connectors**									
5.01.34.10.A	15mm.....................	nr	0.10	9.23	—	2.82	12.05	10.15	—	3.10	13.26
5.01.34.10.B	22mm.....................	nr	0.12	11.52	—	3.75	15.27	12.67	—	4.13	16.79
5.01.34.11.	**Capillary fittings; straight tank connectors**									
5.01.34.11.A	15mm.....................	nr	0.18	17.23	—	2.65	19.88	18.95	—	2.92	21.86
5.01.34.11.B	22mm.....................	nr	0.24	23.03	—	3.26	26.30	25.33	—	3.59	28.93
5.01.34.11.C	28mm.....................	nr	0.30	28.17	—	11.02	39.19	30.99	—	12.13	43.11
5.01.34.14.	**Compression fittings; straight couplings**									
5.01.34.14.A	15mm.....................	nr	0.11	10.37	—	1.71	12.08	11.41	—	1.88	13.29

Hutchins Priced Schedules 2021	Unit	Labour Hours	Labour Net £	Plant Net £	Materials Net £	Unit Rate Net £	Labour Gross £	Plant Gross £	Materials Gross £	Unit Rate Gross £
							(Gross rates include 10% profit)			
5.01.	**NEW WORK**									
5.01.34.	**SERVICE PIPEWORK COPPER**									
5.01.34.14.	**Compression fittings; straight couplings**									
5.01.34.14.B	22mm........................	nr	0.15	13.80	—	2.76	16.56	15.18	—	3.03	18.21
5.01.34.14.C	28mm........................	nr	0.18	17.23	—	3.98	21.20	18.95	—	4.37	23.32
5.01.34.15.	**Compression fittings; elbows**									
5.01.34.15.A	15mm........................	nr	0.11	10.37	—	0.70	11.07	11.41	—	0.77	12.18
5.01.34.15.B	22mm........................	nr	0.15	13.80	—	0.80	14.59	15.18	—	0.87	16.05
5.01.34.15.C	28mm........................	nr	0.18	17.23	—	1.99	19.22	18.95	—	2.19	21.14
5.01.34.16.	**Compression fittings; equal tees**									
5.01.34.16.A	15mm........................	nr	0.16	14.94	—	0.70	15.64	16.44	—	0.77	17.21
5.01.34.16.B	22mm........................	nr	0.22	20.75	—	1.99	22.74	22.82	—	2.19	25.01
5.01.34.16.C	28mm........................	nr	0.27	25.32	—	9.97	35.29	27.85	—	10.97	38.82
5.01.34.18.	**Compression fittings; straight swivel tap adaptors**									
5.01.34.18.A	15mm........................	nr	0.09	8.09	—	3.81	11.89	8.90	—	4.19	13.08
5.01.34.18.B	22mm........................	nr	0.12	10.94	—	9.97	20.92	12.04	—	10.97	23.01
5.01.34.20.	**Compression fittings; flanged tank connector**									
5.01.34.20.A	15mm........................	nr	0.16	14.94	—	3.81	18.75	16.44	—	4.19	20.62
5.01.34.20.B	22mm........................	nr	0.22	20.75	—	4.32	25.07	22.82	—	4.76	27.58
5.01.34.20.C	28mm........................	nr	0.27	25.32	—	11.02	36.34	27.85	—	12.13	39.97
5.01.36.	**SERVICE PIPEWORK GALVANISED STEEL**									
5.01.36.01.	**Pipes; medium grade; screwed and socketed joints; fix with galvanised steel clips plugged and screwed to walls**									
5.01.36.01.A	15mm........................	m	0.22	20.75	—	7.02	27.76	22.82	—	7.72	30.54
5.01.36.01.B	20mm........................	m	0.23	21.89	—	7.23	29.11	24.08	—	7.95	32.03
5.01.36.01.C	25mm........................	m	0.26	24.74	—	7.99	32.73	27.22	—	8.79	36.01
5.01.36.01.D	32mm........................	m	0.30	28.74	—	9.15	37.89	31.62	—	10.07	41.68
5.01.36.02.	**Pipes; heavy grade; screwed and socketed joints; lay in trench**									
5.01.36.02.A	15mm........................	m	0.11	10.37	—	5.17	15.54	11.41	—	5.69	17.10
5.01.36.02.B	20mm........................	m	0.12	10.94	—	6.76	17.71	12.04	—	7.44	19.48
5.01.36.02.C	25mm........................	m	0.13	12.66	—	7.95	20.61	13.92	—	8.74	22.67
5.01.36.04.	**Fittings; elbows**									
5.01.36.04.A	15mm........................	nr	0.18	17.23	—	1.53	18.75	18.95	—	1.68	20.63
5.01.36.04.B	20mm........................	nr	0.24	23.03	—	1.74	24.77	25.33	—	1.91	27.25

Plumbing and Heating

Hutchins Priced Schedules 2021		Unit	Labour Hours	Labour Net £	Plant Net £	Materials Net £	Unit Rate Net £	Labour Gross £	Plant Gross £	Materials Gross £	Unit Rate Gross £
								(Gross rates include 10% profit)			
5.01.	**NEW WORK**										
5.01.36.	**SERVICE PIPEWORK GALVANISED STEEL**										
5.01.36.04.	**Fittings; elbows**										
5.01.36.04.C	25mm.....................	nr	0.30	28.74	—	2.19	30.94	31.62	—	2.41	34.03
5.01.36.04.D	32mm.....................	nr	0.36	34.55	—	3.43	37.98	38.00	—	3.78	41.78
5.01.36.05.	**Fittings; bends**										
5.01.36.05.A	15mm.....................	nr	0.18	17.23	—	2.51	19.74	18.95	—	2.76	21.71
5.01.36.05.B	20mm.....................	nr	0.24	23.03	—	3.38	26.41	25.33	—	3.72	29.05
5.01.36.05.C	25mm.....................	nr	0.30	28.74	—	4.25	32.99	31.62	—	4.68	36.29
5.01.36.05.D	32mm.....................	nr	0.36	34.55	—	6.23	40.78	38.00	—	6.86	44.86
5.01.36.06.	**Fittings; equal tees**										
5.01.36.06.A	15mm.....................	nr	0.27	25.32	—	1.92	27.23	27.85	—	2.11	29.96
5.01.36.06.B	20mm.....................	nr	0.34	32.26	—	2.12	34.38	35.49	—	2.33	37.82
5.01.36.06.C	25mm.....................	nr	0.41	39.11	—	2.88	42.00	43.03	—	3.17	46.20
5.01.36.06.D	32mm.....................	nr	0.48	46.06	—	4.35	50.41	50.67	—	4.78	55.45
5.01.36.07.	**Fittings; longscrews complete with backnut**										
5.01.36.07.A	15mm.....................	nr	0.27	25.32	—	6.39	31.71	27.85	—	7.03	34.88
5.01.36.07.B	20mm.....................	nr	0.34	32.26	—	7.35	39.61	35.49	—	8.08	43.57
5.01.36.07.C	25mm.....................	nr	0.41	39.11	—	9.93	49.05	43.03	—	10.93	53.95
5.01.36.07.D	32mm.....................	nr	0.48	46.06	—	12.50	58.56	50.67	—	13.75	64.42
5.01.38.	**SERVICE PIPEWORK POLYTHENE**										
5.01.38.02.	**Pipes; medium density; black; fix with clips plugged and screwed to walls**										
5.01.38.02.A	20mm.....................	m	0.18	17.23	—	1.31	18.53	18.95	—	1.44	20.39
5.01.38.02.B	20mm in short lengths............	m	0.19	18.46	—	1.62	20.08	20.31	—	1.78	22.09
5.01.38.02.C	25mm.....................	m	0.21	19.61	—	1.31	20.91	21.57	—	1.44	23.00
5.01.38.02.D	25mm in short lengths............	m	0.22	20.75	—	1.62	22.37	22.82	—	1.79	24.61
5.01.38.02.E	32mm.....................	m	0.23	21.89	—	1.90	23.79	24.08	—	2.09	26.17
5.01.38.04.	**Compression fittings; straight couplings**										
5.01.38.04.A	20mm.....................	nr	0.15	13.80	—	2.62	16.42	15.18	—	2.88	18.06
5.01.38.04.B	25mm.....................	nr	0.18	17.23	—	2.66	19.89	18.95	—	2.93	21.88
5.01.38.04.C	32mm.....................	nr	0.22	20.75	—	3.55	24.30	22.82	—	3.91	26.73
5.01.38.05.	**Compression fittings; elbows**										
5.01.38.05.A	20mm.....................	nr	0.15	13.80	—	3.55	17.35	15.18	—	3.91	19.09
5.01.38.05.B	25mm.....................	nr	0.18	17.23	—	3.71	20.94	18.95	—	4.08	23.03
5.01.38.05.C	32mm.....................	nr	0.22	20.75	—	4.45	25.20	22.82	—	4.90	27.72

Hutchins Priced Schedules 2021		Unit	Labour Hours	Labour Net £	Plant Net £	Materials Net £	Unit Rate Net £	Labour Gross £	Plant Gross £	Materials Gross £	Unit Rate Gross £
							(Gross rates include 10% profit)				
5.01.	**NEW WORK**										
5.01.38.	**SERVICE PIPEWORK POLYTHENE**										
5.01.38.06.	**Compression fittings; equal tees**										
5.01.38.06.A	20mm......................	nr	0.22	20.75	—	4.12	24.87	22.82	—	4.54	27.36
5.01.38.06.B	25mm......................	nr	0.27	25.32	—	6.37	31.69	27.85	—	7.01	34.85
5.01.38.06.C	32mm......................	nr	0.30	28.74	—	7.88	36.62	31.62	—	8.66	40.28
5.01.40.	**ANCILLARIES, SCREWED JOINTS AND FITTINGS**										
5.01.40.01.	**Bib tap; brass**										
5.01.40.01.A	1/2in.....................	nr	0.11	10.37	—	5.64	16.01	11.41	—	6.20	17.61
5.01.40.01.B	3/4in.....................	nr	0.15	13.80	—	7.05	20.85	15.18	—	7.75	22.93
5.01.40.02.	**Bib tap; chrome; cross top**										
5.01.40.02.A	1/2in.....................	pair	0.19	17.89	—	36.42	54.31	19.68	—	40.06	59.74
5.01.40.03.	**Bib tap; brass; hose union**										
5.01.40.03.A	1/2in.....................	nr	0.11	10.37	—	5.64	16.01	11.41	—	6.20	17.61
5.01.40.03.B	3/4in.....................	nr	0.15	13.80	—	7.76	21.56	15.18	—	8.54	23.71
5.01.40.04.	**Pillar tap; basin and bath; chrome; cross top**										
5.01.40.04.A	1/2in.....................	pair	0.19	17.89	—	24.88	42.77	19.68	—	27.37	47.05
5.01.40.04.B	3/4in.....................	pair	0.22	20.75	—	28.70	49.45	22.82	—	31.58	54.40
5.01.40.05.	**Single lever basin mixer with pop-up waste with swivel spout; chrome (PC sum £65 per nr)**										
5.01.40.05.A	1/2in.....................	nr	0.24	23.03	—	68.90	91.93	25.33	—	75.79	101.12
5.01.40.06.	**Mono basin mixer tap with waste; chrome (PC sum £50 per nr)**										
5.01.40.06.A	1/2in.....................	nr	0.30	28.74	—	53.00	81.74	31.62	—	58.30	89.92
5.01.40.07.	**Bath shower mixer tap with shower head and hose; chrome (PC sum £75 per nr)**										
5.01.40.07.A	3/4in.....................	nr	0.42	40.26	—	79.50	119.76	44.28	—	87.45	131.73
5.01.40.08.	**Servicing valve; brass; compression joint for copper**										
5.01.40.08.A	15mm.....................	nr	0.13	12.66	—	2.82	15.48	13.92	—	3.10	17.02
5.01.40.09.	**Drain cock; brass**										
5.01.40.09.A	15mm.....................	nr	0.07	6.95	—	1.75	8.70	7.64	—	1.92	9.57
5.01.40.10.	**Spring safety valve; brass**										
5.01.40.10.A	13mm.....................	nr	0.19	17.89	—	3.50	21.39	19.68	—	3.85	23.53
5.01.40.10.B	19mm.....................	nr	0.22	20.75	—	5.38	26.13	22.82	—	5.92	28.75

Plumbing and Heating

	Unit	Labour Hours	Labour Net £	Plant Net £	Materials Net £	Unit Rate Net £	Labour Gross £	Plant Gross £	Materials Gross £	Unit Rate Gross £
								(Gross rates include 10% profit)		
5.01.	**NEW WORK**									
5.01.40.	**ANCILLARIES, SCREWED JOINTS AND FITTINGS**									
5.01.40.11.	**Main gas cock; brass**									
5.01.40.11.A	15mm	nr	0.12	11.52	—	6.53	18.05	12.67	—	7.18	19.85
5.01.40.12.	**High pressure ball valve; brass; with plastic float**									
5.01.40.12.A	13mm	nr	0.18	17.23	—	6.99	24.21	18.95	—	7.68	26.63
5.01.40.12.B	19mm	nr	0.22	20.75	—	21.77	42.52	22.82	—	23.95	46.77
5.01.40.12.C	25mm	nr	0.25	24.17	—	49.52	73.70	26.59	—	54.48	81.07
5.01.40.13.	**Gate valve; brass; wheelhead**									
5.01.40.13.A	13mm	nr	0.13	12.66	—	2.83	15.49	13.92	—	3.11	17.04
5.01.40.13.B	19mm	nr	0.18	16.65	—	6.15	22.80	18.32	—	6.76	25.08
5.01.40.13.C	25mm	nr	0.22	20.75	—	7.78	28.53	22.82	—	8.56	31.38
5.01.40.14.	**Stopcock**									
5.01.40.14.A	15mm	nr	0.15	13.80	—	5.27	19.07	15.18	—	5.80	20.97
5.01.40.14.B	22mm	nr	0.19	18.46	—	8.59	27.05	20.31	—	9.44	29.75
5.01.40.14.C	28mm	nr	0.25	23.60	—	14.18	37.78	25.96	—	15.60	41.56
5.01.42.	**EQUIPMENT**									
5.01.42.01.	**Plastic water storage cistern; 13mm ball valve; holes for pipes; hoisting and placing in position; complete with lid; capacity**									
5.01.42.01.A	18l	nr	0.70	66.14	—	29.57	95.72	72.76	—	32.53	105.29
5.01.42.01.B	114l	nr	0.77	73.66	—	157.34	231.00	81.03	—	173.07	254.10
5.01.42.01.C	227l	nr	0.94	89.17	—	181.00	270.17	98.09	—	199.09	297.19
5.01.42.02.	**Galvanised steel open top water storage cistern; 13mm ball valve; holes for pipes; hoisting and placing in position; capacity**									
5.01.42.02.A	227l	nr	1.08	102.97	—	289.93	392.91	113.27	—	318.92	432.20
5.01.42.02.B	327l	nr	1.13	107.64	—	341.75	449.39	118.40	—	375.93	494.33
5.01.42.06.	**Copper direct cylinder; Grade 3; with four bosses; immersion heater boss; hoisting and placing in position; capacity**									
5.01.42.06.A	96l	nr	0.67	63.29	—	188.25	251.53	69.62	—	207.07	276.69
5.01.42.06.B	120l	nr	0.70	66.14	—	190.26	256.40	72.76	—	209.29	282.04
5.01.42.06.C	166l	nr	0.86	81.75	—	226.12	307.87	89.93	—	248.73	338.66

Hutchins Priced Schedules 2021		Unit	Labour Hours	Labour Net £	Plant Net £	Materials Net £	Unit Rate Net £	Labour Gross £	Plant Gross £	Materials Gross £	Unit Rate Gross £
								(Gross rates include 10% profit)			
5.01.	**NEW WORK**										
5.01.42.	**EQUIPMENT**										
5.01.42.07.	**Copper indirect cylinder; Grade 3; with four bosses; immersion heater boss; hoisting and placing in position; capacity**										
5.01.42.07.A	114l	nr	0.70	66.14	—	199.56	265.70	72.76	—	219.51	292.27
5.01.42.07.B	140l	nr	0.77	73.66	—	211.71	285.38	81.03	—	232.89	313.91
5.01.42.07.C	162l	nr	0.86	81.75	—	243.39	325.14	89.93	—	267.73	357.65
5.01.42.08.	**Copper combination hot water storage units; direct pattern; insulation; four connections; immersion heater boss; drain boss; hoisting and placing in position; capacity**										
5.01.42.08.A	115l hot; 25l cold	nr	0.91	86.32	—	384.47	470.79	94.95	—	422.92	517.87
5.01.42.08.B	115l hot; 45l cold	nr	0.94	89.75	—	303.73	393.48	98.72	—	334.11	432.83
5.01.42.08.C	115l hot; 115l cold	nr	0.98	93.27	—	639.00	732.27	102.59	—	702.90	805.49
5.01.42.09.	**Copper combination hot water storage units; indirect pattern; insulation; four connections; immersion heater boss drain boss; hoisting and placing in position; capacity**										
5.01.42.09.A	115l hot; 25l cold	nr	0.91	86.32	—	458.39	544.71	94.95	—	504.23	599.18
5.01.42.09.B	115l hot; 45l cold	nr	0.94	89.75	—	327.99	417.73	98.72	—	360.78	459.50
5.01.42.09.C	115l hot; 115l cold	nr	0.98	93.27	—	685.00	778.27	102.59	—	753.50	856.10
5.01.42.10.	**Gas fired condensing high efficiency combi boiler; placing in position; output**										
5.01.42.10.A	25kW ErP compliant	nr	8.00	761.36	—	805.22	1566.58	837.50	—	885.75	1723.24
5.01.42.10.B	30kW ErP compliant	nr	8.00	761.36	—	917.33	1678.69	837.50	—	1009.07	1846.56
5.01.42.10.C	32kW ErP compliant	nr	8.00	761.36	—	920.39	1681.75	837.50	—	1012.43	1849.92
5.01.42.10.D	36kW ErP compliant	nr	8.00	761.36	—	1109.73	1871.09	837.50	—	1220.71	2058.20
5.01.42.10.E	38kW ErP compliant	nr	8.00	761.36	—	1273.11	2034.47	837.50	—	1400.43	2237.92
5.01.42.11.	**Gas fired condensing high efficiency regular boiler; placing in position; output**										
5.01.42.11.A	12kW ErP compliant	nr	8.50	808.95	—	706.22	1515.16	889.84	—	776.84	1666.68
5.01.42.11.B	15kW ErP compliant	nr	8.50	808.95	—	721.69	1530.63	889.84	—	793.85	1683.69
5.01.42.11.C	18kW ErP compliant	nr	8.50	808.95	—	768.62	1577.57	889.84	—	845.48	1735.32
5.01.42.11.D	21kW ErP compliant	nr	8.50	808.95	—	834.73	1643.67	889.84	—	918.20	1808.04
5.01.42.11.E	24kW ErP compliant	nr	8.50	808.95	—	814.14	1623.08	889.84	—	895.55	1785.39
5.01.42.11.F	27kW ErP compliant	nr	8.50	808.95	—	907.98	1716.92	889.84	—	998.77	1888.61

Plumbing and Heating

Hutchins Priced Schedules 2021		Unit	Labour Hours	Labour Net £	Plant Net £	Materials Net £	Unit Rate Net £	Labour Gross £	Plant Gross £	Materials Gross £	Unit Rate Gross £
							(Gross rates include 10% profit)				
5.01.	**NEW WORK**										
5.01.42.	**EQUIPMENT**										
5.01.42.12.	**Oil fired condensing high efficiency combi boiler; placing in position; output**										
5.01.42.12.A	25kW ErP compliant; internal	nr	8.00	761.36	—	2168.23	2929.59	837.50	—	2385.05	3222.55
5.01.42.12.B	32kW ErP compliant; internal	nr	8.00	761.36	—	2273.44	3034.80	837.50	—	2500.78	3338.28
5.01.42.12.C	25kW ErP compliant; external	nr	7.50	713.78	—	2295.11	3008.89	785.15	—	2524.62	3309.77
5.01.42.12.D	32kW ErP compliant; external	nr	7.50	713.78	—	2933.27	3647.04	785.15	—	3226.60	4011.75
5.01.42.13.	**Oil fired condensing high efficiency regular boiler; placing in position; output**										
5.01.42.13.A	25kW ErP compliant; internal	nr	8.50	808.95	—	1477.59	2286.54	889.84	—	1625.35	2515.19
5.01.42.13.B	32kW ErP compliant; internal	nr	8.50	808.95	—	1830.16	2639.11	889.84	—	2013.18	2903.02
5.01.42.13.C	25kW ErP compliant; external	nr	8.00	761.36	—	1644.37	2405.73	837.50	—	1808.81	2646.31
5.01.42.13.D	32kW ErP compliant; external	nr	8.00	761.36	—	2012.48	2773.84	837.50	—	2213.73	3051.23
5.01.42.18.	**Circulator pump; domestic type**										
5.01.42.18.A	all connections	nr	0.72	68.52	—	246.66	315.18	75.37	—	271.33	346.70
5.01.42.19.	**Programming control**										
5.01.42.19.A	domestic type; smart combined heating and hot water system; all connections . .	nr	0.80	75.95	—	167.83	243.78	83.54	—	184.61	268.15
5.01.42.20.	**Bunded plastic storage tank; supplied and fitted with oil level monitor; bottom outlet fitting kit; overfill prevention valve; 4in inspection screw lid; 2in vent, 1in BSP bottom outlet (or top outlet connection) and lockable manhole cover**										
5.01.42.20.A	1300l; 1800mm x 1300mm x 1300mm; 180kg	nr	1.85	175.87	—	1212.75	1388.62	193.46	—	1334.03	1527.49
5.01.42.20.B	1900l; 2190mm x 1540mm x 1460mm; 180kg	nr	2.16	205.95	—	1323.00	1528.95	226.54	—	1455.30	1681.84
5.01.42.20.C	2500l; 2310mm x 1550mm x 1600mm; 220kg	nr	2.48	236.02	—	1527.75	1763.77	259.62	—	1680.53	1940.15
5.01.42.20.D	3500l; diameter: 2120mm; height: 1600mm; 200kg	nr	2.79	265.14	—	1974.00	2239.14	291.66	—	2171.40	2463.06
5.01.42.20.E	5000l; diameter: 2150mm; height: 2210mm; 240kg	nr	3.08	293.31	—	2215.50	2508.81	322.65	—	2437.05	2759.70

Hutchins Priced Schedules 2021	Unit	Labour Hours	Labour Net £	Plant Net £	Materials Net £	Unit Rate Net £	Labour Gross £	Plant Gross £	Materials Gross £	Unit Rate Gross £
						(Gross rates include 10% profit)				
5.01. **NEW WORK**										
5.01.43. **RADIATORS**										
5.01.43.01. **Radiators; steel single panel; concealed brackets plugged and screwed to wall length**										
5.01.43.01.A 400mm	nr	1.51	143.90	—	15.91	159.81	158.29	—	17.50	175.79
5.01.43.01.B 800mm	nr	1.51	143.90	—	38.43	182.32	158.29	—	42.27	200.55
5.01.43.01.C 1200mm	nr	1.66	158.27	—	41.95	200.22	174.09	—	46.15	220.24
5.01.43.01.D 1800mm	nr	1.66	158.27	—	69.75	228.02	174.09	—	76.72	250.82
5.01.44. **SANITARY FITTINGS**										
5.01.44.01. **Sink; stainless steel; waste; overflow with chain and plastic plug; fixing to top of standard sink unit (excluding taps and trap)**										
5.01.44.01.A single bowl with single drainer; 1000mm x 500mm (PC sum £140 per nr)	nr	2.24	212.90	—	151.16	364.05	234.18	—	166.27	400.46
5.01.44.01.B single bowl with double drainer; 1500mm x 500mm (PC sum £155 per nr)	nr	2.40	227.93	—	166.91	394.84	250.73	—	183.60	434.32
5.01.44.02. **Sink; fireclay; Belfast pattern; white glazed; waste; chain and plastic plug; wall mounted on pair brackets, screwed to wall (excluding taps and trap)**										
5.01.44.02.A 610mm x 455mm x 255mm (PC sum £210 per nr)	nr	1.95	185.87	—	274.10	459.97	204.45	—	301.51	505.97
5.01.44.02.B 760mm x 455mm x 255mm (PC sum £300 per nr)	nr	1.95	185.87	—	368.60	554.47	204.45	—	405.46	609.92
5.01.44.03. **Bath; white reinforced acrylic; rectangular with waste and overflow (excluding taps and trap)**										
5.01.44.03.A 1700mm long (PC sum £180 per nr) . . .	nr	2.60	247.44	—	193.41	440.85	272.19	—	212.75	484.94
5.01.44.04. **Bath; cast iron; white porcelain enamelled; rectangular; with cradle feet; waste and overflow (excluding taps or trap)**										
5.01.44.04.A 1700mm long (PC sum £800 per nr) . . .	nr	3.00	285.41	—	844.41	1129.82	313.96	—	928.85	1242.81
5.01.44.05. **Bath panel; enamelled hardboard; cutting to required size; fixed with chrome dome headed screws to timber frame**										
5.01.44.05.A end panel (PC sum £16 per nr)	nr	0.28	26.46	—	18.63	45.08	29.10	—	20.49	49.59
5.01.44.05.B side panel (PC sum £32 per nr)	nr	0.44	41.97	—	36.06	78.03	46.17	—	39.66	85.83
5.01.44.06. **Bath panel angle strip; polished aluminium; cut to length**										
5.01.44.06.A fixing with chrome dome headed screws .	nr	0.16	14.94	—	5.04	19.98	16.44	—	5.54	21.98

Plumbing and Heating

Hutchins Priced Schedules 2021	Unit	Labour Hours	Labour Net £	Plant Net £	Materials Net £	Unit Rate Net £	Labour Gross £	Plant Gross £	Materials Gross £	Unit Rate Gross £
							(Gross rates include 10% profit)			
5.01. **NEW WORK**										
5.01.44. **SANITARY FITTINGS**										
5.01.44.07. **Basin; white vitreous china; 560mm x 405mm; waste; overflow with chain and plastic plug**										
5.01.44.07.A wall mounted on fixing kit screwed to wall, (excluding taps or traps) (PC sum £110 per nr)	nr	2.07	197.38	—	141.49	338.88	217.12	—	155.64	372.77
5.01.44.07.B screwed to wall and bedded in mastic on pedestal mounting screwed to floor, (excluding taps or trap) (PC sum £150 per nr)	nr	2.52	239.45	—	161.62	401.07	263.39	—	177.79	441.18
5.01.44.08. **WC suite; white vitreous china; trapped pan screwed to floor; plastic seat and cover; plastic cistern with brackets screwed to wall; ball valve and fittings; connection to cistern and pan**										
5.01.44.08.A high level; plastic flush pipe with clips to wall; (PC sum £180 per nr)	nr	2.58	245.16	—	189.00	434.16	269.67	—	207.90	477.57
5.01.44.08.B low level; flush bend connected (PC sum £140 per nr)	nr	2.66	253.25	—	147.00	400.25	278.57	—	161.70	440.27
5.01.44.08.C close-coupled; washdown; flush bend (PC sum £240 per nr)	nr	1.91	181.87	—	252.00	433.87	200.06	—	277.20	477.26
5.01.44.08.D close-coupled; syphonic; flush bend (PC sum £390 per nr)	nr	2.00	189.96	—	409.50	599.46	208.96	—	450.45	659.41
5.01.44.09. **Urinal; white vitreous china; automatic cistern on wall hangers screwed to wall; stainless steel flush pipe with spreader; domed outlet grating**										
5.01.44.09.A single bowl (PC sum £300 per nr)	nr	1.79	170.35	—	315.00	485.35	187.39	—	346.50	533.89
5.01.44.09.B single stall 610mm x 1065mm high (PC sum £600 per nr)	nr	3.04	288.94	—	630.00	918.94	317.83	—	693.00	1010.83
5.01.46. **INSULATION**										
5.01.46.04. **Hair felt pipe sheath around pipes; pipe external dia**										
5.01.46.04.A 15mm	m	0.15	7.65	—	0.11	7.76	8.42	—	0.12	8.53
5.01.46.04.B 22mm	m	0.16	8.49	—	0.11	8.60	9.34	—	0.12	9.46
5.01.46.05. **Flexible foam pipe lagging around pipes; pipe external dia**										
5.01.46.05.A 13mm	m	0.07	3.88	—	0.63	4.51	4.27	—	0.69	4.96
5.01.46.05.B 19mm	m	0.09	4.61	—	0.65	5.26	5.07	—	0.72	5.79
5.01.46.05.C 25mm	m	0.10	5.35	—	0.92	6.27	5.88	—	1.02	6.90
5.01.46.05.D 38mm	m	0.12	6.50	—	2.22	8.72	7.15	—	2.44	9.59

Hutchins Priced Schedules 2021		Unit	Labour Hours	Labour Net £	Plant Net £	Materials Net £	Unit Rate Net £	Labour Gross £	Plant Gross £	Materials Gross £	Unit Rate Gross £
									(Gross rates include 10% profit)		
5.01.	**NEW WORK**										
5.01.56.	**MISCELLANEOUS PLUMBING AND HEATING WORK**										
5.01.56.01.	**Cut holes for pipes or the like; not exceeding 55mm dia; make good**										
5.01.56.01.A	100mm concrete	nr	0.44	23.06	—	0.36	23.42	25.37	—	0.39	25.76
5.01.56.01.B	150mm concrete	nr	0.73	38.37	—	0.36	38.73	42.21	—	0.39	42.60
5.01.56.01.C	100mm reinforced concrete	nr	0.59	30.72	—	0.36	31.07	33.79	—	0.39	34.18
5.01.56.01.D	150mm reinforced concrete	nr	0.77	40.36	—	0.36	40.72	44.40	—	0.39	44.79
5.01.56.01.E	75mm blockwork	nr	0.24	12.69	—	0.36	13.04	13.95	—	0.39	14.35
5.01.56.01.F	100mm blockwork	nr	0.24	12.69	—	0.36	13.04	13.95	—	0.39	14.35
5.01.56.01.G	102mm brickwork	nr	0.33	17.30	—	0.36	17.65	19.03	—	0.39	19.42
5.01.56.01.H	215mm brickwork	nr	0.59	30.72	—	0.36	31.07	33.79	—	0.39	34.18
5.01.56.01.I	softwood floor boarding	nr	0.11	5.77	—	—	5.77	6.34	—	—	6.34
5.01.56.01.J	softwood floor boarding and plaster board soffit under	nr	0.26	13.42	—	0.39	13.81	14.76	—	0.43	15.19
5.01.56.02.	**Cut holes for pipes or the like; 55mm - 110mm dia; make good**										
5.01.56.02.A	100mm concrete	nr	0.62	32.61	—	0.59	33.20	35.87	—	0.65	36.52
5.01.56.02.B	150mm concrete	nr	0.92	48.02	—	0.59	48.61	52.82	—	0.65	53.47
5.01.56.02.C	100mm reinforced concrete	nr	0.73	38.37	—	0.59	38.96	42.21	—	0.65	42.86
5.01.56.02.D	150mm reinforced concrete	nr	1.03	53.78	—	0.59	54.38	59.16	—	0.65	59.81
5.01.56.02.E	75mm blockwork	nr	0.33	17.30	—	0.59	17.89	19.03	—	0.65	19.68
5.01.56.02.F	100mm blockwork	nr	0.33	17.30	—	0.59	17.89	19.03	—	0.65	19.68
5.01.56.02.G	102mm brickwork	nr	0.44	23.06	—	0.59	23.66	25.37	—	0.65	26.02
5.01.56.02.H	215mm brickwork	nr	0.81	42.25	—	0.59	42.84	46.48	—	0.65	47.13
5.01.56.02.I	softwood floor boarding	nr	0.15	7.65	—	—	7.65	8.42	—	—	8.42
5.01.56.02.J	softwood floor boarding and plaster board soffit under	nr	0.37	19.19	—	0.59	19.77	21.10	—	0.64	21.75
5.01.56.03.	**Framing; sawn softwood; 38mm x 50mm; for bath panel**										
5.01.56.03.A	end panel	nr	0.26	13.84	—	3.35	17.19	15.22	—	3.69	18.91
5.01.56.03.B	side panel	nr	0.53	27.68	—	6.59	34.27	30.45	—	7.25	37.69
5.01.56.04.	**Take up existing softwood floor boarding; 25mm thick; one board wide; cut holes or notches for pipes and refix boards**										
5.01.56.04.A	service cables	m	0.66	34.60	—	—	34.60	38.06	—	—	38.06
5.01.56.04.B	pipes not exceeding 55mm dia	m	0.81	42.25	—	—	42.25	46.48	—	—	46.48
5.01.56.04.C	pipes 55mm - 110mm dia	m	0.95	49.90	—	—	49.90	54.89	—	—	54.89

Plumbing and Heating

Hutchins Priced Schedules 2021	Unit	Labour Hours	Labour Net £	Plant Net £	Materials Net £	Unit Rate Net £	Labour Gross £	Plant Gross £	Materials Gross £	Unit Rate Gross £
										(Gross rates include 10% profit)
5.01. **NEW WORK**										
5.01.56. **MISCELLANEOUS PLUMBING AND HEATING WORK**										
5.01.56.05. Tank bearers nailed to ceiling joists										
5.01.56.05.A 75mm x 50mm	m	0.09	4.61	—	2.63	7.24	5.07	—	2.89	7.96
5.01.56.05.B 100mm x 50mm	m	0.12	6.19	—	3.50	9.68	6.80	—	3.85	10.65
5.01.56.06. Boarded platform; softwood nailed to bearers										
5.01.56.06.A for tank or cistern	m²	1.14	59.55	—	9.03	68.58	65.50	—	9.93	75.44
5.02. **REPAIRS AND ALTERATIONS**										
5.02.01. **REMOVE GUTTERWORK AND PIPEWORK**										
5.02.01.01. Gutterwork and supports										
5.02.01.01.A cement	m	0.21	20.18	—	—	20.18	22.19	—	—	22.19
5.02.01.01.B PVCu	m	0.23	21.89	—	—	21.89	24.08	—	—	24.08
5.02.01.01.C cast iron	m	0.27	25.89	—	—	25.89	28.47	—	—	28.47
5.02.01.02. Rainwater pipework and supports										
5.02.01.02.A cement	m	0.18	17.23	—	—	17.23	18.95	—	—	18.95
5.02.01.02.B PVCu	m	0.20	19.03	—	—	19.03	20.94	—	—	20.94
5.02.01.02.C cast iron	m	0.24	23.03	—	—	23.03	25.33	—	—	25.33
5.02.01.03. Rainwater shoe										
5.02.01.03.A PVCu	nr	0.04	4.00	—	—	4.00	4.40	—	—	4.40
5.02.01.03.B cast iron	nr	0.06	5.81	—	—	5.81	6.39	—	—	6.39
5.02.01.04. Rainwater head and support										
5.02.01.04.A PVCu	nr	0.22	21.32	—	—	21.32	23.45	—	—	23.45
5.02.01.04.B cast iron	nr	0.27	25.89	—	—	25.89	28.47	—	—	28.47
5.02.01.05. Soil and ventilation pipework and supports										
5.02.01.05.A PVCu	m	0.36	34.55	—	—	34.55	38.00	—	—	38.00
5.02.01.05.B cast iron	m	0.41	38.54	—	—	38.54	42.40	—	—	42.40
5.02.01.05.C lead	m	0.45	43.21	—	—	43.21	47.53	—	—	47.53
5.02.01.06. Service, waste and overflow pipework and supports										
5.02.01.06.A PVCu	m	0.08	7.52	—	—	7.52	8.27	—	—	8.27
5.02.01.06.B copper	m	0.10	9.80	—	—	9.80	10.78	—	—	10.78
5.02.01.06.C lead	m	0.10	9.80	—	—	9.80	10.78	—	—	10.78
5.02.01.06.D galvanised steel	m	0.10	9.80	—	—	9.80	10.78	—	—	10.78

Hutchins Priced Schedules 2021	Unit	Labour Hours	Labour Net £	Plant Net £	Materials Net £	Unit Rate Net £	Labour Gross £	Plant Gross £	Materials Gross £	Unit Rate Gross £
								(Gross rates include 10% profit)		
5.02.	**REPAIRS AND ALTERATIONS**									
5.02.01.	**REMOVE GUTTERWORK AND PIPEWORK**									
5.02.01.07.	**Remove sanitary fittings including taps and trap**									
5.02.01.07.A	WC suite	nr	0.25	24.17	—	—	24.17	26.59	—	—	26.59
5.02.01.07.B	wash hand basin	nr	0.22	21.32	—	—	21.32	23.45	—	—	23.45
5.02.01.07.C	bath	nr	0.32	29.98	—	—	29.98	32.98	—	—	32.98
5.02.01.07.D	sink unit	nr	0.22	21.32	—	—	21.32	23.45	—	—	23.45
5.02.01.07.E	shower	nr	0.09	8.66	—	—	8.66	9.53	—	—	9.53
5.02.01.08.	**Remove sanitary fittings including taps, trap and service and waste pipes not exceeding 3.00m girth**									
5.02.01.08.A	WC suite	nr	0.32	29.98	—	—	29.98	32.98	—	—	32.98
5.02.01.08.B	wash hand basin	nr	0.25	24.17	—	—	24.17	26.59	—	—	26.59
5.02.01.08.C	bath	nr	0.38	35.69	—	—	35.69	39.26	—	—	39.26
5.02.01.08.D	sink unit	nr	0.25	24.17	—	—	24.17	26.59	—	—	26.59
5.02.01.08.E	shower	nr	0.13	12.66	—	—	12.66	13.92	—	—	13.92
5.02.01.09.	**Remove bathroom and toilet fittings**									
5.02.01.09.A	toilet roll holder	nr	0.03	2.86	—	—	2.86	3.14	—	—	3.14
5.02.01.09.B	soap dispenser	nr	0.03	2.86	—	—	2.86	3.14	—	—	3.14
5.02.01.09.C	towel rail	nr	0.05	4.57	—	—	4.57	5.02	—	—	5.02
5.02.01.09.D	towel holder	nr	0.10	9.80	—	—	9.80	10.78	—	—	10.78
5.02.01.09.E	mirror	nr	0.10	9.80	—	—	9.80	10.78	—	—	10.78
5.02.01.10.	**Remove equipment; excluding any draining down of system**									
5.02.01.10.A	cold water tank	nr	1.60	152.56	—	—	152.56	167.81	—	—	167.81
5.02.01.10.B	hot water cylinder	nr	0.80	76.52	—	—	76.52	84.17	—	—	84.17
5.02.01.10.C	gas water heater	nr	2.66	253.25	—	—	253.25	278.57	—	—	278.57
5.02.01.10.D	gas fire	nr	1.33	126.58	—	—	126.58	139.23	—	—	139.23
5.02.01.10.E	expansion tank	nr	1.21	115.06	—	—	115.06	126.57	—	—	126.57
5.02.02.	**REPAIRS TO PIPEWORK AND GUTTERWORK**									
5.02.02.01.	**Clean out eaves or parapet gutters, remove rubbish**									
5.02.02.01.A	any height or position	m	0.08	7.52	—	—	7.52	8.27	—	—	8.27
5.02.02.02.	**Clean out rainwater pipes, stack pipes, etc.; remove rubbish**									
5.02.02.02.A	any height or position	m	0.08	7.52	—	—	7.52	8.27	—	—	8.27

Plumbing and Heating

Hutchins Priced Schedules 2021	Unit	Labour Hours	Labour Net £	Plant Net £	Materials Net £	Unit Rate Net £	Labour Gross £	Plant Gross £	Materials Gross £	Unit Rate Gross £
							(Gross rates include 10% profit)			
5.02.	**REPAIRS AND ALTERATIONS**									
5.02.02.	**REPAIRS TO PIPEWORK AND GUTTERWORK**									
5.02.02.03.	**Take down existing gutter; clean and refix to fascia or on brackets**									
5.02.02.03.A	seal joints with non-setting putty and set to proper falls	m	0.42	40.26	—	1.12	41.38	44.28	—	1.23	45.52
5.02.02.04.	**Take down and remove existing 100mm cast iron gutters and provide and fix new gutters**									
5.02.02.04.A	half round	m	0.67	63.29	—	22.88	86.17	69.62	—	25.17	94.78
5.02.02.04.B	ogee	m	0.67	63.29	—	25.07	88.36	69.62	—	27.58	97.20
5.02.02.05.	**Take down existing rainwater pipes and refix to walls**									
5.02.02.05.A	50mm, 63mm, or 75mm	m	0.45	43.21	—	—	43.21	47.53	—	—	47.53
5.02.02.06.	**Take down and remove existing cast iron rain water pipes and provide and fix new**									
5.02.02.06.A	65mm and 75mm pipes	m	0.67	63.29	—	46.61	109.90	69.62	—	51.27	120.89
5.02.02.06.B	rainwater shoe (not extra over)	nr	0.30	28.74	—	27.57	56.31	31.62	—	30.33	61.94
5.02.02.07.	**Cut out and reform caulked lead joints in cast iron soil, vent or waste pipes**									
5.02.02.07.A	50mm	nr	0.42	40.26	—	—	40.26	44.28	—	—	44.28
5.02.02.07.B	75mm	nr	0.48	46.06	—	—	46.06	50.67	—	—	50.67
5.02.02.07.C	100mm	nr	0.70	66.14	—	—	66.14	72.76	—	—	72.76
5.02.02.08.	**Cut and adapt existing 100mm cast iron soil pipe for a new WC by inserting branch and bend (or long junction); connect to pan trap; make good to**									
5.02.02.08.A	wall	nr	2.90	276.28	—	150.64	426.92	303.91	—	165.70	469.61
5.02.02.08.B	extra for access door	nr	—	—	—	151.93	151.93	—	—	167.12	167.12
5.02.06.	**REPAIRS TO PIPEWORK AND PLUMBING**									
5.02.06.01.	**Renew broken cast iron stopcock box with hinged lid**									
5.02.06.01.A	140mm x 115mm x 75mm	nr	0.30	28.74	—	21.57	50.31	31.62	—	23.73	55.34
5.02.06.01.B	152mm x 152mm x 75mm	nr	0.33	31.69	—	29.23	60.92	34.86	—	32.15	67.01
5.02.06.04.	**Cut existing cast iron pipe and insert new tees**									
5.02.06.04.A	15mm	nr	0.42	40.26	—	1.92	42.18	44.28	—	2.11	46.39
5.02.06.04.B	20mm	nr	0.51	48.92	—	2.12	51.04	53.81	—	2.33	56.14

Plumbing and Heating

Hutchins Priced Schedules 2021		Unit	Labour Hours	Labour Net £	Plant Net £	Materials Net £	Unit Rate Net £	Labour Gross £	Plant Gross £	Materials Gross £	Unit Rate Gross £
								(Gross rates include 10% profit)			
5.02.	**REPAIRS AND ALTERATIONS**										
5.02.06.	**REPAIRS TO PIPEWORK AND PLUMBING**										
5.02.06.04.	**Cut existing cast iron pipe and insert new tees**										
5.02.06.04.C	25mm	nr	0.61	57.58	—	2.88	60.46	63.34	—	3.17	66.51
5.02.06.07.	**Cut existing copper pipes and insert new capillary tees**										
5.02.06.07.A	15mm	nr	0.61	57.58	—	0.70	58.28	63.34	—	0.77	64.11
5.02.06.07.B	22mm	nr	0.73	69.09	—	1.99	71.09	76.00	—	2.19	78.19
5.02.06.07.C	28mm	nr	0.91	86.32	—	9.97	96.29	94.95	—	10.97	105.92
5.02.06.09.	**Cut existing polythene pipes and insert new compression tees**										
5.02.06.09.A	20mm	nr	0.91	86.32	—	4.12	90.44	94.95	—	4.54	99.49
5.02.06.09.B	25mm	nr	1.03	97.83	—	6.37	104.21	107.62	—	7.01	114.63
5.02.06.09.C	32mm	nr	1.21	115.06	—	7.88	122.94	126.57	—	8.66	135.23
5.02.06.10.	**Cover iron or copper pipes with hair felt and twine in any position**										
5.02.06.10.A	up to 22mm dia	m	0.42	40.26	—	0.11	40.36	44.28	—	0.12	44.40
5.02.06.12.	**Take off existing bib valves and prepare iron or copper pipes; provide and fix new bib valve**										
5.02.06.12.A	13mm	nr	0.61	57.58	—	5.64	63.22	63.34	—	6.20	69.54
5.02.06.12.B	19mm	nr	0.73	69.09	—	7.05	76.14	76.00	—	7.75	83.76
5.02.06.14.	**Cut into iron or copper pipes and fit new stopcock**										
5.02.06.14.A	13mm	nr	0.79	74.80	—	5.62	80.42	82.28	—	6.18	88.46
5.02.06.14.B	19mm	nr	0.91	86.32	—	11.48	97.80	94.95	—	12.63	107.58
5.02.06.16.	**Rewasher ball valve, tap or indoor stopcock**										
5.02.06.16.A	13mm	nr	0.30	28.74	—	0.14	28.88	31.62	—	0.15	31.77
5.02.06.16.B	19mm	nr	0.36	34.55	—	0.36	34.90	38.00	—	0.39	38.39
5.02.06.18.	**Take off existing copper trap to bath, basin or sink and provide and fit new plastic trap**										
5.02.06.18.A	32mm	nr	0.73	69.09	—	4.12	73.22	76.00	—	4.54	80.54
5.02.06.18.B	38mm	nr	0.79	74.80	—	4.16	78.96	82.28	—	4.57	86.85
5.02.06.20.	**Take off stopcock to iron or copper pipe both ends and provide and fit new stopcock**										
5.02.06.20.A	13mm	nr	0.39	37.40	—	5.62	43.02	41.14	—	6.18	47.32
5.02.06.20.B	19mm	nr	0.51	48.92	—	11.48	60.40	53.81	—	12.63	66.44
5.02.06.20.C	25mm	nr	0.64	60.43	—	21.06	81.50	66.48	—	23.17	89.64

Plumbing and Heating

Hutchins Priced Schedules 2021	Unit	Labour Hours	Labour Net £	Plant Net £	Materials Net £	Unit Rate Net £	Labour Gross £	Plant Gross £	Materials Gross £	Unit Rate Gross £
							(Gross rates include 10% profit)			
5.02. **REPAIRS AND ALTERATIONS**										
5.02.06. **REPAIRS TO PIPEWORK AND PLUMBING**										
5.02.06.22. Take off ball valve to iron or copper pipes; provide and fit new ball valve and ball float complete										
5.02.06.22.A 13mm	nr	0.30	28.74	—	6.99	35.73	31.62	—	7.68	39.30
5.02.06.22.C 19mm	nr	0.39	37.40	—	21.77	59.17	41.14	—	23.95	65.09
5.02.08. **REPAIRS TO SANITARYWARE AND FITTINGS**										
5.02.08.01. Supply WC suite complete (PC sum £180 per nr) and connect to existing services										
5.02.08.01.A copper or iron	nr	3.51	333.76	—	189.00	522.76	367.14	—	207.90	575.04
5.02.08.02. Supply low level WC suite (PC sum £140 per nr) complete; connect to existing services										
5.02.08.02.A copper or iron	nr	3.81	362.60	—	147.00	509.60	398.86	—	161.70	560.56
5.02.08.03. Remove defective WC pan and fix new; make good all connections										
5.02.08.03.A WC pan (PC sum £90 per nr)	nr	2.36	224.41	—	94.50	318.91	246.85	—	103.95	350.80
5.02.08.04. Take off defective seat to pedestal pan; supply and fix new seat										
5.02.08.04.A plastic seat	nr	0.45	43.21	—	17.66	60.87	47.53	—	19.43	66.95
5.02.08.05. Take off WC seat, renew joints to flush pipe including closet and outlet connection and pan										
5.02.08.05.A refix seat	nr	0.70	66.14	—	1.96	68.11	72.76	—	2.16	74.92
5.02.08.07. Disconnect ball valve to WC tank										
5.02.08.07.A re-washer and clean out	nr	0.88	83.46	—	0.36	83.82	91.81	—	0.39	92.20
5.02.08.07.B unscrew ball valve and supply and fit new ball	nr	0.18	17.23	—	2.23	19.45	18.95	—	2.45	21.40
5.02.08.08. Supply and fix flat back basin (PC sum £90) including taps, traps and wall brackets, connect to existing services										
5.02.08.08.A iron or copper	nr	3.51	333.76	—	94.50	428.26	367.14	—	103.95	471.09
5.02.08.09. Supply and fix pedestal basin (PC sum £150 per nr) including taps, traps and wall brackets; connect to existing services										
5.02.08.09.A iron or copper	nr	3.51	333.76	—	157.50	491.26	367.14	—	173.25	540.39

Hutchins Priced Schedules 2021	Unit	Labour Hours	Labour Net £	Plant Net £	Materials Net £	Unit Rate Net £	Labour Gross £	Plant Gross £	Materials Gross £	Unit Rate Gross £
							(Gross rates include 10% profit)			

5.02.   REPAIRS AND ALTERATIONS

5.02.08.   REPAIRS TO SANITARYWARE AND FITTINGS

5.02.08.12.   Supply and fix 1700mm long, white reinforced acrylic rectangular bath (PC sum £180 per nr); with waste and overflow; complete with 3/4in bath shower mixer tap with shower head and hose (PC sum £75 per nr) and trap; connect to existing services

5.02.08.12.A iron or copper	nr	3.51	333.76	—	320.93	654.69	367.14	—	353.02	720.16
5.02.08.14. Clear blockage and flush out; to										
5.02.08.14.A WC pans and traps	nr	0.76	71.95	—	—	71.95	79.14	—	—	79.14
5.02.08.14.B traps and waste pipes of baths, sinks, lavatory basins, etc	nr	0.61	57.58	—	—	57.58	63.34	—	—	63.34

5.02.08.16.   Disconnect all pipework; take out and remove galvanised steel hot water tank; provide and install copper indirect cylinder; allow for cutting holes, tank connectors, make up and all connections to existing pipework; capacity

5.02.08.16.A 114l	nr	1.81	172.64	—	218.25	390.89	189.90	—	240.08	429.98

5.02.08.20.   Turn off water supply; disconnect all pipework; take out and remove galvanised steel cold water storage tank; provide and install plastic tank complete with ball valve, lid and insulation and allow for cutting holes, tank connectors, make up and all connections to existing pipework; capacity

5.02.08.20.A 182l (PC sum £140 per nr)	nr	3.81	362.60	—	175.58	538.18	398.86	—	193.14	591.99

5.02.08.21.   Clean and scour out open-top storage tanks

5.02.08.21.A generally	nr	4.48	425.89	—	—	425.89	468.47	—	—	468.47

Glazing, Painting and Decorating

Hutchins Priced Schedules 2021	Unit	Labour Hours	Labour Net £	Plant Net £	Materials Net £	Unit Rate Net £	Labour Gross £	Plant Gross £	Materials Gross £	Unit Rate Gross £
						(Gross rates include 10% profit)				
6.01.	**NEW WORK**									
6.01.01.	**GLASS IN OPENINGS**									
6.01.01.01.	**2mm clear float glass and glazing to wood with putty; in panes**									
6.01.01.01.A	not exceeding 0.10m2...........	m²	1.20	45.16	—	25.66	70.81	49.67	—	28.22	77.89
6.01.01.01.B	0.10m2 - 0.50m2...............	m²	0.90	33.87	—	25.66	59.52	37.25	—	28.22	65.47
6.01.01.01.C	0.50m2 - 1.00m2...............	m²	0.45	16.93	—	25.66	42.59	18.63	—	28.22	46.85
6.01.01.02.	**3mm clear float glass and glazing to wood with putty; in panes**									
6.01.01.02.A	not exceeding 0.10m2...........	m²	1.20	45.16	—	29.51	74.66	49.67	—	32.46	82.13
6.01.01.02.B	0.10m2 - 0.50m2...............	m²	0.90	33.87	—	29.51	63.38	37.25	—	32.46	69.71
6.01.01.02.C	0.50m2 - 1.00m2...............	m²	0.45	16.93	—	29.51	46.44	18.63	—	32.46	51.09
6.01.01.02.D	exceeding 1.00m2...............	m²	0.35	13.17	—	29.51	42.68	14.49	—	32.46	46.95
6.01.01.03.	**4mm clear float glass and glazing to wood with putty; in panes**									
6.01.01.03.A	0.10m2 - 0.50m2...............	m²	1.00	37.63	—	31.82	69.45	41.39	—	35.00	76.39
6.01.01.03.B	0.50m2 - 1.00m2...............	m²	0.50	18.82	—	32.79	51.60	20.70	—	36.06	56.76
6.01.01.03.C	exceeding 1.00m2...............	m²	0.40	15.05	—	32.79	47.84	16.56	—	36.06	52.62
6.01.01.04.	**2mm clear float glass and glazing to metal with putty; in panes**									
6.01.01.04.A	not exceeding 0.10m2...........	m²	1.75	65.85	—	24.71	90.56	72.44	—	27.18	99.62
6.01.01.04.B	0.10m2 - 0.50m2...............	m²	1.35	50.80	—	24.68	75.48	55.88	—	27.15	83.03
6.01.01.04.C	0.50m2 - 1.00m2...............	m²	0.70	26.34	—	24.65	50.99	28.98	—	27.12	56.09
6.01.01.05.	**3mm clear float glass and glazing to metal with putty; in panes**									
6.01.01.05.A	not exceeding 0.10m2...........	m²	1.75	65.85	—	28.54	94.39	72.44	—	31.39	103.83
6.01.01.05.B	0.10m2 - 0.50m2...............	m²	1.35	50.80	—	28.50	79.31	55.88	—	31.36	87.24
6.01.01.05.C	0.50m2 - 1.00m2...............	m²	0.70	26.34	—	28.48	54.82	28.98	—	31.32	60.30
6.01.01.05.D	exceeding 1.00m2...............	m²	0.55	20.70	—	28.48	49.17	22.77	—	31.32	54.09
6.01.01.06.	**4mm clear float glass and glazing to metal with putty; in panes**									
6.01.01.06.A	0.10m2 - 0.50m2...............	m²	1.50	56.45	—	30.81	87.26	62.09	—	33.90	95.99
6.01.01.06.B	0.50m2 - 1.00m2...............	m²	0.80	30.10	—	30.79	60.89	33.11	—	33.86	66.98
6.01.01.06.C	exceeding 1.00m2...............	m²	0.65	24.46	—	30.79	55.25	26.91	—	33.86	60.77

Glazing, Painting and Decorating

Hutchins Priced Schedules 2021		Unit	Labour Hours	Labour Net £	Plant Net £	Materials Net £	Unit Rate Net £	Labour Gross £	Plant Gross £	Materials Gross £	Unit Rate Gross £
								(Gross rates include 10% profit)			
6.01.	**NEW WORK**										
6.01.01.	**GLASS IN OPENINGS**										
6.01.01.07.	Figured, rolled or cathedral glass and glazing to wood with putty; in panes										
6.01.01.07.A	not exceeding 0.10m2	m²	1.20	45.16	—	57.84	103.00	49.67	—	63.62	113.29
6.01.01.07.B	0.10m2 - 0.50m2	m²	0.90	33.87	—	57.84	91.71	37.25	—	63.62	100.88
6.01.01.07.C	0.50m2 - 1.00m2	m²	0.45	16.93	—	58.90	75.84	18.63	—	64.79	83.42
6.01.01.07.D	exceeding 1.00m2	m²	0.35	13.17	—	56.87	70.04	14.49	—	62.56	77.05
6.01.01.08.	Figured, rolled or cathedral glass and glazing to metal with putty; in panes										
6.01.01.08.A	not exceeding 0.10m2	m²	1.75	65.85	—	55.93	121.78	72.44	—	61.52	133.96
6.01.01.08.B	0.10m2 - 0.50m2	m²	1.35	50.80	—	55.87	106.67	55.88	—	61.45	117.34
6.01.01.08.C	0.50m2 - 1.00m2	m²	0.70	26.34	—	55.84	82.18	28.98	—	61.42	90.40
6.01.01.08.D	exceeding 1.00m2	m²	0.50	18.82	—	55.87	74.68	20.70	—	61.45	82.15
6.01.01.09.	7mm wired cast glass and glazing to wood with putty; in panes										
6.01.01.09.A	not exceeding 0.10m2	m²	1.40	52.68	—	88.49	141.18	57.95	—	97.34	155.29
6.01.01.09.B	0.10m2 - 0.50m2	m²	1.20	45.16	—	87.53	132.68	49.67	—	96.28	145.95
6.01.01.09.C	0.50m2 - 1.00m2	m²	0.60	22.58	—	85.49	108.07	24.84	—	94.04	118.88
6.01.01.09.D	exceeding 1.00m2	m²	0.50	18.82	—	86.46	105.28	20.70	—	95.11	115.80
6.01.01.10.	7mm wired cast glass and glazing to metal with putty; in panes										
6.01.01.10.A	not exceeding 0.10m2	m²	2.10	79.02	—	84.49	163.51	86.93	—	92.94	179.87
6.01.01.10.B	0.10m2 - 0.50m2	m²	1.65	62.09	—	84.55	146.64	68.30	—	93.01	161.30
6.01.01.10.C	0.50m2 - 1.00m2	m²	0.85	31.99	—	84.49	116.48	35.18	—	92.94	128.12
6.01.01.10.D	exceeding 1.00m2	m²	0.75	28.22	—	84.46	112.68	31.04	—	92.91	123.95
6.01.01.11.	7mm wired glass and glazing to wood with putty; roof lights, lantern lights, skylights, etc.; in panes										
6.01.01.11.A	0.10m2 - 0.50m2	m²	2.20	82.79	—	86.46	169.25	91.06	—	95.11	186.17
6.01.01.11.B	0.50m2 - 1.00m2	m²	1.10	41.39	—	85.49	126.89	45.53	—	94.04	139.58
6.01.01.11.C	exceeding 1.00m2	m²	0.90	33.87	—	86.46	120.33	37.25	—	95.11	132.36
6.01.01.12.	7mm wired glass glazing to metal with putty; in panes										
6.01.01.12.A	0.10m2 - 0.50m2	m²	3.00	112.89	—	84.46	197.35	124.18	—	92.91	217.09
6.01.01.12.B	0.50m2 - 1.00m2	m²	1.70	63.97	—	84.49	148.46	70.37	—	92.94	163.31
6.01.01.12.C	exceeding 1.00m2	m²	1.50	56.45	—	84.51	140.95	62.09	—	92.96	155.05
6.01.01.13.	6mm float glass, bedding in wash-leather and beads, both measured separately; in panes										
6.01.01.13.A	not exceeding 4.00m2	m²	3.30	124.18	—	41.94	166.12	136.60	—	46.13	182.73

Glazing, Painting and Decorating

Hutchins Priced Schedules 2021	Unit	Labour Hours	Labour Net £	Plant Net £	Materials Net £	Unit Rate Net £	Labour Gross £	Plant Gross £	Materials Gross £	Unit Rate Gross £
							(Gross rates include 10% profit)			
6.01. **NEW WORK**										
6.01.01. **GLASS IN OPENINGS**										
6.01.01.15. **Curved cutting, including risk, on the following**										
6.01.01.15.A float glass	m	0.25	9.41	—	—	9.41	10.35	—	—	10.35
6.01.01.15.B obscure glass	m	0.33	12.42	—	—	12.42	13.66	—	—	13.66
6.01.01.15.C wire cast glass	m	0.60	22.58	—	—	22.58	24.84	—	—	24.84
6.01.01.15.D 6mm single glaze quality float glass	m	0.40	15.05	—	—	15.05	16.56	—	—	16.56
6.01.03. **DOUBLE GLAZING UNITS**										
6.01.03.01. **Hermetically sealed double glazing unit with non-setting compound and clipped beads; 4+10+4mm; clear low iron/clear low iron**										
6.01.03.01.B 1.00m2 - 2.00m2	m²	2.00	75.26	—	52.23	127.49	82.79	—	57.46	140.24
6.01.03.01.C 0.75m2 - 1.00m2	m²	2.50	94.08	—	54.14	148.21	103.48	—	59.55	163.03
6.01.03.01.D 0.50m2 - 0.75m2	m²	3.00	112.89	—	55.41	168.30	124.18	—	60.95	185.13
6.01.03.01.E 0.35m2 - 0.50m2	m²	3.50	131.71	—	57.95	189.66	144.88	—	63.75	208.62
6.01.03.01.F 0.25m2 - 0.35m2	m²	4.00	150.52	—	61.21	211.73	165.57	—	67.33	232.91
6.01.03.01.G not exceeding 0.25m2	m²	4.50	169.34	—	65.57	234.91	186.27	—	72.13	258.40
6.01.03.02. **Hermetically sealed double glazing unit with non-setting compound and clipped beads; 4+10+4mm; clear low iron/pattern**										
6.01.03.02.B 1.00m2 - 2.00m2	m²	2.00	75.26	—	61.47	136.73	82.79	—	67.62	150.41
6.01.03.02.C 0.75m2 - 1.00m2	m²	2.50	94.08	—	63.38	157.45	103.48	—	69.72	173.20
6.01.03.02.D 0.50m2 - 0.75m2	m²	3.00	112.89	—	64.65	177.54	124.18	—	71.11	195.29
6.01.03.02.E 0.35m2 - 0.50m2	m²	3.50	131.71	—	67.19	198.90	144.88	—	73.91	218.79
6.01.03.02.F 0.25m2 - 0.35m2	m²	4.00	150.52	—	70.45	220.97	165.57	—	77.50	243.07
6.01.03.02.G not exceeding 0.25m2	m²	4.50	169.34	—	74.81	244.15	186.27	—	82.29	268.56
6.01.03.03. **Hermetically sealed double glazing unit with non-setting compound and clipped beads; 4+10+4mm; clear low iron/Planitherm**										
6.01.03.03.B 1.00m2 - 2.00m2	m²	2.00	75.26	—	70.71	145.97	82.79	—	77.78	160.57
6.01.03.03.C 0.75m2 - 1.00m2	m²	2.50	94.08	—	72.62	166.69	103.48	—	79.88	183.36
6.01.03.03.D 0.50m2 - 0.75m2	m²	3.00	112.89	—	73.89	186.78	124.18	—	81.28	205.46
6.01.03.03.E 0.35m2 - 0.50m2	m²	3.50	131.71	—	76.43	208.14	144.88	—	84.07	228.95
6.01.03.03.F 0.25m2 - 0.35m2	m²	4.00	150.52	—	79.69	230.21	165.57	—	87.66	253.23
6.01.03.03.G not exceeding 0.25m2	m²	4.50	169.34	—	84.05	253.39	186.27	—	92.46	278.73

Glazing, Painting and Decorating

Hutchins Priced Schedules 2021		Unit	Labour Hours	Labour Net £	Plant Net £	Materials Net £	Unit Rate Net £	Labour Gross £	Plant Gross £	Materials Gross £	Unit Rate Gross £
										(Gross rates include 10% profit)	
6.01.	**NEW WORK**										
6.01.03.	DOUBLE GLAZING UNITS										
6.01.03.04.	Hermetically sealed double glazing unit with non-setting compound and clipped beads; 4+10+4mm; clear/anti-sun toughened										
6.01.03.04.B	1.00m2 - 2.00m2	m²	2.00	75.26	—	105.36	180.62	82.79	—	115.90	198.69
6.01.03.04.C	0.75m2 - 1.00m2	m²	2.50	94.08	—	107.27	201.34	103.48	—	117.99	221.48
6.01.03.04.D	0.50m2 - 0.75m2	m²	3.00	112.89	—	108.54	221.43	124.18	—	119.39	243.57
6.01.03.04.E	0.35m2 - 0.50m2	m²	3.50	131.71	—	111.08	242.79	144.88	—	122.19	267.06
6.01.03.04.F	0.25m2 - 0.35m2	m²	4.00	150.52	—	114.34	264.86	165.57	—	125.78	291.35
6.01.03.04.G	not exceeding 0.25m2	m²	4.50	169.34	—	118.70	288.04	186.27	—	130.57	316.84
6.01.03.05.	Hermetically sealed double glazing unit with non-setting compound and clipped beads; 4+10+4mm; clear/clear toughened										
6.01.03.05.B	1.00m2 - 2.00m2	m²	2.00	75.26	—	75.33	150.59	82.79	—	82.87	165.65
6.01.03.05.C	0.75m2 - 1.00m2	m²	2.50	94.08	—	77.24	171.31	103.48	—	84.96	188.44
6.01.03.05.D	0.50m2 - 0.75m2	m²	3.00	112.89	—	78.51	191.40	124.18	—	86.36	210.54
6.01.03.05.E	0.35m2 - 0.50m2	m²	3.50	131.71	—	81.05	212.76	144.88	—	89.16	234.03
6.01.03.05.F	0.25m2 - 0.35m2	m²	4.00	150.52	—	84.31	234.83	165.57	—	92.74	258.32
6.01.03.05.G	not exceeding 0.25m2	m²	4.50	169.34	—	88.67	258.01	186.27	—	97.54	283.81
6.01.03.06.	Hermetically sealed double glazing unit with non-setting compound and clipped beads; 4+10+4mm; clear/Pilkington Activ Blue toughened										
6.01.03.06.B	1.00m2 - 2.00m2	m²	2.00	75.26	—	137.70	212.96	82.79	—	151.47	234.26
6.01.03.06.C	0.75m2 - 1.00m2	m²	2.50	94.08	—	139.61	233.68	103.48	—	153.57	257.05
6.01.03.06.D	0.50m2 - 0.75m2	m²	3.00	112.89	—	140.88	253.77	124.18	—	154.97	279.14
6.01.03.06.E	0.35m2 - 0.50m2	m²	3.50	131.71	—	143.42	275.13	144.88	—	157.76	302.64
6.01.03.06.F	0.25m2 - 0.35m2	m²	4.00	150.52	—	146.68	297.20	165.57	—	161.35	326.92
6.01.03.06.G	not exceeding 0.25m2	m²	4.50	169.34	—	151.04	320.38	186.27	—	166.15	352.42
6.01.03.07.	Hermetically sealed double glazing unit with non-setting compound and clipped beads; 4+10+4mm; 4mm clear/Planitherm toughened										
6.01.03.07.B	1.00m2 - 2.00m2	m²	2.00	75.26	—	100.74	176.00	82.79	—	110.82	193.60
6.01.03.07.C	0.75m2 - 1.00m2	m²	2.50	94.08	—	102.65	196.72	103.48	—	112.91	216.40
6.01.03.07.D	0.50m2 - 0.75m2	m²	3.00	112.89	—	103.92	216.81	124.18	—	114.31	238.49
6.01.03.07.E	0.35m2 - 0.50m2	m²	3.50	131.71	—	106.46	238.17	144.88	—	117.11	261.98
6.01.03.07.F	0.25m2 - 0.35m2	m²	4.00	150.52	—	109.72	260.24	165.57	—	120.70	286.27
6.01.03.07.G	not exceeding 0.25m2	m²	4.50	169.34	—	114.08	283.42	186.27	—	125.49	311.76

Hutchins Priced Schedules 2021		Unit	Labour Hours	Labour Net £	Plant Net £	Materials Net £	Unit Rate Net £	Labour Gross £	Plant Gross £	Materials Gross £	Unit Rate Gross £
									(Gross rates include 10% profit)		
6.01.	**NEW WORK**										
6.01.03.	**DOUBLE GLAZING UNITS**										
6.01.03.08.	**Hermetically sealed double glazing unit with non-setting compound and clipped beads; 4+10+4mm; 4mm clear/satin toughened**										
6.01.03.08.B	1.00m2 - 2.00m2	m²	2.00	75.26	—	146.94	222.20	82.79	—	161.64	244.42
6.01.03.08.C	0.75m2 - 1.00m2	m²	2.50	94.08	—	148.85	242.92	103.48	—	163.73	267.22
6.01.03.08.D	0.50m2 - 0.75m2	m²	3.00	112.89	—	150.12	263.01	124.18	—	165.13	289.31
6.01.03.08.E	0.35m2 - 0.50m2	m²	3.50	131.71	—	152.66	284.37	144.88	—	167.93	312.80
6.01.03.08.F	0.25m2 - 0.35m2	m²	4.00	150.52	—	155.92	306.44	165.57	—	171.52	337.09
6.01.03.08.G	not exceeding 0.25m2	m²	4.50	169.34	—	160.28	329.62	186.27	—	176.31	362.58
6.01.03.09.	**Hermetically sealed double glazing unit with non-setting compound and clipped beads; 4+10+4mm; 4mm clear/toughened pattern**										
6.01.03.09.B	1.00m2 - 2.00m2	m²	2.00	75.26	—	98.43	173.69	82.79	—	108.28	191.06
6.01.03.09.C	0.75m2 - 1.00m2	m²	2.50	94.08	—	100.34	194.41	103.48	—	110.37	213.85
6.01.03.09.D	0.50m2 - 0.75m2	m²	3.00	112.89	—	101.61	214.50	124.18	—	111.77	235.95
6.01.03.09.E	0.35m2 - 0.50m2	m²	3.50	131.71	—	104.15	235.86	144.88	—	114.57	259.44
6.01.03.09.F	0.25m2 - 0.35m2	m²	4.00	150.52	—	107.41	257.93	165.57	—	118.15	283.73
6.01.03.09.G	not exceeding 0.25m2	m²	4.50	169.34	—	111.77	281.11	186.27	—	122.95	309.22
6.01.03.10.	**Hermetically sealed double glazing unit with non-setting compound and clipped beads; 4+10+4mm; 4mm K/anti-sun toughened**										
6.01.03.10.B	1.00m2 - 2.00m2	m²	2.00	75.26	—	130.77	206.03	82.79	—	143.85	226.64
6.01.03.10.C	0.75m2 - 1.00m2	m²	2.50	94.08	—	132.68	226.75	103.48	—	145.95	249.43
6.01.03.10.D	0.50m2 - 0.75m2	m²	3.00	112.89	—	133.95	246.84	124.18	—	147.34	271.52
6.01.03.10.E	0.35m2 - 0.50m2	m²	3.50	131.71	—	136.49	268.20	144.88	—	150.14	295.02
6.01.03.10.F	0.25m2 - 0.35m2	m²	4.00	150.52	—	139.75	290.27	165.57	—	153.73	319.30
6.01.03.10.G	not exceeding 0.25m2	m²	4.50	169.34	—	144.11	313.45	186.27	—	158.52	344.79
6.01.03.11.	**Hermetically sealed double glazing unit with non-setting compound and clipped beads; 4+10+4mm; 4mm Planitherm/pattern**										
6.01.03.11.B	1.00m2 - 2.00m2	m²	2.00	75.26	—	84.57	159.83	82.79	—	93.03	175.82
6.01.03.11.C	0.75m2 - 1.00m2	m²	2.50	94.08	—	86.48	180.55	103.48	—	95.13	198.61
6.01.03.11.D	0.50m2 - 0.75m2	m²	3.00	112.89	—	87.75	200.64	124.18	—	96.52	220.70
6.01.03.11.E	0.35m2 - 0.50m2	m²	3.50	131.71	—	90.29	222.00	144.88	—	99.32	244.20
6.01.03.11.F	0.25m2 - 0.35m2	m²	4.00	150.52	—	93.55	244.07	165.57	—	102.91	268.48
6.01.03.11.G	not exceeding 0.25m2	m²	4.50	169.34	—	97.91	267.25	186.27	—	107.70	293.97

Glazing, Painting and Decorating

Hutchins Priced Schedules 2021	Unit	Labour Hours	Labour Net £	Plant Net £	Materials Net £	Unit Rate Net £	Labour Gross £	Plant Gross £	Materials Gross £	Unit Rate Gross £
							(Gross rates include 10% profit)			
6.01.	**NEW WORK**									
6.01.03.	**DOUBLE GLAZING UNITS**									
6.01.03.12.	Hermetically sealed double glazing unit with non-setting compound and clipped beads; 4+10+4mm; 4mm Planitherm/toughened pattern									
6.01.03.12.B	1.00m2 - 2.00m2	m²	2.00	75.26	—	123.84	199.10	82.79	—	136.23	219.01
6.01.03.12.C	0.75m2 - 1.00m2	m²	2.50	94.08	—	125.75	219.82	103.48	—	138.32	241.81
6.01.03.12.D	0.50m2 - 0.75m2	m²	3.00	112.89	—	127.02	239.91	124.18	—	139.72	263.90
6.01.03.12.E	0.35m2 - 0.50m2	m²	3.50	131.71	—	129.56	261.27	144.88	—	142.52	287.39
6.01.03.12.F	0.25m2 - 0.35m2	m²	4.00	150.52	—	132.82	283.34	165.57	—	146.11	311.68
6.01.03.12.G	not exceeding 0.25m2	m²	4.50	169.34	—	137.18	306.52	186.27	—	150.90	337.17
6.01.03.13.	Hermetically sealed double glazing unit with non-setting compound and clipped beads; 4+10+4mm; 4mm Planitherm/Pilkington Activ Blue toughened									
6.01.03.13.B	1.00m2 - 2.00m2	m²	2.00	75.26	—	151.56	226.82	82.79	—	166.72	249.51
6.01.03.13.C	0.75m2 - 1.00m2	m²	2.50	94.08	—	153.47	247.54	103.48	—	168.81	272.30
6.01.03.13.D	0.50m2 - 0.75m2	m²	3.00	112.89	—	154.74	267.63	124.18	—	170.21	294.39
6.01.03.13.E	0.35m2 - 0.50m2	m²	3.50	131.71	—	157.28	288.99	144.88	—	173.01	317.88
6.01.03.13.F	0.25m2 - 0.35m2	m²	4.00	150.52	—	160.54	311.06	165.57	—	176.60	342.17
6.01.03.13.G	not exceeding 0.25m2	m²	4.50	169.34	—	164.90	334.24	186.27	—	181.39	367.66
6.01.03.14.	Hermetically sealed double glazing unit with non-setting compound and clipped beads; 4+10+4mm; 4mm Planitherm/satin									
6.01.03.14.B	1.00m2 - 2.00m2	m²	2.00	75.26	—	118.07	193.33	82.79	—	129.87	212.66
6.01.03.14.C	0.75m2 - 1.00m2	m²	2.50	94.08	—	119.97	214.05	103.48	—	131.97	235.45
6.01.03.14.D	0.50m2 - 0.75m2	m²	3.00	112.89	—	121.24	234.13	124.18	—	133.37	257.55
6.01.03.14.E	0.35m2 - 0.50m2	m²	3.50	131.71	—	123.79	255.49	144.88	—	136.16	281.04
6.01.03.14.F	0.25m2 - 0.35m2	m²	4.00	150.52	—	127.05	277.57	165.57	—	139.75	305.33
6.01.03.14.G	not exceeding 0.25m2	m²	4.50	169.34	—	131.41	300.74	186.27	—	144.55	330.82
6.01.03.15.	Hermetically sealed double glazing unit with non-setting compound and clipped beads; 4+10+4mm; 4mm Planitherm/satin toughened									
6.01.03.15.B	1.00m2 - 2.00m2	m²	2.00	75.26	—	177.09	252.35	82.79	—	194.80	277.58
6.01.03.15.C	0.75m2 - 1.00m2	m²	2.50	94.08	—	178.99	273.07	103.48	—	196.89	300.38
6.01.03.15.D	0.50m2 - 0.75m2	m²	3.00	112.89	—	180.26	293.15	124.18	—	198.29	322.47
6.01.03.15.E	0.35m2 - 0.50m2	m²	3.50	131.71	—	182.81	314.51	144.88	—	201.09	345.96
6.01.03.15.F	0.25m2 - 0.35m2	m²	4.00	150.52	—	186.07	336.59	165.57	—	204.68	370.25

Hutchins Priced Schedules 2021	Unit	Labour Hours	Labour Net £	Plant Net £	Materials Net £	Unit Rate Net £	Labour Gross £	Plant Gross £	Materials Gross £	Unit Rate Gross £
						(Gross rates include 10% profit)				
6.01.	**NEW WORK**									
6.01.03.	**DOUBLE GLAZING UNITS**									
6.01.03.15.	**Hermetically sealed double glazing unit with non-setting compound and clipped beads; 4+10+4mm; 4mm Planitherm/satin toughened**									
6.01.03.15.G	not exceeding 0.25m2	m²	4.50	169.34	—	190.43	359.76	186.27	—	209.47	395.74
6.01.06.	**PAINTING AND DECORATING INTERNALLY**									
6.01.06.01.	**Limewhite**									
6.01.06.01.A	one coat brick walls	m²	0.15	5.64	—	0.13	5.77	6.21	—	0.14	6.35
6.01.06.01.B	two coat brick walls	m²	0.25	9.41	—	0.23	9.64	10.35	—	0.26	10.61
6.01.06.01.C	two coat plaster walls	m²	0.20	7.53	—	0.05	7.57	8.28	—	0.05	8.33
6.01.06.01.D	two coat plaster ceilings	m²	0.22	8.28	—	0.05	8.33	9.11	—	0.05	9.16
6.01.06.02.	**Primer; two coats matt or silk emulsion paint**									
6.01.06.02.A	concrete walls	m²	0.41	15.43	—	1.05	16.48	16.97	—	1.16	18.13
6.01.06.02.B	concrete walls to stairwell	m²	0.50	18.82	—	1.05	19.87	20.70	—	1.16	21.86
6.01.06.02.C	concrete ceilings	m²	0.53	19.94	—	1.05	21.00	21.94	—	1.16	23.10
6.01.06.02.D	concrete ceilings to stairwell	m²	0.59	22.20	—	1.05	23.26	24.42	—	1.16	25.58
6.01.06.02.E	brick walls	m²	0.50	18.82	—	1.35	20.17	20.70	—	1.49	22.19
6.01.06.02.F	brick walls to stairwell	m²	0.56	21.07	—	1.35	22.43	23.18	—	1.49	24.67
6.01.06.02.G	block walls	m²	0.62	23.33	—	1.35	24.68	25.66	—	1.49	27.15
6.01.06.02.H	block walls to stairwell	m²	0.68	25.59	—	1.35	26.94	28.15	—	1.49	29.64
6.01.06.02.I	plastered walls	m²	0.38	14.30	—	0.91	15.21	15.73	—	1.00	16.73
6.01.06.02.J	plastered walls to stairwell	m²	0.44	16.56	—	0.91	17.47	18.21	—	1.00	19.22
6.01.06.02.K	plastered ceilings	m²	0.50	18.82	—	0.91	19.73	20.70	—	1.00	21.70
6.01.06.02.L	plastered ceilings to stairwell	m²	0.53	19.94	—	0.91	20.86	21.94	—	1.00	22.94
6.01.06.02.M	embossed papered or textured walls . . .	m²	0.41	15.43	—	1.05	16.48	16.97	—	1.16	18.13
6.01.06.02.N	embossed papered or textured walls to stairwell	m²	0.50	18.82	—	1.05	19.87	20.70	—	1.16	21.86
6.01.06.02.O	embossed papered or textured ceilings . .	m²	0.53	19.94	—	1.05	21.00	21.94	—	1.16	23.10
6.01.06.02.P	embossed papered or textured ceilings to stairwell	m²	0.59	22.20	—	1.05	23.26	24.42	—	1.16	25.58
6.01.06.03.	**Primer; two coats eggshell paint**									
6.01.06.03.A	concrete walls	m²	0.46	17.31	—	2.88	20.19	19.04	—	3.17	22.21
6.01.06.03.B	concrete walls to stairwell	m²	0.55	20.70	—	2.88	23.58	22.77	—	3.17	25.93
6.01.06.03.C	concrete ceilings	m²	0.58	21.83	—	2.88	24.70	24.01	—	3.17	27.17
6.01.06.03.D	concrete ceilings to stairwell	m²	0.64	24.08	—	2.88	26.96	26.49	—	3.17	29.66
6.01.06.03.E	brick walls	m²	0.54	20.32	—	3.31	23.63	22.35	—	3.64	25.99
6.01.06.03.F	brick walls to stairwell	m²	0.60	22.58	—	3.31	25.88	24.84	—	3.64	28.47
6.01.06.03.G	block walls	m²	0.66	24.84	—	4.12	28.96	27.32	—	4.54	31.86
6.01.06.03.H	block walls to stairwell	m²	0.72	27.09	—	4.12	31.22	29.80	—	4.54	34.34
6.01.06.03.I	plastered walls	m²	0.43	16.18	—	2.49	18.67	17.80	—	2.74	20.54

Glazing, Painting and Decorating

Hutchins Priced Schedules 2021	Unit	Labour Hours	Labour Net £	Plant Net £	Materials Net £	Unit Rate Net £	Labour Gross £	Plant Gross £	Materials Gross £	Unit Rate Gross £
							(Gross rates include 10% profit)			
6.01.	**NEW WORK**									
6.01.06.	**PAINTING AND DECORATING INTERNALLY**									
6.01.06.03.	**Primer; two coats eggshell paint**									
6.01.06.03.J	plastered walls to stairwell	m²	0.50	18.82	—	2.49	21.30	20.70	—	2.74	23.44
6.01.06.03.K	plastered ceilings	m²	0.55	20.70	—	2.49	23.19	22.77	—	2.74	25.50
6.01.06.03.L	plastered ceilings to stairwell	m²	0.59	22.20	—	2.49	24.69	24.42	—	2.74	27.16
6.01.06.03.M	embossed papered or textured walls . . .	m²	0.47	17.69	—	2.88	20.56	19.45	—	3.17	22.62
6.01.06.03.N	embossed papered or textured walls to stairwell	m²	0.56	21.07	—	2.88	23.95	23.18	—	3.17	26.35
6.01.06.03.O	embossed papered or textured ceiling . .	m²	0.59	22.20	—	2.88	25.08	24.42	—	3.17	27.59
6.01.06.03.P	embossed papered or textured ceilings to stairwell	m²	0.66	24.84	—	2.88	27.71	27.32	—	3.17	30.49
6.01.06.04.	**Primer; one undercoat; one coat gloss finishing paint**									
6.01.06.04.A	concrete walls	m²	0.46	17.31	—	1.89	19.20	19.04	—	2.08	21.12
6.01.06.04.B	concrete walls to stairwell	m²	0.55	20.70	—	1.89	22.58	22.77	—	2.08	24.84
6.01.06.04.C	concrete ceilings	m²	0.58	21.83	—	1.89	23.71	24.01	—	2.08	26.08
6.01.06.04.D	concrete ceilings to stairwell	m²	0.64	24.08	—	1.89	25.97	26.49	—	2.08	28.57
6.01.06.04.E	brick walls	m²	0.54	20.32	—	2.45	22.77	22.35	—	2.69	25.04
6.01.06.04.F	brick walls to stairwell	m²	0.60	22.58	—	2.45	25.02	24.84	—	2.69	27.53
6.01.06.04.G	block walls	m²	0.66	24.84	—	2.88	27.71	27.32	—	3.16	30.48
6.01.06.04.H	block walls to stairwell	m²	0.72	27.09	—	2.88	29.97	29.80	—	3.16	32.97
6.01.06.04.I	plastered walls	m²	0.43	16.18	—	1.89	18.07	17.80	—	2.08	19.87
6.01.06.04.J	plastered walls to stairwell	m²	0.50	18.82	—	1.89	20.70	20.70	—	2.08	22.77
6.01.06.04.K	plastered ceilings	m²	0.55	20.70	—	1.89	22.58	22.77	—	2.08	24.84
6.01.06.04.L	plastered ceilings to stairwell	m²	0.59	22.20	—	1.89	24.09	24.42	—	2.08	26.50
6.01.06.04.M	embossed papered or textured walls . . .	m²	0.47	17.69	—	1.89	19.57	19.45	—	2.08	21.53
6.01.06.04.N	embossed papered or textured walls to stairwell	m²	0.56	21.07	—	1.89	22.96	23.18	—	2.08	25.26
6.01.06.04.O	embossed papered or textured ceilings . .	m²	0.59	22.20	—	1.89	24.09	24.42	—	2.08	26.50
6.01.06.04.P	embossed papered or textured ceilings to stairwell	m²	0.66	24.84	—	1.89	26.72	27.32	—	2.08	29.39
6.01.06.04.Q	dado line 25mm wide including cutting in both edges	m	0.30	11.29	—	0.09	11.38	12.42	—	0.10	12.52
6.01.06.05.	**Primer; two undercoats; one coat gloss finishing paint**									
6.01.06.05.A	concrete walls	m²	0.58	21.83	—	3.22	25.05	24.01	—	3.55	27.55
6.01.06.05.B	concrete walls to stairwell	m²	0.70	26.34	—	3.22	29.57	28.98	—	3.55	32.52
6.01.06.05.C	concrete ceilings	m²	0.74	27.85	—	3.22	31.07	30.63	—	3.55	34.18
6.01.06.05.D	concrete ceilings to stairwell	m²	0.82	30.86	—	3.22	34.08	33.94	—	3.55	37.49
6.01.06.05.E	brick walls	m²	0.69	25.96	—	3.04	29.01	28.56	—	3.35	31.91
6.01.06.05.F	brick walls to stairwell	m²	0.77	28.98	—	3.04	32.02	31.87	—	3.35	35.22
6.01.06.05.G	block walls	m²	0.85	31.99	—	3.78	35.77	35.18	—	4.16	39.35

Glazing, Painting and Decorating

Hutchins Priced Schedules 2021		Unit	Labour Hours	Labour Net £	Plant Net £	Materials Net £	Unit Rate Net £	Labour Gross £	Plant Gross £	Materials Gross £	Unit Rate Gross £
								(Gross rates include 10% profit)			
6.01.	**NEW WORK**										
6.01.06.	**PAINTING AND DECORATING INTERNALLY**										
6.01.06.05.	**Primer; two undercoats; one coat gloss finishing paint**										
6.01.06.05.H	block walls to stairwell	m²	0.93	35.00	—	3.78	38.78	38.50	—	4.16	42.66
6.01.06.05.I	plastered walls	m²	0.54	20.32	—	2.20	22.52	22.35	—	2.42	24.77
6.01.06.05.J	plastered walls to stairwell	m²	0.63	23.71	—	2.20	25.91	26.08	—	2.42	28.50
6.01.06.05.K	plastered ceilings	m²	0.70	26.34	—	2.20	28.54	28.98	—	2.42	31.39
6.01.06.05.L	plastered ceilings to stairwell	m²	0.75	28.22	—	2.20	30.42	31.04	—	2.42	33.46
6.01.06.05.M	embossed papered or textured walls . . .	m²	0.59	22.20	—	2.48	24.68	24.42	—	2.73	27.15
6.01.06.05.N	embossed papered or textured walls to stairwell	m²	0.71	26.72	—	2.48	29.20	29.39	—	2.73	32.12
6.01.06.05.O	embossed papered or textured ceilings . .	m²	0.75	28.22	—	2.48	30.71	31.04	—	2.73	33.78
6.01.06.05.P	embossed papered or textured ceilings to stairwell	m²	0.84	31.61	—	2.48	34.09	34.77	—	2.73	37.50
6.01.06.05.Q	dado line 25mm wide included cutting in both edges	m	0.40	15.05	—	0.12	15.17	16.56	—	0.13	16.69
6.01.06.06.	**Knot, stop, prime, one undercoat, one gloss finishing coat gloss oil paint on woodwork**										
6.01.06.06.A	general surfaces; not exceeding 150mm girth	m	0.32	12.04	—	0.22	12.26	13.25	—	0.24	13.48
6.01.06.06.B	general surfaces; 150mm - 300mm girth .	m	0.41	15.43	—	0.48	15.91	16.97	—	0.53	17.50
6.01.06.06.C	general surfaces; over 300mm girth	m²	0.75	28.22	—	1.71	29.93	31.04	—	1.88	32.92
6.01.06.06.G	windows, glazed doors, screens in small panes	m²	1.66	62.47	—	1.40	63.87	68.71	—	1.54	70.25
6.01.06.06.H	windows, glazed doors, screens in medium panes	m²	1.45	54.56	—	1.22	55.79	60.02	—	1.35	61.37
6.01.06.06.I	windows, glazed doors, screens in large panes	m²	1.23	46.28	—	1.10	47.38	50.91	—	1.21	52.12
6.01.06.06.J	windows, glazed doors, screens in extra large panes	m²	1.01	38.01	—	0.87	38.87	41.81	—	0.95	42.76
6.01.06.06.K	frames and sashes (measure over glass)	m²	1.11	41.77	—	1.10	42.87	45.95	—	1.21	47.15
6.01.06.06.L	open balustrade to staircase (measured flat both sides overall)	m²	0.78	29.35	—	1.53	30.88	32.29	—	1.68	33.97
6.01.06.08.	**Knot, stop, prime, two undercoats, one finishing coat gloss oil paint on woodwork**										
6.01.06.08.A	general surfaces; not exceeding 150mm girth	m	0.41	15.43	—	0.30	15.73	16.97	—	0.33	17.30
6.01.06.08.B	general surfaces; 150mm - 300mm girth .	m	0.52	19.57	—	0.65	20.22	21.52	—	0.72	22.24
6.01.06.08.C	general surfaces; over 300mm girth	m²	0.97	36.50	—	2.19	38.69	40.15	—	2.41	42.56
6.01.06.08.G	windows, glazed doors, screens in small panes	m²	2.16	81.28	—	1.80	83.08	89.41	—	1.98	91.39
6.01.06.08.H	windows, glazed doors, screens in medium panes	m²	1.88	70.74	—	1.54	72.28	77.82	—	1.69	79.51

Glazing, Painting and Decorating

Hutchins Priced Schedules 2021		Unit	Labour Hours	Labour Net £	Plant Net £	Materials Net £	Unit Rate Net £	Labour Gross £	Plant Gross £	Materials Gross £	Unit Rate Gross £
									(Gross rates include 10% profit)		
6.01.	**NEW WORK**										
6.01.06.	PAINTING AND DECORATING INTERNALLY										
6.01.06.08.	Knot, stop, prime, two undercoats, one finishing coat gloss oil paint on woodwork										
6.01.06.08.I	windows, glazed doors, screens in large panes	m²	1.60	60.21	—	1.35	61.56	66.23	—	1.49	67.72
6.01.06.08.J	windows, glazed doors, screens in extra large panes	m²	1.31	49.30	—	1.12	50.42	54.22	—	1.23	55.46
6.01.06.08.K	frames and sashes (measure over glass)	m²	1.44	54.19	—	1.35	55.54	59.61	—	1.49	61.09
6.01.06.08.L	open balustrade to staircase (measured flat both sides overall)	m²	1.02	38.38	—	1.98	40.37	42.22	—	2.18	44.40
6.01.06.09.	**Primer, one undercoat, one finishing coat gloss oil paint on metalwork**										
6.01.06.09.A	general surfaces; not exceeding 150mm girth	m	0.30	11.29	—	0.22	11.51	12.42	—	0.24	12.66
6.01.06.09.B	general surfaces; 150mm - 300mm girth	m	0.39	14.68	—	0.49	15.16	16.14	—	0.53	16.68
6.01.06.09.C	general surfaces; over 300mm girth	m²	0.69	25.96	—	1.71	27.68	28.56	—	1.88	30.44
6.01.06.09.D	windows, glazed doors, screens in small panes	m²	1.66	62.47	—	1.40	63.87	68.71	—	1.54	70.26
6.01.06.09.E	windows, glazed doors, screens in medium panes	m²	1.45	54.56	—	1.23	55.79	60.02	—	1.35	61.37
6.01.06.09.F	windows, glazed doors, screens in large panes	m²	1.23	46.28	—	1.10	47.38	50.91	—	1.21	52.12
6.01.06.09.G	windows, glazed doors, screens in extra large panes	m²	1.01	38.01	—	0.87	38.87	41.81	—	0.96	42.76
6.01.06.09.H	corrugated surfaces; over 300mm girth	m²	0.75	28.22	—	1.93	30.15	31.04	—	2.12	33.17
6.01.06.09.I	structural steelwork; over 300mm girth	m²	0.84	31.61	—	1.71	33.32	34.77	—	1.88	36.65
6.01.06.09.J	radiators over; 300mm girth	m²	0.69	25.96	—	1.71	27.68	28.56	—	1.88	30.44
6.01.06.09.K	pipes, ducts, etc.; not exceeding 150mm girth	m	0.25	9.41	—	0.22	9.62	10.35	—	0.24	10.59
6.01.06.09.L	pipes, ducts, etc.; 150mm - 300mm girth	m	0.35	13.17	—	0.49	13.66	14.49	—	0.53	15.02
6.01.06.09.M	pipes, ducts, etc.; over 300mm girth	m²	0.68	25.59	—	1.71	27.30	28.15	—	1.88	30.03
6.01.06.09.O	sundry fittings - casement stays, etc.	nr	0.15	5.64	—	0.13	5.77	6.21	—	0.14	6.35
6.01.06.10.	**Primer, two undercoats, one finishing coat gloss oil paint on metalwork**										
6.01.06.10.A	general surfaces; not exceeding 150mm girth	m	0.38	14.30	—	0.30	14.60	15.73	—	0.33	16.06
6.01.06.10.B	general surfaces; 150mm - 300mm girth	m	0.50	18.82	—	0.66	19.47	20.70	—	0.72	21.42
6.01.06.10.C	general surfaces; over 300mm girth	m²	0.90	33.87	—	2.19	36.06	37.25	—	2.41	39.67
6.01.06.10.D	windows, glazed doors, screens in small panes	m²	2.16	81.28	—	1.80	83.08	89.41	—	1.98	91.39
6.01.06.10.E	windows, glazed doors, screens in medium panes	m²	1.88	70.74	—	1.54	72.28	77.82	—	1.69	79.51
6.01.06.10.F	windows, glazed doors, screens in large panes	m²	1.60	60.21	—	1.35	61.56	66.23	—	1.49	67.72

Hutchins Priced Schedules 2021	Unit	Labour Hours	Labour Net £	Plant Net £	Materials Net £	Unit Rate Net £	Labour Gross £	Plant Gross £	Materials Gross £	Unit Rate Gross £
						(Gross rates include 10% profit)				
6.01.	**NEW WORK**									
6.01.06.	**PAINTING AND DECORATING INTERNALLY**									
6.01.06.10.	**Primer, two undercoats, one finishing coat gloss oil paint on metalwork**									
6.01.06.10.G	windows, glazed doors, screens in extra large panes	m²	1.31	49.30	—	1.12	50.42	54.22	—	1.24	55.46
6.01.06.10.H	corrugated surfaces; over 300mm girth . .	m²	0.98	36.88	—	2.47	39.34	40.57	—	2.71	43.28
6.01.06.10.I	structural steelwork; over 300mm girth . .	m²	1.10	41.39	—	2.19	43.59	45.53	—	2.41	47.95
6.01.06.10.J	radiators over; 300mm girth	m²	0.90	33.87	—	2.19	36.06	37.25	—	2.41	39.67
6.01.06.10.K	pipes, ducts, etc.; not exceeding 150mm girth	m	0.33	12.42	—	0.30	12.72	13.66	—	0.33	13.99
6.01.06.10.L	pipes, ducts, etc.; 150mm - 300mm girth .	m	0.46	17.31	—	0.66	17.97	19.04	—	0.72	19.76
6.01.06.10.M	pipes, ducts, etc.; over 300mm girth	m²	0.89	33.49	—	2.19	35.68	36.84	—	2.41	39.25
6.01.06.10.O	sundry fittings - casement stays, etc.	nr	0.20	7.53	—	0.18	7.71	8.28	—	0.20	8.48
6.01.06.20.	**One coat Cuprinol clear preserver on wrought timber**									
6.01.06.20.A	general surfaces; not exceeding 150mm girth	m	0.09	3.39	—	0.07	3.46	3.73	—	0.08	3.81
6.01.06.20.B	general surfaces; 150mm - 300mm girth .	m	0.14	5.27	—	0.18	5.45	5.80	—	0.20	6.00
6.01.06.20.C	general surfaces; over 300mm girth	m²	0.19	7.15	—	0.59	7.74	7.86	—	0.64	8.51
6.01.06.21.	**Two coats Cuprinol clear preserver on wrought timber**									
6.01.06.21.A	general surfaces; not exceeding 150mm girth	m	0.18	6.77	—	0.18	6.96	7.45	—	0.20	7.65
6.01.06.21.B	general surfaces; 150mm - 300mm girth .	m	0.25	9.41	—	0.37	9.77	10.35	—	0.40	10.75
6.01.06.21.C	general surfaces; over 300mm girth	m²	0.37	13.92	—	1.21	15.13	15.32	—	1.33	16.64
6.01.06.22.	**One coat Cuprinol oak preserver on wrought timber**									
6.01.06.22.A	general surfaces; not exceeding 150mm girth	m	0.08	3.01	—	0.06	3.07	3.31	—	0.07	3.38
6.01.06.22.B	general surfaces; 150mm - 300mm girth .	m	0.12	4.52	—	0.16	4.67	4.97	—	0.17	5.14
6.01.06.22.C	general surfaces; over 300mm girth	m²	0.18	6.77	—	0.53	7.30	7.45	—	0.58	8.03
6.01.06.23.	**Two coats Cuprinol oak preserver wrought timber**									
6.01.06.23.A	general surfaces; not exceeding 150mm girth	m	0.16	6.02	—	0.16	6.18	6.62	—	0.17	6.79
6.01.06.23.B	general surfaces; 150mm - 300mm girth .	m	0.23	8.65	—	0.31	8.97	9.52	—	0.34	9.86
6.01.06.23.C	general surfaces; over 300mm girth	m²	0.36	13.55	—	1.03	14.58	14.90	—	1.13	16.03
6.01.06.24.	**Two coats clear polyurethane on woodwork**									
6.01.06.24.A	general surfaces; not exceeding 150mm girth	m	0.16	6.02	—	0.31	6.33	6.62	—	0.34	6.96
6.01.06.24.B	general surfaces; 150mm - 300mm girth .	m	0.24	9.03	—	0.62	9.65	9.93	—	0.68	10.62

Glazing, Painting and Decorating

Hutchins Priced Schedules 2021		Unit	Labour Hours	Labour Net £	Plant Net £	Materials Net £	Unit Rate Net £	Labour Gross £	Plant Gross £	Materials Gross £	Unit Rate Gross £
							(Gross rates include 10% profit)				
6.01.	**NEW WORK**										
6.01.06.	**PAINTING AND DECORATING INTERNALLY**										
6.01.06.24.	**Two coats clear polyurethane on woodwork**										
6.01.06.24.C	general surfaces; over 300mm girth	m²	0.36	13.55	—	2.05	15.60	14.90	—	2.26	17.16
6.01.06.24.G	windows, glazed doors, screens in small panes......................	m²	0.80	30.10	—	1.63	31.73	33.11	—	1.79	34.90
6.01.06.24.H	windows, glazed doors, screens in medium panes................	m²	0.72	27.09	—	1.43	28.53	29.80	—	1.58	31.38
6.01.06.24.I	windows, glazed doors, screens in large panes......................	m²	0.64	24.08	—	1.24	25.32	26.49	—	1.36	27.86
6.01.06.24.J	windows, glazed doors, screens in extra large panes..................	m²	0.56	21.07	—	1.05	22.12	23.18	—	1.15	24.33
6.01.06.25.	**Three coats clear polyurethane on woodwork**										
6.01.06.25.A	general surfaces; not exceeding 150mm girth......................	m	0.26	9.78	—	0.47	10.25	10.76	—	0.51	11.27
6.01.06.25.B	general surfaces; 150mm - 300mm girth .	m	0.32	12.04	—	0.93	12.97	13.25	—	1.02	14.27
6.01.06.25.C	general surfaces; over 300mm girth	m²	0.54	20.32	—	3.10	23.42	22.35	—	3.41	25.76
6.01.06.25.G	windows, glazed doors, screens in small panes......................	m²	1.10	41.39	—	2.48	43.87	45.53	—	2.73	48.26
6.01.06.25.H	windows, glazed doors, screens in medium panes................	m²	1.02	38.38	—	2.17	40.55	42.22	—	2.39	44.61
6.01.06.25.I	windows, glazed doors, screens in large panes......................	m²	0.94	35.37	—	1.86	37.23	38.91	—	2.05	40.96
6.01.06.25.J	windows, glazed doors, screens in extra large panes..................	m²	0.86	32.36	—	1.55	33.91	35.60	—	1.71	37.30
6.01.06.26.	**Two coats coloured polyurethane on woodwork**										
6.01.06.26.A	general surfaces; not exceeding 150mm girth......................	m	0.16	6.02	—	0.64	6.66	6.62	—	0.71	7.33
6.01.06.26.B	general surfaces; 150mm - 300mm girth .	m	0.24	9.03	—	1.28	10.32	9.93	—	1.41	11.35
6.01.06.26.C	general surfaces; over 300mm girth	m²	0.36	13.55	—	4.25	17.80	14.90	—	4.68	19.58
6.01.06.26.G	windows, glazed doors, screens in small panes......................	m²	0.80	30.10	—	3.37	33.47	33.11	—	3.71	36.82
6.01.06.26.H	windows, glazed doors, screens in medium panes................	m²	0.72	27.09	—	2.97	30.06	29.80	—	3.27	33.07
6.01.06.26.I	windows, glazed doors, screens in large panes......................	m²	0.64	24.08	—	2.57	26.65	26.49	—	2.82	29.32
6.01.06.26.J	windows, glazed doors, screens in extra large panes..................	m²	0.56	21.07	—	2.17	23.24	23.18	—	2.38	25.56
6.01.06.27.	**Three coats coloured polyurethane on woodwork**										
6.01.06.27.A	general surfaces; not exceeding 150mm girth......................	m	0.26	9.78	—	0.96	10.75	10.76	—	1.06	11.82
6.01.06.27.B	general surfaces; 150mm - 300mm girth .	m	0.32	12.04	—	1.93	13.97	13.25	—	2.12	15.36

Hutchins Priced Schedules 2021		Unit	Labour Hours	Labour Net £	Plant Net £	Materials Net £	Unit Rate Net £	Labour Gross £	Plant Gross £	Materials Gross £	Unit Rate Gross £
								(Gross rates include 10% profit)			
6.01.	**NEW WORK**										
6.01.06.	**PAINTING AND DECORATING INTERNALLY**										
6.01.06.27.	**Three coats coloured polyurethane on woodwork**										
6.01.06.27.C	general surfaces; over 300mm girth	m²	0.54	20.32	—	6.42	26.74	22.35	—	7.06	29.41
6.01.06.27.G	windows, glazed doors, screens in small panes	m²	1.10	41.39	—	5.14	46.53	45.53	—	5.65	51.18
6.01.06.27.H	windows, glazed doors, screens in medium panes	m²	1.02	38.38	—	4.49	42.88	42.22	—	4.94	47.16
6.01.06.27.I	windows, glazed doors, screens in large panes	m²	0.94	35.37	—	3.85	39.22	38.91	—	4.24	43.15
6.01.06.27.J	windows, glazed doors, screens in extra large panes	m²	0.86	32.36	—	3.21	35.57	35.60	—	3.53	39.13
6.01.06.28.	**Two coats raw or boiled linseed oil on woodwork**										
6.01.06.28.A	general surfaces; not exceeding 150mm girth	m	0.12	4.52	—	0.19	4.71	4.97	—	0.21	5.18
6.01.06.28.B	general surfaces; 150mm - 300mm girth .	m	0.16	6.02	—	0.35	6.37	6.62	—	0.38	7.01
6.01.06.28.C	general surfaces; over 300mm girth	m²	0.30	11.29	—	1.20	12.48	12.42	—	1.32	13.73
6.01.06.30.	**Seal and wax polish woodwork**										
6.01.06.30.A	general surfaces; not exceeding 150mm girth	m	0.09	3.39	—	0.37	3.75	3.73	—	0.40	4.13
6.01.06.30.B	general surfaces; 150mm - 300mm girth .	m	0.13	4.89	—	0.77	5.66	5.38	—	0.85	6.23
6.01.06.30.C	general surfaces; over 300mm girth	m²	0.20	7.53	—	2.85	10.37	8.28	—	3.13	11.41
6.01.06.30.G	windows, glazed doors, screens in small panes	m²	0.60	22.58	—	2.13	24.71	24.84	—	2.34	27.18
6.01.06.30.H	windows, glazed doors, screens in medium panes	m²	0.50	18.82	—	1.74	20.56	20.70	—	1.92	22.61
6.01.06.30.I	windows, glazed doors, screens in large panes	m²	0.40	15.05	—	1.43	16.49	16.56	—	1.58	18.13
6.01.06.30.J	windows, glazed doors, screens in extra large panes	m²	0.30	11.29	—	0.85	12.14	12.42	—	0.93	13.35
6.01.06.40.	**One coat Artex sealer; one coat Artex standard compound; textured finish**										
6.01.06.40.A	concrete walls	m²	0.42	15.80	—	1.55	17.36	17.39	—	1.71	19.09
6.01.06.40.B	concrete walls to stairwell	m²	0.44	16.56	—	1.55	18.11	18.21	—	1.71	19.92
6.01.06.40.C	brick walls	m²	0.42	15.80	—	1.79	17.60	17.39	—	1.97	19.36
6.01.06.40.D	brick walls to stairwell	m²	0.44	16.56	—	1.79	18.35	18.21	—	1.97	20.18
6.01.06.40.E	block walls	m²	0.42	15.80	—	2.03	17.84	17.39	—	2.24	19.62
6.01.06.40.F	block walls to stairwell	m²	0.44	16.56	—	2.03	18.59	18.21	—	2.24	20.45
6.01.06.40.G	plastered walls	m²	0.37	13.92	—	1.55	15.47	15.32	—	1.71	17.02
6.01.06.40.H	plastered walls to stairwell	m²	0.39	14.68	—	1.55	16.23	16.14	—	1.71	17.85
6.01.06.40.I	plastered ceilings	m²	0.41	15.43	—	1.55	16.98	16.97	—	1.71	18.68
6.01.06.40.J	plastered ceilings to stairwell	m²	0.43	16.18	—	1.55	17.73	17.80	—	1.71	19.50

Glazing, Painting and Decorating

Hutchins Priced Schedules 2021	Unit	Labour Hours	Labour Net £	Plant Net £	Materials Net £	Unit Rate Net £	Labour Gross £	Plant Gross £	Materials Gross £	Unit Rate Gross £
							(Gross rates include 10% profit)			

6.01. NEW WORK

6.01.06. PAINTING AND DECORATING INTERNALLY

6.01.06.40. One coat Artex sealer; one coat Artex standard compound; textured finish

6.01.06.40.K	plasterboard walls	m²	0.37	13.92	—	1.55	15.47	15.32	—	1.71	17.02
6.01.06.40.L	plasterboard walls to stairwell	m²	0.39	14.68	—	1.55	16.23	16.14	—	1.71	17.85
6.01.06.40.M	plasterboard ceiling	m²	0.41	15.43	—	1.55	16.98	16.97	—	1.71	18.68
6.01.06.40.N	plasterboard ceiling to stairwell	m²	0.43	16.18	—	1.55	17.73	17.80	—	1.71	19.50

6.01.08. PAINTING AND DECORATING EXTERNALLY

6.01.08.02. Two coats Snowcem on walls

6.01.08.02.A	concrete	m²	0.45	16.93	—	0.82	17.76	18.63	—	0.91	19.53
6.01.08.02.B	brick	m²	0.48	18.06	—	0.82	18.89	19.87	—	0.91	20.78
6.01.08.02.C	block	m²	0.54	20.32	—	0.82	21.15	22.35	—	0.91	23.26
6.01.08.02.D	cement rendered	m²	0.42	15.80	—	0.82	16.63	17.39	—	0.91	18.29
6.01.08.02.E	rough cast	m²	0.57	21.45	—	2.40	23.85	23.59	—	2.64	26.23

6.01.08.03. Stabilising solution; two coats Sandtex matt on walls

6.01.08.03.A	concrete	m²	0.57	21.45	—	1.42	22.87	23.59	—	1.56	25.16
6.01.08.03.B	brick	m²	0.60	22.58	—	1.67	24.24	24.84	—	1.83	26.67
6.01.08.03.C	block	m²	0.66	24.84	—	2.05	26.89	27.32	—	2.25	29.57
6.01.08.03.D	cement rendered	m²	0.54	20.32	—	1.42	21.74	22.35	—	1.56	23.92
6.01.08.03.E	rough cast	m²	0.69	25.96	—	2.99	28.95	28.56	—	3.29	31.85

6.01.08.04. Two coats Weathershield masonry paint on walls

6.01.08.04.A	concrete	m²	0.50	18.82	—	0.68	19.49	20.70	—	0.74	21.44
6.01.08.04.B	brick	m²	0.61	22.95	—	0.80	23.75	25.25	—	0.88	26.13
6.01.08.04.C	block	m²	0.69	25.96	—	0.96	26.92	28.56	—	1.05	29.61
6.01.08.04.D	cement rendered	m²	0.53	19.94	—	0.80	20.74	21.94	—	0.88	22.82
6.01.08.04.E	rough cast	m²	0.72	27.09	—	1.19	28.29	29.80	—	1.31	31.11

6.01.08.10. Prime only woodwork

6.01.08.10.A	general surfaces; not exceeding 150mm girth	m	0.13	4.89	—	0.11	5.01	5.38	—	0.13	5.51
6.01.08.10.B	general surfaces; 150mm - 300mm girth .	m	0.17	6.40	—	0.23	6.63	7.04	—	0.25	7.29
6.01.08.10.C	general surfaces; over 300mm girth	m²	0.26	9.78	—	0.84	10.63	10.76	—	0.93	11.69
6.01.08.10.G	windows, glazed doors, screens in small panes	m²	0.66	24.84	—	0.69	25.53	27.32	—	0.76	28.08
6.01.08.10.H	windows, glazed doors, screens in medium panes	m²	0.58	21.83	—	0.61	22.44	24.01	—	0.67	24.68
6.01.08.10.I	windows, glazed doors, screens in large panes	m²	0.50	18.82	—	0.54	19.35	20.70	—	0.59	21.29

Hutchins Priced Schedules 2021		Unit	Labour Hours	Labour Net £	Plant Net £	Materials Net £	Unit Rate Net £	Labour Gross £	Plant Gross £	Materials Gross £	Unit Rate Gross £
								(Gross rates include 10% profit)			
6.01.	**NEW WORK**										
6.01.08.	**PAINTING AND DECORATING EXTERNALLY**										
6.01.08.10.	**Prime only woodwork**										
6.01.08.10.J	windows, glazed doors, screens in extra large panes..............	m²	0.42	15.80	—	0.46	16.26	17.39	—	0.51	17.89
6.01.08.10.K	frames and sashes (measured over glass).................	m²	0.44	16.56	—	0.54	17.09	18.21	—	0.59	18.80
6.01.08.20.	**Knot, stop, prime, one undercoat one finishing coat gloss oil paint on woodwork**										
6.01.08.20.A	general surfaces; not exceeding 150mm girth....................	m	0.39	14.68	—	0.22	14.89	16.14	—	0.24	16.38
6.01.08.20.B	general surfaces; 150mm - 300mm girth.	m	0.51	19.19	—	0.48	19.68	21.11	—	0.53	21.64
6.01.08.20.C	general surfaces; over 300mm girth....	m²	0.78	29.35	—	1.71	31.06	32.29	—	1.88	34.17
6.01.08.20.G	windows, glazed doors, screens in small panes...................	m²	1.98	74.51	—	1.40	75.91	81.96	—	1.54	83.50
6.01.08.20.H	windows, glazed doors, screens in medium panes...............	m²	1.74	65.48	—	1.22	66.70	72.02	—	1.35	73.37
6.01.08.20.I	windows, glazed doors, screens in large panes...................	m²	1.50	56.45	—	1.10	57.54	62.09	—	1.21	63.30
6.01.08.20.J	windows, glazed doors, screens in extra large panes................	m²	1.20	45.16	—	0.87	46.02	49.67	—	0.95	50.63
6.01.08.20.K	frames and sashes (measured over glass).................	m²	1.32	49.67	—	1.10	50.77	54.64	—	1.21	55.84
6.01.08.20.L	open balustrade to staircase (measured flat both sides)...............	m²	0.87	32.74	—	1.53	34.27	36.01	—	1.68	37.69
6.01.08.21.	**Knot, stop, prime, two undercoats, one finishing coat gloss oil paint on woodwork**										
6.01.08.21.A	general surfaces; not exceeding 150mm girth....................	m	0.52	19.57	—	0.30	19.87	21.52	—	0.33	21.86
6.01.08.21.B	general surfaces; 150mm - 300mm girth.	m	0.68	25.59	—	0.65	26.24	28.15	—	0.72	28.87
6.01.08.21.C	general surfaces; over 300mm girth....	m²	1.04	39.14	—	2.19	41.33	43.05	—	2.41	45.46
6.01.08.21.G	windows, glazed doors, screens in small panes...................	m²	2.64	99.34	—	1.80	101.14	109.28	—	1.98	111.26
6.01.08.21.H	windows, glazed doors, screens in medium panes...............	m²	2.32	87.30	—	1.54	88.84	96.03	—	1.69	97.72
6.01.08.21.I	windows, glazed doors, screens in large panes...................	m²	2.00	75.26	—	1.35	76.61	82.79	—	1.49	84.27
6.01.08.21.J	windows, glazed doors, screens in extra large panes................	m²	1.60	60.21	—	1.12	61.33	66.23	—	1.23	67.46
6.01.08.21.K	frames and sashes (measured over glass).................	m²	1.76	66.23	—	1.35	67.58	72.85	—	1.49	74.34
6.01.08.21.L	open balustrade to staircase (measured flat both sides)...............	m²	1.16	43.65	—	2.34	46.00	48.02	—	2.58	50.59

Glazing, Painting and Decorating

Hutchins Priced Schedules 2021		Unit	Labour Hours	Labour Net £	Plant Net £	Materials Net £	Unit Rate Net £	Labour Gross £	Plant Gross £	Materials Gross £	Unit Rate Gross £
										(Gross rates include 10% profit)	
6.01.	**NEW WORK**										
6.01.08.	PAINTING AND DECORATING EXTERNALLY										
6.01.08.30.	Prime only metalwork										
6.01.08.30.A	general surfaces; not exceeding 150mm girth	m	0.13	4.89	—	0.12	5.01	5.38	—	0.13	5.51
6.01.08.30.B	general surfaces; 150mm - 300mm girth	m	0.17	6.40	—	0.23	6.63	7.04	—	0.25	7.29
6.01.08.30.C	general surfaces; over 300mm girth	m²	0.26	9.78	—	0.85	10.63	10.76	—	0.93	11.69
6.01.08.30.D	windows, glazed doors, screens in small panes	m²	0.66	24.84	—	0.69	25.53	27.32	—	0.76	28.08
6.01.08.30.E	windows, glazed doors, screens in medium panes	m²	0.58	21.83	—	0.61	22.44	24.01	—	0.68	24.68
6.01.08.30.F	windows, glazed doors, screens in large panes	m²	0.50	18.82	—	0.54	19.35	20.70	—	0.59	21.29
6.01.08.30.G	windows, glazed doors, screens in extra large panes	m²	0.42	15.80	—	0.46	16.27	17.39	—	0.51	17.89
6.01.08.30.H	stairs (measured overall)	m²	0.28	10.54	—	0.73	11.27	11.59	—	0.80	12.39
6.01.08.30.I	corrugated surfaces; over 300mm girth	m²	0.29	10.91	—	0.96	11.87	12.00	—	1.06	13.06
6.01.08.30.J	structural steelwork; over 300mm girth	m²	0.32	12.04	—	0.85	12.89	13.25	—	0.93	14.18
6.01.08.30.K	railings, balustrades (measured flat both sides overall)	m²	0.29	10.91	—	0.73	11.64	12.00	—	0.80	12.81
6.01.08.30.L	eaves gutters inside and out; not exceeding 150mm girth	m	0.15	5.64	—	0.12	5.76	6.21	—	0.13	6.34
6.01.08.30.M	eaves gutters inside and out; 150mm - 300mm girth	m	0.17	6.40	—	0.23	6.63	7.04	—	0.25	7.29
6.01.08.30.N	eaves gutters inside and out; over 300mm girth	m²	0.27	10.16	—	0.85	11.01	11.18	—	0.93	12.11
6.01.08.30.O	pipes, ducts, etc.; not exceeding 150mm girth	m	0.15	5.64	—	0.12	5.76	6.21	—	0.13	6.34
6.01.08.30.P	pipes, ducts, etc.; 150mm - 300mm girth	m	0.19	7.15	—	0.23	7.38	7.86	—	0.25	8.12
6.01.08.30.Q	pipes, ducts, etc.; over 300mm girth	m²	0.26	9.78	—	0.85	10.63	10.76	—	0.93	11.69
6.01.08.30.R	rainwater heads inside and out	nr	0.08	3.01	—	0.12	3.13	3.31	—	0.13	3.44
6.01.08.31.	**Primer, one undercoat, one finishing coat gloss oil paint on metalwork**										
6.01.08.31.A	general surfaces; not exceeding 150mm girth	m	0.39	14.68	—	0.22	14.89	16.14	—	0.24	16.38
6.01.08.31.B	general surfaces; 150mm - 300mm girth	m	0.51	19.19	—	0.49	19.68	21.11	—	0.53	21.64
6.01.08.31.C	general surfaces; over 300mm girth	m²	0.78	29.35	—	1.71	31.06	32.29	—	1.88	34.17
6.01.08.31.D	windows, glazed doors, screens in small panes	m²	1.98	74.51	—	1.40	75.91	81.96	—	1.54	83.50
6.01.08.31.E	windows, glazed doors, screens in medium panes	m²	1.74	65.48	—	1.23	66.70	72.02	—	1.35	73.37
6.01.08.31.F	windows, glazed doors, screens in large panes	m²	1.50	56.45	—	1.10	57.54	62.09	—	1.21	63.30

Glazing, Painting and Decorating

Hutchins Priced Schedules 2021	Unit	Labour Hours	Labour Net £	Plant Net £	Materials Net £	Unit Rate Net £	Labour Gross £	Plant Gross £	Materials Gross £	Unit Rate Gross £
										(Gross rates include 10% profit)

6.01.   **NEW WORK**

6.01.08.   **PAINTING AND DECORATING EXTERNALLY**

6.01.08.31.   **Primer, one undercoat, one finishing coat gloss oil paint on metalwork**

	Unit	Labour Hours	Labour Net £	Plant Net £	Materials Net £	Unit Rate Net £	Labour Gross £	Plant Gross £	Materials Gross £	Unit Rate Gross £
6.01.08.31.G windows, glazed doors, screens in extra large panes	m²	1.20	45.16	—	0.87	46.02	49.67	—	0.96	50.63
6.01.08.31.H stairs (measured overall)	m²	0.84	31.61	—	1.49	33.10	34.77	—	1.64	36.41
6.01.08.31.I corrugated surfaces; over 300mm girth	m²	0.87	32.74	—	1.93	34.67	36.01	—	2.12	38.13
6.01.08.31.J structural steelwork; over 300mm girth	m²	0.96	36.12	—	1.71	37.84	39.74	—	1.88	41.62
6.01.08.31.K railings, balustrades (measured flat both sides overall)	m²	0.87	32.74	—	1.34	34.08	36.01	—	1.48	37.49
6.01.08.31.L eaves gutters inside and out; not exceeding 150mm girth	m	0.45	16.93	—	0.22	17.15	18.63	—	0.24	18.87
6.01.08.31.M eaves gutters inside and out; 150mm - 300mm girth	m	0.51	19.19	—	0.49	19.68	21.11	—	0.53	21.64
6.01.08.31.N eaves gutters inside and out; over 300mm girth	m²	0.81	30.48	—	1.71	32.19	33.53	—	1.88	35.41
6.01.08.31.O pipes, ducts, etc.; not exceeding 150mm girth	m	0.45	16.93	—	0.22	17.15	18.63	—	0.24	18.87
6.01.08.31.P pipes, ducts etc.; 150mm - 300mm girth	m	0.51	19.19	—	0.49	19.68	21.11	—	0.53	21.64
6.01.08.31.Q pipes, ducts, etc.; over 300mm girth	m²	0.78	29.35	—	1.71	31.06	32.29	—	1.88	34.17
6.01.08.31.R rainwater heads inside and out	nr	0.24	9.03	—	0.22	9.25	9.93	—	0.24	10.17

6.01.08.32.   **Primer, two undercoats, one finishing coat gloss oil finishing paint on metalwork**

	Unit	Labour Hours	Labour Net £	Plant Net £	Materials Net £	Unit Rate Net £	Labour Gross £	Plant Gross £	Materials Gross £	Unit Rate Gross £
6.01.08.32.A general surfaces; not exceeding 150mm girth	m	0.52	19.57	—	0.30	19.87	21.52	—	0.33	21.86
6.01.08.32.B general surfaces; 150mm - 300mm girth	m	0.68	25.59	—	0.66	26.24	28.15	—	0.72	28.87
6.01.08.32.C general surfaces; over 300mm girth	m²	1.04	39.14	—	2.19	41.33	43.05	—	2.41	45.46
6.01.08.32.D windows, glazed doors, screens in small panes	m²	2.64	99.34	—	1.80	101.15	109.28	—	1.98	111.26
6.01.08.32.E windows, glazed doors, screens in medium panes	m²	2.32	87.30	—	1.54	88.84	96.03	—	1.69	97.72
6.01.08.32.F windows, glazed doors, screens in large panes	m²	2.00	75.26	—	1.35	76.61	82.79	—	1.49	84.27
6.01.08.32.G windows, glazed doors, screens in extra large panes	m²	1.60	60.21	—	1.12	61.33	66.23	—	1.24	67.47
6.01.08.32.H stairs (measured overall)	m²	1.12	42.15	—	1.89	44.04	46.36	—	2.08	48.44
6.01.08.32.I corrugated surfaces; over 300mm girth	m²	1.16	43.65	—	2.47	46.12	48.02	—	2.71	50.73
6.01.08.32.J structural steelwork; over 300mm girth	m²	1.28	48.17	—	2.19	50.36	52.98	—	2.41	55.40
6.01.08.32.K railings, balustrades (measured flat both sides overall)	m²	1.16	43.65	—	1.65	45.30	48.02	—	1.82	49.83
6.01.08.32.L eaves gutters inside and out; not exceeding 150mm girth	m	0.60	22.58	—	0.30	22.88	24.84	—	0.33	25.17
6.01.08.32.M eaves gutters inside and out; 150mm - 300mm girth	m	0.68	25.59	—	0.66	26.24	28.15	—	0.72	28.87

Glazing, Painting and Decorating

Hutchins Priced Schedules 2021	Unit	Labour Hours	Labour Net £	Plant Net £	Materials Net £	Unit Rate Net £	Labour Gross £	Plant Gross £	Materials Gross £	Unit Rate Gross £
							(Gross rates include 10% profit)			

6.01.   **NEW WORK**

6.01.08.   PAINTING AND DECORATING EXTERNALLY

6.01.08.32.   **Primer, two undercoats, one finishing coat gloss oil finishing paint on metalwork**

6.01.08.32.N eaves gutters inside and out; over 300mm girth	m²	1.08	40.64	—	2.19	42.83	44.70	—	2.41	47.12
6.01.08.32.O pipes, ducts, etc.; not exceeding 150mm girth	m	0.60	22.58	—	0.30	22.88	24.84	—	0.33	25.17
6.01.08.32.P pipes, ducts, etc.; 150mm - 300mm girth	m	0.68	25.59	—	0.66	26.24	28.15	—	0.72	28.87
6.01.08.32.Q pipes, ducts, etc.; over 300mm girth	m²	1.04	39.14	—	2.19	41.33	43.05	—	2.41	45.46
6.01.08.32.R rainwater heads inside and out	nr	0.32	12.04	—	0.30	12.34	13.25	—	0.33	13.58

6.01.08.40.   **One coat bituminous paint on metalwork**

6.01.08.40.A water storage cisterns, tanks, inside and out	m²	0.40	15.05	—	0.13	15.19	16.56	—	0.15	16.70
6.01.08.40.B gutters (inside surfaces)	m	0.10	3.76	—	0.05	3.81	4.14	—	0.05	4.19

6.01.08.41.   **Two coats bituminous paint on metalwork**

6.01.08.41.A water storage cisterns, tanks, inside and out	m²	0.80	30.10	—	0.28	30.39	33.11	—	0.31	33.43
6.01.08.41.B corrugated surfaces; over 300mm girth	m²	1.00	37.63	—	0.28	37.91	41.39	—	0.31	41.70
6.01.08.41.C gutters (inside surfaces)	m	0.18	6.77	—	0.08	6.86	7.45	—	0.09	7.54
6.01.08.41.D pipes; not exceeding 150mm girth	m	0.13	4.89	—	0.05	4.94	5.38	—	0.05	5.44
6.01.08.41.E pipes; 150mm - 300mm girth	m	0.24	9.03	—	0.08	9.11	9.93	—	0.09	10.03
6.01.08.41.F pipes; over 300mm girth	m²	0.34	12.79	—	0.28	13.08	14.07	—	0.31	14.38

6.01.08.42.   **Two coats timber preserver on wrought timber**

6.01.08.42.A general surfaces; not exceeding 150mm girth	m	0.20	7.53	—	0.07	7.60	8.28	—	0.08	8.36
6.01.08.42.B general surfaces; 150mm - 300mm girth	m	0.26	9.78	—	0.13	9.92	10.76	—	0.15	10.91
6.01.08.42.C general surfaces; over 300mm girth	m²	0.38	14.30	—	0.47	14.77	15.73	—	0.51	16.24

6.01.08.43.   **Two coats timber preserver on sawn timber**

6.01.08.43.A general surfaces; not exceeding 150mm girth	m	0.21	7.90	—	0.13	8.04	8.69	—	0.15	8.84
6.01.08.43.B general surfaces; 150mm - 300mm girth	m	0.27	10.16	—	0.25	10.41	11.18	—	0.28	11.45
6.01.08.43.C general surfaces; over 300mm girth	m²	0.40	15.05	—	0.60	15.65	16.56	—	0.66	17.22

6.01.08.50.   **Two coats fungicidal and insecticidal preservative on wrought timber**

6.01.08.50.A general surfaces; not exceeding 150mm girth	m	0.19	7.15	—	0.08	7.23	7.86	—	0.09	7.95
6.01.08.50.B general surfaces; 150mm - 300mm girth	m	0.25	9.41	—	0.16	9.56	10.35	—	0.17	10.52
6.01.08.50.C general surfaces; over 300mm girth	m²	0.36	13.55	—	0.51	14.05	14.90	—	0.56	15.46

Hutchins Priced Schedules 2021	Unit	Labour Hours	Labour Net £	Plant Net £	Materials Net £	Unit Rate Net £	Labour Gross £	Plant Gross £	Materials Gross £	Unit Rate Gross £
						(Gross rates include 10% profit)				

6.01.   NEW WORK

6.01.08.   PAINTING AND DECORATING EXTERNALLY

6.01.08.51.   **Two coats fungicidal and insecticidal preservative on sawn timber**

	Unit	Labour Hours	Labour Net £	Plant Net £	Materials Net £	Unit Rate Net £	Labour Gross £	Plant Gross £	Materials Gross £	Unit Rate Gross £
6.01.08.51.A general surfaces; not exceeding 150mm girth	m	0.20	7.53	—	0.10	7.63	8.28	—	0.11	8.39
6.01.08.51.B general surfaces; 150mm - 300mm girth	m	0.26	9.78	—	0.23	10.02	10.76	—	0.26	11.02
6.01.08.51.C general surfaces; over 300mm girth	m²	0.38	14.30	—	0.75	15.05	15.73	—	0.83	16.56

6.01.08.55.   **Two coats Cuprinol preserver on wrought timber (green)**

	Unit	Labour Hours	Labour Net £	Plant Net £	Materials Net £	Unit Rate Net £	Labour Gross £	Plant Gross £	Materials Gross £	Unit Rate Gross £
6.01.08.55.A general surfaces; not exceeding 150mm girth	m	0.20	7.53	—	0.25	7.78	8.28	—	0.27	8.55
6.01.08.55.B general surfaces; 150mm - 300mm girth	m	0.26	9.78	—	0.50	10.28	10.76	—	0.55	11.31
6.01.08.55.C general surfaces; over 300mm girth	m²	0.38	14.30	—	1.65	15.95	15.73	—	1.82	17.55

6.01.08.56.   **Two coats Cuprinol preserver on sawn timber (green)**

	Unit	Labour Hours	Labour Net £	Plant Net £	Materials Net £	Unit Rate Net £	Labour Gross £	Plant Gross £	Materials Gross £	Unit Rate Gross £
6.01.08.56.A general surfaces; not exceeding 150mm girth	m	0.21	7.90	—	0.34	8.25	8.69	—	0.38	9.07
6.01.08.56.B general surfaces; 150mm - 300mm girth	m	0.27	10.16	—	0.66	10.82	11.18	—	0.72	11.90
6.01.08.56.C general surfaces; over 300mm girth	m²	0.40	15.05	—	2.22	17.27	16.56	—	2.44	18.99

6.01.08.57.   **Two coats Cuprinol preserver on wrought timber (clear)**

	Unit	Labour Hours	Labour Net £	Plant Net £	Materials Net £	Unit Rate Net £	Labour Gross £	Plant Gross £	Materials Gross £	Unit Rate Gross £
6.01.08.57.A general surfaces; not exceeding 150mm girth	m	0.20	7.53	—	0.29	7.82	8.28	—	0.32	8.60
6.01.08.57.B general surfaces; 150mm - 300mm girth	m	0.26	9.78	—	0.59	10.37	10.76	—	0.64	11.41
6.01.08.57.C general surfaces; over 300mm girth	m²	0.38	14.30	—	1.94	16.24	15.73	—	2.13	17.86

6.01.08.58.   **Two coats Cuprinol preserver on sawn timber (clear)**

	Unit	Labour Hours	Labour Net £	Plant Net £	Materials Net £	Unit Rate Net £	Labour Gross £	Plant Gross £	Materials Gross £	Unit Rate Gross £
6.01.08.58.A general surfaces; not exceeding 150mm girth	m	0.21	7.90	—	0.40	8.30	8.69	—	0.44	9.14
6.01.08.58.B general surfaces; 150mm - 300mm girth	m	0.27	10.16	—	0.77	10.93	11.18	—	0.85	12.02
6.01.08.58.C general surfaces; over 300mm girth	m²	0.40	15.05	—	2.60	17.65	16.56	—	2.86	19.41

6.01.08.59.   **Two coats raw or boiled linseed oil on woodwork**

	Unit	Labour Hours	Labour Net £	Plant Net £	Materials Net £	Unit Rate Net £	Labour Gross £	Plant Gross £	Materials Gross £	Unit Rate Gross £
6.01.08.59.A general surfaces; not exceeding 150mm girth	m	0.14	5.27	—	0.42	5.69	5.80	—	0.47	6.26
6.01.08.59.B general surfaces; 150mm - 300mm girth	m	0.18	6.77	—	0.83	7.60	7.45	—	0.91	8.36
6.01.08.59.C general surfaces; over 300mm girth	m²	0.32	12.04	—	2.78	14.83	13.25	—	3.06	16.31

Glazing, Painting and Decorating

Hutchins Priced Schedules 2021		Unit	Labour Hours	Labour Net £	Plant Net £	Materials Net £	Unit Rate Net £	Labour Gross £	Plant Gross £	Materials Gross £	Unit Rate Gross £
								(Gross rates include 10% profit)			
6.01.	**NEW WORK**										
6.01.08.	**PAINTING AND DECORATING EXTERNALLY**										
6.01.08.60.	**Two coats clear polyurethane on woodwork**										
6.01.08.60.A	general surfaces; not exceeding 150mm girth	m	0.20	7.53	—	0.31	7.84	8.28	—	0.34	8.62
6.01.08.60.B	general surfaces; 150mm - 300mm girth	m	0.26	9.78	—	0.66	10.44	10.76	—	0.72	11.49
6.01.08.60.C	general surfaces; over 300mm girth	m²	0.38	14.30	—	2.17	16.47	15.73	—	2.39	18.12
6.01.08.60.G	windows, glazed doors, screens in small panes	m²	0.84	31.61	—	1.63	33.24	34.77	—	1.79	36.56
6.01.08.60.H	windows, glazed doors, screens in medium panes	m²	0.76	28.60	—	1.40	29.99	31.46	—	1.53	32.99
6.01.08.60.I	windows, glazed doors, screens in large panes	m²	0.68	25.59	—	1.12	26.71	28.15	—	1.24	29.38
6.01.08.60.J	windows, glazed doors, screens in extra large panes	m²	0.60	22.58	—	0.66	23.24	24.84	—	0.72	25.56
6.01.08.61.	**Three coats clear polyurethane on woodwork**										
6.01.08.61.A	general surfaces; not exceeding 150mm girth	m	0.28	10.54	—	0.50	11.04	11.59	—	0.55	12.14
6.01.08.61.B	general surfaces; 150mm - 300mm girth	m	0.34	12.79	—	0.97	13.76	14.07	—	1.07	15.14
6.01.08.61.C	general surfaces; over 300mm girth	m²	0.56	21.07	—	3.29	24.37	23.18	—	3.62	26.80
6.01.08.61.G	windows, glazed doors, screens in small panes	m²	1.14	42.90	—	2.48	45.38	47.19	—	2.73	49.92
6.01.08.61.H	windows, glazed doors, screens in medium panes	m²	1.06	39.89	—	2.09	41.98	43.88	—	2.30	46.18
6.01.08.61.I	windows, glazed doors, screens in large panes	m²	0.98	36.88	—	1.63	38.50	40.57	—	1.79	42.36
6.01.08.61.J	windows, glazed doors, screens in extra large panes	m²	0.90	33.87	—	0.97	34.84	37.25	—	1.07	38.32
6.01.09.	**SIGNWRITING**										
6.01.09.01.	**Gloss oil paint; per coat; per 25mm high**										
6.01.09.01.A	capital letters	nr	0.16	6.02	—	0.05	6.07	6.62	—	0.05	6.67
6.01.09.01.B	lower case letters	nr	0.14	5.27	—	0.02	5.29	5.80	—	0.02	5.82
6.01.09.01.C	numbers	nr	0.16	6.02	—	0.05	6.07	6.62	—	0.05	6.67
6.01.09.01.D	stops, commas and hyphens	nr	0.05	1.88	—	0.02	1.90	2.07	—	0.02	2.09
6.01.09.02.	**Gilt or gold leaf; per 25mm high**										
6.01.09.02.A	capital letters	nr	0.16	6.02	—	0.09	6.11	6.62	—	0.10	6.72
6.01.09.02.B	lower case letters	nr	0.14	5.27	—	0.07	5.34	5.80	—	0.08	5.87
6.01.09.02.C	numbers	nr	0.16	6.02	—	0.09	6.11	6.62	—	0.10	6.72

Glazing, Painting and Decorating

Hutchins Priced Schedules 2021	Unit	Labour Hours	Labour Net £	Plant Net £	Materials Net £	Unit Rate Net £	Labour Gross £	Plant Gross £	Materials Gross £	Unit Rate Gross £
							(Gross rates include 10% profit)			

6.01. **NEW WORK**

6.01.10. **PRESSURE SPRAY PAINTING**

6.01.10.01. **One coat emulsion paint**

6.01.10.01.A	brick.....................	m²	0.07	4.57	—	0.30	4.87	5.03	—	0.33	5.36
6.01.10.01.B	block.....................	m²	0.08	5.65	—	0.38	6.03	6.21	—	0.42	6.63
6.01.10.01.C	concrete..................	m²	0.05	3.56	—	0.30	3.86	3.92	—	0.33	4.25
6.01.10.01.D	plaster...................	m²	0.05	3.56	—	0.26	3.82	3.92	—	0.28	4.20

6.01.10.02. **One coat oil colour**

6.01.10.02.B	brick.....................	m²	0.10	6.38	—	0.47	6.86	7.02	—	0.52	7.54
6.01.10.02.C	block.....................	m²	0.12	8.13	—	0.61	8.74	8.95	—	0.67	9.61
6.01.10.02.D	concrete..................	m²	0.12	8.13	—	0.47	8.60	8.95	—	0.52	9.47
6.01.10.02.E	plaster...................	m²	0.08	5.31	—	0.41	5.71	5.84	—	0.45	6.29

6.01.10.03. **One basecoat by brush; one coat multicolour**

6.01.10.03.A	brick.....................	m²	0.18	11.96	—	3.55	15.52	13.16	—	3.91	17.07
6.01.10.03.C	block.....................	m²	0.18	11.96	—	3.55	15.52	13.16	—	3.91	17.07
6.01.10.03.D	concrete..................	m²	0.18	11.96	—	3.55	15.52	13.16	—	3.91	17.07
6.01.10.03.E	plaster...................	m²	0.16	10.55	—	3.55	14.10	11.61	—	3.91	15.51

6.01.11. **PAPERHANGING**

6.01.11.01. **Strip off one layer woodchip paper; stop cracks and rub down**

6.01.11.01.A	walls.....................	m²	0.18	6.77	—	0.11	6.89	7.45	—	0.13	7.58
6.01.11.01.B	walls to stairwell..........	m²	0.20	7.53	—	0.11	7.64	8.28	—	0.13	8.40
6.01.11.01.C	ceilings..................	m²	0.22	8.28	—	0.11	8.39	9.11	—	0.13	9.23
6.01.11.01.D	ceilings to stairwell........	m²	0.24	9.03	—	0.11	9.15	9.93	—	0.13	10.06

6.01.11.02. **Strip off one layer of standard patterned or ready-pasted paper; stop cracks and rub down**

6.01.11.02.A	walls.....................	m²	0.21	7.90	—	0.11	8.02	8.69	—	0.13	8.82
6.01.11.02.B	walls to stairwell..........	m²	0.23	8.65	—	0.11	8.77	9.52	—	0.13	9.65
6.01.11.02.C	ceilings..................	m²	0.25	9.41	—	0.11	9.52	10.35	—	0.13	10.47
6.01.11.02.D	ceilings to stairwell........	m²	0.27	10.16	—	0.11	10.27	11.18	—	0.13	11.30

6.01.11.03. **Strip off one layer of vinyl paper; stop cracks and rub down**

6.01.11.03.A	walls.....................	m²	0.23	8.65	—	0.11	8.77	9.52	—	0.13	9.65
6.01.11.03.B	walls to stairwell..........	m²	0.25	9.41	—	0.11	9.52	10.35	—	0.13	10.47
6.01.11.03.C	ceilings..................	m²	0.27	10.16	—	0.11	10.27	11.18	—	0.13	11.30
6.01.11.03.D	ceilings to stairwell........	m²	0.29	10.91	—	0.11	11.03	12.00	—	0.13	12.13

Glazing, Painting and Decorating

Hutchins Priced Schedules 2021	Unit	Labour Hours	Labour Net £	Plant Net £	Materials Net £	Unit Rate Net £	Labour Gross £	Plant Gross £	Materials Gross £	Unit Rate Gross £
						(Gross rates include 10% profit)				
6.01.	**NEW WORK**									
6.01.11.	**PAPERHANGING**									
6.01.11.04.	**Strip off one layer of embossed paper; stop cracks and rub down**									
6.01.11.04.A	walls	m²	0.23	8.65	—	0.11	8.77	9.52	—	0.13	9.65
6.01.11.04.B	walls to stairwell	m²	0.25	9.41	—	0.11	9.52	10.35	—	0.13	10.47
6.01.11.04.C	ceilings	m²	0.27	10.16	—	0.11	10.27	11.18	—	0.13	11.30
6.01.11.04.D	ceilings to stairwell	m²	0.29	10.91	—	0.11	11.03	12.00	—	0.13	12.13
6.01.11.05.	**Strip off one layer of Lincrusta or Anaglypta paper; stop cracks and rub down**									
6.01.11.05.A	walls	m²	0.28	10.54	—	0.11	10.65	11.59	—	0.13	11.72
6.01.11.05.B	walls to stairwell	m²	0.30	11.29	—	0.11	11.40	12.42	—	0.13	12.54
6.01.11.05.C	ceilings	m²	0.33	12.42	—	0.11	12.53	13.66	—	0.13	13.79
6.01.11.05.D	ceilings to stairwell	m²	0.38	14.30	—	0.11	14.41	15.73	—	0.13	15.86
6.01.11.06.	**Strip off two layers of woodchip paper; stop cracks and rub down**									
6.01.11.06.A	walls	m²	0.28	10.54	—	0.11	10.65	11.59	—	0.13	11.72
6.01.11.06.B	walls to stairwell	m²	0.30	11.29	—	0.11	11.40	12.42	—	0.13	12.54
6.01.11.06.C	ceilings	m²	0.35	13.17	—	0.11	13.29	14.49	—	0.13	14.61
6.01.11.06.D	ceilings to stairwell	m²	0.38	14.30	—	0.11	14.41	15.73	—	0.13	15.86
6.01.11.07.	**Strip off two layers of standard patterned or ready-pasted paper; stop cracks and rub down**									
6.01.11.07.A	walls	m²	0.32	12.04	—	0.11	12.16	13.25	—	0.13	13.37
6.01.11.07.B	walls to stairwell	m²	0.35	13.17	—	0.11	13.29	14.49	—	0.13	14.61
6.01.11.07.C	ceilings	m²	0.38	14.30	—	0.11	14.41	15.73	—	0.13	15.86
6.01.11.07.D	ceilings to stairwell	m²	0.42	15.80	—	0.11	15.92	17.39	—	0.13	17.51
6.01.11.08.	**Strip off two layers of vinyl paper; stop cracks and rub down**									
6.01.11.08.A	walls	m²	0.35	13.17	—	0.11	13.29	14.49	—	0.13	14.61
6.01.11.08.B	walls to stairwell	m²	0.38	14.30	—	0.11	14.41	15.73	—	0.13	15.86
6.01.11.08.C	ceilings	m²	0.40	15.05	—	0.11	15.17	16.56	—	0.13	16.68
6.01.11.08.D	ceilings to stairwell	m²	0.45	16.93	—	0.11	17.05	18.63	—	0.13	18.75
6.01.11.09.	**Strip off two layers of embossed paper; stop cracks and rub down**									
6.01.11.09.A	walls	m²	0.34	12.79	—	0.11	12.91	14.07	—	0.13	14.20
6.01.11.09.B	walls to stairwell	m²	0.37	13.92	—	0.11	14.04	15.32	—	0.13	15.44
6.01.11.09.C	ceilings	m²	0.42	15.80	—	0.11	15.92	17.39	—	0.13	17.51
6.01.11.09.D	ceilings to stairwell	m²	0.45	16.93	—	0.11	17.05	18.63	—	0.13	18.75

Hutchins Priced Schedules 2021	Unit	Labour Hours	Labour Net £	Plant Net £	Materials Net £	Unit Rate Net £	Labour Gross £	Plant Gross £	Materials Gross £	Unit Rate Gross £

(Gross rates include 10% profit)

6.01. NEW WORK

6.01.11. PAPERHANGING

6.01.11.10. Strip off two layers of Lincrusta or Anaglypta paper; stop cracks and rub down

	Unit	Labour Hours	Labour Net £	Plant Net £	Materials Net £	Unit Rate Net £	Labour Gross £	Plant Gross £	Materials Gross £	Unit Rate Gross £
6.01.11.10.A walls	m²	0.38	14.30	—	0.11	14.41	15.73	—	0.13	15.86
6.01.11.10.B walls to stairwell	m²	0.42	15.80	—	0.11	15.92	17.39	—	0.13	17.51
6.01.11.10.C ceilings	m²	0.50	18.82	—	0.11	18.93	20.70	—	0.13	20.82
6.01.11.10.D ceilings to stairwell	m²	0.53	19.94	—	0.11	20.06	21.94	—	0.13	22.06
6.01.11.11. Prepare and hang lining paper										
6.01.11.11.A walls	m²	0.25	9.41	—	0.38	9.79	10.35	—	0.42	10.76
6.01.11.11.B walls to stairwell	m²	0.28	10.54	—	0.38	10.91	11.59	—	0.42	12.01
6.01.11.11.C ceilings	m²	0.30	11.29	—	0.38	11.67	12.42	—	0.42	12.83
6.01.11.11.D ceilings to stairwell	m²	0.32	12.04	—	0.38	12.42	13.25	—	0.42	13.66
6.01.11.13. Prepare and hang ready-pasted paper										
6.01.11.13.A walls	m²	0.26	9.78	—	3.53	13.32	10.76	—	3.89	14.65
6.01.11.13.B walls to stairwell	m²	0.29	10.91	—	3.53	14.44	12.00	—	3.89	15.89
6.01.11.13.C ceilings	m²	0.31	11.67	—	3.53	15.20	12.83	—	3.89	16.72
6.01.11.13.D ceilings to stairwell	m²	0.33	12.42	—	3.53	15.95	13.66	—	3.89	17.54
6.01.11.14. Prepare and hang standard patterned paper (PC sum £16 per roll)										
6.01.11.14.A walls	m²	0.26	9.78	—	3.39	13.18	10.76	—	3.73	14.49
6.01.11.14.B walls to stairwell	m²	0.29	10.91	—	3.39	14.30	12.00	—	3.73	15.74
6.01.11.14.C ceilings	m²	0.31	11.67	—	3.39	15.06	12.83	—	3.73	16.56
6.01.11.14.D ceilings to stairwell	m²	0.33	12.42	—	3.39	15.81	13.66	—	3.73	17.39
6.01.11.15. Prepare and hang vinyl surface paper										
6.01.11.15.A walls	m²	0.26	9.78	—	3.53	13.32	10.76	—	3.89	14.65
6.01.11.15.B walls to stairwell	m²	0.29	10.91	—	3.53	14.45	12.00	—	3.89	15.89
6.01.11.15.C ceilings	m²	0.31	11.67	—	3.53	15.20	12.83	—	3.89	16.72
6.01.11.15.D ceilings to stairwell	m²	0.33	12.42	—	3.53	15.95	13.66	—	3.89	17.55
6.01.11.16. Prepare and hang Anaglypta paper										
6.01.11.16.A walls	m²	0.27	10.16	—	1.24	11.40	11.18	—	1.36	12.54
6.01.11.16.B walls to stairwell	m²	0.30	11.29	—	1.24	12.52	12.42	—	1.36	13.78
6.01.11.16.C ceilings	m²	0.32	12.04	—	1.24	13.28	13.25	—	1.36	14.61
6.01.11.16.D ceilings to stairwell	m²	0.33	12.42	—	1.24	13.65	13.66	—	1.36	15.02
6.01.11.17. Prepare and hang flock paper (PC sum £39 per roll)										
6.01.11.17.A walls	m²	0.27	10.16	—	8.27	18.43	11.18	—	9.09	20.27
6.01.11.17.B walls to stairwell	m²	0.30	11.29	—	8.27	19.56	12.42	—	9.09	21.51

Glazing, Painting and Decorating

Hutchins Priced Schedules 2021	Unit	Labour Hours	Labour Net £	Plant Net £	Materials Net £	Unit Rate Net £	Labour Gross £	Plant Gross £	Materials Gross £	Unit Rate Gross £
						(Gross rates include 10% profit)				
6.01. **NEW WORK**										
6.01.11. PAPERHANGING										
6.01.11.17. Prepare and hang flock paper (PC sum £39 per roll)										
6.01.11.17.C ceilings	m²	0.32	12.04	—	8.27	20.31	13.25	—	9.09	22.34
6.01.11.17.D ceilings to stairwell	m²	0.33	12.42	—	8.27	20.69	13.66	—	9.09	22.75
6.01.11.18. Cut and hang standard border strip										
6.01.11.18.A 75mm - 150mm wide	m	0.09	3.39	—	1.26	4.65	3.73	—	1.39	5.11
6.01.12. POLISHING BY SPECIALIST										
6.01.12.01. Body in and polish										
6.01.12.01.A general surfaces	m²	1.05	39.51	—	4.58	44.09	43.46	—	5.04	48.50
6.01.12.01.B general surfaces in narrow widths; not exceeding 150mm	m	0.25	9.41	—	0.68	10.09	10.35	—	0.75	11.09
6.01.12.01.C general surfaces in narrow widths; 150mm - 300mm	m	0.40	15.05	—	1.37	16.43	16.56	—	1.51	18.07
6.01.12.02. Seal and wax polish										
6.01.12.02.A general surfaces	m²	0.45	16.93	—	1.21	18.15	18.63	—	1.34	19.96
6.01.12.02.B general surfaces in narrow widths; not exceeding 150mm	m	0.11	4.14	—	0.18	4.32	4.55	—	0.20	4.75
6.01.12.02.C general surfaces in narrow widths; 150mm - 300mm	m	0.18	6.77	—	0.36	7.14	7.45	—	0.40	7.85
6.01.12.03. Body in and wax polish										
6.01.12.03.A general surfaces	m²	0.68	25.59	—	1.21	26.80	28.15	—	1.34	29.48
6.01.12.03.B general surfaces in narrow widths; not exceeding 150mm	m	0.17	6.40	—	0.18	6.58	7.04	—	0.20	7.24
6.01.12.03.C general surfaces in narrow widths; 150mm - 300mm	m	0.27	10.16	—	0.36	10.52	11.18	—	0.40	11.58
6.01.12.04. Set and dry shine										
6.01.12.04.A general surfaces	m²	0.60	22.58	—	—	22.58	24.84	—	—	24.84
6.01.12.04.B general surfaces in narrow widths; not exceeding 150mm	m	0.15	5.64	—	—	5.64	6.21	—	—	6.21
6.01.12.04.C general surfaces in narrow widths; 150mm - 300mm	m	0.20	7.53	—	—	7.53	8.28	—	—	8.28
6.01.12.05. Body in with polish and two coats yacht varnish										
6.01.12.05.A general surfaces	m²	1.11	41.77	—	4.23	46.00	45.95	—	4.65	50.60
6.01.12.05.B general surfaces in narrow widths; not exceeding 150mm	m	0.26	9.78	—	0.63	10.41	10.76	—	0.69	11.45
6.01.12.05.C general surfaces in narrow widths; 150mm - 300mm	m	0.42	15.80	—	1.27	17.07	17.39	—	1.40	18.78

Glazing, Painting and Decorating

| | | | | | | | | | (Gross rates include 10% profit) | |

	Unit	Labour Hours	Labour Net £	Plant Net £	Materials Net £	Unit Rate Net £	Labour Gross £	Plant Gross £	Materials Gross £	Unit Rate Gross £
6.01. **NEW WORK**											
6.01.12. **POLISHING BY SPECIALIST**											
6.01.12.06. **Grain fill, body in and spirit off piano type finish**											
6.01.12.06.A general surfaces	m²	2.75	103.48	—	5.26	108.74	113.83	—	5.78	119.61
6.01.12.06.B general surfaces in narrow widths; not exceeding 150mm	m	0.65	24.46	—	0.72	25.18	26.91	—	0.79	27.69
6.01.12.06.C general surfaces in narrow widths; 150mm - 300mm	m	1.05	39.51	—	1.62	41.13	43.46	—	1.78	45.24
6.01.13. **FIRE RETARDANT FINISHES**											
6.01.13.01. **Zeroflame**											
6.01.13.01.A paint; class 0/1	m²	0.65	24.46	—	33.60	58.06	26.91	—	36.96	63.87
6.01.13.01.C treatment; class 0	m²	0.65	24.46	—	39.28	63.74	26.91	—	43.21	70.12
6.01.13.02. **Nullifire**											
6.01.13.02.A intumescent coating for steelwork	m²	0.60	22.58	—	28.27	50.85	24.84	—	31.10	55.94
6.01.13.03. **Thermoguard Timbercoat**											
6.01.13.03.A class 0/1	m²	0.60	22.58	—	28.84	51.42	24.84	—	31.72	56.56
6.01.13.04. **Envirograf**											
6.01.13.04.A class 0/1	m²	0.60	22.58	—	37.02	59.60	24.84	—	40.72	65.56
6.01.14. **LACQUERS AND POLYURETHANE BY SPECIALIST**											
6.01.14.01. **Prepare and seal**											
6.01.14.01.A general surfaces	m²	0.24	9.03	—	0.80	9.83	9.93	—	0.88	10.82
6.01.14.01.B general surfaces in narrow widths; not exceeding 150mm girth	m	0.06	2.26	—	0.13	2.39	2.48	—	0.14	2.62
6.01.14.01.C general surfaces in narrow widths; 150mm - 300mm girth	m	0.08	3.01	—	0.25	3.26	3.31	—	0.27	3.59
6.01.14.02. **Each additional coat of lacquer**											
6.01.14.02.A general surfaces	m²	0.20	7.53	—	0.80	8.33	8.28	—	0.88	9.16
6.01.14.02.B general surfaces in narrow widths; not exceeding 150mm girth	m	0.05	1.88	—	0.13	2.01	2.07	—	0.14	2.21
6.01.14.02.C general surfaces in narrow widths; 150mm - 300mm girth	m	0.07	2.63	—	0.25	2.88	2.90	—	0.27	3.17
6.01.14.03. **Wire wool and burnish with wax**											
6.01.14.03.A general surfaces	m²	0.55	20.70	—	1.23	21.92	22.77	—	1.35	24.12
6.01.14.03.B general surfaces in narrow widths; not exceeding 150mm girth	m	0.15	5.64	—	0.20	5.84	6.21	—	0.22	6.43
6.01.14.03.C general surfaces in narrow widths; 150mm - 300mm girth	m	0.20	7.53	—	0.39	7.91	8.28	—	0.43	8.71

Glazing, Painting and Decorating

Hutchins Priced Schedules 2021		Unit	Labour Hours	Labour Net £	Plant Net £	Materials Net £	Unit Rate Net £	Labour Gross £	Plant Gross £	Materials Gross £	Unit Rate Gross £
							(Gross rates include 10% profit)				
6.01.	**NEW WORK**										
6.01.15.	**PREPARE EXISTING WORK BY SPECIALIST**										
6.01.15.01.	**Wash down to degrease and repolish**										
6.01.15.01.A	general surfaces	m²	0.60	22.58	—	1.23	23.81	24.84	—	1.35	26.19
6.01.15.01.B	general surfaces in narrow widths; not exceeding 150mm girth	m	0.15	5.64	—	0.20	5.84	6.21	—	0.22	6.43
6.01.15.01.C	general surfaces in narrow widths; 150mm - 300mm girth	m	0.20	7.53	—	0.39	7.91	8.28	—	0.43	8.71
6.01.15.02.	**Wash down, degrease and clean to revive**										
6.01.15.02.A	general surfaces	m²	0.55	20.70	—	—	20.70	22.77	—	—	22.77
6.01.15.02.B	general surfaces in narrow widths; not exceeding 150mm girth	m	0.14	5.27	—	—	5.27	5.80	—	—	5.80
6.01.15.02.C	general surfaces in narrow widths; 150mm - 300mm girth	m	0.18	6.77	—	—	6.77	7.45	—	—	7.45
6.01.15.03.	**Wash down to degrease and rewax**										
6.01.15.03.A	general surfaces	m²	0.46	17.31	—	0.96	18.27	19.04	—	1.06	20.10
6.01.15.03.B	general surfaces in narrow widths; not exceeding 150mm girth	m	0.12	4.52	—	0.14	4.66	4.97	—	0.16	5.13
6.01.15.03.C	general surfaces in narrow widths; 150mm - 300mm girth	m	0.16	6.02	—	0.29	6.31	6.62	—	0.32	6.94
6.01.15.04.	**Strip off paint and varnish with remover**										
6.01.15.04.A	general surfaces	m²	0.75	28.22	—	1.26	29.48	31.04	—	1.38	32.43
6.01.15.04.B	general surfaces in narrow widths; not exceeding 150mm girth	m	0.19	7.15	—	0.19	7.34	7.86	—	0.21	8.07
6.01.15.04.C	general surfaces in narrow widths; 150mm - 300mm girth	m	0.23	8.65	—	0.38	9.03	9.52	—	0.41	9.94
6.02.	**REPAIRS AND ALTERATIONS**										
6.02.01.	**GLAZING**										
6.02.01.01.	**Hack out**										
6.02.01.01.A	broken glass other than plate	m²	2.34	88.05	—	—	88.05	96.86	—	—	96.86
6.02.01.01.B	plate glass	m²	3.55	133.59	—	—	133.59	146.95	—	—	146.95
6.02.01.02.	**Take out**										
6.02.01.02.A	all types of glass other than plate and set aside for re-use	m²	3.23	121.54	—	—	121.54	133.70	—	—	133.70
6.02.01.02.B	plate glass and set aside for re-use	m²	4.68	176.11	—	—	176.11	193.72	—	—	193.72
6.02.01.03.	**Remove old putty**										
6.02.01.03.A	paint rebate one coat oil colour ready to receive new glass	m	0.16	6.02	—	0.31	6.33	6.62	—	0.34	6.97

Glazing, Painting and Decorating

Hutchins Priced Schedules 2021		Unit	Labour Hours	Labour Net £	Plant Net £	Materials Net £	Unit Rate Net £	Labour Gross £	Plant Gross £	Materials Gross £	Unit Rate Gross £
								(Gross rates include 10% profit)			
6.02.	**REPAIRS AND ALTERATIONS**										
6.02.01.	**GLAZING**										
6.02.01.04.	**Glaze, sprig and putty to wood**										
6.02.01.04.A	sashes; average 0.40m2; 4mm clear . . .	m²	1.10	41.39	—	31.82	73.21	45.53	—	35.00	80.53
6.02.01.04.B	sashes; average 0.40m2; 4mm obscured	m²	1.10	41.39	—	48.49	89.89	45.53	—	53.34	98.87
6.02.01.04.C	sashes; average 0.40m2; 7mm wired cast	m²	1.20	45.16	—	85.49	130.65	49.67	—	94.04	143.72
6.02.01.05.	**Glaze, putty and sprig to metal**										
6.02.01.05.A	sashes extra over foregoing	m²	0.40	15.05	—	—	15.05	16.56	—	—	16.56
6.02.01.06.	**Wired cast glass in rooflights**										
6.02.01.06.A	panes up to 0.70m2	m²	0.90	33.87	—	85.49	119.36	37.25	—	94.04	131.30
6.02.01.06.B	panes exceeding 0.70m2	m²	1.00	37.63	—	85.49	123.12	41.39	—	94.04	135.44
6.02.01.07.	**Float glass**										
6.02.01.07.A	6mm	m²	3.50	131.71	—	38.94	170.64	144.88	—	42.83	187.71
6.02.01.09.	**Remove temporary coverings to sashes**										
6.02.01.09.A	stopping up nail holes, etc.	m²	1.10	41.39	—	—	41.39	45.53	—	—	45.53
6.02.02.	**INTERNAL DECORATING**										
6.02.02.01.	**Brush down and apply two coats lime white on brick walls, plaster walls and ceilings**										
6.02.02.01.A	new work	m²	0.33	12.42	—	0.17	12.59	13.66	—	0.19	13.85
6.02.02.01.B	old work	m²	0.38	14.30	—	0.17	14.47	15.73	—	0.19	15.92
6.02.02.02.	**Brush down and apply two coats emulsion to plaster**										
6.02.02.02.A	new work	m²	0.46	17.31	—	0.51	17.82	19.04	—	0.56	19.61
6.02.02.02.B	old work	m²	0.51	19.19	—	0.51	19.70	21.11	—	0.56	21.67
6.02.02.03.	**Brush down and apply two coats emulsion paint on brick walls**										
6.02.02.03.A	new work	m²	0.58	21.83	—	0.95	22.78	24.01	—	1.05	25.06
6.02.02.03.B	old work	m²	0.63	23.71	—	0.95	24.66	26.08	—	1.05	27.13
6.02.02.04.	**Wash down plaster surfaces, fill cracks nail holes, etc. with filler**										
6.02.02.04.A	bring forward for new decoration	m²	0.20	7.53	—	0.49	8.02	8.28	—	0.54	8.82
6.02.02.05.	**Prepare and apply oil colour on plaster ceilings**										
6.02.02.05.A	one coat	m²	0.27	10.16	—	0.70	10.86	11.18	—	0.77	11.95
6.02.02.05.B	two coats	m²	0.55	20.70	—	1.40	22.10	22.77	—	1.54	24.31

Glazing, Painting and Decorating

Hutchins Priced Schedules 2021	Unit	Labour Hours	Labour Net £	Plant Net £	Materials Net £	Unit Rate Net £	Labour Gross £	Plant Gross £	Materials Gross £	Unit Rate Gross £
							(Gross rates include 10% profit)			
6.02. **REPAIRS AND ALTERATIONS**										
6.02.02. **INTERNAL DECORATING**										
6.02.02.06. **Prepare and apply oil colour on walls**										
6.02.02.06.A plaster walls; one coat	m²	0.21	7.90	—	0.70	8.60	8.69	—	0.77	9.46
6.02.02.06.B plaster walls; two coats	m²	0.43	16.18	—	1.40	17.58	17.80	—	1.54	19.34
6.02.02.06.C brick walls; one coat	m²	0.27	10.16	—	1.40	11.56	11.18	—	1.54	12.72
6.02.02.06.D brick walls; two coats	m²	0.54	20.32	—	2.80	23.12	22.35	—	3.08	25.43
6.02.02.07. **Wash down, touch up and two coats oil**										
6.02.02.07.A general surfaces of wood	m²	0.65	24.46	—	0.71	25.17	26.91	—	0.78	27.69
6.02.02.07.B add for each extra coat applied or deduct for one coat	m²	0.24	9.03	—	0.37	9.40	9.93	—	0.41	10.34
6.02.02.07.C general surfaces of windows frames and sashes (measured over glass)	m²	1.00	37.63	—	1.08	38.71	41.39	—	1.19	42.58
6.02.02.07.D add for each extra coat applied or deduct for one coat	m²	0.40	15.05	—	0.54	15.59	16.56	—	0.59	17.15
6.02.02.07.E on surfaces exceeding 150mm girth	m	0.14	5.27	—	0.14	5.41	5.80	—	0.16	5.95
6.02.02.07.F on surfaces 150mm - 300mm girth	m	0.24	9.03	—	0.31	9.34	9.93	—	0.34	10.28
6.02.02.07.G add for each extra coat applied or deduct for one coat; not exceeding 150mm girth .	m	0.07	2.63	—	0.09	2.72	2.90	—	0.09	2.99
6.02.02.07.H add for extra coat applied or deduct for one coat; 150mm - 300mm girth	m	0.11	4.14	—	0.14	4.28	4.55	—	0.16	4.71
6.02.02.08. **Clean down and apply two coats oil colour to metal frames and sashes; bring forward bare patches**										
6.02.02.08.A measured over glass	m²	1.11	41.77	—	1.08	42.85	45.95	—	1.19	47.13
6.02.02.08.B general surfaces; over 300mm girth	m²	0.69	25.96	—	1.02	26.99	28.56	—	1.12	29.69
6.02.02.08.C general surfaces; not exceeding 150mm girth	m	0.30	11.29	—	0.28	11.57	12.42	—	0.31	12.73
6.02.02.08.D general surfaces; 150mm - 300mm girth .	m	0.39	14.68	—	0.54	15.22	16.14	—	0.59	16.74
6.02.02.09. **Clean down and one coat gloss oil paint**										
6.02.02.09.A to fireplace jambs, stoves, mantel registers and the like	nr	1.00	37.63	—	0.61	38.24	41.39	—	0.67	42.06
6.02.02.11. **Clean and apply one coat gloss paint**										
6.02.02.11.A on casement stays, fasteners, bolts, rimlocks and sundry fittings	nr	0.20	7.53	—	0.07	7.59	8.28	—	0.07	8.35
6.02.02.12. **Prepare and repolish existing wood surfaces**										
6.02.02.12.A general surfaces	m²	2.10	79.02	—	0.29	79.31	86.93	—	0.32	87.25
6.02.02.12.B handrails	m	0.60	22.58	—	0.11	22.69	24.84	—	0.12	24.96

Glazing, Painting and Decorating

Hutchins Priced Schedules 2021	Unit	Labour Hours	Labour Net £	Plant Net £	Materials Net £	Unit Rate Net £	Labour Gross £	Plant Gross £	Materials Gross £	Unit Rate Gross £
								(Gross rates include 10% profit)		
6.02.	**REPAIRS AND ALTERATIONS**									
6.02.02.	**INTERNAL DECORATING**									
6.02.02.13.	**Strip, body in and repolish wood surfaces**									
6.02.02.13.A	general surfaces	m²	4.70	176.86	—	0.40	177.26	194.55	—	0.44	194.98
6.02.02.13.B	handrails	m	1.10	41.39	—	0.12	41.51	45.53	—	0.13	45.66
6.02.02.14.	**Prepare and wax polish floors**									
6.02.02.14.A	general surfaces	m²	0.50	18.82	—	0.31	19.13	20.70	—	0.35	21.04
6.02.02.16.	**Strip paper from walls or ceilings**									
6.02.02.16.A	stop, size ready for new paper; first layer .	m²	0.30	11.29	—	0.39	11.68	12.42	—	0.43	12.85
6.02.02.16.B	add for each extra layer stripped	m²	0.12	4.52	—	—	4.52	4.97	—	—	4.97
6.02.02.17.	**Strip varnished paper from walls or ceilings**									
6.02.02.17.A	stop, size ready for new paper; first layer .	m²	0.60	22.58	—	0.39	22.97	24.84	—	0.43	25.27
6.02.02.17.B	add for each extra layer stripped	m²	0.22	8.28	—	—	8.28	9.11	—	—	9.11
6.02.02.18.	**Cut, trim and hang paper to walls**									
6.02.02.18.B	standard (PC sum £16 per roll)	m²	0.26	9.78	—	3.50	13.29	10.76	—	3.85	14.61
6.02.02.19.	**Cut, trim and hang paper to ceilings**									
6.02.02.19.A	lining	m²	0.30	11.29	—	0.49	11.78	12.42	—	0.54	12.95
6.02.03.	**EXTERNAL DECORATING**									
6.02.03.01.	**Wash down and apply two coats of masonry paint on**									
6.02.03.01.A	rendered walls	m²	0.53	19.94	—	0.74	20.69	21.94	—	0.82	22.76
6.02.03.02.	**Wash down and apply two coats Sandtex Matt masonry paint on**									
6.02.03.02.A	rendered walls	m²	0.54	20.32	—	0.83	21.15	22.35	—	0.91	23.26
6.02.03.03.	**Wash down and apply two coats Snowcem masonry paint on**									
6.02.03.03.A	rendered walls	m²	0.42	15.80	—	1.42	17.23	17.39	—	1.57	18.95
6.02.03.05.	**Wash down, touch up and two coats oil**									
6.02.03.05.A	general surfaces wood	m²	0.68	25.59	—	1.02	26.61	28.15	—	1.12	29.27
6.02.03.05.B	add for each extra coat applied	m²	0.27	10.16	—	0.51	10.67	11.18	—	0.56	11.74
6.02.03.05.C	window frames and sashes (over glass) .	m²	1.21	45.53	—	1.08	46.61	50.09	—	1.19	51.27
6.02.03.05.D	add for each extra coat applied	m²	0.43	16.18	—	0.54	16.72	17.80	—	0.59	18.39
6.02.03.06.	**Burn off paint to woodwork general surfaces**									
6.02.03.06.A	and prepare for priming	m²	0.80	30.10	—	—	30.10	33.11	—	—	33.11

Glazing, Painting and Decorating

Hutchins Priced Schedules 2021		Unit	Labour Hours	Labour Net £	Plant Net £	Materials Net £	Unit Rate Net £	Labour Gross £	Plant Gross £	Materials Gross £	Unit Rate Gross £
								(Gross rates include 10% profit)			
6.02.	**REPAIRS AND ALTERATIONS**										
6.02.03.	**EXTERNAL DECORATING**										
6.02.03.07.	**Strip paint to wood with paint remover**										
6.02.03.07.A	and prepare for priming	m²	0.50	18.82	—	0.07	18.89	20.70	—	0.08	20.78
6.02.03.08.	**Prime and two coats oil paint to putties**										
6.02.03.08.A	after reglazing	m	0.18	6.77	—	0.07	6.84	7.45	—	0.07	7.52
6.02.03.13.	**Clean down wire brush metal surfaces; bring forward bare patches and apply oil colour on previously painted surfaces**										
6.02.03.13.A	general surfaces; over 300mm girth	m²	0.78	29.35	—	0.77	30.12	32.29	—	0.84	33.13
6.02.03.13.B	general surfaces; not exceeding 150mm girth	m	0.39	14.68	—	0.23	14.90	16.14	—	0.25	16.39
6.02.03.13.C	general surfaces; 150mm - 300mm girth .	m	0.51	19.19	—	0.48	19.67	21.11	—	0.53	21.64
6.02.03.13.D	corrugated iron surfaces; (measured flat)	m²	0.87	32.74	—	0.85	33.59	36.01	—	0.94	36.95
6.02.03.13.E	structural steelwork	m²	0.96	36.12	—	0.77	36.89	39.74	—	0.84	40.58
6.02.03.13.G	railings, balustrades (measured flat overall)	m²	0.87	32.74	—	0.65	33.39	36.01	—	0.72	36.73
6.02.03.13.I	stairs (measured overall)	m²	0.84	31.61	—	0.54	32.15	34.77	—	0.59	35.36
6.02.03.13.J	windows, glazed doors in small panes . .	m²	1.98	74.51	—	0.54	75.05	81.96	—	0.59	82.55
6.02.03.13.K	windows, glazed doors in medium panes .	m²	1.74	65.48	—	0.48	65.96	72.02	—	0.53	72.55
6.02.03.13.L	windows, glazed doors in large panes . .	m²	1.50	56.45	—	0.43	56.87	62.09	—	0.47	62.56
6.02.03.13.M	eaves gutters; inside and outside	m	0.51	19.19	—	0.28	19.48	21.11	—	0.31	21.42
6.02.03.13.N	rainwater pipes, soil pipes, etc.	m	0.51	19.19	—	0.11	19.30	21.11	—	0.12	21.24
6.02.03.13.O	pipes, bars, straps, etc. up to 150mm girth	m	0.45	16.93	—	0.11	17.05	18.63	—	0.12	18.75
6.02.04.	**WOOD PRESERVATIVES**										
6.02.04.01.	**Two coats timber preserver**										
6.02.04.01.A	sawn surfaces	m²	0.40	15.05	—	0.60	15.65	16.56	—	0.66	17.22
6.02.04.01.B	wrought surfaces	m²	0.38	14.30	—	0.47	14.77	15.73	—	0.51	16.24
6.02.04.02.	**Two coats Cuprinol clear preserver**										
6.02.04.02.A	sawn surfaces	m²	0.40	15.05	—	1.94	16.99	16.56	—	2.13	18.69
6.02.04.02.B	wrought surfaces	m²	0.38	14.30	—	1.54	15.84	15.73	—	1.69	17.42
6.02.04.03.	**Two coats fungicidal and insecticidal preservative**										
6.02.04.03.A	sawn surfaces	m²	0.38	14.30	—	0.69	14.99	15.73	—	0.75	16.48
6.02.04.03.B	wrought surfaces	m²	0.36	13.55	—	0.54	14.09	14.90	—	0.60	15.50

Hutchins Priced Schedules 2021	Unit	Labour Hours	Labour Net £	Plant Net £	Materials Net £	Unit Rate Net £	Labour Gross £	Plant Gross £	Materials Gross £	Unit Rate Gross £
									(Gross rates include 10% profit)	
7.01.	**NEW WORK**									
7.01.01.	**RECONSTRUCTED STONE BY SPECIALIST**									
7.01.01.01.	**Plain cladding**									
7.01.01.01.A	50mm	m²	2.26	151.89	—	44.61	196.51	167.08	—	49.08	216.16
7.01.01.01.B	75mm	m²	2.51	168.76	—	49.69	218.45	185.64	—	54.66	240.30
7.01.01.02.	**Plain face ashlar**									
7.01.01.02.A	100mm	m²	2.76	185.63	—	60.26	245.89	204.20	—	66.28	270.48
7.01.01.03.	**Plain string**									
7.01.01.03.A	100mm x 300mm	m	0.75	50.41	—	47.18	97.59	55.45	—	51.90	107.35
7.01.01.04.	**Sills, jambs and heads**									
7.01.01.04.A	slip sill 140mm x 150mm	m	0.55	36.97	—	81.20	118.17	40.66	—	89.33	129.99
7.01.01.04.B	200mm x 75mm	m	0.60	40.33	—	26.90	67.23	44.36	—	29.59	73.95
7.01.01.04.C	plain head 215mm x 100mm	m	0.85	57.13	—	89.85	146.98	62.84	—	98.84	161.68
7.01.01.04.D	plain head 140mm x 100mm	m	0.80	53.77	—	75.25	129.01	59.14	—	82.77	141.92
7.01.01.05.	**Coping; bedded, jointed and pointed in cement mortar (1:6)**									
7.01.01.05.A	225mm x 75mm; twice weathered and throated	m	0.27	14.05	—	44.36	58.41	15.45	—	48.79	64.25
7.01.01.05.B	305mm x 75mm; twice weathered and throated	m	0.34	17.61	—	44.49	62.10	19.37	—	48.93	68.31
7.01.01.05.C	361mm x 75mm; twice weathered and throated	m	0.47	24.64	—	60.05	84.69	27.10	—	66.06	93.16
7.01.01.05.D	417mm x 75mm; twice weathered and throated	m	0.54	28.10	—	82.27	110.36	30.91	—	90.49	121.40
7.01.01.05.E	473mm x 75mm; twice weathered and throated	m	0.67	35.12	—	93.38	128.50	38.63	—	102.72	141.35
7.01.01.05.F	529mm x 75mm; twice weathered and throated	m	0.67	35.12	—	104.35	139.47	38.63	—	114.79	153.42
7.01.01.05.G	225mm x 75mm; once weathered and throated	m	0.27	14.05	—	44.36	58.41	15.45	—	48.79	64.25
7.01.01.05.H	305mm x 75mm; once weathered and throated	m	0.40	21.07	—	60.18	81.26	23.18	—	66.20	89.38
7.01.01.05.I	361mm x 75mm; once weathered and throated	m	0.54	28.10	—	71.30	99.39	30.91	—	78.42	109.33
7.01.01.05.J	417mm x 75mm; once weathered and throated	m	0.54	28.10	—	82.27	110.36	30.91	—	90.49	121.40
7.01.01.05.K	473mm x 75mm; once weathered and throated	m	0.67	35.12	—	93.38	128.50	38.63	—	102.72	141.35

Masonry

Hutchins Priced Schedules 2021	Unit	Labour Hours	Labour Net £	Plant Net £	Materials Net £	Unit Rate Net £	Labour Gross £	Plant Gross £	Materials Gross £	Unit Rate Gross £
							(Gross rates include 10% profit)			
7.01.	**NEW WORK**									
7.01.01.	**RECONSTRUCTED STONE BY SPECIALIST**									
7.01.01.05.	**Coping; bedded, jointed and pointed in cement mortar (1:6)**									
7.01.01.05.L	529mm x 75mm; once weathered and throated	m	0.67	35.12	—	104.35	139.47	38.63	—	114.79	153.42
7.01.01.05.M	305mm x 75mm; flat topped and throated	m	0.40	21.07	—	60.18	81.26	23.18	—	66.20	89.38
7.01.01.05.N	361mm x 75mm; flat topped and throated	m	0.54	28.10	—	71.30	99.39	30.91	—	78.42	109.33
7.01.01.05.O	417mm x 75mm; flat topped and throated	m	0.54	28.10	—	82.27	110.36	30.91	—	90.49	121.40
7.01.01.05.P	473mm x 75mm; flat topped and throated	m	0.67	35.12	—	93.38	128.50	38.63	—	102.72	141.35
7.01.01.05.Q	330mm x 100mm; moulded twice weathered	m	0.40	21.07	—	89.75	110.82	23.18	—	98.73	121.91
7.01.01.05.R	428mm x 100mm; moulded twice weathered	m	0.54	28.10	—	130.11	158.20	30.91	—	143.12	174.02
7.01.01.05.S	305mm x 75mm; twice chamfered and throated	m	0.40	21.07	—	60.18	81.26	23.18	—	66.20	89.38
7.01.01.05.T	361mm x 75mm; twice chamfered and throated	m	0.54	28.10	—	71.30	99.39	30.91	—	78.42	109.33
7.01.01.05.U	417mm x 75mm; twice chamfered and throated	m	0.54	28.10	—	82.27	110.36	30.91	—	90.49	121.40
7.01.01.05.V	473mm x 75mm; twice chamfered and throated	m	0.67	35.12	—	93.38	128.50	38.63	—	102.72	141.35
7.01.01.06.	**Chimney cap; weathered and throated all round**									
7.01.01.06.A	940mm x 600mm x 100mm holed for two 225mm x 225mm flues	nr	2.25	151.22	—	150.02	301.24	166.34	—	165.02	331.37
7.01.01.07.	**Square pier cap; weathered and throated all round**									
7.01.01.07.A	500mm	nr	0.60	40.33	—	84.08	124.40	44.36	—	92.48	136.84
7.01.01.07.B	400mm	nr	0.60	40.33	—	67.28	107.60	44.36	—	74.00	118.36
7.01.02.	**SOFT STONE BY SPECIALIST**									
7.01.02.01.	**Cut out to a depth of approximately 25mm, properly key, dowel and reinforce as necessary with non-ferrous metal; make good in plastic artificial stone to match existing**									
7.01.02.01.A	ashlar	m²	9.75	655.30	—	305.11	960.41	720.83	—	335.62	1056.45
7.01.02.01.B	moulding; 75mm girth	m	3.25	218.43	—	68.13	286.56	240.28	—	74.94	315.21
7.01.02.01.C	moulding; 75mm girth; increasing per 25mm of girth	m	0.60	40.33	—	5.37	45.69	44.36	—	5.90	50.26
7.01.02.01.D	plain or weathered coping; 300mm wide	m	4.40	295.72	—	94.45	390.17	325.30	—	103.89	429.19
7.01.02.01.E	plain or weathered coping; 300mm wide; increasing per 25mm of width	m	1.25	84.01	—	7.77	91.78	92.41	—	8.54	100.96
7.01.02.01.F	tracery	m²	18.00	1209.78	—	310.17	1519.95	1330.76	—	341.19	1671.95
7.01.02.01.G	mullion front from glazing; 150mm wide	m	4.75	319.25	—	94.45	413.70	351.17	—	103.89	455.06

Hutchins Priced Schedules 2021	Unit	Labour Hours	Labour Net £	Plant Net £	Materials Net £	Unit Rate Net £	Labour Gross £	Plant Gross £	Materials Gross £	Unit Rate Gross £
							(Gross rates include 10% profit)			
7.01.	**NEW WORK**									
7.01.02.	**SOFT STONE BY SPECIALIST**									
7.01.02.01.	Cut out to a depth of approximately 25mm, properly key, dowel and reinforce as necessary with non-ferrous metal; make good in plastic artificial stone to match existing									
7.01.02.01.H	circular labels or hoods; to 225mm girth	m	6.60	443.59	—	141.57	585.15	487.94	—	155.72	643.67
7.01.02.01.I	circular columns; plain	m	10.00	672.10	—	303.35	975.45	739.31	—	333.69	1073.00
7.01.02.01.J	circular columns; fluted	m	18.00	1209.78	—	303.35	1513.13	1330.76	—	333.69	1664.44
7.01.02.01.K	stooling to jambs or mullions; moulded	nr	2.65	178.11	—	24.89	203.00	195.92	—	27.38	223.30
7.01.02.01.L	stooling to jambs or mullions; plain	nr	1.40	94.09	—	24.22	118.31	103.50	—	26.64	130.14
7.01.03.	**PORTLAND STONE BY SPECIALIST**									
7.01.03.01.	Cut out to a depth of approximately 25mm, properly key, dowel and reinforce as necessary with non-ferrous metal; make good in plastic artificial stone to match existing									
7.01.03.01.A	ashlar	m²	12.50	840.13	—	305.11	1145.24	924.14	—	335.62	1259.76
7.01.03.01.B	moulding; 75mm girth	m	4.15	278.92	—	68.13	347.05	306.81	—	74.94	381.75
7.01.03.01.C	moulding; 75mm girth; increasing per 25mm of girth	m	0.83	55.78	—	7.77	63.55	61.36	—	8.54	69.91
7.01.03.01.D	plain or weathered coping; 300mm wide	m	5.50	369.66	—	94.45	464.10	406.62	—	103.89	510.51
7.01.03.01.E	plain or weathered coping; 300mm wide; per 25mm of width	m	1.50	100.82	—	7.77	108.58	110.90	—	8.54	119.44
7.01.03.01.F	tracery	m²	22.25	1495.42	—	310.17	1805.60	1644.96	—	341.19	1986.15
7.01.03.01.G	mullion front from glazing; 150mm wide	m	5.95	399.90	—	94.45	494.35	439.89	—	103.89	543.78
7.01.03.01.H	circular labels or hoods; to 225mm girth	m	8.25	554.48	—	141.57	696.05	609.93	—	155.72	765.65
7.01.03.01.I	circular columns; plain	m	12.85	863.65	—	303.35	1167.00	950.01	—	333.69	1283.70
7.01.03.01.J	circular columns; fluted	m	22.50	1512.23	—	303.35	1815.58	1663.45	—	333.69	1997.13
7.01.03.01.K	stooling to jambs or mullions; moulded	nr	3.35	225.15	—	24.89	250.05	247.67	—	27.38	275.05
7.01.03.01.L	stooling to jambs or mullions; plain	nr	1.75	117.62	—	24.89	142.51	129.38	—	27.38	156.76
7.01.04.	**YORK STONE BY SPECIALIST**									
7.01.04.01.	Cut out to a depth of approximately 25mm, properly key, dowel and reinforce as necessary with non-ferrous metal; make good in plastic artificial stone to match existing									
7.01.04.01.A	ashlar	m²	13.75	924.14	—	305.11	1229.25	1016.55	—	335.62	1352.17
7.01.04.01.B	moulding; 75mm girth	m	4.65	312.53	—	68.13	380.65	343.78	—	74.94	418.72
7.01.04.01.C	moulding; 75mm girth; increasing per 25mm of girth	m	1.00	67.21	—	7.77	74.98	73.93	—	8.54	82.48
7.01.04.01.D	plain or weathered coping; 300mm wide	m	6.20	416.70	—	94.45	511.15	458.37	—	103.89	562.26
7.01.04.01.E	plain or weathered coping; 300mm wide; increasing per 25mm of width	m	1.70	114.26	—	7.77	122.02	125.68	—	8.54	134.23
7.01.04.01.F	tracery	m²	25.00	1680.25	—	310.17	1990.42	1848.28	—	341.19	2189.46
7.01.04.01.G	mullion front from glazing; 150mm wide	m	6.65	446.95	—	94.45	541.39	491.64	—	103.89	595.53

Masonry

	Unit	Labour Hours	Labour Net £	Plant Net £	Materials Net £	Unit Rate Net £	Labour Gross £	Plant Gross £	Materials Gross £	Unit Rate Gross £
						(Gross rates include 10% profit)				
7.01.	**NEW WORK**									
7.01.04.	**YORK STONE BY SPECIALIST**									
7.01.04.01.	Cut out to a depth of approximately 25mm, properly key, dowel and reinforce as necessary with non-ferrous metal; make good in plastic artificial stone to match existing									
7.01.04.01.H	circular labels or hoods; to 225mm girth	m	9.25	621.69	—	141.57	763.26	683.86	—	155.72	839.58
7.01.04.01.I	circular columns; plain	m	14.50	974.55	—	303.35	1277.90	1072.00	—	333.69	1405.68
7.01.04.01.J	circular columns; fluted	m	25.00	1680.25	—	303.35	1983.60	1848.28	—	333.69	2181.96
7.01.04.01.K	stooling to jambs or mullions; moulded	nr	3.65	245.32	—	24.89	270.21	269.85	—	27.38	297.23
7.01.04.01.L	stooling to jambs or mullions; plain	nr	1.95	131.06	—	24.89	155.95	144.17	—	27.38	171.55
7.01.05.	**CLEANING STONEWORK OR BRICKWORK BY SPECIALIST**									
7.01.05.01.	Clean by nebulous cold water									
7.01.05.01.A	spray process assisted by suitable graded brushes	m²	0.43	28.56	16.89	—	45.45	31.42	18.58	—	50.00
7.01.05.03.	Clean by hydrochloric acid based cleaner									
7.01.05.03.A	and high pressure water process	m²	0.16	10.75	6.36	0.05	17.16	11.83	6.99	0.05	18.88
7.01.06.	**REPOINTING BY SPECIALIST**									
7.01.06.01.	Rake out and repoint									
7.01.06.01.A	rubble walling	m²	1.34	70.35	—	1.58	71.93	77.38	—	1.74	79.13
7.01.06.01.B	flint walling	m²	4.53	237.36	—	2.64	240.00	261.09	—	2.91	264.00
7.01.06.01.C	ashlar walling	m²	1.01	52.73	—	0.92	53.66	58.01	—	1.02	59.02
7.01.07.	**STONE FIXING BY SPECIALIST**									
7.01.07.01.	Labours on stonework									
7.01.07.01.A	fixing only natural or reconstructed stonework	m³	80.49	4219.29	—	3.30	4222.59	4641.21	—	3.63	4644.85
7.01.07.01.B	cut out and piece into existing work	m³	251.53	13185.31	—	1.58	13186.89	14503.84	—	1.74	14505.58
7.01.07.01.C	fixing only rubble walling average 125mm thick faced one side only	m²	8.22	430.79	—	1.72	432.50	473.87	—	1.89	475.75
7.01.07.01.D	fixing only flint walling	m²	19.28	1010.87	—	1.72	1012.58	1111.95	—	1.89	1113.84
7.01.07.01.E	lay paving stones	m²	2.41	126.54	—	0.66	127.20	139.20	—	0.73	139.92
7.01.10.	**CAST STONEWORK**									
7.01.10.01.	Bradstone walling blocks 100mm thick in gauged mortar (1:1:6) flush pointed on exposed faces as the work proceeds; walls and skins of hollow walls; finish									
7.01.10.01.A	tooled	m²	0.92	48.02	—	61.90	109.92	52.82	—	68.09	120.91
7.01.10.01.B	rough dressed	m²	0.92	48.02	—	63.27	111.29	52.82	—	69.60	122.42
7.01.10.01.C	square coursed rubble	m²	0.92	48.02	—	63.87	111.89	52.82	—	70.26	123.07
7.01.10.01.D	masonry block	m²	0.92	48.02	—	63.27	111.29	52.82	—	69.60	122.42

Hutchins Priced Schedules 2021	Unit	Labour Hours	Labour Net £	Plant Net £	Materials Net £	Unit Rate Net £	Labour Gross £	Plant Gross £	Materials Gross £	Unit Rate Gross £
						(Gross rates include 10% profit)				
7.01.	**NEW WORK**									
7.01.10.	**CAST STONEWORK**									
7.01.10.01.	Bradstone walling blocks 100mm thick in gauged mortar (1:1:6) flush pointed on exposed faces as the work proceeds; walls and skins of hollow walls; finish									
7.01.10.01.E	traditional	m²	0.92	48.02	—	63.40	111.42	52.82	—	69.74	122.56
7.01.10.01.F	square dressed	m²	0.92	48.02	—	63.40	111.42	52.82	—	69.74	122.56
7.01.10.02.	**Extra for quoin blocks**									
7.01.10.02.A	masonry finish	m	0.37	19.34	—	121.60	140.94	21.28	—	133.76	155.04
7.01.10.02.B	tooled finish	m	0.37	19.34	—	121.60	140.94	21.28	—	133.76	155.04
7.01.10.03.	**Fair returns; 100mm wide**									
7.01.10.03.A	masonry finish; masonry ends	m	0.35	18.35	—	—	18.35	20.18	—	—	20.18
7.01.10.03.B	masonry finish; dressed ends	m	0.35	18.35	—	4.79	23.14	20.18	—	5.27	25.45
7.01.10.03.C	squared course rubble finish	m	0.70	36.59	—	—	36.59	40.25	—	—	40.25
7.01.10.03.D	rough hewn finish	m	0.76	39.73	—	—	39.73	43.71	—	—	43.71
7.01.10.04.	**Lintel; bedded, jointed and pointed in cement mortar (1:3); in stock lengths**									
7.01.10.04.A	100mm x 140mm x 1200mm	nr	0.54	28.20	—	46.82	75.02	31.02	—	51.50	82.52
7.01.10.04.B	100mm x 140mm x 1675mm	nr	0.67	35.12	—	65.37	100.49	38.63	—	71.91	110.54
7.01.10.04.C	100mm x 140mm x 2000mm	nr	0.94	49.27	—	77.98	127.26	54.20	—	85.78	139.98
7.01.10.04.D	100mm x 215mm x 1525mm	nr	0.80	42.15	—	80.20	122.35	46.36	—	88.22	134.58
7.01.10.04.E	100mm x 215mm x 2000mm	nr	0.94	49.27	—	105.15	154.43	54.20	—	115.67	169.87
7.01.10.04.F	100mm x 215mm x 2450mm	nr	1.07	56.30	—	128.80	185.10	61.93	—	141.68	203.61
7.01.10.04.G	100mm x 215mm x 2900mm	nr	1.21	63.32	—	152.45	215.77	69.66	—	167.70	237.35
7.01.10.05.	**Sill including stooling; bedded, jointed and pointed in cement mortar (1:1:6); in stock lengths**									
7.01.10.05.A	140mm x 150mm; 665mm - 2545mm . . .	m	0.40	21.07	—	80.47	101.54	23.18	—	88.51	111.69
7.01.10.07.	**Traditional window surround components for non-standard windows in cement mortar (1:1:6)**									
7.01.10.07.A	146mm x 143mm; head	m	0.54	28.20	—	37.75	65.95	31.02	—	41.52	72.54
7.01.10.07.B	146mm x 143mm; continuous jamb	m	0.54	28.20	—	37.82	66.02	31.02	—	41.60	72.63
7.01.10.07.C	146mm x 143mm; sill	m	0.54	28.20	—	37.75	65.95	31.02	—	41.52	72.54
7.01.10.07.D	146mm x 108mm; mullion	m	0.54	28.20	—	28.69	56.89	31.02	—	31.55	62.58
7.01.10.07.E	162mm x 102mm; label mould	m	0.54	28.20	—	30.01	58.21	31.02	—	33.01	64.03
7.01.10.07.F	162mm x 102mm; kneeler	nr	0.20	10.59	—	30.01	40.60	11.65	—	33.01	44.66
7.01.12.	**NATURAL STONE RUBBLE WORK**									
7.01.12.01.	**Random walling in Yorkshire limestone (PC sum £330 per m3); in mortar; unfaced; average thickness**									
7.01.12.01.A	300mm	m²	2.35	123.08	—	119.74	242.83	135.39	—	131.72	267.11

Masonry

(Gross rates include 10% profit)

		Unit	Labour Hours	Labour Net £	Plant Net £	Materials Net £	Unit Rate Net £	Labour Gross £	Plant Gross £	Materials Gross £	Unit Rate Gross £
7.01.	**NEW WORK**										
7.01.12.	**NATURAL STONE RUBBLE WORK**										
7.01.12.01.	**Random walling in Yorkshire limestone (PC sum £330 per m3); in mortar; unfaced; average thickness**										
7.01.12.01.B	450mm	m²	3.52	184.62	—	180.39	365.01	203.09	—	198.43	401.51
7.01.12.01.C	600mm	m²	4.70	246.16	—	239.49	485.65	270.78	—	263.44	534.22
7.01.12.01.D	extra for one fair face	m²	0.67	35.12	—	—	35.12	38.63	—	—	38.63
7.01.12.01.E	extra for two fair faces	m²	1.27	66.78	—	—	66.78	73.46	—	—	73.46
7.01.12.02.	**Square rubble walling in Yorkshire limestone (PC sum £330 per m3); average thickness**										
7.01.12.02.A	300mm	m²	3.35	175.82	—	118.46	294.27	193.40	—	130.30	323.70
7.01.12.02.B	450mm	m²	5.10	267.24	—	177.56	444.79	293.96	—	195.31	489.27
7.01.12.02.C	600mm	m²	6.71	351.63	—	236.79	588.42	386.80	—	260.46	647.26
7.01.12.02.D	extra for one fair face	m²	0.67	35.12	—	—	35.12	38.63	—	—	38.63
7.01.12.02.E	extra for two fair faces	m²	1.27	66.78	—	—	66.78	73.46	—	—	73.46
7.01.20.	**NATURAL FLINTWORK**										
7.01.20.10.	**Random flintwork; in cement/lime mortar (1:2:9); face pointed one side; to backing blockwork (measured separately); thickness**										
7.01.20.10.A	100mm	m²	1.84	96.66	—	46.59	143.26	106.33	—	51.25	157.58
7.01.20.10.B	215mm	m²	3.52	184.62	—	85.31	269.93	203.09	—	93.84	296.93
7.02.	**REPAIRS AND ALTERATIONS**										
7.02.01.	**LABOURS**										
7.02.01.01.	**Take down masonry, clean and set aside**										
7.02.01.01.A	ashlar walling	m²	2.01	105.47	—	—	105.47	116.02	—	—	116.02
7.02.01.01.B	cornices, etc.	m	0.80	42.15	—	—	42.15	46.36	—	—	46.36
7.02.01.01.C	arches	nr	0.54	28.20	—	—	28.20	31.02	—	—	31.02
7.02.01.01.D	steps, sills, etc.	m	1.01	52.73	—	—	52.73	58.01	—	—	58.01
7.02.01.02.	**Take down masonry, clean and reset**										
7.02.01.02.A	ashlar walling	m²	5.20	272.48	—	11.58	284.06	299.73	—	12.74	312.47
7.02.01.02.B	cornices, etc.	m	2.21	116.06	—	2.70	118.76	127.66	—	2.97	130.64
7.02.01.02.C	arches	nr	1.48	77.37	—	4.76	82.13	85.11	—	5.24	90.35
7.02.01.02.D	steps, sills, etc.	m	1.34	70.35	—	3.48	73.82	77.38	—	3.82	81.21

Hutchins Priced Schedules 2021	Unit	Labour Hours	Labour Net £	Plant Net £	Materials Net £	Unit Rate Net £	Labour Gross £	Plant Gross £	Materials Gross £	Unit Rate Gross £
							(Gross rates include 10% profit)			
7.02.	**REPAIRS AND ALTERATIONS**									
7.02.01.	LABOURS									
7.02.01.03.	**Cut out decayed Portland (or similar) stone (PC sum £2580 per m3) in**									
7.02.01.03.A	facings of wall built in lime mortar in adjacent stones; prepare for and supply and fix new stones, average 50mm thick, point and clean down on completion . . .	m²	2.01	105.47	—	148.60	254.07	116.02	—	163.46	279.47
7.02.01.03.B	facings of wall built in lime mortar in separate stones; prepare for and supply and fix new stones, average 50mm thick, point and clean down on completion . . .	m²	2.41	126.54	—	155.42	281.96	139.20	—	170.96	310.16
7.02.01.04.	**Rake out joints of**									
7.02.01.04.A	ashlar stonework and repoint	m²	0.94	49.27	—	0.90	50.18	54.20	—	0.99	55.19
7.02.01.04.B	squared rubble and repoint	m²	1.07	56.30	—	0.90	57.20	61.93	—	0.99	62.92
7.02.01.05.	**Redress face of walling where decayed**									
7.02.01.05.A	with picked face and repoint	m²	5.37	281.29	—	0.51	281.80	309.41	—	0.57	309.98
7.02.01.06.	**Take up and reset 50mm thick Yorkstone slabs; any size; in landings, hearths, cover stones, pavings, etc.; in**									
7.02.01.06.A	lime mortar	m²	1.34	70.35	—	3.48	73.82	77.38	—	3.82	81.21
7.02.01.06.B	cement mortar	m²	1.61	84.40	—	3.20	87.60	92.84	—	3.52	96.36
7.02.01.07.	**Cut and form toothing in old masonry for**									
7.02.01.07.A	new brick or stone	m²	1.34	70.35	—	—	70.35	77.38	—	—	77.38
7.02.01.08.	**Take down and reset blocking courses, cornices, strings, plinths, apexes, kneelers, etc.; in**									
7.02.01.08.A	lime mortar	m	0.80	42.15	—	0.51	42.66	46.36	—	0.57	46.93
7.02.01.08.B	cement mortar	m	0.94	49.27	—	0.47	49.75	54.20	—	0.52	54.72
7.02.01.09.	**Take down and reset window sills, steps, etc.; including cutting away and making good**									
7.02.01.09.A	lime mortar	m	1.34	70.35	—	0.26	70.61	77.38	—	0.28	77.67
7.02.01.09.B	cement mortar	m	1.48	77.37	—	0.24	77.61	85.11	—	0.26	85.37
7.02.01.10.	**Repair with granite chippings concrete, including cutting out to a depth of at least 19mm; finish concrete fair and flush with original surface**									
7.02.01.10.A	treads	m²	1.74	91.42	—	3.10	94.52	100.56	—	3.41	103.97
7.02.01.10.B	landing	m²	1.07	56.30	—	3.10	59.40	61.93	—	3.41	65.33

Hutchins Priced Schedules 2021	Unit	Labour Hours	Labour Net £	Plant Net £	Materials Net £	Unit Rate Net £	Labour Gross £	Plant Gross £	Materials Gross £	Unit Rate Gross £
						(Gross rates include 10% profit)				

8.01.   **SLATE ROOFING**

8.01.01.   **WELSH SLATE**

8.01.01.01.   **Welsh slates; size 600mm x 300mm; 75mm lap; 50mm x 25mm treated sawn softwood battens; reinforced slaters underlining felt (formerly type 1F)**

Hutchins Priced Schedules 2021	Unit	Labour Hours	Labour Net £	Plant Net £	Materials Net £	Unit Rate Net £	Labour Gross £	Plant Gross £	Materials Gross £	Unit Rate Gross £
8.01.01.01.A sloping	m²	0.23	12.27	—	64.99	77.26	13.49	—	71.49	84.98
8.01.01.01.B vertical or mansard	m²	0.37	19.40	—	64.99	84.39	21.33	—	71.49	92.83
8.01.01.01.C extra for double eaves course	m	0.27	14.05	—	19.86	33.90	15.45	—	21.84	37.29
8.01.01.01.D extra for mitred hips; both sides measured	m	1.61	84.40	—	29.78	114.18	92.84	—	32.76	125.60
8.01.01.01.E extra for cutting to valleys; both sides measured	m	0.97	50.95	—	19.86	70.81	56.05	—	21.84	77.89
8.01.01.01.F hole for small pipe	nr	0.40	21.07	—	—	21.07	23.18	—	—	23.18
8.01.01.01.G hole for large pipe	nr	0.54	28.10	—	—	28.10	30.91	—	—	30.91
8.01.01.01.H fix only lead soakers	nr	0.03	1.78	—	—	1.78	1.96	—	—	1.96
8.01.01.01.I fix only hip irons	nr	0.17	8.81	—	—	8.81	9.69	—	—	9.69

8.01.01.02.   **Welsh slates; size 500mm x 250mm; 75mm lap; 47mm x 25mm treated sawn softwood battens; reinforced slaters underlining felt (formerly type 1F)**

Hutchins Priced Schedules 2021	Unit	Labour Hours	Labour Net £	Plant Net £	Materials Net £	Unit Rate Net £	Labour Gross £	Plant Gross £	Materials Gross £	Unit Rate Gross £
8.01.01.02.A sloping	m²	0.34	17.61	—	45.71	63.32	19.37	—	50.28	69.65
8.01.01.02.B vertical or mansard	m²	0.54	28.10	—	45.71	73.80	30.91	—	50.28	81.18
8.01.01.02.C extra for double eaves course	m	0.30	15.83	—	9.92	25.75	17.41	—	10.91	28.33
8.01.01.02.D extra for mitred hips; both sides measured	m	1.81	94.99	—	19.84	114.83	104.48	—	21.82	126.31
8.01.01.02.E extra for cutting to valleys; both sides measured	m	1.17	61.54	—	19.84	81.38	67.70	—	21.82	89.52
8.01.01.02.F hole for small pipe	nr	0.40	21.07	—	—	21.07	23.18	—	—	23.18
8.01.01.02.G hole for large pipe	nr	0.54	28.10	—	—	28.10	30.91	—	—	30.91
8.01.01.02.H fix only lead soakers	nr	0.03	1.78	—	—	1.78	1.96	—	—	1.96
8.01.01.02.I fix only hip irons	nr	0.17	8.81	—	—	8.81	9.69	—	—	9.69

8.01.01.03.   **Welsh slates; size 400mm x 200mm; 75mm lap; 47mm x 25mm treated sawn softwood battens; reinforced slaters underlining felt (formerly type 1F)**

Hutchins Priced Schedules 2021	Unit	Labour Hours	Labour Net £	Plant Net £	Materials Net £	Unit Rate Net £	Labour Gross £	Plant Gross £	Materials Gross £	Unit Rate Gross £
8.01.01.03.A sloping	m²	0.57	29.88	—	31.90	61.78	32.87	—	35.09	67.95
8.01.01.03.B vertical or mansard	m²	0.84	43.93	—	31.90	75.82	48.32	—	35.09	83.41
8.01.01.03.C extra for double eaves course	m	0.40	21.07	—	5.76	26.83	23.18	—	6.33	29.51

Roofing

|---|---|---|---|---|---|---|---|---|---|---|
| | | | | | | | | | | (Gross rates include 10% profit) |

8.01.   **SLATE ROOFING**

8.01.01.   **WELSH SLATE**

8.01.01.03.   **Welsh slates; size 400mm x 200mm; 75mm lap; 47mm x 25mm treated sawn softwood battens; reinforced slaters underlining felt (formerly type 1F)**

8.01.01.03.D	extra for mitred hips; both sides measured	m	2.35	123.08	—	9.60	132.68	135.39	—	10.55	145.94
8.01.01.03.E	extra for cutting to valleys; both sides measured	m	1.34	70.35	—	9.60	79.94	77.38	—	10.55	87.94
8.01.01.03.F	hole for small pipe	nr	0.40	21.07	—	—	21.07	23.18	—	—	23.18
8.01.01.03.G	hole for large pipe	nr	0.54	28.10	—	—	28.10	30.91	—	—	30.91
8.01.01.03.H	fix only lead soakers	nr	0.03	1.78	—	—	1.78	1.96	—	—	1.96
8.01.01.03.I	fix only hip irons	nr	0.17	8.81	—	—	8.81	9.69	—	—	9.69

8.01.02.   **ARTIFICIAL SLATES**

8.01.02.01.   **Artificial slates; blue / black; Eternit Thrutone Slate or the like; size 600mm x 300mm; 75mm lap; 47mm x 25mm treated sawn softwood battens; reinforced slaters underlining felt (formerly type 1F)**

8.01.02.01.A	sloping	m²	0.34	17.61	—	27.41	45.02	19.37	—	30.15	49.52
8.01.02.01.B	vertical or mansard	m²	0.44	22.86	—	27.41	50.26	25.14	—	30.15	55.29
8.01.02.01.C	extra for double eaves course	m	0.27	14.05	—	4.71	18.76	15.45	—	5.18	20.63
8.01.02.01.D	extra for mitred hips or cutting to valleys; both sides measured	m	0.97	50.95	—	4.71	55.66	56.05	—	5.18	61.23

8.01.03.   **FORTICRETE HARDROW SLATES**

8.01.03.01.   **Forticrete Hardrow slates; size 457mm x 305mm; 75mm lap; 38mm x 25mm treated softwood battens; reinforced slaters underlining felt (formerly type 1F)**

8.01.03.01.A	sloping	m²	0.34	17.61	—	45.22	62.84	19.37	—	49.74	69.12
8.01.03.01.B	vertical or mansard	m²	0.44	22.86	—	45.22	68.08	25.14	—	49.74	74.89
8.01.03.01.C	extra for eaves course	m	0.13	7.02	—	6.08	13.10	7.73	—	6.68	14.41
8.01.03.01.D	extra for ridge slates	m	0.20	10.59	—	28.66	39.25	11.65	—	31.53	43.17
8.01.03.01.E	extra for mitred hips or cutting to valleys; both sides measured	m	0.97	50.95	—	10.13	61.08	56.05	—	11.14	67.18

Hutchins Priced Schedules 2021	Unit	Labour Hours	Labour Net £	Plant Net £	Materials Net £	Unit Rate Net £	Labour Gross £	Plant Gross £	Materials Gross £	Unit Rate Gross £
						(Gross rates include 10% profit)				
8.02. **TILE ROOFING**										
8.02.01. **PLAIN TILES**										
8.02.01.01. Clay plain tiles; machine-made; smooth red; size 265mm x 165mm (PC sum £600 per 1000); 64mm lap; 19mm x 38mm treated sawn softwood battens; reinforced slaters underlining felt (formerly type 1F)										
8.02.01.01.A sloping	m²	0.94	49.27	—	46.30	95.57	54.20	—	50.93	105.13
8.02.01.01.B vertical or mansard	m²	1.07	56.30	—	46.30	102.60	61.93	—	50.93	112.86
8.02.01.01.C extra for verges	m	0.13	7.02	—	1.42	8.45	7.73	—	1.57	9.29
8.02.01.01.D extra for double eaves course	m	0.13	7.02	—	3.60	10.62	7.73	—	3.96	11.69
8.02.01.01.E extra for half round ridge tiles	m	0.57	29.88	—	13.17	43.04	32.87	—	14.48	47.35
8.02.01.01.F extra for half round hip tiles; cutting both sides	m	0.67	35.12	—	13.17	48.29	38.63	—	14.48	53.12
8.02.01.01.G extra for angle ridge tiles	m	0.57	29.88	—	18.43	48.31	32.87	—	20.27	53.14
8.02.01.01.H extra for bonnet hip tiles; cutting both sides	m	0.67	35.12	—	53.30	88.42	38.63	—	58.63	97.26
8.02.01.01.I extra for valley tiles; cutting both sides	m	0.67	35.12	—	53.50	88.42	38.63	—	58.63	97.26
8.02.01.01.J extra for intersection of ridge and hip	nr	0.47	24.64	—	—	24.64	27.10	—	—	27.10
8.02.01.03. Clay plain tiles; hand-made; 265mm x 165mm (PC sum £640 per 1000); 64mm lap; 19mm x 38mm treated sawn softwood battens; reinforced slaters underlining felt (formerly type 1F)										
8.02.01.03.A sloping	m²	0.94	49.27	—	48.94	98.21	54.20	—	53.83	108.04
8.02.01.03.B vertical or mansard	m²	1.07	56.30	—	48.94	105.24	61.93	—	53.83	115.76
8.02.01.03.C extra for verges	m	0.13	7.02	—	1.42	8.45	7.73	—	1.57	9.29
8.02.01.03.D extra for double eaves course	m	0.13	7.02	—	4.48	11.50	7.73	—	4.93	12.65
8.02.01.03.E extra for half round ridge tiles	m	0.57	29.88	—	13.91	43.79	32.87	—	15.30	48.16
8.02.01.03.F extra for hip tiles; cutting both sides	m	0.67	35.12	—	13.91	49.03	38.63	—	15.30	53.93
8.02.01.03.G extra for angle ridge tiles	m	0.57	29.88	—	13.91	43.79	32.87	—	15.30	48.16
8.02.01.03.H extra for bonnet hip tiles; cutting both sides	m	0.67	35.12	—	56.30	91.42	38.63	—	61.93	100.56
8.02.01.03.I extra for valley tiles; cutting both sides	m	0.67	35.12	—	56.30	91.42	38.63	—	61.93	100.56
8.02.01.03.J extra for intersection of ridge and hip	nr	0.47	24.64	—	—	24.64	27.10	—	—	27.10
8.02.01.05. Concrete plain tiles; BS EN 490:2011+A1:2017; size 267mm x 168mm; 64mm lap; 19mm x 38mm treated sawn softwood battens; reinforced slaters underlining felt (formerly type 1F)										
8.02.01.05.A sloping	m²	0.94	49.27	—	27.82	77.09	54.20	—	30.60	84.80
8.02.01.05.B vertical or mansard	m²	1.07	56.30	—	27.82	84.12	61.93	—	30.60	92.53
8.02.01.05.C extra for verges	m	0.11	5.98	—	1.42	7.40	6.57	—	1.57	8.14
8.02.01.05.D extra for double eaves course	m	0.13	7.02	—	1.92	8.94	7.73	—	2.11	9.84
8.02.01.05.E extra for segmental ridge tiles	m	0.57	29.88	—	8.13	38.01	32.87	—	8.94	41.81
8.02.01.05.F extra for segmental hip tiles; cutting both sides	m	0.67	35.12	—	8.84	43.96	38.63	—	9.73	48.36

Roofing

Hutchins Priced Schedules 2021	Unit	Labour Hours	Labour Net £	Plant Net £	Materials Net £	Unit Rate Net £	Labour Gross £	Plant Gross £	Materials Gross £	Unit Rate Gross £

(Gross rates include 10% profit)

	Unit	Labour Hours	Labour Net £	Plant Net £	Materials Net £	Unit Rate Net £	Labour Gross £	Plant Gross £	Materials Gross £	Unit Rate Gross £
8.02.	**TILE ROOFING**										
8.02.01.	**PLAIN TILES**										
8.02.01.05.	**Concrete plain tiles; BS EN 490:2011+A1:2017; size 267mm x 168mm; 64mm lap; 19mm x 38mm treated sawn softwood battens; reinforced slaters underlining felt (formerly type 1F)**										
8.02.01.05.G	extra for bonnet hip tiles; cutting both sides	m	0.67	35.12	—	60.86	95.98	38.63	—	66.95	105.58
8.02.01.05.H	extra for valley tiles; cutting both sides . .	m	0.67	35.12	—	60.86	95.98	38.63	—	66.95	105.58
8.02.01.05.I	extra for intersection of ridge and hip . . .	m	0.47	24.64	—	—	24.64	27.10	—	—	27.10
8.02.02.	**INTERLOCKING TILES**										
8.02.02.01.	**Concrete interlocking tiles; smooth finish; size 387mm x 230mm; 75mm lap 22mm x 38mm treated sawn softwood battens; reinforced underlining felt (formerly type 1F)**										
8.02.02.01.A	sloping	m²	0.22	11.64	—	16.48	28.12	12.80	—	18.13	30.93
8.02.02.01.B	vertical or mansard	m²	0.30	15.83	—	16.48	32.31	17.41	—	18.13	35.54
8.02.02.01.C	extra for verges; 150mm fibre reinforced cement strip undercloak	m	0.23	12.27	—	3.93	16.19	13.49	—	4.32	17.81
8.02.02.01.D	extra for ridge tiles	m	0.43	22.54	—	8.13	30.67	24.79	—	8.94	33.73
8.02.02.01.E	extra for hip tiles; cutting both sides	m	0.56	29.15	—	8.84	37.99	32.06	—	9.73	41.79
8.02.02.03.	**Concrete interlocking tiles; smooth finish; size 418mm x 330mm; 75mm lap; 22mm x 38mm treated sawn softwood battens; reinforced slaters underlining felt (formerly type 1F)**										
8.02.02.03.A	sloping	m²	0.15	7.76	—	14.07	21.83	8.53	—	15.48	24.01
8.02.02.03.B	vertical or mansard	m²	0.20	10.59	—	14.07	24.66	11.65	—	15.48	27.13
8.02.02.03.C	extra for verges; 150mm fibre reinforced cement strip undercloak	m	0.23	12.27	—	3.93	16.19	13.49	—	4.32	17.81
8.02.02.03.D	extra for ridge tiles	m	0.43	22.54	—	8.13	30.67	24.79	—	8.94	33.73
8.02.02.03.E	extra for hip tiles; cutting both sides	m	0.56	29.15	—	8.84	37.99	32.06	—	9.73	41.79

Roofing

Hutchins Priced Schedules 2021		Unit	Labour Hours	Labour Net £	Plant Net £	Materials Net £	Unit Rate Net £	Labour Gross £	Plant Gross £	Materials Gross £	Unit Rate Gross £
								(Gross rates include 10% profit)			
8.03.	**COUNTER BATTENS AND UNDERFELT**										
8.03.01.	**COUNTER BATTENS**										
8.03.01.01.	**Treated sawn softwood counter battens; fixed with galvanised nails to softwood**										
8.03.01.01.A	19mm x 38mm; 450mm centres	m²	0.05	2.83	—	0.92	3.75	3.11	—	1.01	4.13
8.03.01.01.B	19mm x 38mm; 600mm centres	m²	0.05	2.52	—	0.71	3.23	2.77	—	0.78	3.55
8.03.01.01.C	19mm x 38mm; 750mm centres	m²	0.04	2.10	—	0.58	2.68	2.31	—	0.64	2.95
8.03.01.01.D	25mm x 38mm; 450mm centres	m²	0.07	3.56	—	1.08	4.65	3.92	—	1.19	5.11
8.03.01.01.E	25mm x 38mm; 600mm centres	m²	0.06	3.15	—	0.83	3.98	3.46	—	0.92	4.38
8.03.01.01.F	25mm x 38mm; 750mm centres	m²	0.05	2.83	—	0.68	3.51	3.11	—	0.75	3.86
8.03.01.01.G	38mm x 38mm; 450mm centres	m²	0.09	4.93	—	1.25	6.18	5.42	—	1.37	6.79
8.03.01.01.H	38mm x 38mm; 600mm centres	m²	0.08	4.19	—	0.96	5.15	4.61	—	1.05	5.67
8.03.01.01.I	38mm x 38mm; 750mm centres	m²	0.07	3.56	—	0.78	4.34	3.92	—	0.86	4.78
8.03.01.01.J	38mm x 50mm; 450mm centres	m²	0.11	5.66	—	2.80	8.46	6.23	—	3.08	9.31
8.03.01.01.K	38mm x 50mm; 600mm centres	m²	0.09	4.61	—	2.13	6.74	5.07	—	2.34	7.41
8.03.01.01.L	38mm x 50mm; 750mm centres	m²	0.07	3.88	—	1.71	5.59	4.27	—	1.88	6.15
8.03.03.	**UNDERFELT**										
8.03.03.01.	**Reinforced slaters underlining felt; formerly type 1F; 150mm laps; secured with galvanised nails**										
8.03.03.01.A	standard	m²	0.05	2.52	—	2.92	5.43	2.77	—	3.21	5.98

Roofing

Hutchins Priced Schedules 2021		Unit	Labour Hours	Labour Net £	Plant Net £	Materials Net £	Unit Rate Net £	Labour Gross £	Plant Gross £	Materials Gross £	Unit Rate Gross £
								(Gross rates include 10% profit)			
8.04.	**ASPHALT BITUMEN FELT**										
8.04.01.	**MASTIC ASPHALT**										
8.04.01.01.	**20mm two coat coverings; felt isolating membrane**										
8.04.01.01.A	over 300mm wide	m²	0.54	28.10	0.98	54.62	83.69	30.91	1.07	60.08	92.06
8.04.01.01.B	not exceeding 150mm wide	m	0.19	10.17	0.64	8.21	19.01	11.19	0.70	9.03	20.91
8.04.01.01.C	150mm - 300mm wide	m	0.25	13.31	1.06	16.39	30.76	14.65	1.17	18.03	33.84
8.04.01.02.	**Extra over two coat coverings for**										
8.04.01.02.A	turning nibs into grooves	m	0.14	7.55	—	—	7.55	8.30	—	—	8.30
8.04.01.02.B	working to metal flashings	m	0.20	10.27	—	—	10.27	11.30	—	—	11.30
8.04.01.02.C	working into outlets	nr	1.36	71.40	—	—	71.40	78.54	—	—	78.54
8.04.01.03.	**13mm two coat skirtings; internal angle fillet; top edge turned into groove**										
8.04.01.03.A	150mm high	m	0.52	27.05	0.85	5.00	32.90	29.75	0.93	5.50	36.19
8.04.01.03.B	250mm high	m	0.64	33.44	0.68	8.34	42.46	36.79	0.75	9.17	46.71
8.04.01.04.	**13mm two coat aprons; undercut drip edge and rounded arris; including angles**										
8.04.01.04.A	75mm high	m	0.47	24.64	0.68	2.51	27.82	27.10	0.75	2.76	30.61
8.04.01.04.B	100mm high	m	0.60	31.66	0.30	3.34	35.29	34.83	0.33	3.67	38.82
8.04.01.05.	**13mm two coat linings to gutter; two rounded arrises; two internal angle fillets; including angles and intersections**										
8.04.01.05.A	300mm	m	1.31	68.57	0.76	8.69	78.02	75.42	0.84	9.56	85.83
8.04.01.05.B	extra for end	nr	0.49	25.48	—	—	25.48	28.02	—	—	28.02
8.04.01.05.C	extra for outlet	nr	1.34	70.35	—	—	70.35	77.38	—	—	77.38

Hutchins Priced Schedules 2021	Unit	Labour Hours	Labour Net £	Plant Net £	Materials Net £	Unit Rate Net £	Labour Gross £	Plant Gross £	Materials Gross £	Unit Rate Gross £
							(Gross rates include 10% profit)			

8.04.   **ASPHALT BITUMEN FELT**

8.04.01.   **MASTIC ASPHALT**

8.04.01.07.   **13mm two coat collar; 100mm high; internal angle fillet at junction with covering; fair edge and arris**

	Unit	Labour Hours	Labour Net £	Plant Net £	Materials Net £	Unit Rate Net £	Labour Gross £	Plant Gross £	Materials Gross £	Unit Rate Gross £
8.04.01.07.A small pipes and the like	nr	0.62	32.50	0.64	1.73	34.86	35.75	0.70	1.90	38.35
8.04.01.07.B large pipes and the like	nr	0.92	48.33	1.06	3.16	52.55	53.16	1.17	3.48	57.81

8.04.01.08.   **Accessories for asphalt roofing**

	Unit	Labour Hours	Labour Net £	Plant Net £	Materials Net £	Unit Rate Net £	Labour Gross £	Plant Gross £	Materials Gross £	Unit Rate Gross £
8.04.01.08.A aluminium edge trims; 64mm face x 60mm leg; including fixing with galvanised nails at not more than 300mm centres; working asphalt to trim	m	0.34	17.61	—	4.65	22.27	19.37	—	5.12	24.49
8.04.01.08.B extra for right angle corner piece	nr	0.57	29.88	—	5.47	35.35	32.87	—	6.02	38.89
8.04.01.08.C aluminium edge trims; 100mm x 63mm; including fixing with galvanised nails at not more than 300mm centres; working asphalt to trim	m	0.34	17.61	—	6.30	23.92	19.37	—	6.93	26.31
8.04.01.08.D extra for right angle corner piece	nr	0.57	29.88	—	7.42	37.30	32.87	—	8.16	41.03
8.04.01.08.E aluminium pressure release ventilators including asphalt collars	nr	0.81	42.25	—	26.06	68.31	46.48	—	28.67	75.14

8.04.02.   **BITUMEN FELT ROOFING**

8.04.02.01.   **Glassfibre felt coverings; stone chippings surfacing**

	Unit	Labour Hours	Labour Net £	Plant Net £	Materials Net £	Unit Rate Net £	Labour Gross £	Plant Gross £	Materials Gross £	Unit Rate Gross £
8.04.02.01.A two layer; flat coverings; over 300mm wide	m²	0.40	21.07	—	4.55	25.62	23.18	—	5.01	28.19
8.04.02.01.B extra for working into outlet	nr	0.67	35.12	—	—	35.12	38.63	—	—	38.63
8.04.02.01.C three layer; flat coverings; over 300mm wide	m²	0.54	28.10	—	6.65	34.75	30.91	—	7.32	38.22
8.04.02.01.D extra for working into outlet	nr	0.67	35.12	—	—	35.12	38.63	—	—	38.63

8.04.02.03.   **IKO glassfibre felt mineral surface coverings**

	Unit	Labour Hours	Labour Net £	Plant Net £	Materials Net £	Unit Rate Net £	Labour Gross £	Plant Gross £	Materials Gross £	Unit Rate Gross £
8.04.02.03.A two layer; sloping coverings; over 300mm wide	m²	0.54	28.10	—	4.20	32.30	30.91	—	4.62	35.53
8.04.02.03.B extra for working into outlet	nr	0.67	35.12	—	—	35.12	38.63	—	—	38.63
8.04.02.03.C three layer; sloping coverings; over 300mm wide	m²	0.67	35.12	—	6.30	41.42	38.63	—	6.93	45.56
8.04.02.03.D extra for working into outlet	nr	0.67	35.12	—	—	35.12	38.63	—	—	38.63
8.04.02.03.E 75mm wide aprons; fair drip edge at eaves or verges	m	0.25	13.00	—	0.16	13.16	14.30	—	0.17	14.47
8.04.02.03.F 150mm wide aprons; fair drip edge at eaves or verges	m	0.36	19.08	—	0.32	19.40	20.99	—	0.35	21.34

Roofing

| --- | --- | --- | --- | --- | --- | --- | --- | --- | --- | --- |
							(Gross rates include 10% profit)			
8.04.	**ASPHALT BITUMEN FELT**									
8.04.02.	**BITUMEN FELT ROOFING**									
8.04.02.03.	**IKO glassfibre felt mineral surface coverings**									
8.04.02.03.G	150mm girth skirtings; dressed over angle fillet	m	0.29	14.99	—	0.32	15.31	16.49	—	0.35	16.84
8.04.02.03.H	300mm girth skirtings; dressed over angle fillet	m	0.37	19.19	—	0.70	19.89	21.10	—	0.77	21.87
8.04.02.03.I	200mm three coat linings to gutters dressed over two angle fillets	m	0.27	14.26	—	1.26	15.52	15.68	—	1.39	17.07
8.04.02.03.J	extra for end	nr	0.51	26.63	—	—	26.63	29.29	—	—	29.29
8.04.02.03.K	extra for outlet	nr	0.67	35.12	—	—	35.12	38.63	—	—	38.63
8.04.02.03.L	300mm three coat linings to gutters dressed over two angle fillets	m	0.41	21.39	—	1.89	23.28	23.53	—	2.08	25.61
8.04.02.03.M	extra for end	nr	0.51	26.63	—	—	26.63	29.29	—	—	29.29
8.04.02.03.N	extra for outlet	nr	0.67	35.12	—	—	35.12	38.63	—	—	38.63
8.04.02.03.O	collars around small pipes	nr	1.01	52.73	—	—	52.73	58.01	—	—	58.01
8.04.02.03.P	collars around large pipes	nr	1.34	70.35	—	—	70.35	77.38	—	—	77.38
8.04.02.04.	**Fibre insulation board; bedded in hot bitumen**									
8.04.02.04.A	19mm thick	m²	0.61	25.02	1.27	8.54	34.83	27.52	1.40	9.40	38.32
8.04.02.04.B	extra for forming holes for pipe	nr	0.19	10.17	—	—	10.17	11.19	—	—	11.19
8.04.02.05.	**Resin bonded glassfibre slabs; bedded in hot bitumen**									
8.04.02.05.A	25mm thick	m²	1.07	49.03	1.27	4.52	54.82	53.93	1.40	4.97	60.30
8.04.02.05.B	extra for forming holes for pipe	nr	0.20	10.38	—	—	10.38	11.42	—	—	11.42
8.04.02.06.	**Vapour barrier; bedded in hot bitumen**									
8.04.02.06.A	felt	m²	0.60	24.70	1.27	1.84	27.82	27.18	1.40	2.03	30.60
8.04.02.08.	**Accessories for felt roofing**									
8.04.02.08.A	aluminium edge trims; 45mm face x 60mm leg; including fixing with galvanised nails at not more than 300mm centres; working feltwork to trim	m	0.23	11.95	—	3.88	15.83	13.15	—	4.26	17.41
8.04.02.08.B	extra for right angle corner piece	nr	0.43	22.65	—	5.12	27.76	24.91	—	5.63	30.54
8.04.02.08.C	aluminium edge trims; 100mm face x 63mm leg; including fixing with galvanised nails at not more than 300mm centres; working feltwork to trim	m	0.14	7.13	—	6.30	13.43	7.84	—	6.93	14.77
8.04.02.08.D	extra for right angle corner piece	nr	0.40	21.07	—	7.42	28.49	23.18	—	8.16	31.34

Roofing

Hutchins Priced Schedules 2021	Unit	Labour Hours	Labour Net £	Plant Net £	Materials Net £	Unit Rate Net £	Labour Gross £	Plant Gross £	Materials Gross £	Unit Rate Gross £
									(Gross rates include 10% profit)	
8.05.	**CORRUGATED SHEETING**									
8.05.01.	**REINFORCED CEMENT**									
8.05.01.01.	**Coverings; horizontal; 1090mm nominal width; 51mm corrugations; 100mm side and 152mm end laps; straight cutting**									
8.05.01.01.A	fixed to wood with galvanised drive screws and washers............	m²	0.34	17.61	—	10.04	27.66	19.37	—	11.05	30.42
8.05.01.01.B	fixed to steel purlins with hook bolts....	m²	0.47	24.64	—	10.07	34.70	27.10	—	11.07	38.17
8.05.01.02.	**Coverings; mansard and vertical; 1090mm nominal width; 51mm corrugations; 100mm side and 152mm end laps; straight cutting**									
8.05.01.02.A	fixed to wood with galvanised drive screws and washers............	m²	0.47	24.64	—	10.72	35.36	27.10	—	11.80	38.90
8.05.01.02.B	fixed to steel framing with hook bolts...	m²	0.67	35.12	—	10.76	45.88	38.63	—	11.83	50.46
8.05.01.03.	**Extra over cement sheets for vinyl translucent sheets**									
8.05.01.03.A	762mm wide............	m²	—	—	—	-1.40	-1.40	—	—	-1.53	-1.53
8.05.01.04.	**Sundry labours and finishings**									
8.05.01.04.A	raking cutting.............	m	0.23	12.27	—	—	12.27	13.49	—	—	13.49
8.05.01.04.B	barge boards.............	m	0.23	12.27	—	15.18	27.45	13.49	—	16.70	30.19
8.05.01.04.C	two piece close fitting ridge.........	m	0.37	19.40	—	23.82	43.22	21.33	—	26.21	47.54
8.05.01.04.E	eaves filler pieces.............	m	0.23	12.27	—	12.62	24.89	13.49	—	13.88	27.38
8.05.01.04.F	apron flashings.............	m	0.23	12.27	—	17.00	29.27	13.49	—	18.70	32.19
8.05.02.	**GALVANISED STEEL**									
8.05.02.01.	**0.7mm galvanised corrugated steel sheeting; profiled 32/1000; including one corrugation side lap and 150mm end lap**									
8.05.02.01.A	to wood purlins with drive screws and washers.............	m²	0.23	12.27	—	7.96	20.23	13.49	—	8.76	22.25
8.05.02.01.B	to steel purlins with hook bolts.......	m²	0.34	17.61	—	7.99	25.60	19.37	—	8.79	28.16
8.05.02.02.	**Sundry labours and finishings**									
8.05.02.02.A	raking cutting.............	m	0.47	24.64	—	—	24.64	27.10	—	—	27.10
8.05.02.02.B	24G corrugated steel ridge 375mm girth galvanised roll top roofing ridge 130mm x 130mm.............	m	0.20	10.59	—	8.19	18.78	11.65	—	9.01	20.66

Roofing

Hutchins Priced Schedules 2021	Unit	Labour Hours	Labour Net £	Plant Net £	Materials Net £	Unit Rate Net £	Labour Gross £	Plant Gross £	Materials Gross £	Unit Rate Gross £
							(Gross rates include 10% profit)			
8.08.	**SHEET METAL: ROOF DECKING**									
8.08.01.	COVERINGS LEAD; 1.80MM (CODE 4)									
8.08.01.01.	Coverings									
8.08.01.01.A	flat roof....................	m²	2.32	155.86	—	52.36	208.22	171.45	—	57.60	229.05
8.08.01.01.B	gutters, valleys, dormer roofs and cheeks	m²	2.90	194.84	—	52.36	247.21	214.33	—	57.60	271.93
8.08.01.01.C	aprons; cappings to ridges or hips	m²	3.00	201.90	—	52.36	254.26	222.09	—	57.60	279.69
8.08.01.01.D	damp proof course..............	m²	1.16	77.96	—	52.36	130.33	85.76	—	57.60	143.36
8.08.01.02.	Flashings; wedging into groove									
8.08.01.02.A	150mm girth.................	m	0.23	15.59	—	8.33	23.92	17.15	—	9.16	26.31
8.08.01.02.B	200mm girth.................	m	0.32	21.24	—	11.12	32.36	23.36	—	12.23	35.60
8.08.01.02.C	300mm girth.................	m	0.46	31.19	—	16.66	47.84	34.30	—	18.32	52.63
8.08.01.03.	Stepped flashings; wedging into groove									
8.08.01.03.A	150mm girth.................	m	0.29	19.49	—	8.33	27.82	21.44	—	9.16	30.60
8.08.01.03.B	200mm girth.................	m	0.50	33.27	—	11.12	44.39	36.60	—	12.23	48.83
8.08.01.03.C	300mm girth.................	m	0.61	41.07	—	16.66	57.72	45.17	—	18.32	63.49
8.08.01.04.	Soakers and hand to roofer									
8.08.01.04.A	200mm x 200mm	nr	0.17	11.69	—	2.24	13.94	12.86	—	2.47	15.33
8.08.01.04.B	300mm x 300mm	nr	0.20	13.44	—	4.99	18.43	14.79	—	5.49	20.27
8.08.01.05.	Slate; 400mm x 400mm with collar 200mm high around pipe									
8.08.01.05.A	100mm dia...................	nr	0.87	58.47	—	12.82	71.29	64.32	—	14.10	78.42
8.08.01.05.B	150mm dia...................	nr	0.88	59.14	—	14.41	73.56	65.06	—	15.85	80.91
8.08.01.06.	Slate; dressed through 225mm wall into rainwater head									
8.08.01.06.A	600mm x 450mm	nr	1.32	88.58	—	15.56	104.14	97.44	—	17.12	114.56
8.08.02.	COVERINGS LEAD; 2.24MM (CODE 5)									
8.08.02.01.	Coverings									
8.08.02.01.A	flat roof....................	m²	2.44	163.66	—	64.12	227.78	180.02	—	70.54	250.56
8.08.02.01.B	gutters, valleys, dormer roofs and cheeks	m²	3.05	204.72	—	64.12	268.85	225.19	—	70.54	295.73
8.08.02.01.C	aprons; cappings to ridges or hips	m²	3.15	211.85	—	64.12	275.97	233.03	—	70.54	303.57
8.08.02.01.D	damp proof course..............	m²	1.22	81.79	—	64.12	145.92	89.97	—	70.54	160.51
8.08.02.02.	Flashings; wedging into groove									
8.08.02.02.A	150mm girth.................	m	0.26	17.34	—	10.20	27.54	19.07	—	11.22	30.29
8.08.02.02.B	200mm girth.................	m	0.35	23.39	—	13.62	37.01	25.73	—	14.98	40.71
8.08.02.02.C	300mm girth.................	m	0.48	32.26	—	20.40	52.66	35.49	—	22.44	57.92
8.08.02.03.	Stepped flashings; wedging into groove									
8.08.02.03.A	150mm girth.................	m	0.32	21.64	—	10.20	31.84	23.81	—	11.22	35.02

Roofing

Hutchins Priced Schedules 2021		Unit	Labour Hours	Labour Net £	Plant Net £	Materials Net £	Unit Rate Net £	Labour Gross £	Plant Gross £	Materials Gross £	Unit Rate Gross £
								(Gross rates include 10% profit)			
8.08.	**SHEET METAL: ROOF DECKING**										
8.08.02.	**COVERINGS LEAD; 2.24MM (CODE 5)**										
8.08.02.03.	**Stepped flashings; wedging into groove**										
8.08.02.03.B	200mm girth	m	0.52	35.08	—	13.62	48.70	38.59	—	14.98	53.57
8.08.02.03.C	300mm girth	m	0.64	42.88	—	20.40	63.28	47.17	—	22.44	69.61
8.08.02.04.	**Soakers and hand to roofer**										
8.08.02.04.A	200mm x 200mm	nr	0.23	15.26	—	27.18	42.43	16.78	—	29.89	46.68
8.08.02.04.B	300mm x 300mm	nr	0.25	17.00	—	6.11	23.11	18.70	—	6.72	25.42
8.08.02.05.	**Slate; 400mm x 400mm with collar 200mm high around pipe**										
8.08.02.05.A	100mm dia	nr	0.93	62.37	—	15.69	78.07	68.61	—	17.26	85.87
8.08.02.05.B	200mm dia	nr	1.04	70.17	—	17.65	87.82	77.18	—	19.41	96.60
8.08.02.06.	**Slate; dressed through 225mm wall into rainwater head**										
8.08.02.06.A	600mm x 450mm	nr	1.37	92.08	—	19.05	111.13	101.29	—	20.96	122.24
8.08.03.	**SUNDRIES; LEAD SHEET COVERINGS**										
8.08.03.01.	**Coverings**										
8.08.03.01.A	beaded edges	m	0.32	21.24	—	—	21.24	23.36	—	—	23.36
8.08.03.01.B	soldered seam	m	1.58	106.26	—	1.10	107.36	116.88	—	1.21	118.09
8.08.03.01.C	soldered dots with brass screws	nr	0.53	35.42	—	0.31	35.73	38.96	—	0.34	39.30
8.08.03.01.D	bossed ends to rolls	nr	0.63	42.48	—	—	42.48	46.72	—	—	46.72
8.08.03.01.E	bossed angles to rolls	nr	0.79	53.16	—	—	53.16	58.48	—	—	58.48
8.08.03.01.F	bossed intersections to rolls	nr	0.92	61.97	—	—	61.97	68.16	—	—	68.16
8.08.04.	**COPPER SHEET COVERINGS; 0.55MM (24 SWG)**										
8.08.04.01.	**Coverings**										
8.08.04.01.A	flat roof covering; rolls and laps	m²	1.86	124.67	—	83.08	207.75	137.14	—	91.38	228.53
8.08.04.01.B	gutters, valleys, dormer roofs and cheeks	m²	2.32	155.86	—	83.08	238.94	171.45	—	91.38	262.83
8.08.04.01.C	aprons; cappings to ridges or hips	m²	2.42	162.92	—	83.08	245.99	179.21	—	91.38	270.59
8.08.04.02.	**Flashings; wedging into groove**										
8.08.04.02.A	150mm girth	m	0.29	19.49	—	13.21	32.70	21.44	—	14.53	35.97
8.08.04.02.B	200mm girth	m	0.35	23.39	—	17.64	41.03	25.73	—	19.41	45.14
8.08.04.02.C	300mm girth	m	0.52	35.08	—	26.43	61.51	38.59	—	29.07	67.66
8.08.04.03.	**Stepped flashings; wedging into groove**										
8.08.04.03.A	150mm girth	m	0.35	23.39	—	13.21	36.60	25.73	—	14.53	40.26
8.08.04.03.B	200mm girth	m	0.41	27.29	—	17.64	44.93	30.02	—	19.41	49.42
8.08.04.03.C	300mm girth	m	0.58	38.98	—	26.43	65.41	42.88	—	29.07	71.95

Roofing

Hutchins Priced Schedules 2021	Unit	Labour Hours	Labour Net £	Plant Net £	Materials Net £	Unit Rate Net £	Labour Gross £	Plant Gross £	Materials Gross £	Unit Rate Gross £
									(Gross rates include 10% profit)	
8.08.	**SHEET METAL: ROOF DECKING**									
8.08.05.	**ZINC SHEET COVERINGS; 0.6MM (12G)**									
8.08.05.01.	Coverings									
8.08.05.01.A	flat roof covering; rolls and laps	m²	1.74	116.88	—	33.68	150.56	128.57	—	37.05	165.62
8.08.05.01.B	gutters, valleys, dormer roofs and cheeks	m²	2.18	146.32	—	33.68	180.00	160.95	—	37.05	198.00
8.08.05.01.C	aprons; cappings to ridges or hips	m²	2.28	153.37	—	33.68	187.06	168.71	—	37.05	205.76
8.08.05.02.	**Flashings; wedging to groove**									
8.08.05.02.A	150mm girth	m	0.23	15.59	—	5.36	20.95	17.15	—	5.89	23.05
8.08.05.02.B	200mm girth	m	0.29	19.49	—	7.15	26.64	21.44	—	7.87	29.31
8.08.05.02.C	300mm girth	m	0.46	31.19	—	10.71	41.90	34.30	—	11.79	46.09
8.08.05.03.	**Stepped flashings; wedging into groove**									
8.08.05.03.A	150mm girth	m	0.29	19.49	—	5.36	24.85	21.44	—	5.89	27.33
8.08.05.03.B	200mm girth	m	0.35	23.39	—	7.15	30.54	25.73	—	7.87	33.60
8.08.05.03.C	300mm girth	m	0.52	35.08	—	10.71	45.80	38.59	—	11.79	50.38
8.08.06.	**ZINC SHEET COVERINGS; 0.8MM (14G)**									
8.08.06.01.	Coverings									
8.08.06.01.A	flat roof covering; rolls and laps	m²	1.74	116.88	—	45.60	162.48	128.57	—	50.16	178.73
8.08.06.01.B	gutters, valleys, dormer roofs and cheeks	m²	2.18	146.32	—	45.60	191.92	160.95	—	50.16	211.11
8.08.06.01.C	aprons; cappings to ridges or hips	m²	2.28	153.37	—	45.60	198.97	168.71	—	50.16	218.87
8.08.06.02.	**Flashings; wedging to groove**									
8.08.06.02.A	150mm girth	m	0.23	15.59	—	7.25	22.85	17.15	—	7.98	25.13
8.08.06.02.B	200mm girth	m	0.29	19.49	—	9.68	29.18	21.44	—	10.65	32.09
8.08.06.02.C	300mm girth	m	0.46	31.19	—	14.51	45.69	34.30	—	15.96	50.26
8.08.06.03.	**Stepped flashings; wedging into groove**									
8.08.06.03.A	150mm girth	m	0.29	19.49	—	7.25	26.74	21.44	—	7.98	29.42
8.08.06.03.B	200mm girth	m	0.35	23.39	—	9.68	33.07	25.73	—	10.65	36.38
8.08.06.03.C	300mm girth	m	0.52	35.08	—	14.51	49.59	38.59	—	15.96	54.55
8.08.08.	**ALUMINIUM SHEET COVERINGS; 1MM COMMERCIAL GRADE**									
8.08.08.01.	Coverings									
8.08.08.01.A	flat roof covering; rolls and laps	m²	1.86	124.67	—	21.03	145.71	137.14	—	23.13	160.28
8.08.08.01.B	gutters, valleys, dormer roofs and cheeks	m²	2.32	155.86	—	21.03	176.89	171.45	—	23.13	194.58
8.08.08.01.C	aprons; cappings to ridges or hips	m²	2.42	162.92	—	21.03	183.95	179.21	—	23.13	202.34
8.08.08.02.	**Flashings; wedging to groove**									
8.08.08.02.A	150mm girth	m	0.29	19.49	—	3.35	22.84	21.44	—	3.68	25.12
8.08.08.02.B	200mm girth	m	0.35	23.39	—	4.47	27.86	25.73	—	4.91	30.64
8.08.08.02.C	300mm girth	m	0.52	35.08	—	6.69	41.77	38.59	—	7.36	45.95

Hutchins Priced Schedules 2021	Unit	Labour Hours	Labour Net £	Plant Net £	Materials Net £	Unit Rate Net £	Labour Gross £	Plant Gross £	Materials Gross £	Unit Rate Gross £
							(Gross rates include 10% profit)			

8.08. **SHEET METAL: ROOF DECKING**

8.08.08. **ALUMINIUM SHEET COVERINGS; 1MM COMMERCIAL GRADE**

8.08.08.03. Stepped flashings; wedging into groove

8.08.08.03.A	150mm girth	m	0.35	23.39	—	3.35	26.73	25.73	—	3.68	29.41
8.08.08.03.B	200mm girth	m	0.41	27.29	—	4.47	31.75	30.02	—	4.91	34.93
8.08.08.03.C	300mm girth	m	0.58	38.98	—	6.69	45.67	42.88	—	7.36	50.24

8.09. **REPAIRS AND ALTERATIONS**

8.09.01. **REMOVE COVERINGS AND LOAD INTO SKIP**

8.09.01.01. Roof coverings

8.09.01.01.A	slates	m²	0.55	16.27	—	—	16.27	17.90	—	—	17.90
8.09.01.01.B	nibbed tiles	m²	0.45	13.31	—	—	13.31	14.64	—	—	14.64
8.09.01.01.C	corrugated metal sheeting	m²	0.45	13.31	—	—	13.31	14.64	—	—	14.64
8.09.01.01.D	underfelt and nails	m²	0.07	2.07	—	—	2.07	2.28	—	—	2.28
8.09.01.01.E	three layers felt	m²	0.28	8.28	—	—	8.28	9.11	—	—	9.11
8.09.01.01.F	sheet metal	m²	0.55	16.27	—	—	16.27	17.90	—	—	17.90

8.09.01.02. Metal finishings

8.09.01.02.A	horizontal	m	0.22	6.51	—	—	6.51	7.16	—	—	7.16
8.09.01.02.B	stepped	m	0.28	8.28	—	—	8.28	9.11	—	—	9.11

8.09.01.03. Battens including withdrawing nails

8.09.01.03.A	tile or slate	m²	0.10	2.96	—	—	2.96	3.25	—	—	3.25

8.09.01.04. Remove coverings, carefully handling and disposing of toxic or other special waste by approved method

8.09.01.04.A	asbestos cement sheeting (including active waste landfill tax standard rate 1 April 2020 to 31 March 2021)	m²	1.48	77.37	4.99	—	82.36	85.11	5.49	—	90.60
8.09.01.04.B	asbestos cement sheeting (including active waste landfill tax standard rate 1 April 2021 to 31 March 2022)	m²	1.48	77.37	5.13	—	82.50	85.11	5.64	—	90.75

8.09.01.05. Strip, clean and set aside sound slates or tiles for re-use

8.09.01.05.A	slates	m²	0.40	11.83	—	—	11.83	13.02	—	—	13.02
8.09.01.05.B	tiles	m²	0.30	8.87	—	—	8.87	9.76	—	—	9.76
8.09.01.05.C	cement slates	m²	0.40	11.83	—	—	11.83	13.02	—	—	13.02

8.09.01.06. Strip, clean and set aside for removal

8.09.01.06.A	slates	m²	0.30	8.87	—	—	8.87	9.76	—	—	9.76
8.09.01.06.B	tiles	m²	0.24	7.10	—	—	7.10	7.81	—	—	7.81
8.09.01.06.C	cement slates	m²	0.30	8.87	—	—	8.87	9.76	—	—	9.76

Roofing

Hutchins Priced Schedules 2021		Unit	Labour Hours	Labour Net £	Plant Net £	Materials Net £	Unit Rate Net £	Labour Gross £	Plant Gross £	Materials Gross £	Unit Rate Gross £
								(Gross rates include 10% profit)			
8.09.	REPAIRS AND ALTERATIONS										
8.09.02.	RENEWALS										
8.09.02.01.	Tile battens to 100mm gauge										
8.09.02.01.A	area not exceeding 8.50m2	m²	0.23	11.95	—	5.06	17.01	13.15	—	5.57	18.71
8.09.02.01.B	area exceeding 8.50m2	m²	0.17	9.12	—	5.06	14.18	10.03	—	5.57	15.60
8.09.02.02.	Slate battens to 205mm gauge										
8.09.02.02.A	area not exceeding 8.50m2	m²	0.17	9.12	—	2.96	12.08	10.03	—	3.26	13.29
8.09.02.02.B	area exceeding 8.50m2	m²	0.11	5.98	—	2.96	8.94	6.57	—	3.26	9.83
8.09.02.03.	Roofing felt										
8.09.02.03.A	per m2	m²	0.09	4.93	—	3.30	8.23	5.42	—	3.63	9.05
8.09.02.04.	Single slates with clips; including taking off; first slate										
8.09.02.04.A	400mm x 200mm	nr	0.36	18.66	—	1.00	19.66	20.53	—	1.10	21.63
8.09.02.04.B	500mm x 250mm	nr	0.38	19.71	—	1.72	21.43	21.68	—	1.89	23.57
8.09.02.04.C	600mm x 300mm	nr	0.39	20.44	—	2.48	22.92	22.49	—	2.72	25.21
8.09.02.05.	Single slates with clips; including taking off; second and subsequent slate up to 30nr										
8.09.02.05.A	400mm x 200mm	nr	0.15	7.76	—	1.00	8.76	8.53	—	1.10	9.63
8.09.02.05.B	500mm x 250mm	nr	0.15	8.07	—	1.72	9.79	8.88	—	1.89	10.77
8.09.02.05.C	600mm x 300mm	nr	0.16	8.49	—	2.48	10.97	9.34	—	2.72	12.06
8.09.02.06.	Single tiles, including taking off; first tile										
8.09.02.06.A	clay (PC sum £600 per 1000)	nr	0.24	12.69	—	0.60	13.29	13.95	—	0.66	14.61
8.09.02.06.B	concrete	nr	0.24	12.69	—	0.32	13.01	13.95	—	0.35	14.31
8.09.02.07.	Single tiles, including taking off; second and subsequent tiles up to 50nr										
8.09.02.07.A	clay (PC sum £600 per 1000)	nr	0.15	7.76	—	0.60	8.36	8.53	—	0.66	9.19
8.09.02.07.B	concrete	nr	0.15	7.76	—	0.32	8.08	8.53	—	0.35	8.89
8.09.02.10.	Cement mortar (1:3) angle fillet to slate or tile roof at abutment to										
8.09.02.10.A	walls or chimney stacks	m	0.27	14.05	—	0.36	14.40	15.45	—	0.39	15.84
8.09.02.11.	Strip roof slating; sort slates and reslate roof using 50% salvaged slates										
8.09.02.11.A	400mm x 200mm	m²	0.77	40.47	—	13.10	53.57	44.52	—	14.41	58.93
8.09.02.11.B	500mm x 250mm	m²	0.54	28.10	—	13.13	41.22	30.91	—	14.44	45.34
8.09.02.11.C	600mm x 300mm	m²	0.34	17.61	—	13.28	30.89	19.37	—	14.60	33.98

Roofing

Hutchins Priced Schedules 2021	Unit	Labour Hours	Labour Net £	Plant Net £	Materials Net £	Unit Rate Net £	Labour Gross £	Plant Gross £	Materials Gross £	Unit Rate Gross £
									(Gross rates include 10% profit)	
8.09.	**REPAIRS AND ALTERATIONS**									
8.09.02.	**RENEWALS**									
8.09.02.12.	**Strip 267mm x 165mm plain roof tiles and retile using 50% salvaged tiles**									
8.09.02.12.A	clay (PC sum for new £600 per 1000) . . .	m²	1.14	59.76	—	19.80	79.56	65.73	—	21.78	87.51
8.09.02.12.B	concrete	m²	1.14	59.76	—	10.56	70.32	65.73	—	11.62	77.35
8.09.03.	**REPAIRS TO SHEET METAL**									
8.09.03.01.	**Take up old lead; any position or weight**									
8.09.03.01.A	and set aside	m²	0.21	14.18	—	—	14.18	15.60	—	—	15.60
8.09.03.01.B	and relay to boarded flats including dressing over rolls and drips, with new bossed ends, etc.	m²	2.27	152.30	—	—	152.30	167.53	—	—	167.53
8.09.03.02.	**Redress lead flashings**									
8.09.03.02.A	and rewedge and repoint	m	0.42	28.36	—	—	28.36	31.20	—	—	31.20
8.09.03.03.	**Take up old zinc; in any position**									
8.09.03.03.A	and remove from site	m²	0.53	35.42	—	—	35.42	38.96	—	—	38.96
8.09.03.04.	**Take up existing zinc and supply and lay new zinc including rolls, laps, etc.**									
8.09.03.04.A	0.4mm thick (10G)	m²	2.11	141.68	—	25.87	167.55	155.85	—	28.46	184.31
8.09.03.04.B	0.6mm thick (12G)	m²	2.37	159.42	—	33.68	193.11	175.36	—	37.05	212.42
8.09.03.04.C	0.8mm thick (14G)	m²	2.64	177.10	—	45.60	222.70	194.81	—	50.16	244.97
8.09.03.05.	**Prepare zinc roofs and supply and lay bitumenised fabric**									
8.09.03.05.A	bedded in bitumen, apply two coats of bitumen paint	m²	1.00	67.28	—	6.84	74.12	74.00	—	7.53	81.53
8.09.03.06.	**Seal crack in holes in zinc or asphalt flats**									
8.09.03.06.A	and apply two coats of bitumen paint . . .	m²	0.47	31.86	—	1.67	33.52	35.04	—	1.83	36.87
8.09.03.07.	**Clean and treat defective sheet zinc with**									
8.09.03.07.A	zinc repair compound	m²	0.58	38.98	—	3.41	42.39	42.88	—	3.75	46.63
8.09.03.08.	**Remove slates, take out defective box gutter linings or valleys and**									
8.09.03.08.A	renew wood linings and line with 0.6mm (12G) zinc and replace slates	m²	4.22	283.36	—	90.89	374.25	311.69	—	99.98	411.68

Roofing

Hutchins Priced Schedules 2021	Unit	Labour Hours	Labour Net £	Plant Net £	Materials Net £	Unit Rate Net £	Labour Gross £	Plant Gross £	Materials Gross £	Unit Rate Gross £
							(Gross rates include 10% profit)			
8.09.	**REPAIRS AND ALTERATIONS**									
8.09.04.	PROVISION OF LADDERS									
8.09.04.01.	Labour and transport up to five miles each way, and set up and remove ladders up to									
8.09.04.01.A	two storeys high	nr	2.00	59.16	4.27	—	63.43	65.08	4.70	—	69.77
8.09.04.01.B	four storeys high	nr	2.50	73.95	4.27	—	78.22	81.35	4.70	—	86.04
8.09.04.01.C	extra over last two items if access is difficult e.g. one house in a terrace with no side entrance	nr	2.00	59.16	4.27	—	63.43	65.08	4.70	—	69.77
8.09.05.	REMOVAL OF ASPHALT									
8.09.05.01.	Remove waterproofing finishes and load into skip									
8.09.05.01.A	asphalt paving	m²	0.67	19.82	—	—	19.82	21.80	—	—	21.80
8.09.05.01.B	asphalt roofing	m²	1.10	32.54	—	—	32.54	35.79	—	—	35.79
8.09.05.01.C	asphalt skirting	m	0.17	5.03	—	—	5.03	5.53	—	—	5.53
8.09.06.	ROOF TREATMENTS BY SPECIALISTS									
8.09.06.01.	Clean, prepare and apply bitumen roofing compound									
8.09.06.01.A	to sound surfaces; asphalt, concrete, felt, slate or tile	m²	0.55	16.27	—	6.47	22.74	17.90	—	7.11	25.01
8.09.06.01.B	and fungicide solution to surfaces likely to support fungal or algae growth; asphalt, concrete, felt, slate or tile	m²	0.60	17.75	—	7.08	24.83	19.52	—	7.79	27.31

Sundries

Hutchins Priced Schedules 2021		Unit	Labour Hours	Labour Net £	Plant Net £	Materials Net £	Unit Rate Net £	Labour Gross £	Plant Gross £	Materials Gross £	Unit Rate Gross £
								(Gross rates include 10% profit)			
9.01.	**STEELWORK AND METALWORK**										
9.01.01.	**UNFRAMED STEELWORK**										
9.01.01.01.	**Rolled joist beams**										
9.01.01.01.A	127mm x 76mm x 13kg	m	0.15	10.15	—	15.00	25.15	11.16	—	16.50	27.66
9.01.01.01.B	152mm x 89mm x 16kg	m	0.21	13.85	—	20.00	33.85	15.23	—	22.00	37.23
9.01.01.01.C	178mm x 102mm x 19kg	m	0.24	16.20	—	23.00	39.20	17.82	—	25.30	43.12
9.01.01.01.D	203mm x 102mm x 23kg	m	0.29	19.56	—	29.00	48.56	21.51	—	31.90	53.41
9.01.01.02.	**Universal beams**										
9.01.01.02.A	203mm x 133mm x 25kg	m	0.29	19.56	—	26.60	46.16	21.51	—	29.26	50.77
9.01.01.02.B	203mm x 133mm x 30kg	m	0.35	23.32	—	33.00	56.32	25.65	—	36.30	61.95
9.01.01.02.C	254mm x 146mm x 31kg	m	0.36	23.99	—	33.25	57.24	26.39	—	36.58	62.97
9.01.01.02.D	254mm x 146mm x 37kg	m	0.43	28.70	—	38.95	67.65	31.57	—	42.85	74.41
9.01.01.02.E	254mm x 146mm x 43kg	m	0.50	33.40	—	45.60	79.00	36.74	—	50.16	86.90
9.01.01.02.F	305mm x 165mm x 40kg	m	0.46	31.05	—	42.75	73.80	34.16	—	47.03	81.18
9.01.01.02.G	305mm x 165mm x 46kg	m	0.53	35.76	—	48.45	84.21	39.33	—	53.30	92.63
9.01.01.02.H	305mm x 165mm x 54kg	m	0.62	41.87	—	57.00	98.87	46.06	—	62.70	108.76
9.01.01.02.I	254mm x 102mm x 25kg	m	0.29	19.56	—	28.00	47.56	21.51	—	30.80	52.31
9.01.01.02.J	305mm x 102mm x 33kg	m	0.38	25.67	—	37.00	62.67	28.24	—	40.70	68.94
9.01.01.02.K	356mm x 171mm x 51kg	m	0.59	39.52	—	54.15	93.67	43.47	—	59.57	103.04
9.01.02.	**METALWORK**										
9.01.02.01.	**Mild steel flats and plates**										
9.01.02.01.A	50mm x 6mm	m	0.14	9.48	—	2.45	11.92	10.42	—	2.69	13.12
9.01.02.01.B	80mm x 6mm	m	0.20	13.51	—	4.58	18.09	14.86	—	5.04	19.90
9.01.02.01.C	80mm x 12mm	m	0.20	13.51	—	8.65	22.16	14.86	—	9.52	24.38
9.01.02.01.D	100mm x 6mm	m	0.24	16.20	—	4.88	21.08	17.82	—	5.37	23.19
9.01.02.01.E	100mm x 12mm	m	0.24	16.20	—	9.78	25.97	17.82	—	10.75	28.57
9.01.02.01.F	125mm x 6mm	m	0.29	19.56	—	7.28	26.83	21.51	—	8.00	29.52
9.01.02.01.G	150mm x 6mm	m	0.34	22.99	—	8.15	31.13	25.28	—	8.96	34.25
9.01.02.01.H	150mm x 12mm	m	0.34	22.99	—	16.83	39.82	25.28	—	18.51	43.80
9.01.02.01.I	200mm x 6mm	m	0.44	29.71	—	11.36	41.07	32.68	—	12.50	45.17
9.01.02.01.J	254mm x 6mm	m	0.55	37.10	—	14.11	51.21	40.81	—	15.52	56.33
9.01.02.01.K	254mm x 12mm	m	0.55	37.10	—	27.71	64.81	40.81	—	30.48	71.29

Sundries

Hutchins Priced Schedules 2021		Unit	Labour Hours	Labour Net £	Plant Net £	Materials Net £	Unit Rate Net £	Labour Gross £	Plant Gross £	Materials Gross £	Unit Rate Gross £
								(Gross rates include 10% profit)			
9.01.	**STEELWORK AND METALWORK**										
9.01.03.	**CATNIC GALVANISED STEEL LINTELS**										
9.01.03.01.	CG90/100; combined lintel; for cavity wall; length										
9.01.03.01.A	750mm	nr	0.24	12.69	—	19.50	32.19	13.95	—	21.45	35.41
9.01.03.01.B	900mm	nr	0.25	13.31	—	22.84	36.16	14.65	—	25.13	39.77
9.01.03.01.C	1050mm	nr	0.27	14.05	—	26.12	40.17	15.45	—	28.73	44.18
9.01.03.01.D	1200mm	nr	0.28	14.78	—	30.45	45.24	16.26	—	33.50	49.76
9.01.03.01.E	1350mm	nr	0.29	15.41	—	34.28	49.69	16.95	—	37.71	54.66
9.01.03.02.	CG90/100; combined lintel; for cavity wall; length										
9.01.03.02.A	1500mm	nr	0.31	16.15	—	38.08	54.22	17.76	—	41.88	59.64
9.01.03.02.B	1650mm	nr	0.32	16.88	—	41.97	58.84	18.57	—	46.16	64.73
9.01.03.02.C	1800mm	nr	0.34	17.61	—	45.83	63.45	19.37	—	50.42	69.79
9.01.03.02.E	1950mm	nr	0.35	18.35	—	51.56	69.91	20.18	—	56.71	76.90
9.01.03.02.F	2100mm	nr	0.36	18.98	—	54.52	73.49	20.87	—	59.97	80.84
9.01.03.02.G	2250mm	nr	0.38	19.71	—	61.91	81.62	21.68	—	68.11	89.79
9.01.03.02.H	2400mm	nr	0.39	20.44	—	65.73	86.17	22.49	—	72.30	94.79
9.01.03.02.I	2550mm	nr	0.40	21.07	—	72.21	93.28	23.18	—	79.43	102.61
9.01.03.02.J	2700mm	nr	0.42	21.81	—	78.71	100.51	23.99	—	86.58	110.56
9.01.03.02.K	2850mm	nr	0.43	22.44	—	93.45	115.89	24.68	—	102.80	127.47
9.01.03.02.L	3000mm	nr	0.44	23.12	—	101.42	124.54	25.43	—	111.56	136.99
9.01.03.02.M	3300mm	nr	0.46	23.85	—	114.96	138.82	26.24	—	126.46	152.70
9.01.03.02.N	3600mm	nr	0.47	24.53	—	126.16	150.69	26.99	—	138.77	165.76
9.01.03.03.	CN81C; combined lintel; for cavity wall; length										
9.01.03.03.A	2100mm	nr	0.36	18.98	—	141.36	160.34	20.87	—	155.50	176.37
9.01.03.03.B	2250mm	nr	0.38	19.71	—	152.89	172.60	21.68	—	168.18	189.86
9.01.03.03.C	2400mm	nr	0.39	20.44	—	156.62	177.06	22.49	—	172.28	194.76
9.01.03.03.D	2550mm	nr	0.40	21.07	—	162.02	183.09	23.18	—	178.22	201.40
9.01.03.03.E	2700mm	nr	0.42	21.81	—	166.32	188.13	23.99	—	182.96	206.94
9.01.03.03.F	2850mm	nr	0.43	22.54	—	179.80	202.34	24.79	—	197.78	222.57
9.01.03.03.G	3000mm	nr	0.44	23.17	—	189.02	212.19	25.49	—	207.92	233.41
9.01.03.03.H	3300mm	nr	0.46	23.90	—	195.61	219.52	26.29	—	215.17	241.47
9.01.03.03.I	3600mm	nr	0.47	24.64	—	212.40	237.04	27.10	—	233.64	260.74
9.01.03.03.J	3900mm	nr	0.48	25.27	—	229.26	254.52	27.79	—	252.18	279.98
9.01.03.03.K	4200mm	nr	0.50	26.00	—	243.27	269.27	28.60	—	267.60	296.20
9.01.03.03.L	4575mm	nr	0.51	26.73	—	255.00	281.74	29.41	—	280.50	309.91
9.01.03.03.M	4800mm	nr	0.52	27.47	—	271.81	299.27	30.21	—	298.99	329.20
9.01.03.04.	CN71A; single lintel for external solid walls; length										
9.01.03.04.A	750mm	nr	0.23	11.95	—	28.81	40.76	13.15	—	31.69	44.84

Hutchins Priced Schedules 2021		Unit	Labour Hours	Labour Net £	Plant Net £	Materials Net £	Unit Rate Net £	Labour Gross £	Plant Gross £	Materials Gross £	Unit Rate Gross £
							(Gross rates include 10% profit)				
9.01.	**STEELWORK AND METALWORK**										
9.01.03.	**CATNIC GALVANISED STEEL LINTELS**										
9.01.03.04.	**CN71A; single lintel for external solid walls; length**										
9.01.03.04.B	900mm	nr	0.24	12.69	—	28.80	41.49	13.95	—	31.68	45.63
9.01.03.04.C	1050mm	nr	0.25	13.31	—	33.55	46.86	14.65	—	36.90	51.55
9.01.03.04.D	1200mm	nr	0.27	14.05	—	37.20	51.24	15.45	—	40.91	56.37
9.01.03.04.E	1350mm	nr	0.28	14.78	—	43.03	57.81	16.26	—	47.33	63.59
9.01.03.04.F	1500mm	nr	0.29	15.41	—	48.74	64.15	16.95	—	53.61	70.57
9.01.03.04.G	1650mm	nr	0.31	16.15	—	53.78	69.93	17.76	—	59.16	76.92
9.01.03.04.H	1800mm	nr	0.32	16.88	—	58.83	75.71	18.57	—	64.71	83.28
9.01.03.04.I	1950mm	nr	0.34	17.61	—	60.57	78.18	19.37	—	66.63	86.00
9.01.03.04.J	2100mm	nr	0.35	18.35	—	68.55	86.90	20.18	—	75.41	95.59
9.01.03.04.K	2250mm	nr	0.36	18.98	—	80.24	99.22	20.87	—	88.27	109.14
9.01.03.04.L	2400mm	nr	0.38	19.71	—	85.37	105.08	21.68	—	93.91	115.59
9.01.03.04.M	2550mm	nr	0.39	20.44	—	88.20	108.65	22.49	—	97.02	119.51
9.01.03.04.N	2700mm	nr	0.40	21.07	—	93.36	114.44	23.18	—	102.70	125.88
9.01.03.05.	**CN92; single lintel; for 75mm internal partitions and non-loadbearing walls; length**										
9.01.03.05.B	1050mm	nr	0.27	14.05	—	3.96	18.01	15.45	—	4.36	19.81
9.01.03.05.C	1200mm	nr	0.28	14.78	—	4.48	19.27	16.26	—	4.93	21.19
9.01.03.06.	**CN102; single lintel; for 100mm internal partitions and non-loadbearing walls; length**										
9.01.03.06.B	1050mm	nr	0.27	14.05	—	5.01	19.06	15.45	—	5.52	20.97
9.01.03.06.C	1200mm	nr	0.28	14.78	—	5.53	20.32	16.26	—	6.09	22.35
9.01.04.	**STRUCTURAL STEELWORK BY SPECIALISTS**										
9.01.04.01.	**Portal framework; where members are in the range of**										
9.01.04.01.A	18kg - 30kg per linear metre	t	14.56	978.78	—	1413.48	2392.26	1076.66	—	1554.83	2631.48
9.01.04.01.B	30kg - 45kg per linear metre	t	12.68	852.22	—	1465.85	2318.08	937.45	—	1612.44	2549.88
9.01.04.01.C	45kg - 70kg per linear metre	t	11.93	801.55	—	1518.20	2319.74	881.70	—	1670.01	2551.72
9.01.04.01.D	70kg - 100kg per linear metre	t	11.30	759.41	—	1570.55	2329.95	835.35	—	1727.60	2562.95
9.01.04.01.E	exceeding 100kg per linear metre	t	10.67	717.20	—	1622.89	2340.09	788.92	—	1785.18	2574.10
9.01.04.01.F	purlins	t	13.56	911.23	—	1388.16	2299.40	1002.36	—	1526.98	2529.34
9.01.04.02.	**Beam and post work with rigid connection where members are in the range of**										
9.01.04.02.A	18kg - 30kg per linear metre	t	13.81	928.10	—	1413.48	2341.58	1020.91	—	1554.83	2575.74
9.01.04.02.B	30kg - 45kg per linear metre	t	13.06	877.49	—	1465.85	2343.35	965.24	—	1612.44	2577.68

Sundries

		Unit	Labour Hours	Labour Net £	Plant Net £	Materials Net £	Unit Rate Net £	Labour Gross £	Plant Gross £	Materials Gross £	Unit Rate Gross £
								(Gross rates include 10% profit)			
9.01.	**STEELWORK AND METALWORK**										
9.01.04.	**STRUCTURAL STEELWORK BY SPECIALISTS**										
9.01.04.02.	**Beam and post work with rigid connection where members are in the range of**										
9.01.04.02.C	45kg - 70kg per linear metre	t	12.55	843.75	—	1518.20	2361.95	928.13	—	1670.01	2598.14
9.01.04.02.D	70kg - 100kg per linear metre	t	11.05	742.54	—	1570.55	2313.08	816.79	—	1727.60	2544.39
9.01.04.02.E	exceeding 100kg per linear metre	t	9.54	641.25	—	1622.89	2264.14	705.38	—	1785.18	2490.56
9.01.04.03.	**Beam and post work with simple web cleated connections where members are in the range of**										
9.01.04.03.A	18kg - 30kg per linear metre	t	12.68	852.22	—	1413.48	2265.70	937.45	—	1554.83	2492.27
9.01.04.03.B	30kg - 45kg per linear metre	t	11.44	768.88	—	1465.85	2234.73	845.77	—	1612.44	2458.21
9.01.04.03.C	45kg - 70kg per linear metre	t	10.04	674.99	—	1518.20	2193.19	742.49	—	1670.01	2412.50
9.01.04.03.D	70kg - 100kg per linear metre	t	9.54	641.25	—	1570.55	2211.80	705.38	—	1727.60	2432.98
9.01.04.03.E	exceeding 100kg per linear metre	t	8.54	573.77	—	1622.89	2196.66	631.15	—	1785.18	2416.33
9.02.	**WINDOWS**										
9.02.01.	**PVCU WINDOWS**										
9.02.01.01.	**PVCu double glazed casement window; finished in white; including frame; side hung; including trickle vent; fitted with polished chrome handles; sealed both sides**										
9.02.01.01.A	height 750mm x 630mm; 1-lite with left opener	nr	0.88	33.11	—	88.97	122.09	36.43	—	97.87	134.30
9.02.01.01.B	height 900mm x 630mm; 1-lite with left opener	nr	0.99	37.25	—	100.17	137.42	40.98	—	110.19	151.16
9.02.01.01.C	height 1050mm x 630mm; 1-lite with left opener	nr	1.15	43.27	—	106.71	149.99	47.60	—	117.38	164.99
9.02.01.01.D	height 1050mm x 915mm; 2-lite with left opener	nr	1.25	47.04	—	144.99	192.03	51.74	—	159.49	211.23
9.02.01.01.E	height 1050mm x 1200mm; 2-lite with left and right opener	nr	1.40	52.68	—	190.58	243.27	57.95	—	209.64	267.59
9.02.01.01.F	height 1200mm x 1200mm; 2-lite with left and right opener	nr	1.50	56.45	—	199.65	256.09	62.09	—	219.61	281.70
9.02.01.01.G	height 1050mm x 1770mm; 3-lite with left and right opener	nr	1.55	58.33	—	243.05	301.38	64.16	—	267.35	331.51
9.02.01.01.H	height 1200mm x 1770mm; 3-lite with left and right opener	nr	2.00	75.26	—	256.46	331.72	82.79	—	282.11	364.89
9.02.01.01.I	height 900mm x 2339mm; 4-lite with left and right opener	nr	2.15	80.90	—	275.68	356.58	88.99	—	303.24	392.24
9.02.01.01.J	height 1050mm x 2339mm; 4-lite with left and right opener	nr	2.20	82.79	—	297.30	380.08	91.06	—	327.03	418.09
9.02.01.01.K	height 1200mm x 2339mm; 4-lite with left and right opener	nr	2.25	84.67	—	315.08	399.74	93.13	—	346.58	439.72

Hutchins Priced Schedules 2021		Unit	Labour Hours	Labour Net £	Plant Net £	Materials Net £	Unit Rate Net £	Labour Gross £	Plant Gross £	Materials Gross £	Unit Rate Gross £
									(Gross rates include 10% profit)		
9.02.	**WINDOWS**										
9.02.01.	**PCVU WINDOWS**										
9.02.01.02.	**PVCu double glazed casement window; finished in light oak; including frame; side hung; including trickle vent; fitted with polished chrome handles; sealed both sides**										
9.02.01.02.A	height 750mm x 630mm; 1-lite with left opener....................	nr	0.88	33.11	—	127.93	161.04	36.43	—	140.72	177.15
9.02.01.02.B	height 900mm x 630mm; 1-lite with left opener....................	nr	0.99	37.25	—	133.75	171.00	40.98	—	147.12	188.10
9.02.01.02.C	height 1050mm x 630mm; 1-lite with left opener....................	nr	1.15	43.27	—	141.44	184.71	47.60	—	155.58	203.18
9.02.01.02.D	height 1050mm x 915mm; 2-lite with left opener....................	nr	1.25	47.04	—	191.36	238.40	51.74	—	210.49	262.24
9.02.01.02.E	height 1050mm x 1200mm; 2-lite with left and right opener...............	nr	1.40	52.68	—	251.83	304.51	57.95	—	277.01	334.96
9.02.01.02.F	height 1200mm x 1200mm; 2-lite with left and right opener...............	nr	1.50	56.45	—	261.77	318.21	62.09	—	287.94	350.03
9.02.01.02.G	height 1050mm x 1770mm; 3-lite with left and right opener...............	nr	1.55	58.33	—	317.87	376.20	64.16	—	349.66	413.82
9.02.01.02.H	height 1200mm x 1770mm; 3-lite with left and right opener...............	nr	2.00	75.26	—	332.60	407.86	82.79	—	365.86	448.64
9.02.01.02.I	height 900mm x 2339mm; 4-lite with left and right opener...............	nr	2.15	80.90	—	360.86	441.77	88.99	—	396.95	485.94
9.02.01.02.J	height 1050mm x 2339mm; 4-lite with left and right opener...............	nr	2.20	82.79	—	385.80	468.59	91.06	—	424.38	515.45
9.02.01.02.K	height 1200mm x 2339mm; 4-lite with left and right opener...............	nr	2.25	84.67	—	405.25	489.92	93.13	—	445.78	538.91
9.02.01.03.	**PVCu triple glazed casement window; finished in white; including frame; side hung; including trickle vent; fitted with polished chrome handles; sealed both sides**										
9.02.01.03.A	height 750mm x 630mm; 1-lite with left opener....................	nr	0.88	33.11	—	113.21	146.32	36.43	—	124.53	160.95
9.02.01.03.B	height 900mm x 630mm; 1-lite with left opener....................	nr	0.99	37.25	—	122.78	160.03	40.98	—	135.05	176.03
9.02.01.03.C	height 1050mm x 630mm; 1-lite with left opener....................	nr	1.15	43.27	—	133.58	176.86	47.60	—	146.94	194.54
9.02.01.03.D	height 1050mm x 915mm; 2-lite with left opener....................	nr	1.25	47.04	—	185.16	232.20	51.74	—	203.68	255.42
9.02.01.03.E	height 1050mm x 1200mm; 2-lite with left and right opener...............	nr	1.40	52.68	—	241.15	293.83	57.95	—	265.27	323.22
9.02.01.03.F	height 1200mm x 1200mm; 2-lite with left and right opener...............	nr	1.50	56.45	—	258.09	314.54	62.09	—	283.90	345.99

Sundries

Hutchins Priced Schedules 2021	Unit	Labour Hours	Labour Net £	Plant Net £	Materials Net £	Unit Rate Net £	Labour Gross £	Plant Gross £	Materials Gross £	Unit Rate Gross £
								(Gross rates include 10% profit)		

9.02. WINDOWS

9.02.01. PVCU WINDOWS

9.02.01.03. PVCu triple glazed casement window; finished in white; including frame; side hung; including trickle vent; fitted with polished chrome handles; sealed both sides

Hutchins Priced Schedules 2021	Unit	Labour Hours	Labour Net £	Plant Net £	Materials Net £	Unit Rate Net £	Labour Gross £	Plant Gross £	Materials Gross £	Unit Rate Gross £
9.02.01.03.G height 1050mm x 1770mm; 3-lite with left and right opener	nr	1.55	58.33	—	318.34	376.67	64.16	—	350.18	414.34
9.02.01.03.H height 1200mm x 1770mm; 3-lite with left and right opener	nr	2.00	75.26	—	343.42	418.68	82.79	—	377.76	460.55
9.02.01.03.I height 900mm x 2339mm; 4-lite with left and right opener	nr	2.15	80.90	—	361.76	442.67	88.99	—	397.94	486.94
9.02.01.03.J height 1050mm x 2339mm; 4-lite with left and right opener	nr	2.20	82.79	—	399.05	481.84	91.06	—	438.96	530.02
9.02.01.03.K height 1200mm x 2339mm; 4-lite with left and right opener	nr	2.25	84.67	—	432.60	517.27	93.13	—	475.86	569.00

9.02.01.04. PVCu triple glazed casement window; finished in light oak; including frame; side hung; including trickle vent; fitted with polished chrome handles; sealed both sides

Hutchins Priced Schedules 2021	Unit	Labour Hours	Labour Net £	Plant Net £	Materials Net £	Unit Rate Net £	Labour Gross £	Plant Gross £	Materials Gross £	Unit Rate Gross £
9.02.01.04.A height 750mm x 630mm; 1-lite with left opener	nr	0.88	33.11	—	146.21	179.32	36.43	—	160.83	197.26
9.02.01.04.B height 900mm x 630mm; 1-lite with left opener	nr	0.99	37.25	—	156.36	193.62	40.98	—	172.00	212.98
9.02.01.04.C height 1050mm x 630mm; 1-lite with left opener	nr	1.15	43.27	—	168.32	211.59	47.60	—	185.15	232.75
9.02.01.04.D height 1050mm x 915mm; 2-lite with left opener	nr	1.25	47.04	—	221.08	268.12	51.74	—	243.19	294.93
9.02.01.04.E height 1050mm x 1200mm; 2-lite with left and right opener	nr	1.40	52.68	—	302.39	355.07	57.95	—	332.63	390.58
9.02.01.04.F height 1200mm x 1200mm; 2-lite with left and right opener	nr	1.50	56.45	—	320.20	376.64	62.09	—	352.22	414.31
9.02.01.04.G height 1050mm x 1770mm; 3-lite with left and right opener	nr	1.55	58.33	—	393.17	451.49	64.16	—	432.48	496.64
9.02.01.04.H height 1200mm x 1770mm; 3-lite with left and right opener	nr	2.00	75.26	—	419.56	494.82	82.79	—	461.51	544.30
9.02.01.04.I height 900mm x 2339mm; 4-lite with left and right opener	nr	2.15	80.90	—	446.94	527.85	88.99	—	491.63	580.63
9.02.01.04.J height 1050mm x 2339mm; 4-lite with left and right opener	nr	2.20	82.79	—	487.59	570.38	91.06	—	536.35	627.41
9.02.01.04.K height 1200mm x 2339mm; 4-lite with left and right opener	nr	2.25	84.67	—	522.78	607.44	93.13	—	575.06	668.19

9.02.01.05. PVCu tilt and turn double glazed window; finished in white; including frame; side hung; including trickle vent; fitted with polished chrome handles; sealed both sides

Hutchins Priced Schedules 2021	Unit	Labour Hours	Labour Net £	Plant Net £	Materials Net £	Unit Rate Net £	Labour Gross £	Plant Gross £	Materials Gross £	Unit Rate Gross £
9.02.01.05.A height 750mm x 630mm; 1-lite with left opener	nr	0.88	33.11	—	150.71	183.83	36.43	—	165.78	202.21

Hutchins Priced Schedules 2021		Unit	Labour Hours	Labour Net £	Plant Net £	Materials Net £	Unit Rate Net £	Labour Gross £	Plant Gross £	Materials Gross £	Unit Rate Gross £
									(Gross rates include 10% profit)		
9.02.	**WINDOWS**										
9.02.01.	PVCU WINDOWS										
9.02.01.05.	PVCu tilt and turn double glazed window; finished in white; including frame; side hung; including trickle vent; fitted with polished chrome handles; sealed both sides										
9.02.01.05.B	height 900mm x 630mm; 1-lite with left opener......................	nr	0.99	37.25	—	155.96	193.21	40.98	—	171.55	212.53
9.02.01.05.C	height 1050mm x 630mm; 1-lite with left opener......................	nr	1.15	43.27	—	162.50	205.77	47.60	—	178.75	226.35
9.02.01.05.D	height 1050mm x 915mm; 2-lite with left opener......................	nr	1.25	47.04	—	191.74	238.77	51.74	—	210.91	262.65
9.02.01.05.E	height 1050mm x 1200mm; 2-lite with left and right opener................	nr	1.40	52.68	—	297.10	349.78	57.95	—	326.81	384.76
9.02.01.05.F	height 1200mm x 1200mm; 2-lite with left and right opener................	nr	1.50	56.45	—	306.16	362.61	62.09	—	336.78	398.87
9.02.01.05.G	height 1050mm x 1770mm; 3-lite with left and right opener................	nr	1.55	58.33	—	349.56	407.89	64.16	—	384.52	448.68
9.02.01.05.H	height 1200mm x 1770mm; 3-lite with left and right opener................	nr	2.00	75.26	—	362.96	438.22	82.79	—	399.26	482.04
9.02.01.05.I	height 900mm x 2339mm; 4-lite with left and right opener................	nr	2.15	80.90	—	382.19	463.09	88.99	—	420.41	509.40
9.02.01.05.J	height 1050mm x 2339mm; 4-lite with left and right opener................	nr	2.20	82.79	—	403.80	486.59	91.06	—	444.18	535.24
9.02.01.05.K	height 1200mm x 2339mm; 4-lite with left and right opener................	nr	2.25	84.67	—	421.58	506.25	93.13	—	463.74	556.87
9.02.01.06.	**PVCu tilt and turn double glazed window; finished in light oak; including frame; side hung; including trickle vent; fitted with polished chrome handles; sealed both sides**										
9.02.01.06.A	height 750mm x 630mm; 1-lite with left opener......................	nr	0.88	33.11	—	198.33	231.44	36.43	—	218.16	254.59
9.02.01.06.B	height 900mm x 630mm; 1-lite with left opener......................	nr	0.99	37.25	—	203.89	241.14	40.98	—	224.28	265.26
9.02.01.06.C	height 1050mm x 630mm; 1-lite with left opener......................	nr	1.15	43.27	—	211.24	254.52	47.60	—	232.36	279.97
9.02.01.06.D	height 1050mm x 915mm; 2-lite with left opener......................	nr	1.25	47.04	—	258.89	305.93	51.74	—	284.78	336.53
9.02.01.06.E	height 1050mm x 1200mm; 2-lite with left and right opener................	nr	1.40	52.68	—	404.12	456.81	57.95	—	444.54	502.49
9.02.01.06.F	height 1200mm x 1200mm; 2-lite with left and right opener................	nr	1.50	56.45	—	414.06	470.50	62.09	—	455.46	517.55
9.02.01.06.G	height 1050mm x 1770mm; 3-lite with left and right opener................	nr	1.55	58.33	—	470.17	528.50	64.16	—	517.19	581.35
9.02.01.06.H	height 1200mm x 1770mm; 3-lite with left and right opener................	nr	2.00	75.26	—	484.89	560.15	82.79	—	533.38	616.16
9.02.01.06.I	height 900mm x 2339mm; 4-lite with left and right opener................	nr	2.15	80.90	—	513.16	594.07	88.99	—	564.48	653.48

Sundries

|---|---|---|---|---|---|---|---|---|---|---|
| | | | | | | | (Gross rates include 10% profit) | | | |

9.02. **WINDOWS**

9.02.01. **PVCU WINDOWS**

9.02.01.06. PVCu tilt and turn double glazed window; finished in light oak; including frame; side hung; including trickle vent; fitted with polished chrome handles; sealed both sides

9.02.01.06.J	height 1050mm x 2339mm; 4-lite with left and right opener	nr	2.20	82.79	—	538.10	620.89	91.06	—	591.92	682.98
9.02.01.06.K	height 1200mm x 2339mm; 4-lite with left and right opener	nr	2.25	84.67	—	557.55	642.22	93.13	—	613.31	706.44

9.02.01.07. PVCu tilt and turn triple glazed window; finished in white; including frame; side hung; including trickle vent; fitted with polished chrome handles; sealed both sides

9.02.01.07.A	height 750mm x 630mm; 1-lite with left opener	nr	0.88	33.11	—	168.99	202.11	36.43	—	185.89	222.32
9.02.01.07.B	height 900mm x 630mm; 1-lite with left opener	nr	0.99	37.25	—	178.56	215.82	40.98	—	196.42	237.40
9.02.01.07.C	height 1050mm x 630mm; 1-lite with left opener	nr	1.15	43.27	—	189.37	232.64	47.60	—	208.31	255.91
9.02.01.07.D	height 1050mm x 915mm; 2-lite with left opener	nr	1.25	47.04	—	230.08	277.12	51.74	—	253.09	304.83
9.02.01.07.E	height 1050mm x 1200mm; 2-lite with left and right opener	nr	1.40	52.68	—	347.66	400.35	57.95	—	382.43	440.38
9.02.01.07.F	height 1200mm x 1200mm; 2-lite with left and right opener	nr	1.50	56.45	—	364.60	421.05	62.09	—	401.06	463.15
9.02.01.07.G	height 1050mm x 1770mm; 3-lite with left and right opener	nr	1.55	58.33	—	424.85	483.17	64.16	—	467.33	531.49
9.02.01.07.H	height 1200mm x 1770mm; 3-lite with left and right opener	nr	2.00	75.26	—	449.92	525.18	82.79	—	494.91	577.70
9.02.01.07.I	height 900mm x 2339mm; 4-lite with left and right opener	nr	2.15	80.90	—	468.27	549.17	88.99	—	515.09	604.09
9.02.01.07.J	height 1050mm x 2339mm; 4-lite with left and right opener	nr	2.20	82.79	—	505.58	588.36	91.06	—	556.13	647.20
9.02.01.07.K	height 1200mm x 2339mm; 4-lite with left and right opener	nr	2.25	84.67	—	539.10	623.77	93.13	—	593.02	686.15

9.02.01.08. PVCu tilt and turn triple glazed window; finished in light oak; including frame; side hung; including trickle vent; fitted with polished chrome handles; sealed both sides

9.02.01.08.A	height 750mm x 630mm; 1-lite with left opener	nr	0.88	33.11	—	215.78	248.90	36.43	—	237.36	273.79
9.02.01.08.B	height 900mm x 630mm; 1-lite with left opener	nr	0.99	37.25	—	225.47	262.72	40.98	—	248.01	288.99
9.02.01.08.C	height 1050mm x 630mm; 1-lite with left opener	nr	1.15	43.27	—	236.89	280.17	47.60	—	260.58	308.18
9.02.01.08.D	height 1050mm x 915mm; 2-lite with left opener	nr	1.25	47.04	—	297.24	344.28	51.74	—	326.96	378.71

Hutchins Priced Schedules 2021		Unit	Labour Hours	Labour Net £	Plant Net £	Materials Net £	Unit Rate Net £	Labour Gross £	Plant Gross £	Materials Gross £	Unit Rate Gross £
									(Gross rates include 10% profit)		
9.02.	**WINDOWS**										
9.02.01.	**PCVU WINDOWS**										
9.02.01.08.	**PVCu tilt and turn triple glazed window; finished in light oak; including frame; side hung; including trickle vent; fitted with polished chrome handles; sealed both sides**										
9.02.01.08.E	height 1050mm x 1200mm; 2-lite with left and right opener..............	nr	1.40	52.68	—	454.69	507.37	57.95	—	500.16	558.11
9.02.01.08.F	height 1200mm x 1200mm; 2-lite with left and right opener..............	nr	1.50	56.45	—	472.50	528.95	62.09	—	519.75	581.84
9.02.01.08.G	height 1050mm x 1770mm; 3-lite with left and right opener..............	nr	1.55	58.33	—	545.47	603.80	64.16	—	600.02	664.18
9.02.01.08.H	height 1200mm x 1770mm; 3-lite with left and right opener..............	nr	2.00	75.26	—	571.86	647.12	82.79	—	629.04	711.83
9.02.01.08.I	height 900mm x 2339mm; 4-lite with left and right opener..............	nr	2.15	80.90	—	599.23	680.14	88.99	—	659.16	748.15
9.02.01.08.J	height 1050mm x 2339mm; 4-lite with left and right opener..............	nr	2.20	82.79	—	639.88	722.67	91.06	—	703.87	794.93
9.02.01.08.K	height 1200mm x 2339mm; 4-lite with left and right opener..............	nr	2.25	84.67	—	675.08	759.75	93.13	—	742.59	835.72
9.02.02.	**TIMBER WINDOWS**										
9.02.02.01.	**Stormsure standard double glazed casement window; finished in Hi-Build white paint; including frames; side hung; including trickle vent; fitted with polished chrome handles; sealed both sides**										
9.02.02.01.A	height 750mm x 630mm; 1-lite with left opener.....................	nr	0.88	33.11	—	324.71	357.82	36.43	—	357.18	393.61
9.02.02.01.B	height 900mm x 630mm; 1-lite with left opener.....................	nr	0.99	37.25	—	340.10	377.36	40.98	—	374.11	415.09
9.02.02.01.C	height 1050mm x 630mm; 1-lite with left opener.....................	nr	1.15	43.27	—	356.11	399.38	47.60	—	391.72	439.32
9.02.02.01.D	height 1050mm x 915mm; 2-lite with left opener.....................	nr	1.25	47.04	—	561.83	608.87	51.74	—	618.01	669.75
9.02.02.01.E	height 1050mm x 1200mm; 2-lite with left and right opener..............	nr	1.40	52.68	—	614.85	667.53	57.95	—	676.33	734.28
9.02.02.01.F	height 1200mm x 1200mm; 2-lite with left and right opener..............	nr	1.50	56.45	—	647.04	703.49	62.09	—	711.75	773.84
9.02.02.01.G	height 1050mm x 1770mm; 3-lite with left and right opener..............	nr	1.55	58.33	—	787.37	845.70	64.16	—	866.11	930.26
9.02.02.01.H	height 1200mm x 1770mm; 3-lite with left and right opener..............	nr	2.00	75.26	—	829.63	904.89	82.79	—	912.60	995.38
9.02.02.01.I	height 900mm x 2339mm; 4-lite with left and right opener..............	nr	2.15	80.90	—	907.40	988.30	88.99	—	998.14	1087.13
9.02.02.01.J	height 1050mm x 2339mm; 4-lite with left and right opener..............	nr	2.20	82.79	—	957.55	1040.33	91.06	—	1053.30	1144.37
9.02.02.01.K	height 1200mm x 2339mm; 4-lite with left and right opener..............	nr	2.25	84.67	—	1009.89	1094.56	93.13	—	1110.88	1204.02

Sundries

Hutchins Priced Schedules 2021	Unit	Labour Hours	Labour Net £	Plant Net £	Materials Net £	Unit Rate Net £	Labour Gross £	Plant Gross £	Materials Gross £	Unit Rate Gross £
						(Gross rates include 10% profit)				
9.02.	**WINDOWS**									
9.02.02.	**TIMBER WINDOWS**									
9.02.02.02.	**Stormsure Energy+ high performance triple glazed casement window; finished in Hi-Build white paint; including frames; side hung; including trickle vent; fitted with polished chrome handles; sealed both sides**									
9.02.02.02.A	height 750mm x 630mm; 1-lite with left opener	nr	0.88	33.11	—	508.52	541.64	36.43	—	559.37	595.80
9.02.02.02.B	height 900mm x 630mm; 1-lite with left opener	nr	0.99	37.25	—	567.06	604.32	40.98	—	623.77	664.75
9.02.02.02.C	height 1050mm x 630mm; 1-lite with left opener	nr	1.15	43.27	—	626.18	669.45	47.60	—	688.80	736.40
9.02.02.02.D	height 1050mm x 915mm; 2-lite with left opener	nr	1.25	47.04	—	949.67	996.71	51.74	—	1044.64	1096.38
9.02.02.02.E	height 1050mm x 1200mm; 2-lite with left and right opener	nr	1.40	52.68	—	1154.21	1206.89	57.95	—	1269.63	1327.58
9.02.02.02.F	height 1200mm x 1200mm; 2-lite with left and right opener	nr	1.50	56.45	—	1272.84	1329.29	62.09	—	1400.13	1462.22
9.02.02.02.G	height 1050mm x 1770mm; 3-lite with left and right opener	nr	1.55	58.33	—	1627.13	1685.45	64.16	—	1789.84	1854.00
9.02.02.02.H	height 1200mm x 1770mm; 3-lite with left and right opener	nr	2.00	75.26	—	1801.68	1876.94	82.79	—	1981.85	2064.64
9.02.02.02.I	height 900mm x 2339mm; 4-lite with left and right opener	nr	2.15	80.90	—	1869.90	1950.81	88.99	—	2056.89	2145.89
9.02.02.02.J	height 1050mm x 2339mm; 4-lite with left and right opener	nr	2.20	82.79	—	2098.17	2180.96	91.06	—	2307.99	2399.06
9.02.02.02.K	height 1200mm x 2339mm; 4-lite with left and right opener	nr	2.25	84.67	—	2328.78	2413.44	93.13	—	2561.66	2654.79
9.02.02.03.	**Stormsure oak casement double glazed window; finished in golden oak stain; including frames; side hung; including trickle vent; fitted with polished chrome handles; sealed both sides**									
9.02.02.03.A	height 750mm x 630mm; 1-lite with left opener	nr	0.88	33.11	—	652.52	685.63	36.43	—	717.77	754.20
9.02.02.03.B	height 900mm x 630mm; 1-lite with left opener	nr	0.99	37.25	—	694.49	731.74	40.98	—	763.94	804.92
9.02.02.03.C	height 1050mm x 630mm; 1-lite with left opener	nr	1.15	43.27	—	737.16	780.44	47.60	—	810.88	858.48
9.02.02.03.D	height 1050mm x 915mm; 2-lite with left opener	nr	1.25	47.04	—	1172.21	1219.24	51.74	—	1289.43	1341.17
9.02.02.03.E	height 1050mm x 1200mm; 2-lite with left and right opener	nr	1.40	52.68	—	1302.00	1354.68	57.95	—	1432.20	1490.15

Sundries

Hutchins Priced Schedules 2021	Unit	Labour Hours	Labour Net £	Plant Net £	Materials Net £	Unit Rate Net £	Labour Gross £	Plant Gross £	Materials Gross £	Unit Rate Gross £
							(Gross rates include 10% profit)			

9.02. **WINDOWS**

9.02.02. **TIMBER WINDOWS**

9.02.02.03. Stormsure oak casement double glazed window; finished in golden oak stain; including frames; side hung; including trickle vent; fitted with polished chrome handles; sealed both sides

9.02.02.03.F height 1200mm x 1200mm; 2-lite with left and right opener	nr	1.50	56.45	—	1145.51	1201.96	62.09	—	1260.06	1322.15
9.02.02.03.G height 1050mm x 1770mm; 3-lite with left and right opener	nr	1.55	58.33	—	1592.36	1650.69	64.16	—	1751.60	1815.76
9.02.02.03.H height 1200mm x 1770mm; 3-lite with left and right opener	nr	2.00	75.26	—	1683.21	1758.47	82.79	—	1851.53	1934.32
9.02.02.03.I height 900mm x 2339mm; 4-lite with left and right opener	nr	2.15	80.90	—	1879.40	1960.31	88.99	—	2067.34	2156.34
9.02.02.03.J height 1050mm x 2339mm; 4-lite with left and right opener	nr	2.20	82.79	—	1992.30	2075.09	91.06	—	2191.53	2282.60
9.02.02.03.K height 1200mm x 2339mm; 4-lite with left and right opener	nr	2.25	84.67	—	2107.76	2192.43	93.13	—	2318.54	2411.67

9.02.02.04. Velux roof windows; centre pivot; laminated Nordic red pine frame and sash; sealed unit double pre-glazing; 3mm clear float glass; exterior aluminium cladding; natural brownish-grey finish; recessed flashings for installation in tiles up to 90mm in profile; roof pitch 20 deg to 90 deg; including insulation collar

9.02.02.04.A height 780mm x 550mm	nr	6.05	227.66	—	269.65	497.31	250.43	—	296.62	547.05
9.02.02.04.B height 980mm x 550mm	nr	6.92	260.44	—	283.32	543.75	286.48	—	311.65	598.13
9.02.02.04.C height 1180mm x 660mm	nr	7.31	275.00	—	320.21	595.21	302.50	—	352.23	654.73
9.02.02.04.D height 980mm x 780mm	nr	7.29	274.13	—	311.08	585.21	301.55	—	342.18	643.73
9.02.02.04.E height 1400mm x 780mm	nr	7.81	294.04	—	371.83	665.87	323.44	—	409.02	732.46
9.02.02.04.F height 1600mm x 940mm	nr	8.48	319.06	—	445.53	764.60	350.97	—	490.08	841.06
9.02.02.04.G height 1180mm x 1140mm	nr	8.22	309.43	—	432.59	742.02	340.37	—	475.85	816.22
9.02.02.04.H height 980mm x 1340mm	nr	8.17	307.47	—	426.34	733.82	338.22	—	468.98	807.20
9.02.02.04.I height 1400mm x 1340mm	nr	9.08	341.68	—	497.61	839.29	375.85	—	547.37	923.22

9.02.02.05. Velux roof windows; top hung; laminated Nordic red pine frame and sash; sealed unit double pre-glazing; 3mm clear float glass; exterior aluminium cladding; natural brownish-grey finish; recessed flashings for installation in tiles up to 90mm in profile; roof pitch 15 deg to 75 deg; including insulation collar

9.02.02.05.A height 780mm x 550mm	nr	6.05	227.66	—	316.89	544.55	250.43	—	348.58	599.01
9.02.02.05.B height 980mm x 550mm	nr	6.92	260.44	—	353.29	613.72	286.48	—	388.62	675.10
9.02.02.05.C height 1180mm x 660mm	nr	7.31	275.00	—	424.29	699.29	302.50	—	466.71	769.21
9.02.02.05.D height 980mm x 780mm	nr	7.29	274.13	—	417.18	691.31	301.55	—	458.90	760.44

Sundries

|---|---|---|---|---|---|---|---|---|---|---|
						(Gross rates include 10% profit)				
9.02.	**WINDOWS**									
9.02.02.	**TIMBER WINDOWS**									
9.02.02.05.	Velux roof windows; top hung; laminated Nordic red pine frame and sash; sealed unit double pre-glazing; 3mm clear float glass; exterior aluminium cladding; natural brownish-grey finish; recessed flashings for installation in tiles up to 90mm in profile; roof pitch 15 deg to 75 deg; including insulation collar									
9.02.02.05.E	height 1400mm x 780mm	nr	7.81	294.04	—	487.06	781.10	323.44	—	535.77	859.21
9.02.02.05.F	height 1600mm x 940mm	nr	8.48	319.06	—	607.16	926.22	350.97	—	667.87	1018.84
9.02.02.05.G	height 1180mm x 1140mm	nr	8.22	309.43	—	588.74	898.18	340.37	—	647.62	987.99
9.02.02.05.H	height 980mm x 1340mm	nr	8.17	307.47	—	569.66	877.13	338.22	—	626.62	964.85
9.02.02.05.I	height 1400mm x 1340mm	nr	9.08	341.68	—	650.06	991.74	375.85	—	715.07	1090.91
9.03.	**EXTERNAL DOORS**									
9.03.01.	**PVCU DOORS**									
9.03.01.01.	PVCu door including frame fitted with 155mm external sill, 15mm aluminium threshold, letterbox, knocker, spyhole, polished chrome handle and clear glass panel squares if applicable; sealed both sides									
9.03.01.01.A	854mm x 2086mm; 6 panel in white	nr	2.15	144.50	—	392.05	536.55	158.95	—	431.26	590.21
9.03.01.01.B	943mm x 2086mm; 6 panel in white	nr	2.20	147.86	—	395.80	543.66	162.65	—	435.38	598.03
9.03.01.01.C	1006mm x 2086mm; 6 panel in white . . .	nr	2.25	151.22	—	399.39	550.61	166.34	—	439.32	605.67
9.03.01.01.D	854mm x 2086mm; 6 panel in light oak . .	nr	2.15	144.50	—	545.39	689.89	158.95	—	599.93	758.88
9.03.01.01.E	943mm x 2086mm; 6 panel in light oak . .	nr	2.20	147.86	—	550.70	698.57	162.65	—	605.77	768.42
9.03.01.01.F	1006mm x 2086mm; 6 panel in light oak .	nr	2.25	151.22	—	555.76	706.98	166.34	—	611.34	777.68
9.03.01.01.G	854mm x 2086mm; 2 panel 4 square in white	nr	2.15	144.50	—	406.94	551.44	158.95	—	447.63	606.59
9.03.01.01.H	943mm x 2086mm; 2 panel 4 square in white	nr	2.20	147.86	—	410.69	558.55	162.65	—	451.76	614.40
9.03.01.01.I	1006mm x 2086mm; 2 panel 4 square in white	nr	2.25	151.22	—	414.27	565.50	166.34	—	455.70	622.05
9.03.01.01.J	854mm x 2086mm; 2 panel 4 square in light oak	nr	2.15	144.50	—	560.27	704.77	158.95	—	616.30	775.25
9.03.01.01.K	943mm x 2086mm; 2 panel 4 square in light oak	nr	2.20	147.86	—	565.58	713.44	162.65	—	622.14	784.79
9.03.01.01.L	1006mm x 2086mm; 2 panel 4 square in light oak	nr	2.25	151.22	—	570.64	721.86	166.34	—	627.70	794.05
9.03.01.02.	PVCu sliding patio door, double glazed, 2 pane with 1 sliding and 1 fixed including frame; fitted with 155mm external sill, polished chrome handles; sealed both sides									
9.03.01.02.A	2000mm x 2086mm in white	nr	2.69	180.66	—	591.77	772.43	198.73	—	650.95	849.67
9.03.01.02.B	2500mm x 2086mm in white	nr	2.91	195.31	—	662.80	858.11	214.84	—	729.08	943.92
9.03.01.02.C	3000mm x 2086mm in white	nr	3.13	210.03	—	1012.49	1222.52	231.03	—	1113.74	1344.78
9.03.01.02.D	2000mm x 2086mm in light oak or anthracite grey	nr	2.69	180.66	—	737.40	918.06	198.73	—	811.14	1009.87

Hutchins Priced Schedules 2021	Unit	Labour Hours	Labour Net £	Plant Net £	Materials Net £	Unit Rate Net £	Labour Gross £	Plant Gross £	Materials Gross £	Unit Rate Gross £
						(Gross rates include 10% profit)				

9.03.   **EXTERNAL DOORS**

9.03.01.   **PVCU DOORS**

9.03.01.02.   **PVCu sliding patio door, double glazed, 2 pane with 1 sliding and 1 fixed including frame; fitted with 155mm external sill, polished chrome handles; sealed both sides**

9.03.01.02.E 2500mm x 2086mm in light oak or anthracite grey...............	nr	2.91	195.31	—	820.86	1016.18	214.84	—	902.95	1117.79
9.03.01.02.F 3000mm x 2086mm in light oak or anthracite grey...............	nr	3.13	210.03	—	1182.97	1393.00	231.03	—	1301.27	1532.30

9.03.01.03.   **PVCu French door, 2 pane including frame, fitted with 155mm external sill, 15mm aluminium threshold and polished chrome handles; sealed both sides**

9.03.01.03.A 1000mm x 2086mm in white; double glazed.....................	nr	2.25	151.22	—	483.94	635.16	166.34	—	532.33	698.67
9.03.01.03.B 1500mm x 2086mm in white; double glazed.....................	nr	2.47	165.94	—	551.08	717.02	182.54	—	606.18	788.72
9.03.01.03.C 1900mm x 2086mm in white; double glazed.....................	nr	2.64	177.70	—	605.76	783.46	195.47	—	666.33	861.81
9.03.01.03.D 1000mm x 2086mm in light oak or anthracite grey; double glazed.......	nr	2.25	151.22	—	624.47	775.69	166.34	—	686.91	853.26
9.03.01.03.E 1500mm x 2086mm in light oak or anthracite grey; double glazed.......	nr	2.47	165.94	—	698.81	864.75	182.54	—	768.69	951.23
9.03.01.03.F 1900mm x 2086mm in light oak or anthracite grey; double glazed.......	nr	2.64	177.70	—	759.26	936.96	195.47	—	835.18	1030.66
9.03.01.03.G 1000mm x 2086mm in white; triple glazed.....................	nr	2.25	151.22	—	528.12	679.34	166.34	—	580.93	747.28
9.03.01.03.H 1500mm x 2086mm in white; triple glazed.....................	nr	2.47	165.94	—	635.31	801.25	182.54	—	698.84	881.37
9.03.01.03.I 1900mm x 2086mm in white; triple glazed.....................	nr	2.64	177.70	—	722.80	900.50	195.47	—	795.08	990.55
9.03.01.03.J 1000mm x 2086mm in light oak or anthracite grey; triple glazed........	nr	2.25	151.22	—	668.65	819.87	166.34	—	735.52	901.86
9.03.01.03.K 1500mm x 2086mm in light oak or anthracite grey; triple glazed........	nr	2.47	165.94	—	783.04	948.98	182.54	—	861.35	1043.88
9.03.01.03.L 1900mm x 2086mm in light oak or anthracite grey; triple glazed........	nr	2.64	177.70	—	876.29	1053.99	195.47	—	963.92	1159.39

9.03.02.   **COMPOSITE DOORS**

9.03.02.01.   **Composite timber core door including frame fitted with 150mm external sill, 15mm aluminium threshold, letterbox, knocker, spyhole, polished chrome handle and clear glass panel squares if applicable; sealed both sides**

9.03.02.01.A 854mm x 2086mm; 4 panel 2 square in white....................	nr	2.15	144.50	—	480.37	624.87	158.95	—	528.40	687.36

Sundries

Hutchins Priced Schedules 2021		Unit	Labour Hours	Labour Net £	Plant Net £	Materials Net £	Unit Rate Net £	Labour Gross £	Plant Gross £	Materials Gross £	Unit Rate Gross £
									(Gross rates include 10% profit)		
9.03.	**EXTERNAL DOORS**										
9.03.02.	**COMPOSITE DOORS**										
9.03.02.01.	**Composite timber core door including frame fitted with 150mm external sill, 15mm aluminium threshold, letterbox, knocker, spyhole, polished chrome handle and clear glass panel squares if applicable; sealed both sides**										
9.03.02.01.B	943mm x 2086mm; 4 panel 2 square in white	nr	2.20	147.86	—	481.02	628.88	162.65	—	529.12	691.77
9.03.02.01.C	1006mm x 2086mm; 4 panel 2 square in white	nr	2.25	151.22	—	481.48	632.71	166.34	—	529.63	695.98
9.03.02.01.D	854mm x 2086mm; 4 panel 2 square in light oak	nr	2.15	144.50	—	723.26	867.76	158.95	—	795.59	954.54
9.03.02.01.E	943mm x 2086mm; 4 panel 2 square in light oak	nr	2.20	147.86	—	724.01	871.87	162.65	—	796.41	959.06
9.03.02.01.F	1006mm x 2086mm; 4 panel 2 square in light oak	nr	2.25	151.22	—	723.79	875.02	166.34	—	796.17	962.52
9.03.02.01.G	854mm x 2086mm; 2 panel 4 square in white	nr	2.15	144.50	—	511.24	655.74	158.95	—	562.36	721.31
9.03.02.01.H	943mm x 2086mm; 2 panel 4 square in white	nr	2.20	147.86	—	511.89	659.75	162.65	—	563.07	725.72
9.03.02.01.I	1006mm x 2086mm; 2 panel 4 square in white	nr	2.25	151.22	—	512.35	663.58	166.34	—	563.59	729.93
9.03.02.01.J	854mm x 2086mm; 2 panel 4 square in light oak	nr	2.15	144.50	—	754.23	898.73	158.95	—	829.65	988.60
9.03.02.01.K	943mm x 2086mm; 2 panel 4 square in light oak	nr	2.20	147.86	—	754.88	902.74	162.65	—	830.36	993.01
9.03.02.01.L	1006mm x 2086mm; 2 panel 4 square in light oak	nr	2.25	151.22	—	755.35	906.57	166.34	—	830.88	997.22
9.03.03.	**ALUMINIUM DOORS**										
9.03.03.01.	**Aluminium French door; double glazed, 2 pane including frame, fitted with 105mm external sill, 15mm aluminium threshold and polished chrome handles; sealed both sides**										
9.03.03.01.A	1300mm x 2086mm in white, black or anthracite grey	nr	2.25	151.22	—	1340.13	1491.36	166.34	—	1474.15	1640.49
9.03.03.01.B	1800mm x 2086mm in white, black or anthracite grey	nr	2.47	165.94	—	1401.19	1567.13	182.54	—	1541.30	1723.84
9.03.03.01.C	2300mm x 2086mm in white, black or anthracite grey	nr	2.64	177.70	—	1480.93	1658.63	195.47	—	1629.02	1824.49
9.03.03.02.	**Aluminium bi-fold door; double glazed; including frame fitted with 150mm external sill, 15mm aluminium threshold and polished chrome handles; all sections fold left with no external access; sealed both sides**										
9.03.03.02.A	2000mm x 2086mm; 2 section; in white, black or anthracite grey	nr	2.69	180.66	—	1425.42	1606.08	198.73	—	1567.96	1766.68

Hutchins Priced Schedules 2021	Unit	Labour Hours	Labour Net £	Plant Net £	Materials Net £	Unit Rate Net £	Labour Gross £	Plant Gross £	Materials Gross £	Unit Rate Gross £
							(Gross rates include 10% profit)			

9.03. **EXTERNAL DOORS**

9.03.03. **ALUMINIUM DOORS**

9.03.03.02. Aluminium bi-fold door; double glazed; including frame fitted with 150mm external sill, 15mm aluminium threshold and polished chrome handles; all sections fold left with no external access; sealed both sides

	Unit	Labour Hours	Labour Net £	Plant Net £	Materials Net £	Unit Rate Net £	Labour Gross £	Plant Gross £	Materials Gross £	Unit Rate Gross £
9.03.03.02.B 2500mm x 2086mm; 3 section; in white, black or anthracite grey	nr	2.91	195.31	—	2010.68	2205.99	214.84	—	2211.75	2426.59
9.03.03.02.C 3000mm x 2086mm; 3 section; in white, black or anthracite grey	nr	3.13	210.03	—	2074.93	2284.97	231.03	—	2282.43	2513.46
9.03.03.02.D 3000mm x 2086mm; 4 section; in white, black or anthracite grey	nr	3.13	210.03	—	2478.07	2688.10	231.03	—	2725.88	2956.91
9.03.03.02.E 3500mm x 2086mm; 4 section; in white, black or anthracite grey	nr	3.34	224.75	—	2524.70	2749.45	247.23	—	2777.17	3024.39
9.03.03.02.F 4000mm x 2086mm; 4 section; in white, black or anthracite grey	nr	3.56	239.47	—	2593.73	2833.20	263.42	—	2853.10	3116.52
9.03.03.02.G 4000mm x 2086mm; 5 section; in white, black or anthracite grey	nr	3.56	239.47	—	3059.45	3298.92	263.42	—	3365.40	3628.82
9.03.03.02.H 4500mm x 2086mm; 5 section; in white, black or anthracite grey	nr	3.78	254.12	—	3127.38	3381.50	279.53	—	3440.12	3719.65
9.03.03.02.I 5000mm x 2086mm; 5 section; in white, black or anthracite grey	nr	4.00	268.84	—	3208.18	3477.02	295.72	—	3528.99	3824.72

9.04. **ASPHALT WORK**

9.04.01. **MASTIC ASPHALT TANKING; BS 6925:1988**

9.04.01.01. 13mm one coat horizontal covering on concrete

	Unit	Labour Hours	Labour Net £	Plant Net £	Materials Net £	Unit Rate Net £	Labour Gross £	Plant Gross £	Materials Gross £	Unit Rate Gross £
9.04.01.01.A over 300mm wide	m²	0.44	22.86	0.91	33.35	57.11	25.14	1.00	36.69	62.82
9.04.01.01.B not exceeding 150mm wide	m	0.16	8.18	0.34	5.00	13.52	9.00	0.37	5.50	14.87
9.04.01.01.C 150mm - 300mm wide	m	0.22	11.64	0.45	10.01	22.09	12.80	0.49	11.01	24.30

9.04.01.02. 20mm two coat horizontal covering on concrete

9.04.01.02.A over 300mm wide	m²	0.48	25.37	0.98	51.31	77.66	27.91	1.07	56.44	85.43
9.04.01.02.B not exceeding 150mm wide	m	0.19	9.75	0.42	7.69	17.87	10.73	0.47	8.46	19.65
9.04.01.02.C 150mm - 300mm wide	m	0.26	13.42	0.55	15.39	29.36	14.76	0.61	16.93	32.30

9.04.01.03. 30mm three coat horizontal covering on concrete

9.04.01.03.A over 300mm wide	m²	0.88	46.02	1.96	76.96	124.95	50.63	2.16	84.65	137.44
9.04.01.03.B not exceeding 150mm wide	m	0.32	16.88	0.70	11.55	29.12	18.57	0.77	12.70	32.04
9.04.01.03.C 150mm - 300mm wide	m	0.46	24.22	0.98	23.09	48.29	26.64	1.07	25.40	53.11

9.04.01.04. 13mm two coat vertical covering on brickwork

9.04.01.04.A over 300mm wide	m²	1.59	83.35	3.39	33.35	120.09	91.68	3.73	36.69	132.10
9.04.01.04.B not exceeding 150mm wide	m	0.48	24.95	1.06	5.00	31.01	27.45	1.17	5.50	34.12
9.04.01.04.C 150mm - 300mm wide	m	0.72	37.64	1.65	10.01	49.30	41.40	1.82	11.01	54.23

Sundries

Hutchins Priced Schedules 2021	Unit	Labour Hours	Labour Net £	Plant Net £	Materials Net £	Unit Rate Net £	Labour Gross £	Plant Gross £	Materials Gross £	Unit Rate Gross £
						(Gross rates include 10% profit)				
9.04.	**ASPHALT WORK**									
9.04.01.	**MASTIC ASPHALT TANKING; BS 6925:1988**									
9.04.01.05.	**20mm three coat vertical covering on brickwork**									
9.04.01.05.A	over 300mm wide	m²	2.05	107.25	4.66	51.31	163.23	117.98	5.13	56.44	179.55
9.04.01.05.B	not exceeding 150mm wide	m	0.62	32.71	1.44	7.69	41.85	35.98	1.59	8.46	46.03
9.04.01.05.C	150mm - 300mm wide	m	0.93	48.86	2.12	15.39	66.37	53.74	2.33	16.93	73.01
9.04.01.06.	**Labours**									
9.04.01.06.A	internal angle fillets	m	0.28	14.47	—	—	14.47	15.91	—	—	15.91
9.04.01.06.B	turning nibs into grooves	m	0.16	8.18	—	—	8.18	9.00	—	—	9.00
9.04.01.06.C	working into outlets	nr	1.64	86.18	—	—	86.18	94.80	—	—	94.80
9.04.01.06.D	collars and internal angle fillets around small pipes (not exceeding 55mm)	nr	1.34	70.03	—	—	70.03	77.04	—	—	77.04
9.04.01.06.E	collars and internal angle fillets around large pipes (55mm to 110mm)	nr	1.95	102.32	—	—	102.32	112.56	—	—	112.56
9.04.03.	**MASTIC ASPHALT FLOORING; BS 6925:1988**									
9.04.03.01.	**15mm one coat light duty flooring, including isolating membrane**									
9.04.03.01.A	over 300mm wide	m²	0.54	28.10	0.98	41.81	70.88	30.91	1.07	45.99	77.97
9.04.03.01.B	not exceeding 150mm wide	m	0.19	10.17	0.40	6.27	16.85	11.19	0.44	6.90	18.53
9.04.03.01.C	150mm - 300mm wide	m	0.27	14.05	0.68	12.55	27.27	15.45	0.75	13.80	30.00
9.04.03.02.	**20mm one coat medium duty flooring, including isolating membrane**									
9.04.03.02.A	over 300mm wide	m²	0.54	28.10	0.98	54.62	83.69	30.91	1.07	60.08	92.06
9.04.03.02.B	not exceeding 150mm wide	m	0.19	10.17	0.64	8.69	19.50	11.19	0.70	9.56	21.45
9.04.03.02.C	150mm - 300mm wide	m	0.25	13.31	1.06	16.39	30.76	14.65	1.17	18.03	33.84
9.04.03.03.	**30mm one coat heavy duty flooring, including isolating membrane**									
9.04.03.03.A	over 300mm wide	m²	0.67	35.12	1.10	80.27	116.49	38.63	1.21	88.29	128.14
9.04.03.03.B	not exceeding 150mm wide	m	0.24	12.79	1.17	12.05	26.00	14.07	1.28	13.25	28.60
9.04.03.03.C	150mm - 300mm wide	m	0.33	17.40	1.26	24.08	42.74	19.14	1.39	26.49	47.02
9.04.03.04.	**Labours**									
9.04.03.04.A	working against metal frames	m	0.07	3.88	—	—	3.88	4.27	—	—	4.27
9.04.03.04.B	extra for working flooring into recessed cover; not exceeding 1.00m2	nr	0.60	31.66	—	—	31.66	34.83	—	—	34.83
9.04.03.05.	**Skirtings; 13mm two coat; fair edge, angles, coved angle fillet and nib turned into groove**									
9.04.03.05.A	150mm high	m	0.52	27.05	0.85	5.00	32.90	29.75	0.93	5.50	36.19

Hutchins Priced Schedules 2021		Unit	Labour Hours	Labour Net £	Plant Net £	Materials Net £	Unit Rate Net £	Labour Gross £	Plant Gross £	Materials Gross £	Unit Rate Gross £
										(Gross rates include 10% profit)	
9.05.	**ELECTRICAL WORK**										
9.05.01.	**LIGHTING POINTS**										
9.05.01.01.	**PVCu insulated and sheathed cables in residential property; installed in floor cavities or roof voids; protected by PVCu channel in walls**										
9.05.01.01.A	lighting point controlled by 1 switch	nr	1.24	52.14	—	5.26	57.40	57.35	—	5.79	63.14
9.05.01.01.B	lighting point controlled by 2 switches . . .	nr	1.58	66.70	—	8.86	75.57	73.38	—	9.75	83.12
9.05.01.01.C	lighting point controlled by 3 switches . . .	nr	1.73	72.98	—	12.70	85.69	80.28	—	13.97	94.25
9.05.01.01.D	2 lighting points controlled by 1 switch . .	nr	1.32	55.59	—	9.34	64.92	61.15	—	10.27	71.42
9.05.01.02.	**PVCu insulated cables contained in black enamelled screwed, welded conduit in commercial property**										
9.05.01.02.A	lighting point controlled by 1 switch	nr	1.46	61.52	—	33.14	94.66	67.67	—	36.46	104.13
9.05.01.02.B	lighting point controlled by 2 switches . . .	nr	1.87	78.71	—	34.32	113.03	86.58	—	37.75	124.33
9.05.01.02.C	lighting point controlled by 3 switches . . .	nr	2.05	86.11	—	83.63	169.74	94.72	—	91.99	186.72
9.05.01.02.D	2 lighting points controlled by 1 switch . .	nr	1.56	65.61	—	57.86	123.47	72.17	—	63.64	135.82
9.05.01.03.	**PVCu insulated cables contained in galvanised screwed, welded conduit in industrial property**										
9.05.01.03.A	lighting point controlled by 1 switch	nr	1.72	72.59	—	37.62	110.21	79.85	—	41.38	121.23
9.05.01.03.B	lighting point controlled by 2 switches . . .	nr	2.21	92.91	—	73.58	166.48	102.20	—	80.93	183.13
9.05.01.03.C	lighting point controlled by 3 switches . . .	nr	2.41	101.65	—	96.77	198.42	111.81	—	106.44	218.26
9.05.01.03.D	2 lighting points controlled by 1 switch . .	nr	1.84	77.40	—	66.65	144.05	85.14	—	73.32	158.46
9.05.01.04.	**MICS cables in commercial property**										
9.05.01.04.A	lighting point controlled by 1 switch	nr	1.46	61.52	—	24.95	86.47	67.67	—	27.45	95.12
9.05.01.04.B	lighting point controlled by 2 switches . . .	nr	1.87	78.71	—	26.13	104.84	86.58	—	28.74	115.32
9.05.01.04.C	lighting point controlled by 3 switches . . .	nr	2.05	86.11	—	64.10	150.21	94.72	—	70.51	165.23
9.05.01.04.D	2 lighting points controlled by 3 switches .	nr	2.18	91.80	—	78.48	170.28	100.98	—	86.33	187.31
9.05.01.05.	**MICS cables in industrial property**										
9.05.01.05.A	lighting point controlled by 1 switch	nr	1.72	72.59	—	37.62	110.21	79.85	—	41.38	121.23
9.05.01.05.B	lighting point controlled by 2 switches . . .	nr	2.21	92.91	—	73.58	166.48	102.20	—	80.93	183.13
9.05.01.05.C	lighting point controlled by 3 switches . . .	nr	2.41	101.65	—	96.77	198.42	111.81	—	106.44	218.26
9.05.01.05.D	2 lighting points controlled by 1 switch . .	nr	1.84	77.40	—	54.84	132.24	85.14	—	60.32	145.46

Sundries

	Unit	Labour Hours	Labour Net £	Plant Net £	Materials Net £	Unit Rate Net £	Labour Gross £	Plant Gross £	Materials Gross £	Unit Rate Gross £
									(Gross rates include 10% profit)	
9.05.	**ELECTRICAL WORK**									
9.05.02.	**SWITCHED SOCKET OUTLETS**									
9.05.02.01.	**PVCu insulated and sheathed cable in residential property, installed**									
9.05.02.01.A	13A single switch socket outlet	nr	1.36	57.35	—	7.61	64.96	63.09	—	8.37	71.46
9.05.02.01.B	1A - 13A dual switch socket outlet	nr	1.36	57.35	—	8.02	65.37	63.09	—	8.83	71.91
9.05.02.02.	**PVCu insulated cables contained in black enamelled screwed welded conduit in commercial property**									
9.05.02.02.A	1A - 13A single switch socket outlet	nr	1.44	60.76	—	24.34	85.10	66.84	—	26.77	93.61
9.05.02.02.B	13A dual switch socket outlet	nr	1.44	60.76	—	24.75	85.51	66.84	—	27.23	94.06
9.05.02.03.	**PVCu insulated cables contained in galvanised screwed welded conduit in industrial property**									
9.05.02.03.A	1A - 13A single switch socket outlet	nr	1.73	72.98	—	28.81	101.79	80.28	—	31.69	111.97
9.05.02.03.B	1A - 13A dual switch socket outlet	nr	1.86	78.20	—	29.23	107.42	86.02	—	32.15	118.16
9.05.02.04.	**MICS cables in commercial property**									
9.05.02.04.A	1A - 13A single switch socket outlet	nr	1.24	52.14	—	42.09	94.23	57.35	—	46.30	103.65
9.05.02.04.B	1A - 13A dual switch socket outlet	nr	1.49	62.54	—	42.50	105.04	68.79	—	46.75	115.54
9.05.04.	**COOKER POINTS**									
9.05.04.01.	**PVCu insulated and sheathed cable in residential property, installed**									
9.05.04.01.B	cooker point 45A with cooker panel	nr	1.61	67.80	—	30.33	98.13	74.58	—	33.36	107.94
9.05.04.02.	**PVCu insulated cables contained in enamelled screwed welded conduit in commercial property**									
9.05.04.02.B	cooker point 45A with cooker panel	nr	1.73	72.98	—	45.70	118.68	80.28	—	50.26	130.55
9.05.04.03.	**MICS cables in commercial properties**									
9.05.04.03.B	cooker point 45A with cooker panel	nr	1.86	78.20	—	48.28	126.48	86.02	—	53.11	139.12
9.05.06.	**IMMERSION HEATER POINTS**									
9.05.06.01.	**Immersion heater point with control switch (excluding heater)**									
9.05.06.01.A	PVCu insulated cables in PVCu conduit (domestic).	nr	1.49	62.54	—	8.21	70.75	68.79	—	9.03	77.82
9.05.06.01.B	in enamelled steel welded conduit, etc. . .	nr	1.86	78.20	—	26.24	104.44	86.02	—	28.86	114.88
9.05.06.02.	**MICS cables in commercial property**									
9.05.06.02.A	immersion heater point with control switch (excluding heater)	nr	1.65	69.57	—	36.90	106.47	76.53	—	40.58	117.11
9.05.06.02.B	supply and connect 3kW immersion heater complete with fixed thermostat to new cylinder.	nr	1.36	57.35	—	18.43	75.78	63.09	—	20.27	83.36

Sundries

Hutchins Priced Schedules 2021	Unit	Labour Hours	Labour Net £	Plant Net £	Materials Net £	Unit Rate Net £	Labour Gross £	Plant Gross £	Materials Gross £	Unit Rate Gross £
							(Gross rates include 10% profit)			
9.05. **ELECTRICAL WORK**										
9.05.08. **INFRARED HEATER POINTS**										
9.05.08.01. **PVCu insulated and sheathed cables installed in residential properties**										
9.05.08.01.A infrared heater with control switch (excluding heater and earth bonding)...	nr	1.36	57.35	—	8.17	65.52	63.09	—	8.99	72.07
9.05.08.02. **PVCu insulated cables in black enamelled screwed conduit in commercial property**										
9.05.08.02.A infrared heater point with control switch (excluding heater and earth bonding)...	nr	1.36	57.35	—	35.44	92.79	63.09	—	38.98	102.07
9.05.08.03. **MICS cables in commercial property**										
9.05.08.03.A infrared heater point with control switch (excluding heater and earth bonding)...	nr	1.36	57.35	—	28.85	86.20	63.09	—	31.73	94.82
9.05.09. **BELL INSTALLATIONS**										
9.05.09.01. **PVCu insulated cable in residential property**										
9.05.09.01.A bell controlled by front door push including transformer............	nr	1.26	51.34	—	21.91	73.26	56.48	—	24.10	80.58
9.05.09.01.B bell controlled by front door push and back door push including transformer...	nr	1.61	67.80	—	25.37	93.17	74.58	—	27.90	102.48
9.05.09.02. **PVCu insulated cables contained in black enamelled screwed welded conduit in commercial property**										
9.05.09.02.A bell controlled by front door push including transformer............	nr	1.49	62.54	—	51.39	113.92	68.79	—	56.52	125.31
9.05.09.02.B bell controlled by front door push and back door push including transformer...	nr	1.73	72.98	—	79.76	152.74	80.28	—	87.73	168.01
9.05.09.03. **MICS cables in commercial property**										
9.05.09.03.A bell controlled by front door push including transformer............	nr	1.49	62.54	—	59.99	122.53	68.79	—	65.99	134.78
9.05.09.03.B bell controlled by front door push and back door push including transformer...	nr	1.61	67.80	—	88.99	156.79	74.58	—	97.89	172.47
9.05.10. **TV OUTLETS**										
9.05.10.01. **Low loss coaxial cable in residential property protected by PVCu conduit in walls; terminating with a single TV outlet; allow 3 metres of cable in roof void**										
9.05.10.01.A for connection to aerial by others.....	nr	1.73	72.98	—	5.39	78.37	80.28	—	5.93	86.21

Sundries

Hutchins Priced Schedules 2021	Unit	Labour Hours	Labour Net £	Plant Net £	Materials Net £	Unit Rate Net £	Labour Gross £	Plant Gross £	Materials Gross £	Unit Rate Gross £
						(Gross rates include 10% profit)				
9.05. **ELECTRICAL WORK**										
9.05.12. **ELECTRICAL SHAVER POINT**										
9.05.12.01. PVCu insulated and sheathed cable in residential properties										
9.05.12.01.A dual voltage type, isolated for bathrooms, etc.	nr	1.11	46.90	—	17.97	64.88	51.60	—	19.77	71.37
9.05.12.02. PVCu insulated cables contained in black enamelled screwed and welded conduit in commercial property										
9.05.12.02.A dual voltage type, isolated for bathrooms, etc.	nr	1.11	46.90	—	34.01	80.91	51.60	—	37.41	89.00
9.05.14. **STORAGE HEATING**										
9.05.14.01. PVCu insulated and sheathed cable in residential properties										
9.05.14.01.A 3kW heater point controlled by switch adjacent to heater position (from independent fuse)	nr	1.44	60.76	—	6.44	67.20	66.84	—	7.09	73.92
9.05.14.02. PVCu insulated cables contained in black enamelled screwed and welded conduit in commercial property										
9.05.14.02.A 3kW heater point controlled by switch adjacent to heater position (from independent fuse)	nr	1.73	72.98	—	22.49	95.47	80.28	—	24.74	105.02
9.05.16. **LIGHTING FITTINGS**										
9.05.16.02. LED tube fitting; ceiling mounted; single										
9.05.16.02.A 637mm; 9W	nr	0.50	20.85	—	9.96	30.81	22.93	—	10.96	33.89
9.05.16.02.B 1237mm; 18W	nr	0.50	20.85	—	12.86	33.71	22.93	—	14.15	37.08
9.05.16.02.C 1537mm; 22W	nr	0.50	20.85	—	17.42	38.27	22.93	—	19.17	42.10
9.05.16.02.D 1837mm; 30W	nr	0.62	26.06	—	29.24	55.30	28.67	—	32.17	60.83

Sundries

Hutchins Priced Schedules 2021	Unit	Labour Hours	Labour Net £	Plant Net £	Materials Net £	Unit Rate Net £	Labour Gross £	Plant Gross £	Materials Gross £	Unit Rate Gross £
									(Gross rates include 10% profit)	
9.05.	**ELECTRICAL WORK**									
9.05.18.	**REWIRING**									
9.05.18.01.	**To an average dwelling house containing**									
9.05.18.01.A	15nr 13A twin power points and 14nr 5A lighting points; 3nr two-way lighting points complete with switches; 150mm pendants sets; all in two circuits and 3kw immersion heater and cooker feed complete and 1nr external 15W LED bulkhead with sensor including consumer unit with MCBs	nr	45.47	1914.91	—	308.62	2223.53	2106.40	—	339.49	2445.88
9.06.	**DAMP-PROOFING**									
9.06.01.	**GENERALLY**									
9.06.01.01.	**Chase out mortar joint in brickwork with mechanical saw; insert flexible membrane of lead, zinc, copper, bituminous material or low density polythene and force in new mortar (material cost of membrane not included)**									
9.06.01.01.A	to half brick walls	m	1.05	54.83	2.14	0.13	57.10	60.31	2.35	0.14	62.80
9.06.01.01.B	to one brick walls	m	2.09	109.66	4.27	0.26	114.19	120.63	4.70	0.28	125.61
9.06.01.01.C	to one and a half brick walls	m	3.14	164.60	6.41	0.39	171.39	181.06	7.05	0.42	188.53
9.06.01.01.D	extra over last for material cost of membrane in one brick wall in copper; 0.45mm (26 SWG)	m	—	—	—	14.80	14.80	—	—	16.28	16.28
9.06.01.01.E	extra over for material cost of membrane in one brick wall in 1.80mm lead (Code 4)	m	—	—	—	11.22	11.22	—	—	12.34	12.34
9.06.01.01.F	extra over for pitch polymer damp proof course; 1.25mm thick	m	—	—	—	1.19	1.19	—	—	1.31	1.31
9.06.01.02.	**Drill brickwork and provide Dryzone damp proof course cream to**									
9.06.01.02.A	half brick walls	m	0.15	7.86	—	27.14	35.01	8.65	—	29.86	38.51
9.06.01.02.B	one brick walls	m	0.28	14.89	—	45.24	60.12	16.38	—	49.76	66.14
9.06.01.02.C	one and half brick wall	m	0.40	21.07	—	74.03	95.10	23.18	—	81.43	104.61
9.06.01.03.	**Drill brickwork and provide chemical damp-proof course injected under pressure to**									
9.06.01.03.A	half brick wall	m	0.14	7.13	—	4.50	11.63	7.84	—	4.95	12.79
9.06.01.03.B	one brick wall	m	0.26	13.42	—	9.00	22.42	14.76	—	9.90	24.66
9.06.01.03.C	one and half brick wall	m	0.37	19.50	—	13.50	33.00	21.45	—	14.85	36.30

Sundries

Hutchins Priced Schedules 2021		Unit	Labour Hours	Labour Net £	Plant Net £	Materials Net £	Unit Rate Net £	Labour Gross £	Plant Gross £	Materials Gross £	Unit Rate Gross £
								(Gross rates include 10% profit)			
9.06.	**DAMP-PROOFING**										
9.06.01.	**GENERALLY**										
9.06.01.04.	**Hack off existing plastering or rendering to a height of one metre above the damp-proof course and later replace with**										
9.06.01.04.A	Renderguard Gold with Thistle multi-purpose finish plaster	m²	1.08	44.10	—	3.01	47.10	48.51	—	3.31	51.81
9.06.01.05.	**Tanking slurry**										
9.06.01.05.A	Vandex BB75; two layers; minimum 2mm thickness per layer	m²	0.75	22.19	—	18.32	40.50	24.40	—	20.15	44.55
9.06.01.06.	**Clean, prepare and apply bitumen roofing compound to**										
9.06.01.06.A	sound surfaces, asphalt, concrete, felt, slate or tile	m²	0.55	16.27	—	6.47	22.74	17.90	—	7.11	25.01
9.06.01.07.	**Clean, prepare and apply bitumen roofing compound and fungicide solution to**										
9.06.01.07.A	surfaces likely to support fungal or algae growth, asphalt, concrete, felt, slate or tile	m²	0.60	17.75	—	6.74	24.48	19.52	—	7.41	26.93
9.07.	**UNDERPINNING**										
9.07.01.	**GENERALLY**										
9.07.01.01.	**Break up concrete paving and hardcore under; level and ram hardcore on completion; 150mm concrete paving with screeded finish**										
9.07.01.01.A	150mm concrete paving	m²	2.60	76.91	—	15.23	92.14	84.60	—	16.76	101.35
9.07.01.02.	**Excavate for access trench, part backfill**										
9.07.01.02.A	dispose of surplus in skip	m³	7.00	207.06	—	—	207.06	227.77	—	—	227.77
9.07.01.03.	**Excavate under existing wall or foundation**										
9.07.01.03.A	and get out	m³	5.40	159.73	—	—	159.73	175.71	—	—	175.71
9.07.01.04.	**Hire of skip; delivery to site; removing when full; dispose of contents; payment of tipping charges**										
9.07.01.04.A	skip size; 4.50m3	m³	—	—	43.60	—	43.60	—	47.96	—	47.96
9.07.01.06.	**Break out existing foundations in short lengths and get out**										
9.07.01.06.A	concrete	m³	32.00	946.56	—	—	946.56	1041.22	—	—	1041.22

Sundries

Hutchins Priced Schedules 2021	Unit	Labour Hours	Labour Net £	Plant Net £	Materials Net £	Unit Rate Net £	Labour Gross £	Plant Gross £	Materials Gross £	Unit Rate Gross £
								(Gross rates include 10% profit)		

9.07.   UNDERPINNING

9.07.01.   GENERALLY

9.07.01.06.   Break out existing foundations in short lengths and get out

	Unit	Labour Hours	Labour Net £	Plant Net £	Materials Net £	Unit Rate Net £	Labour Gross £	Plant Gross £	Materials Gross £	Unit Rate Gross £
9.07.01.06.B brickwork	m³	27.00	798.66	—	—	798.66	878.53	—	—	878.53
9.07.01.07. Concrete underpinning in short lengths including necessary formwork; mix										
9.07.01.07.A GEN1 mix	m³	13.00	384.54	—	99.14	483.68	422.99	—	109.05	532.05
9.07.01.07.B GEN2 mix	m³	13.00	384.54	—	101.85	486.39	422.99	—	112.03	535.03
9.07.01.08. Concrete in projecting pier bases including necessary formwork; mix										
9.07.01.08.A GEN1 mix	m³	15.50	458.49	—	99.14	557.63	504.34	—	109.05	613.39
9.07.01.08.B GEN2 mix	m³	15.50	458.49	—	101.85	560.34	504.34	—	112.03	616.37
9.07.01.09. Reinforced concrete in ground beams including one 13mm mild steel bar for each 0.02m2 sectional area of beam; including necessary formwork; mix										
9.07.01.09.A GEN2 mix	m³	9.86	516.86	—	138.51	655.37	568.55	—	152.36	720.90
9.07.01.10. Brickwork one brick thick in short lengths; in underpinning										
9.07.01.10.A class B engineering bricks (PC sum £500 per 1000)	m²	5.37	281.29	—	71.38	352.67	309.41	—	78.52	387.93
9.07.01.10.B common brickwork (PC sum £450 per 1000)	m²	5.37	281.29	—	64.83	346.12	309.41	—	71.31	380.73
9.07.01.11. Wedge and pin up new brickwork to underside of existing with slates; thickness										
9.07.01.11.A 215mm	m	0.67	35.12	—	8.50	43.62	38.63	—	9.35	47.98
9.07.01.11.B 327mm	m	0.94	49.27	—	12.70	61.97	54.20	—	13.97	68.17

House Renovation Grants, Repairs and Alterations

Hutchins Priced Schedules 2021	Unit	Labour Hours	Labour Net £	Plant Net £	Materials Net £	Unit Rate Net £	Labour Gross £	Plant Gross £	Materials Gross £	Unit Rate Gross £
									(Gross rates include 10% profit)	

A.01. **GRANTWORK**

A.01.01. **HOUSE RENOVATION**

A.01.01.01. **Take up old wood flooring in basement including joists and plates, excavate for, breaking up any obstructions; lay 150mm hardcore, 100mm concrete and vinyl tile floor**

	Unit	Labour Hours	Labour Net £	Plant Net £	Materials Net £	Unit Rate Net £	Labour Gross £	Plant Gross £	Materials Gross £	Unit Rate Gross £
A.01.01.01.A take up flooring and excavate to required depth	m²	6.00	177.48	—	—	177.48	195.23	—	—	195.23
A.01.01.01.B lay 150mm hardcore	m²	0.30	8.87	—	9.95	18.83	9.76	—	10.95	20.71
A.01.01.01.C lay 100mm concrete	m²	1.00	29.58	—	9.84	39.42	32.54	—	10.82	43.36
A.01.01.01.E inclusive cost for preparing	m²	7.30	215.93	—	19.79	235.72	237.53	—	21.77	259.29
A.01.01.01.F vinyl tile flooring	m²	0.28	10.54	—	7.72	18.25	11.59	—	8.49	20.08
A.01.01.01.G total cost	m²	7.58	226.47	—	27.50	253.97	249.12	—	30.25	279.37

A.01.01.02. **Break up existing concrete area paving; excavate oversite 150mm deep; prepare subsoil and lay new concrete paving**

	Unit	Labour Hours	Labour Net £	Plant Net £	Materials Net £	Unit Rate Net £	Labour Gross £	Plant Gross £	Materials Gross £	Unit Rate Gross £
A.01.01.02.A break up concrete paving	m²	2.25	66.56	—	—	66.56	73.21	—	—	73.21
A.01.01.02.B excavate oversite	m²	0.42	12.42	—	—	12.42	13.67	—	—	13.67
A.01.01.02.C lay 100mm concrete	m²	1.00	29.58	—	9.84	39.42	32.54	—	10.82	43.36
A.01.01.02.D total cost	m²	3.67	108.56	—	9.84	118.39	119.41	—	10.82	130.23

A.01.01.03. **Build concrete block and cavity lining to existing external and party walls; including plastering and two coats emulsion paint**

	Unit	Labour Hours	Labour Net £	Plant Net £	Materials Net £	Unit Rate Net £	Labour Gross £	Plant Gross £	Materials Gross £	Unit Rate Gross £
A.01.01.03.A 100mm concrete block walling with four ties per m2	m²	0.80	42.15	—	16.29	58.43	46.36	—	17.91	64.27
A.01.01.03.B render and set walls	m²	0.43	28.70	—	2.89	31.59	31.57	—	3.18	34.74
A.01.01.03.C emulsion paint	m²	0.32	12.04	—	0.54	12.58	13.25	—	0.60	13.84
A.01.01.03.D total cost	m²	1.55	82.89	—	19.71	102.60	91.17	—	21.68	112.86

A.01.01.04. **Rake out existing brickwork joints to form key; plaster walls and emulsion paint**

	Unit	Labour Hours	Labour Net £	Plant Net £	Materials Net £	Unit Rate Net £	Labour Gross £	Plant Gross £	Materials Gross £	Unit Rate Gross £
A.01.01.04.A rake out brickwork	m²	0.27	14.05	—	—	14.05	15.45	—	—	15.45
A.01.01.04.B render and set walls	m²	0.43	28.70	—	2.89	31.59	31.57	—	3.18	34.74
A.01.01.04.C emulsion paint	m²	0.32	12.04	—	0.54	12.58	13.25	—	0.60	13.84
A.01.01.04.D total cost	m²	1.02	54.79	—	3.43	58.22	60.27	—	3.77	64.04

House Renovation Grants, Repairs and Alterations

Hutchins Priced Schedules 2021	Unit	Labour Hours	Labour Net £	Plant Net £	Materials Net £	Unit Rate Net £	Labour Gross £	Plant Gross £	Materials Gross £	Unit Rate Gross £
							(Gross rates include 10% profit)			
A.01.	**GRANTWORK**									
A.01.01.	HOUSE RENOVATION									
A.01.01.05.	**Build 100mm concrete block partition wall off site; plaster and emulsion paint both sides; tie into existing walls and pin to soffit**									
A.01.01.05.A	100mm concrete block walling including tying in	m²	0.80	42.15	—	18.73	60.88	46.36	—	20.61	66.97
A.01.01.05.B	render and set both sides	m²	0.85	57.40	—	2.89	60.28	63.14	—	3.18	66.31
A.01.01.05.C	emulsion paint both sides	m²	0.84	31.61	—	1.08	32.69	34.77	—	1.19	35.96
A.01.01.05.D	total cost	m²	2.50	131.15	—	22.70	153.86	144.27	—	24.97	169.24
A.01.01.06.	**Take out and remove kitchen range and mantel-shelf; build up half brick thick wall across opening flush with existing walls and render and set**									
A.01.01.06.A	take out and remove	nr	6.00	177.48	—	—	177.48	195.23	—	—	195.23
A.01.01.06.B	take out and remove mantel-shelf	nr	0.80	23.66	—	—	23.66	26.03	—	—	26.03
A.01.01.06.C	half brick wall including tying; 3.00m2 (common brickwork PC sum £450 per 1000)	nr	4.70	246.16	—	122.74	368.91	270.78	—	135.02	405.80
A.01.01.06.D	render and set including jointing to existing; 3.00m2	nr	1.51	101.29	—	9.34	110.62	111.41	—	10.27	121.68
A.01.01.06.E	total cost	nr	13.00	548.59	—	132.08	680.67	603.45	—	145.29	748.74
A.01.01.08.	**Take out existing window frame, size 1.20m x 1.00m. Cut away stone sill and one brick apron below, about 1.20m x 0.45m. Reform brick jambs and build in brick-on-edge sill. Provide and fix new standard casement window size 1.20m x 1.35m. Glaze, paint both sides; make good all plaster. No alteration work to lintel or arch over. (Common brickwork PC sum £450 per 1000; facing brickwork PC sum £550 per 1000)**									
A.01.01.08.A	take out window sill	nr	1.60	47.33	—	—	47.33	52.06	—	—	52.06
A.01.01.08.B	cut away brickwork; 0.50m2	nr	1.90	56.20	—	—	56.20	61.82	—	—	61.82
A.01.01.08.C	reform jambs; 1.00m	nr	1.68	87.96	—	8.42	96.38	96.76	—	9.27	106.02
A.01.01.08.D	brick sill; 1.00m	nr	0.87	45.71	—	9.92	55.63	50.28	—	10.92	61.20
A.01.01.08.E	window casement	nr	1.21	63.32	—	106.36	169.69	69.66	—	117.00	186.66
A.01.01.08.F	6mm float glass; 1.50m2	nr	1.65	62.09	—	61.30	123.39	68.30	—	67.43	135.73
A.01.01.08.G	paint window casement; 3.50m2	nr	3.50	131.71	—	1.93	133.64	144.88	—	2.13	147.00
A.01.01.08.H	make good plaster	nr	0.45	30.38	—	1.29	31.67	33.42	—	1.42	34.84
A.01.01.08.I	total cost	nr	12.86	524.70	—	189.23	713.93	577.17	—	208.16	785.33

House Renovation Grants, Repairs and Alterations

Hutchins Priced Schedules 2021		Unit	Labour Hours	Labour Net £	Plant Net £	Materials Net £	Unit Rate Net £	Labour Gross £	Plant Gross £	Materials Gross £	Unit Rate Gross £
								(Gross rates include 10% profit)			
A.01.	**GRANTWORK**										
A.01.01.	HOUSE RENOVATION										
A.01.01.09.	Take out old door frame and wing light 1.80m x 2.00m (extreme). Build up one brick wall in old door opening up to sill level; take out old stone sill and build up new brick sill. Provide and fix new purpose made window frame 1.80m x 1.00m. Glaze, paint both sides; make good all plaster										
A.01.01.09.A	take out existing frame	nr	1.85	54.72	—	—	54.72	60.20	—	—	60.20
A.01.01.09.B	take out old sill	nr	0.40	11.83	—	—	11.83	13.02	—	—	13.02
A.01.01.09.C	build up one-brick wall (PC sum £550 per 1000) in old opening 0.75m x 0.90m . . .	nr	2.52	131.89	—	58.12	190.01	145.08	—	63.94	209.01
A.01.01.09.D	build new brick sill (PC sum £550 per 1000) 2.00m	nr	1.74	91.42	—	25.78	117.20	100.56	—	28.36	128.92
A.01.01.09.E	window frame and sashes	nr	7.38	386.75	—	73.21	459.97	425.43	—	80.53	505.96
A.01.01.09.F	6mm float glass in window frame and sashes; 1.60m2	nr	5.20	195.68	—	65.79	261.46	215.24	—	72.37	287.61
A.01.01.09.G	painting window frame and sashes; 4.00m2	nr	5.20	195.68	—	1.58	197.25	215.24	—	1.74	216.98
A.01.01.09.H	make good wall plaster	nr	0.45	30.38	—	2.58	32.96	33.42	—	2.84	36.25
A.01.01.09.I	total cost	nr	24.74	1098.35	—	227.06	1325.41	1208.18	—	249.77	1457.95
A.01.01.10.	Make and fix cupboard front and return end to form linen cupboard, internal size 750mm x 750mm, full height floor to ceiling 2.40m. Return wall and over door wall to be 75mm x 50mm fir stud covered both sides with expanded metal lathing and plastered. Hardboard flush door hung on 75mm light steel hinges, fitted with bow handle and Bales catch. Standard door lining with 50mm x 18mm architrave outside. Four open slat shelves, comprising 50mm x 25mm in softwood slats spaced 25mm apart and including bearers full width and depth of cupboard. New wall plaster and existing plaster inside cupboard to be painted in two coats emulsion. Door, lining and architrave to be primed and painted two coats of oil colour										
A.01.01.10.A	stud partition wall with expanded metal lathing and plaster both sides; 3.00m2 . .	nr	10.00	376.30	—	104.26	480.56	413.93	—	114.69	528.62
A.01.01.10.B	shelving; 2.50m2 including bearers	nr	5.00	188.15	—	28.62	216.77	206.97	—	31.49	238.45
A.01.01.10.C	door lining	nr	0.75	28.22	—	17.15	45.37	31.04	—	18.86	49.91

House Renovation Grants, Repairs and Alterations

Hutchins Priced Schedules 2021	Unit	Labour Hours	Labour Net £	Plant Net £	Materials Net £	Unit Rate Net £	Labour Gross £	Plant Gross £	Materials Gross £	Unit Rate Gross £
							(Gross rates include 10% profit)			

A.01.   **GRANTWORK**

A.01.01.   **HOUSE RENOVATION**

A.01.01.10. Make and fix cupboard front and return end to form linen cupboard, internal size 750mm x 750mm, full height floor to ceiling 2.40m. Return wall and over door wall to be 75mm x 50mm fir stud covered both sides with expanded metal lathing and plastered. Hardboard flush door hung on 75mm light steel hinges, fitted with bow handle and Bales catch. Standard door lining with 50mm x 18mm architrave outside. Four open slat shelves, comprising 50mm x 25mm in softwood slats spaced 25mm apart and including bearers full width and depth of cupboard. New wall plaster and existing plaster inside cupboard to be painted in two coats emulsion. Door, lining and architrave to be primed and painted two coats of oil colour

Code	Description	Unit	Labour Hours	Labour Net £	Plant Net £	Materials Net £	Unit Rate Net £	Labour Gross £	Plant Gross £	Materials Gross £	Unit Rate Gross £
A.01.01.10.D	moulded 6 panel door; white primed; 1981mm x 610mm x 35mm	nr	0.75	28.22	—	44.11	72.33	31.04	—	48.52	79.57
A.01.01.10.E	architrave; 5.00m	nr	0.75	28.22	—	14.31	42.54	31.04	—	15.75	46.79
A.01.01.10.F	hinges; chrome plated	nr	0.45	16.93	—	2.02	18.95	18.63	—	2.22	20.85
A.01.01.10.G	handle	nr	0.30	11.29	—	7.05	18.34	12.42	—	7.75	20.17
A.01.01.10.H	catch	nr	0.60	22.58	—	4.09	26.67	24.84	—	4.50	29.34
A.01.01.10.I	emulsion paint; 8.00m2	nr	2.60	97.84	—	3.52	101.36	107.62	—	3.87	111.49
A.01.01.10.J	painting; 4.00m2	nr	2.75	103.48	—	3.41	106.89	113.83	—	3.75	117.58
A.01.01.10.K	total cost	nr	23.95	901.24	—	228.55	1129.79	991.36	—	251.40	1242.77

A.01.01.12. Take out old door frame; provide and fix 44mm cottage style softwood entrance door and frame, including cylinder night latch and ketter plate; and decorate

Code	Description	Unit	Labour Hours	Labour Net £	Plant Net £	Materials Net £	Unit Rate Net £	Labour Gross £	Plant Gross £	Materials Gross £	Unit Rate Gross £
A.01.01.12.A	take out door and frame	nr	1.50	44.37	—	—	44.37	48.81	—	—	48.81
A.01.01.12.B	new door and frame	nr	3.00	112.89	—	261.81	374.70	124.18	—	288.00	412.17
A.01.01.12.C	cylinder latch	nr	1.60	60.21	—	22.47	82.68	66.23	—	24.72	90.95
A.01.01.12.D	letter plate	nr	1.75	65.85	—	14.13	79.98	72.44	—	15.54	87.98
A.01.01.12.E	painting; 3.50m2	nr	3.00	112.89	—	2.09	114.98	124.18	—	2.30	126.47
A.01.01.12.F	total cost	nr	10.85	396.21	—	300.50	696.71	435.83	—	330.55	766.38

A.01.01.13. Take down old timber stair flight and remove; fill over stair opening with 47mm x 150mm joists and board over to form extension to upper floor. Plaster baseboard and set soffit and emulsion paint; make good wall plaster of spandrel and emulsion paint on wall

Code	Description	Unit	Labour Hours	Labour Net £	Plant Net £	Materials Net £	Unit Rate Net £	Labour Gross £	Plant Gross £	Materials Gross £	Unit Rate Gross £
A.01.01.13.A	take out stairs	nr	9.40	278.05	—	—	278.05	305.86	—	—	305.86
A.01.01.13.B	floor joists and boarding over 2.25m2 . . .	nr	3.25	122.30	—	130.73	253.02	134.53	—	143.80	278.33
A.01.01.13.C	ceiling plaster; 2.60m2	nr	1.21	80.99	—	3.92	84.90	89.09	—	4.31	93.39
A.01.01.13.D	emulsion paint; 2.60m2	nr	1.05	39.51	—	0.50	40.01	43.46	—	0.55	44.01
A.01.01.13.E	wall plaster; 3.50m2	nr	1.71	114.73	—	11.24	125.97	126.20	—	12.36	138.56
A.01.01.13.F	emulsion paint; 3.50m2	nr	1.12	42.15	—	1.92	44.07	46.36	—	2.12	48.48
A.01.01.13.G	total cost	nr	17.73	677.72	—	148.30	826.03	745.49	—	163.13	908.63

A.01.01.14. Take down existing door and lining and set aside and fix in new partition wall including architraves both sides; oil and adjust lock and decorate woodwork

Code	Description	Unit	Labour Hours	Labour Net £	Plant Net £	Materials Net £	Unit Rate Net £	Labour Gross £	Plant Gross £	Materials Gross £	Unit Rate Gross £
A.01.01.14.A	take out door and lining	nr	1.50	44.37	—	—	44.37	48.81	—	—	48.81
A.01.01.14.B	reset door and lining in new opening . . .	nr	3.75	141.11	—	—	141.11	155.22	—	—	155.22
A.01.01.14.C	architrave; 10.00m	nr	1.80	67.73	—	13.22	80.95	74.51	—	14.54	89.05

Hutchins Priced Schedules 2021		Unit	Labour Hours	Labour Net £	Plant Net £	Materials Net £	Unit Rate Net £	Labour Gross £	Plant Gross £	Materials Gross £	Unit Rate Gross £
							(Gross rates include 10% profit)				
A.01.	**GRANTWORK**										
A.01.01.	**HOUSE RENOVATION**										
A.01.01.14.	Take down existing door and lining and set aside and fix in new partition wall including architraves both sides; oil and adjust lock and decorate woodwork										
A.01.01.14.D	painting; 3.50m2	nr	3.30	124.18	—	3.10	127.28	136.60	—	3.42	140.01
A.01.01.14.E	oil and adjust lock	nr	0.25	9.41	—	—	9.41	10.35	—	—	10.35
A.01.01.14.F	total cost	nr	10.60	386.80	—	16.32	403.13	425.48	—	17.96	443.44
A.01.01.15.	**Excavate oversite 225mm deep to annexe building for bathroom and WC or kitchen and deposit soil over adjacent garden**										
A.01.01.15.A	excavate and deposit soil	m²	1.00	29.58	—	—	29.58	32.54	—	—	32.54
A.01.01.16.	**Excavate foundation trench 450mm x 600mm deep; lay concrete foundation 450mm x 300mm thick; new 275mm cavity brickwork 450mm high up to and including dampcourse**										
A.01.01.16.A	trench excavation	m	1.20	35.50	—	—	35.50	39.05	—	—	39.05
A.01.01.16.B	foundation concrete	m	1.00	29.58	—	13.75	43.33	32.54	—	15.13	47.67
A.01.01.16.C	common brickwork (PC sum £450 per 1000)	m	1.14	59.76	—	29.75	89.51	65.73	—	32.73	98.46
A.01.01.16.D	slate dampcourse	m	0.15	8.07	—	7.10	15.17	8.88	—	7.81	16.69
A.01.01.16.E	total cost	m	3.49	132.91	—	50.61	183.51	146.20	—	55.67	201.86
A.01.01.18.	**Lay 100mm hardcore and 100mm concrete oversite and wood flooring on and including 47mm x 150mm joists (no sleeper walls)**										
A.01.01.18.A	hardcore	m²	0.25	7.40	—	6.62	14.01	8.13	—	7.28	15.41
A.01.01.18.B	concrete	m²	1.00	29.58	—	10.18	39.76	32.54	—	11.20	43.74
A.01.01.18.C	flooring and joists	m²	1.10	41.39	—	51.79	93.19	45.53	—	56.97	102.51
A.01.01.18.D	total cost	m²	2.35	78.37	—	68.60	146.96	86.20	—	75.46	161.66
A.01.01.19.	**Build 275mm cavity brick wall in common brickwork (PC sum £450 per 1000) with facing brickwork (PC sum £550 per 1000) externally, plastered and emulsion paint internally**										
A.01.01.19.A	cavity wall	m²	2.85	149.50	—	70.50	220.01	164.45	—	77.56	242.01
A.01.01.19.B	wall plaster	m²	0.43	28.70	—	2.89	31.59	31.57	—	3.18	34.74
A.01.01.19.C	emulsion paint	m²	0.32	12.04	—	0.50	12.54	13.25	—	0.55	13.79
A.01.01.19.D	total cost	m²	3.60	190.24	—	73.89	264.13	209.27	—	81.28	290.55
A.01.01.20.	**Form lean-to roof of concrete tiles on battens including felt, 125mm x 50mm rafters and 100mm x 75mm plate**										
A.01.01.20.A	plates and rafters	m²	0.80	30.10	—	22.21	52.31	33.11	—	24.43	57.55

House Renovation Grants, Repairs and Alterations

Hutchins Priced Schedules 2021	Unit	Labour Hours	Labour Net £	Plant Net £	Materials Net £	Unit Rate Net £	Labour Gross £	Plant Gross £	Materials Gross £	Unit Rate Gross £
							(Gross rates include 10% profit)			

A.01.   **GRANTWORK**

A.01.01.   **HOUSE RENOVATION**

A.01.01.20.   **Form lean-to roof of concrete tiles on battens including felt, 125mm x 50mm rafters and 100mm x 75mm plate**

Hutchins Priced Schedules 2021	Unit	Labour Hours	Labour Net £	Plant Net £	Materials Net £	Unit Rate Net £	Labour Gross £	Plant Gross £	Materials Gross £	Unit Rate Gross £
A.01.01.20.B battens and tiles	m²	0.94	49.27	—	13.61	62.88	54.20	—	14.97	69.17
A.01.01.20.C felt	m²	0.09	4.61	—	3.30	7.92	5.07	—	3.63	8.71
A.01.01.20.D total cost	m²	1.83	83.99	—	39.12	123.11	92.39	—	43.03	135.42
A.01.01.21. **150mm x 25mm fascia and painting and PVCu half round gutter**										
A.01.01.21.A fascia board	m	0.25	9.41	—	3.00	12.40	10.35	—	3.30	13.64
A.01.01.21.B paint fascia boards	m	0.30	11.29	—	0.32	11.61	12.42	—	0.35	12.77
A.01.01.21.C PVCu gutter including stop ends and outlets	m	0.30	28.74	—	15.79	44.53	31.62	—	17.37	48.98
A.01.01.21.D total cost	m	0.85	49.44	—	19.10	68.54	54.38	—	21.01	75.39
A.01.01.22. **PVCu downpipe including swan neck and shoe**										
A.01.01.22.A 68mm PVCu downpipe 3.00m	nr	0.91	86.32	—	51.55	137.87	94.95	—	56.71	151.66
A.01.01.22.B swan neck comprising two bends	nr	0.30	28.74	—	14.25	42.99	31.62	—	15.67	47.29
A.01.01.22.C shoe	nr	0.20	19.03	—	7.28	26.31	20.94	—	8.00	28.94
A.01.01.22.D total cost	nr	1.41	134.09	—	73.08	207.17	147.50	—	80.38	227.89
A.01.01.23. **Renew ceiling joists and plasterboard and emulsion paint ceiling**										
A.01.01.23.A joists 100mm x 50mm	m²	0.20	7.53	—	9.27	16.79	8.28	—	10.19	18.47
A.01.01.23.B plasterboard and set ceiling	m²	0.45	30.38	—	3.57	33.95	33.42	—	3.93	37.34
A.01.01.23.C emulsion paint plasterboard and ceiling	m²	0.40	15.05	—	0.58	15.64	16.56	—	0.64	17.20
A.01.01.23.D total cost	m²	1.05	52.96	—	13.42	66.38	58.25	—	14.76	73.02
A.01.01.24. **Provide and fix standard casement window size 630mm x 1050mm including lintel, brick sill, glazing and painting**										
A.01.01.24.A steel lintel for external solid wall; 1050mm long	nr	0.85	25.14	—	21.03	46.17	27.66	—	23.14	50.79
A.01.01.24.B facing bricks sill (PC sum £550 per 1000)	nr	0.87	45.71	—	45.80	91.51	50.28	—	50.38	100.66
A.01.01.24.C casement window	nr	1.20	45.16	—	149.18	194.34	49.67	—	164.10	213.77
A.01.01.24.D 6mm float glass 0.60m2	nr	0.80	30.10	—	24.86	54.96	33.11	—	27.34	60.46
A.01.01.24.E paint both sides 1.30m2	nr	2.00	75.26	—	0.89	76.15	82.79	—	0.98	83.77
A.01.01.24.F total cost	nr	5.72	221.37	—	241.77	463.14	243.51	—	265.95	509.46
A.01.01.25. **Cut through one brick external wall for new door opening including lintel over; provide and fix new door and frame, size 825mm x 2025mm overall**										
A.01.01.25.A cut opening through one brick wall 2m2	nr	7.50	221.85	—	—	221.85	244.04	—	—	244.04

House Renovation Grants, Repairs and Alterations

Hutchins Priced Schedules 2021		Unit	Labour Hours	Labour Net £	Plant Net £	Materials Net £	Unit Rate Net £	Labour Gross £	Plant Gross £	Materials Gross £	Unit Rate Gross £
								(Gross rates include 10% profit)			

A.01. GRANTWORK

A.01.01. HOUSE RENOVATION

A.01.01.25. Cut through one brick external wall for new door opening including lintel over; provide and fix new door and frame, size 825mm x 2025mm overall

A.01.01.25.B	steel lintel for external solid wall; 143mm high, 1200mm long	nr	1.00	29.58	—	38.23	67.81	32.54	—	42.05	74.59
A.01.01.25.C	door frame	nr	1.75	65.85	—	73.84	139.70	72.44	—	81.23	153.67
A.01.01.25.D	door; 762mm x 1981mm including lock, hinges, etc.	nr	4.00	150.52	—	226.19	376.71	165.57	—	248.81	414.39
A.01.01.25.E	paint door and frame overall both sides; 4.00m2	nr	3.60	135.47	—	4.44	139.91	149.01	—	4.88	153.90
A.01.01.25.F	make good wall plaster	nr	0.50	33.74	—	5.77	39.51	37.11	—	6.35	43.46
A.01.01.25.G	total cost	nr	18.35	637.01	—	348.48	985.49	700.71	—	383.32	1084.04
A.01.01.26.	**Break up concrete paving; excavate drain trench; average 0.60m deep; lay 3.00m of 100mm clayware pipe including hole through existing manhole and make good; drain bend and vertical pipe set in concrete floor to receive WC pan; refill trench and reinstate paving on completion**										
A.01.01.26.A	break up concrete paving; 0.60m wide, 3.00m long	nr	3.00	88.74	—	—	88.74	97.61	—	—	97.61
A.01.01.26.B	excavate trench; 3.00m	nr	2.80	82.82	—	—	82.82	91.11	—	—	91.11
A.01.01.26.C	drain; 100mm; 3.00m long	nr	1.65	48.81	—	33.06	81.86	53.69	—	36.36	90.05
A.01.01.26.D	reinstate concrete paving; 3.00m	nr	3.60	106.49	—	24.89	131.38	117.14	—	27.38	144.52
A.01.01.26.E	end of drain into side of manhole and reform channel benching	nr	1.80	67.73	—	28.67	96.41	74.51	—	31.54	106.05
A.01.01.26.F	drain bend and vertical pipe	nr	3.60	106.49	—	58.44	164.93	117.14	—	64.28	181.42
A.01.01.26.G	total cost	nr	16.45	501.08	—	145.69	646.77	551.19	—	160.26	711.45
A.01.01.28.	**Break up concrete paving; excavate drain trench; average 0.60m deep; lay 3.00m of 100mm clayware pipe including hole through existing manhole and make good; drain for surface and / or disposal waste pipes; 100mm clayware gully trap in paving with gully surround**										
A.01.01.28.A	break up concrete paving; 0.60m wide; 3.00m long	nr	3.00	88.74	—	—	88.74	97.61	—	—	97.61
A.01.01.28.B	excavate trench; 3.00m long	nr	2.80	82.82	—	—	82.82	91.11	—	—	91.11
A.01.01.28.C	drain; 100mm; 3.00m long	nr	1.65	48.81	—	33.06	81.86	53.69	—	36.36	90.05
A.01.01.28.D	reinstate concrete paving; 3.00m	nr	3.60	106.49	—	24.89	131.38	117.14	—	27.38	144.52
A.01.01.28.E	end of drain into side of manhole and reform channel benching	nr	1.80	67.73	—	28.67	96.41	74.51	—	31.54	106.05
A.01.01.28.F	clayware gully and kerb	nr	1.10	32.54	—	93.05	125.59	35.79	—	102.36	138.15
A.01.01.28.G	total cost	nr	13.95	427.13	—	179.68	606.81	469.84	—	197.64	667.49

House Renovation Grants, Repairs and Alterations

Hutchins Priced Schedules 2021

Hutchins Priced Schedules 2021		Unit	Labour Hours	Labour Net £	Plant Net £	Materials Net £	Unit Rate Net £	Labour Gross £	Plant Gross £	Materials Gross £	Unit Rate Gross £
							(Gross rates include 10% profit)				
A.01.	**GRANTWORK**										
A.01.01.	HOUSE RENOVATION										
A.01.01.30.	**Take down existing partition wall 2.40m high (brick, clinker concrete or stud) and clear away; make good ceiling plaster along top of wall; make good wood flooring at base of wall**										
A.01.01.30.A	take down and clear away wall	m	0.70	20.71	—	—	20.71	22.78	—	—	22.78
A.01.01.30.B	make good ceiling plaster and emulsion paint; 0.30m wide	m	0.25	16.87	—	0.27	17.14	18.56	—	0.29	18.85
A.01.01.30.C	make good wood flooring and emulsion paint; 0.30m wide	m	0.65	24.46	—	2.94	27.40	26.91	—	3.24	30.14
A.01.01.30.D	total cost	m	1.60	62.04	—	3.21	65.24	68.24	—	3.53	71.77
A.01.01.32.	**Make good existing wall plaster 2.40m high after removal of partition wall; piece in skirting and picture rail and paint**										
A.01.01.32.A	make good wall plaster 450mm wide and joint to existing 3.00m	nr	0.95	64.12	—	4.79	68.91	70.53	—	5.27	75.80
A.01.01.32.B	piece in 200mm wood skirting about 300mm long and paint	nr	1.10	41.39	—	0.93	42.33	45.53	—	1.03	46.56
A.01.01.32.C	piece in wood picture rail about 300mm long and paint	nr	0.50	18.82	—	0.61	19.42	20.70	—	0.67	21.36
A.01.01.32.D	total cost	nr	2.55	124.33	—	6.33	130.66	136.76	—	6.97	143.73
A.01.01.33.	**Build half brick partition 2.40m high; plaster and emulsion paint both sides; fix 150mm softwood skirting and decorate both sides**										
A.01.01.33.A	half brick wall in commons (PC sum £450 per 1000); 2.40m high	m	4.02	210.94	—	78.95	289.89	232.03	—	86.85	318.88
A.01.01.33.B	plaster both sides half brick wall; 2.40m high	m	2.13	143.43	—	16.77	160.19	157.77	—	18.44	176.21
A.01.01.33.C	emulsion paint both sides	m	1.53	57.57	—	2.24	59.81	63.33	—	2.46	65.79
A.01.01.33.D	skirting both sides	m	0.70	26.34	—	11.89	38.23	28.98	—	13.08	42.05
A.01.01.33.E	paint skirting both sides	m	0.40	15.05	—	1.45	16.50	16.56	—	1.60	18.16
A.01.01.33.F	total cost	m	8.79	453.33	—	111.29	564.63	498.66	—	122.42	621.09
A.01.01.34.	**Build 100mm concrete block wall 2.40m high; plaster and emulsion paint both sides; fix 150mm softwood skirting both sides and paint**										
A.01.01.34.A	concrete block wall; 100mm	m	2.01	105.47	—	33.14	138.60	116.02	—	36.45	152.46
A.01.01.34.B	plaster both sides	m	2.13	143.43	—	16.77	160.19	157.77	—	18.44	176.21
A.01.01.34.C	emulsion paint both sides	m	1.53	57.57	—	2.24	59.81	63.33	—	2.46	65.79
A.01.01.34.D	skirting both sides	m	0.70	26.34	—	11.89	38.23	28.98	—	13.08	42.05

House Renovation Grants, Repairs and Alterations

Hutchins Priced Schedules 2021	Unit	Labour Hours	Labour Net £	Plant Net £	Materials Net £	Unit Rate Net £	Labour Gross £	Plant Gross £	Materials Gross £	Unit Rate Gross £
							(Gross rates include 10% profit)			

A.01. **GRANTWORK**

A.01.01. **HOUSE RENOVATION**

A.01.01.34. **Build 100mm concrete block wall 2.40m high; plaster and emulsion paint both sides; fix 150mm softwood skirting both sides and paint**

A.01.01.34.E	paint both sides	m	0.40	15.05	—	1.45	16.50	16.56	—	1.60	18.16
A.01.01.34.F	total cost	m	6.78	347.86	—	65.48	413.34	382.65	—	72.03	454.67
A.01.01.35.	**Construct stud partition 2.40m high of 100mm x 50mm studs clad both sides with plasterboard set in plaster and emulsion painted; fix 150mm wood skirting both sides and paint**										
A.01.01.35.A	stud partition	m	0.50	33.54	—	30.67	64.20	36.89	—	33.73	70.62
A.01.01.35.B	plasterboard and set both sides	m	2.18	146.79	—	15.01	161.80	161.47	—	16.51	177.98
A.01.01.35.C	emulsion paint both sides	m	1.53	57.57	—	2.24	59.81	63.33	—	2.46	65.79
A.01.01.35.D	skirting both sides	m	0.70	26.34	—	11.89	38.23	28.98	—	13.08	42.05
A.01.01.35.E	paint both sides	m	0.40	15.05	—	1.45	16.50	16.56	—	1.60	18.16
A.01.01.35.F	total cost	m	5.31	279.29	—	61.25	340.54	307.22	—	67.38	374.60
A.01.01.38.	**Take down door and frame and set aside; fill opening with half brick walling and plaster both sides, including jointing to existing; emulsion paint both sides; fix 150mm skirting both sides and paint**										
A.01.01.38.A	take down door and frame	nr	1.50	44.37	—	—	44.37	48.81	—	—	48.81
A.01.01.38.B	half brick wall in commons (PC sum £450 per 1000); 2.00m2	nr	3.22	168.79	—	63.71	232.51	185.67	—	70.09	255.76
A.01.01.38.C	plaster both sides; 4.00m2	nr	2.13	143.43	—	7.06	150.49	157.77	—	7.77	165.54
A.01.01.38.D	emulsion paint both sides; 4.00m2	nr	1.28	48.17	—	1.85	50.02	52.98	—	2.04	55.02
A.01.01.38.E	skirting; 2.00m	nr	0.70	26.34	—	11.89	38.23	28.98	—	13.08	42.05
A.01.01.38.F	paint; 2.00m	nr	0.40	15.05	—	1.45	16.50	16.56	—	1.60	18.16
A.01.01.38.G	total cost	nr	9.23	446.15	—	85.97	532.12	490.76	—	94.57	585.33
A.01.01.40.	**Take down door and frame and set aside; fill opening with 47mm x 100mm studs clad both sides with plaster baseboard and set in plaster; including making good to existing; emulsion paint both sides; fix 150mm skirting both sides and paint**										
A.01.01.40.A	stud partition in opening; 2.00m2	nr	2.00	75.26	—	33.45	108.71	82.79	—	36.80	119.58
A.01.01.40.B	plaster baseboard and set both sides; 4.00m2	nr	1.71	114.73	—	21.56	136.29	126.20	—	23.72	149.92
A.01.01.40.C	emulsion paint both sides; 4.00m2	nr	1.28	48.17	—	1.85	50.02	52.98	—	2.04	55.02
A.01.01.40.D	take down door and frame	nr	1.50	44.37	—	—	44.37	48.81	—	—	48.81
A.01.01.40.E	skirting; 2.00m	nr	0.70	26.34	—	11.89	38.23	28.98	—	13.08	42.05
A.01.01.40.F	paint; 2.00m	nr	0.40	15.05	—	1.45	16.50	16.56	—	1.60	18.16

House Renovation Grants, Repairs and Alterations

Hutchins Priced Schedules 2021	Unit	Labour Hours	Labour Net £	Plant Net £	Materials Net £	Unit Rate Net £	Labour Gross £	Plant Gross £	Materials Gross £	Unit Rate Gross £
									(Gross rates include 10% profit)	

A.01.   **GRANTWORK**

A.01.01.   **HOUSE RENOVATION**

A.01.01.40. Take down door and frame and set aside; fill opening with 47mm x 100mm studs clad both sides with plaster baseboard and set in plaster; including making good to existing; emulsion paint both sides; fix 150mm skirting both sides and paint

	Unit	Labour Hours	Labour Net £	Plant Net £	Materials Net £	Unit Rate Net £	Labour Gross £	Plant Gross £	Materials Gross £	Unit Rate Gross £
A.01.01.40.G total cost	nr	7.59	323.92	—	70.21	394.12	356.31	—	77.23	433.53

A.01.01.42. Existing door frame and door re-used in door opening in new partition; fix new architraves both sides; paint door frame and architrave both sides

	Unit	Labour Hours	Labour Net £	Plant Net £	Materials Net £	Unit Rate Net £	Labour Gross £	Plant Gross £	Materials Gross £	Unit Rate Gross £
A.01.01.42.A fix existing door frame in new opening . .	nr	1.20	45.16	—	—	45.16	49.67	—	—	49.67
A.01.01.42.B architraves; 10.00m	nr	1.50	56.45	—	28.05	84.49	62.09	—	30.85	92.94
A.01.01.42.C rehang door	nr	0.50	18.82	—	—	18.82	20.70	—	—	20.70
A.01.01.42.D paint door and frame overall sides; 4.00m2	nr	3.75	141.11	—	4.44	145.55	155.22	—	4.88	160.11
A.01.01.42.E total cost	nr	6.95	261.53	—	32.49	294.02	287.68	—	35.74	323.42

A.01.01.43. Provide and hang 762mm x 1981mm x 40mm standard hardboard flush door and lining with new architraves both sides in new door opening; door to be hung on 75mm butt hinges and fitted with mortice lock, all to be primed and painted two coats oil colour

	Unit	Labour Hours	Labour Net £	Plant Net £	Materials Net £	Unit Rate Net £	Labour Gross £	Plant Gross £	Materials Gross £	Unit Rate Gross £
A.01.01.43.A door lining	nr	0.68	25.59	—	20.77	46.36	28.15	—	22.85	51.00
A.01.01.43.B door	nr	0.50	18.82	—	57.49	76.30	20.70	—	63.24	83.93
A.01.01.43.C architraves; 10.00m	nr	1.50	56.45	—	28.05	84.49	62.09	—	30.85	92.94
A.01.01.43.D butt hinges 75mm; chrome plated	nr	0.80	30.10	—	1.28	31.39	33.11	—	1.41	34.53
A.01.01.43.E mortice lock and furniture	nr	2.00	75.26	—	18.26	93.52	82.79	—	20.09	102.88
A.01.01.43.F paint; 4.00m2	nr	3.75	141.11	—	4.44	145.55	155.22	—	4.88	160.11
A.01.01.43.G total cost	nr	9.23	347.32	—	130.30	477.62	382.06	—	143.33	525.38

A.01.01.44. Cut opening through one brick wall for window frame, 1200mm x 1200mm; build in steel lintel and roofing tile sill and reform brick jambs; provide and fix new casement window and frame; glaze and paint both sides; plaster reveals inside; make good internal wall plaster; fix 32mm x 150mm softwood window board and paint

	Unit	Labour Hours	Labour Net £	Plant Net £	Materials Net £	Unit Rate Net £	Labour Gross £	Plant Gross £	Materials Gross £	Unit Rate Gross £
A.01.01.44.A cut opening through one-brick wall for window 1200mm x 1200mm and lintel over (1.50m2)	nr	7.20	212.98	—	—	212.98	234.27	—	—	234.27
A.01.01.44.B steel lintel for solid wall; 143mm high; 1500mm long	nr	1.50	44.37	—	49.38	93.75	48.81	—	54.32	103.13
A.01.01.44.C roofing tile sill 1.50m long	nr	0.90	26.62	—	6.15	32.77	29.28	—	6.76	36.05
A.01.01.44.D reform brick jambs to opening; 3.00m; in commons (PC sum £450 per 1000) and facings (PC sum £550 per 1000)	nr	7.50	282.23	—	18.06	300.28	310.45	—	19.87	330.31

Hutchins Priced Schedules 2021	Unit	Labour Hours	Labour Net £	Plant Net £	Materials Net £	Unit Rate Net £	Labour Gross £	Plant Gross £	Materials Gross £	Unit Rate Gross £
								(Gross rates include 10% profit)		

A.01.   **GRANTWORK**

A.01.01.   **HOUSE RENOVATION**

A.01.01.44.   **Cut opening through one brick wall for window frame, 1200mm x 1200mm; build in steel lintel and roofing tile sill and reform brick jambs; provide and fix new casement window and frame; glaze and paint both sides; plaster reveals inside; make good internal wall plaster; fix 32mm x 150mm softwood window board and paint**

Ref	Description	Unit	Labour Hours	Labour Net £	Plant Net £	Materials Net £	Unit Rate Net £	Labour Gross £	Plant Gross £	Materials Gross £	Unit Rate Gross £
A.01.01.44.E	casement window and frame	nr	1.80	67.73	—	210.65	278.39	74.51	—	231.72	306.23
A.01.01.44.F	6mm float glass window and frame; 1.00m2; small squares	nr	1.20	45.16	—	38.86	84.02	49.67	—	42.75	92.42
A.01.01.44.G	painting window sashes and both side; 3.50m2	nr	4.50	169.34	—	3.09	172.43	186.27	—	3.40	189.67
A.01.01.44.H	plaster reveals; make good around opening	nr	0.20	13.51	—	4.79	18.30	14.86	—	5.27	20.13
A.01.01.44.I	window board; 1.20m long; including painting	nr	0.50	18.82	—	5.86	24.67	20.70	—	6.44	27.14
A.01.01.44.J	total cost	nr	25.30	880.74	—	336.84	1217.59	968.82	—	370.53	1339.34

A.01.01.46.   **Take off tiles or slates and form opening in roof size 1200mm x 1200mm for dormer window size 1200mm x 900mm frame and board dormer flat and cheeks; provide and fix standard window, glaze and paint internally; plasterboard and set to soffit and dormer cheeks; make good existing plaster**

Ref	Description	Unit	Labour Hours	Labour Net £	Plant Net £	Materials Net £	Unit Rate Net £	Labour Gross £	Plant Gross £	Materials Gross £	Unit Rate Gross £
A.01.01.46.A	strip tiles or slates and battens; 1.70m2	nr	1.30	38.45	—	—	38.45	42.30	—	—	42.30
A.01.01.46.B	trim opening	nr	0.45	16.93	—	37.15	54.08	18.63	—	40.87	59.49
A.01.01.46.C	carcassing timber 100mm x 47mm to sides and roof of dormer; 1.50m	nr	3.60	135.47	—	51.99	187.46	149.01	—	57.19	206.21
A.01.01.46.D	boarding to cheeks and roof of dormer; 2.50m2	nr	4.35	163.69	—	22.79	186.48	180.06	—	25.07	205.13
A.01.01.46.E	2.27kg lead covering to roof and cheeks; 3.00m2	nr	10.34	984.15	—	194.20	1178.36	1082.57	—	213.62	1296.19
A.01.01.46.F	casement window and frame	nr	1.55	58.33	—	181.65	239.97	64.16	—	199.81	263.97
A.01.01.46.G	6mm float glass window and frame; 1.00m2	nr	1.10	41.39	—	38.79	80.18	45.53	—	42.66	88.20
A.01.01.46.H	painting window sashes and frame both sides; 3.50m2	nr	4.50	169.34	—	0.42	169.76	186.27	—	0.47	186.74
A.01.01.46.I	plasterboard and set to soffit and cheeks; 3.50m2	nr	1.71	114.73	—	11.52	126.25	126.20	—	12.68	138.88
A.01.01.46.J	make good new plaster to existing	nr	0.25	16.87	—	2.27	19.14	18.56	—	2.50	21.06
A.01.01.46.K	emulsion paint new plastering; 3.50m2	nr	1.13	42.52	—	2.01	44.53	46.77	—	2.21	48.98
A.01.01.46.L	total cost	nr	30.28	1781.87	—	542.80	2324.67	1960.06	—	597.08	2557.14

House Renovation Grants, Repairs and Alterations

Hutchins Priced Schedules 2021	Unit	Labour Hours	Labour Net £	Plant Net £	Materials Net £	Unit Rate Net £	Labour Gross £	Plant Gross £	Materials Gross £	Unit Rate Gross £
								(Gross rates include 10% profit)		

A.01. GRANTWORK

A.01.01. HOUSE RENOVATION

A.01.01.47. Take down length of stair handrail and balustrade to stair flights, 3.00m long x 2.40m high; fill in triangular stair spandrel between stair tread and soffit with stud partition; cover with expanded metal lathing and plaster both sides; 25mm x 225mm cut wall string to be fitted over tread risers; fix 63mm mopstick hand rail to full length of wall

Hutchins Priced Schedules 2021	Unit	Labour Hours	Labour Net £	Plant Net £	Materials Net £	Unit Rate Net £	Labour Gross £	Plant Gross £	Materials Gross £	Unit Rate Gross £
A.01.01.47.A take down and remove handrail and balustrade	nr	3.50	103.53	—	—	103.53	113.88	—	—	113.88
A.01.01.47.B triangular stud spandrel wall with expanded metal lathing and plaster both sides; average 4.00m2	nr	7.19	483.24	—	100.86	584.10	531.56	—	110.95	642.51
A.01.01.47.C emulsion paint wall both sides; 8.00m2	nr	2.56	96.33	—	2.19	98.53	105.97	—	2.41	108.38
A.01.01.47.D 225mm x 25mm cut wall string planted on; 3.60m; paint	nr	4.30	161.81	—	17.71	179.52	177.99	—	19.48	197.47
A.01.01.47.E mopstick handrail; 3.60m	nr	2.60	97.84	—	29.48	127.31	107.62	—	32.42	140.05
A.01.01.47.F total cost	nr	20.15	942.75	—	150.25	1092.99	1037.02	—	165.27	1202.29

A.01.01.48. Take down length of stair handrail and balustrade to stair flights, 3.00m long x 2.40m high; fill in triangular stair spandrel between stair tread and soffit with stud partition; cover with expanded metal lathing and plaster both sides; 25mm x 225mm cut wall string to be fitted over tread risers; fix 63mm mopstick hand rail to full length of wall

Hutchins Priced Schedules 2021	Unit	Labour Hours	Labour Net £	Plant Net £	Materials Net £	Unit Rate Net £	Labour Gross £	Plant Gross £	Materials Gross £	Unit Rate Gross £
A.01.01.48.A take down handrail and balustrade	nr	3.50	103.53	—	—	103.53	113.88	—	—	113.88
A.01.01.48.B spandrel wall as last described but with parallel raking sides; 7.50m2	nr	28.20	1061.17	—	190.06	1251.23	1167.28	—	209.07	1376.35
A.01.01.48.C emulsion paint wall both sides; 15.00m2	nr	4.80	180.62	—	4.10	184.73	198.69	—	4.51	203.20
A.01.01.48.D 225mm x 25mm cut wall string planted on; 3.60m; paint	nr	4.30	161.81	—	17.71	179.52	177.99	—	19.48	197.47
A.01.01.48.E mopstick handrail; 3.60m	nr	3.10	116.65	—	29.48	146.13	128.32	—	32.42	160.74
A.01.01.48.F total cost	nr	43.90	1623.78	—	241.35	1865.14	1786.16	—	265.49	2051.65

House Renovation Grants, Repairs and Alterations

Hutchins Priced Schedules 2021	Unit	Labour Hours	Labour Net £	Plant Net £	Materials Net £	Unit Rate Net £	Labour Gross £	Plant Gross £	Materials Gross £	Unit Rate Gross £
							(Gross rates include 10% profit)			

A.01.   GRANTWORK

A.01.01.   HOUSE RENOVATION

A.01.01.49. Take down and remove handrail and balustrade to landing; form new wall 2.40m high in studwork covered with metal lath and plastered both sides; provide and fix 25mm x 150mm skirting one side; decorate wall and skirting

A.01.01.49.A take down handrail and balustrade	m	1.00	29.58	—	—	29.58	32.54	—	—	32.54
A.01.01.49.B stud partition wall as described	m	7.85	295.40	—	87.34	382.74	324.94	—	96.08	421.01
A.01.01.49.C emulsion paint wall both sides	m	1.53	57.57	—	1.23	58.80	63.33	—	1.35	64.68
A.01.01.49.D skirting one side; 150mm x 25mm	m	0.35	13.17	—	5.61	18.78	14.49	—	6.17	20.66
A.01.01.49.E painting skirting one side; 150mm x 25mm....................	m	0.20	7.53	—	0.42	7.95	8.28	—	0.47	8.75
A.01.01.49.F total cost....................	m	10.93	403.25	—	94.60	497.85	443.57	—	104.06	547.63

A.01.01.50. Make and fix cupboard front in 38mm softwood framing and plywood panelling including 750mm x 1950mm door across recess 1.20m x 2.40m high to form food store; supply and fit three shelves 200mm x 25mm and one shelf 300mm x 25mm

A.01.01.50.A cupboard front 38mm, 1.20m x 2.40mm .	nr	20.00	752.60	—	56.22	808.82	827.86	—	61.84	889.70
A.01.01.50.B extra over for hanging door on 75mm butt hinges chrome plated, with Bales catch and bow handle	nr	2.00	75.26	—	12.42	87.68	82.79	—	13.67	96.45
A.01.01.50.C set of three 200mm x 25mm shelves and one 300mm x 25mm shelf	nr	3.20	120.42	—	30.61	151.03	132.46	—	33.67	166.13
A.01.01.50.F painting cupboard front both sides; 5.75m2....................	nr	4.60	173.10	—	5.29	178.39	190.41	—	5.82	196.23
A.01.01.50.G total cost....................	nr	29.80	1121.37	—	104.55	1225.92	1233.51	—	115.00	1348.52

A.01.01.52. Stainless steel sink complete with single drainer fixed complete on brackets built into wall; fit and connect waste trap to waste pipe, discharging into external hopper head or gully; mixer tap with swivel spout; service pipes measured separately

A.01.01.52.A single bowl with single drainer; 1000mm x 500mm (PC sum £140 per nr) and fix ..	nr	2.48	235.93	—	147.00	382.93	259.52	—	161.70	421.22
A.01.01.52.B plastic waste trap; 40mm	nr	0.54	51.77	—	22.64	74.41	56.95	—	24.91	81.86
A.01.01.52.C plastic waste pipe 40mm dia x 2.70m long....................	nr	0.91	86.32	—	17.16	103.48	94.95	—	18.88	113.83
A.01.01.52.D hole through wall for plastic waste pipe; 40mm dia x 2.70m long; make good ...	nr	0.51	26.94	—	0.36	27.30	29.64	—	0.39	30.03
A.01.01.52.E mixer tap swivel spout (PC sum £50 per nr).....................	nr	0.24	23.03	—	52.50	75.53	25.33	—	57.75	83.08
A.01.01.52.F total cost....................	nr	4.69	423.99	—	239.66	663.65	466.39	—	263.62	730.01

House Renovation Grants, Repairs and Alterations

Hutchins Priced Schedules 2021		Unit	Labour Hours	Labour Net £	Plant Net £	Materials Net £	Unit Rate Net £	Labour Gross £	Plant Gross £	Materials Gross £	Unit Rate Gross £
								(Gross rates include 10% profit)			
A.01.	**GRANTWORK**										
A.01.01.	HOUSE RENOVATION										
A.01.01.57.	**1700mm bath; white reinforced acrylic complete with mixer tap, fixed spout, plug and chain, side panel and fixing; fit and connect waste trap to waste pipe discharging into external hopper head or gully; service pipes measured separately**										
A.01.01.57.A	bath (PC sum £180 per nr) and fix	nr	3.87	368.31	—	251.03	619.34	405.14	—	276.14	681.28
A.01.01.57.B	plastic waste trap; 40mm	nr	0.54	51.77	—	22.64	74.41	56.95	—	24.91	81.86
A.01.01.57.C	plastic waste pipe; 40mm dia x 2.70m long	nr	0.91	86.32	—	17.16	103.48	94.95	—	18.88	113.83
A.01.01.57.D	hole through wall for plastic waste pipe 40mm dia x 2.70m long; make good . . .	nr	0.51	26.94	—	0.36	27.30	29.64	—	0.39	30.03
A.01.01.57.E	plastic overflow pipe; 21.5mm dia x 0.60m long	nr	0.39	37.40	—	2.84	40.24	41.14	—	3.13	44.27
A.01.01.57.F	hole through wall for plastic overflow pipe; 21.5mm dia x 0.60m long; make good	nr	0.51	26.94	—	0.36	27.30	29.64	—	0.39	30.03
A.01.01.57.G	total cost	nr	6.74	597.69	—	294.39	892.08	657.46	—	323.83	981.29
A.01.01.58.	**560mm x 405mm white vitreous china wash basin complete with pair of taps, plug and chain; fit and connect waste trap to waste pipe, discharging into external hopper head or gully; service pipes measured separately**										
A.01.01.58.A	wash basin (PC sum £110 per nr) and fix	nr	2.90	276.28	—	115.50	391.78	303.91	—	127.05	430.96
A.01.01.58.B	plastic waste trap; 32mm	nr	0.39	37.40	—	15.12	52.52	41.14	—	16.63	57.77
A.01.01.58.C	plastic waste pipe; 32mm dia x 2.70m long	nr	0.76	71.95	—	14.78	86.73	79.14	—	16.26	95.40
A.01.01.58.D	hole through wall for plastic waste pipe; 32mm dia x 2.70m long; make good . . .	nr	0.51	26.94	—	0.36	27.30	29.64	—	0.39	30.03
A.01.01.58.E	total cost	nr	4.57	412.57	—	145.75	558.32	453.83	—	160.32	614.15
A.01.01.59.	**Extra over waste pipe in last three items for cutting and fitting end of waste pipe into existing 100mm cast iron soil stack pipe; including cutting in and jointing 100mm cast iron boss pipe connector**										
A.01.01.59.A	labour and material	nr	3.27	310.83	—	45.35	356.17	341.91	—	49.88	391.79

Hutchins Priced Schedules 2021	Unit	Labour Hours	Labour Net £	Plant Net £	Materials Net £	Unit Rate Net £	Labour Gross £	Plant Gross £	Materials Gross £	Unit Rate Gross £
								(Gross rates include 10% profit)		

A.01. GRANTWORK

A.01.01. HOUSE RENOVATION

A.01.01.60. WC suite low level complete and fixing at upper floor level including 100mm cast iron bend, long arm and junction to existing stack pipe; service pipes measured separately

	Unit	Labour Hours	Labour Net £	Plant Net £	Materials Net £	Unit Rate Net £	Labour Gross £	Plant Gross £	Materials Gross £	Unit Rate Gross £
A.01.01.60.A WC suite complete and fix.........	nr	3.27	310.83	—	164.49	475.32	341.91	—	180.94	522.85
A.01.01.60.B cast iron bend; 100mm...........	nr	0.61	57.58	—	38.51	96.09	63.34	—	42.36	105.70
A.01.01.60.C cast iron bend junction; 100mm......	nr	1.21	115.06	—	60.07	175.13	126.57	—	66.08	192.64
A.01.01.60.D cut 100mm cast iron soil pipe to receive cast iron bend junction...........	nr	2.06	195.67	—	—	195.67	215.24	—	—	215.24
A.01.01.60.E cut hole through one brick wall for 100mm pipe; make good..........	nr	0.81	42.25	—	0.71	42.96	46.48	—	0.78	47.26
A.01.01.60.F plastic overflow pipe; 21.5mm dia x 1.00m long...................	nr	0.45	43.21	—	4.44	47.65	47.53	—	4.89	52.41
A.01.01.60.G cut hole through one brick wall for small pipe; make good...............	nr	0.51	26.94	—	0.36	27.30	29.64	—	0.39	30.03
A.01.01.60.H total cost.................	nr	8.91	791.53	—	268.58	1060.12	870.69	—	295.44	1166.13

A.01.01.61. WC suite low level complete and fix at ground level including jointing to clayware drain bend (measured separately); service pipes measured separately

	Unit	Labour Hours	Labour Net £	Plant Net £	Materials Net £	Unit Rate Net £	Labour Gross £	Plant Gross £	Materials Gross £	Unit Rate Gross £
A.01.01.61.A WC suite complete and fix.........	nr	3.27	310.83	—	164.49	475.32	341.91	—	180.94	522.85
A.01.01.61.B plastic overflow pipe; 21.5mm dia x 1.00m long...................	nr	0.45	43.21	—	4.44	47.65	47.53	—	4.89	52.41
A.01.01.61.C cut hole through one brick wall for small pipe; make good...............	nr	0.51	26.94	—	0.36	27.30	29.64	—	0.39	30.03
A.01.01.61.D total cost.................	nr	4.23	380.98	—	169.29	550.27	419.07	—	186.22	605.29

A.01.01.62. 227l plastic cold water cistern in loft space, including 13mm ball valve 21.5mm overflow pipe, tank bearers and hardboard tank casing; service pipes measured separately

	Unit	Labour Hours	Labour Net £	Plant Net £	Materials Net £	Unit Rate Net £	Labour Gross £	Plant Gross £	Materials Gross £	Unit Rate Gross £
A.01.01.62.A cistern, ball valve and float.........	nr	2.09	198.52	—	181.00	379.52	218.38	—	199.09	417.47
A.01.01.62.B plastic overflow pipe; 21.5mm dia x 3.00m long...................	nr	1.36	129.53	—	13.32	142.85	142.48	—	14.66	157.14
A.01.01.62.C cut hole through wall for plastic overflow pipe; 21.5mm dia x 3.00m long; make good...........................	nr	0.51	26.94	—	0.36	27.30	29.64	—	0.39	30.03
A.01.01.62.D pair tank bearers 100mm x 50mm each 1.80m long...................	nr	0.12	4.52	—	10.04	14.55	4.97	—	11.04	16.01
A.01.01.62.E hardboard tank casing.............	nr	2.15	80.90	—	13.31	94.21	88.99	—	14.64	103.63
A.01.01.62.F total cost.................	nr	6.23	440.41	—	218.02	658.44	484.46	—	239.82	724.28

House Renovation Grants, Repairs and Alterations

Hutchins Priced Schedules 2021		Unit	Labour Hours	Labour Net £	Plant Net £	Materials Net £	Unit Rate Net £	Labour Gross £	Plant Gross £	Materials Gross £	Unit Rate Gross £
							(Gross rates include 10% profit)				
A.01.	**GRANTWORK**										
A.01.01.	HOUSE RENOVATION										
A.01.01.63.	**Cut through existing ceiling and trim joists to form opening for 750mm x 750mm loft hatch; softwood lining with planted stop and architraves; loft hatch comprises plywood face on 75mm x 25mm framed backing, hinged to lining and fitted with bow handle; all exposed woodwork painted**										
A.01.01.63.A	cut opening and trim joists	nr	0.60	22.58	—	20.97	43.55	24.84	—	23.07	47.90
A.01.01.63.B	loft hatch and lining; 150mm x 25mm x 3.00m long	nr	2.00	75.26	—	14.90	90.16	82.79	—	16.39	99.17
A.01.01.63.C	architrave x 3.50m	nr	0.45	16.93	—	8.52	25.45	18.63	—	9.37	28.00
A.01.01.63.D	bow handle	nr	0.25	9.41	—	7.05	16.46	10.35	—	7.75	18.10
A.01.01.63.E	paint; 0.85m2	nr	0.80	30.10	—	1.06	31.16	33.11	—	1.17	34.28
A.01.01.63.F	make good plaster	nr	0.18	11.83	—	0.65	12.48	13.01	—	0.72	13.73
A.01.01.63.G	total cost	nr	4.28	166.11	—	53.15	219.26	182.72	—	58.46	241.19
A.01.01.66.	**The following rates for water service pipes including bends, sockets and an average allowance of one tee (or junction) for every 4.00m of pipe; holes through walls for pipes and making good plaster included; one hole for every 4.00m of pipe**										
A.01.01.66.A	galvanised steel water pipe; 15mm dia, including extra cost of tees and holes and through partition walls, etc.; all as described	m	0.67	63.29	—	8.29	71.57	69.62	—	9.12	78.73
A.01.01.66.B	galvanised steel water pipe; 20mm dia, including extra cost of tees and holes and through partition walls, etc.; all as described	m	0.73	69.09	—	8.74	77.83	76.00	—	9.61	85.61
A.01.01.66.C	copper water pipe; 15mm dia, including extra cost of tees and holes and through partition walls, etc.; all as described	m	0.54	51.77	—	8.38	60.15	56.95	—	9.22	66.17
A.01.01.66.D	copper water pipe; 22mm dia, including extra cost of tees and holes and through partition walls, etc.; all as described	m	0.58	54.72	—	7.44	62.17	60.20	—	8.19	68.38
A.01.01.66.E	polythene normal gauge water pipe; 20mm dia, including extra cost of tees and holes through partition walls, etc.; all as described	m	0.76	71.95	—	3.19	75.14	79.14	—	3.51	82.65
A.01.01.66.F	polythene normal gauge water pipe; 25mm dia, including extra cost of tees and holes through partition walls, etc.; all as described	m	0.97	92.12	—	4.05	96.18	101.34	—	4.46	105.79

House Renovation Grants, Repairs and Alterations

Hutchins Priced Schedules 2021		Unit	Labour Hours	Labour Net £	Plant Net £	Materials Net £	Unit Rate Net £	Labour Gross £	Plant Gross £	Materials Gross £	Unit Rate Gross £
								(Gross rates include 10% profit)			
A.02.	**REPAIRS AND ALTERATIONS**										
A.02.01.	**EXCAVATION**										
A.02.01.01.	**Excavate by hand oversite area; wheel 18.00m; deposit in skip**										
A.02.01.01.A	average 300mm deep	m²	1.86	55.02	—	—	55.02	60.52	—	—	60.52
A.02.01.02.	**Excavate by hand for trenches to receive foundations; wheel 18.00m; deposit in skip**										
A.02.01.02.A	not exceeding 1.00m deep	m³	6.52	192.86	—	—	192.86	212.15	—	—	212.15
A.02.01.02.B	over 1.00m deep; not exceeding 2.00m deep	m³	8.02	237.23	—	—	237.23	260.95	—	—	260.95
A.02.01.03.	**Excavate by hand for basement; wheel 18.00m; deposit in skip**										
A.02.01.03.A	not exceeding 1.00m deep	m³	6.00	177.48	—	—	177.48	195.23	—	—	195.23
A.02.01.03.B	over 1.00m deep; not exceeding 2.00m deep	m³	6.72	198.78	—	—	198.78	218.66	—	—	218.66
A.02.01.04.	**Excavated material as filling to excavations; deposited and compacted by hand**										
A.02.01.04.A	in 250mm layers	m³	2.00	59.16	—	—	59.16	65.08	—	—	65.08
A.02.01.05.	**Extra over excavation for breaking up brickwork by hand**										
A.02.01.05.A	in old foundations	m³	7.09	209.72	—	—	209.72	230.69	—	—	230.69
A.02.01.06.	**Hire of skip; delivery to site; remove when full; dispose of contents; pay tipping charges**										
A.02.01.06.A	size 4.50m3	m³	—	—	43.60	—	43.60	—	47.96	—	47.96
A.02.02.	**EARTHWORK SUPPORT AND HARDCORE**										
A.02.02.01.	**In firm ground (four uses assumed) to opposing faces not exceeding 2.00m apart; maximum depth not exceeding**										
A.02.02.01.A	1.00m	m²	0.85	25.14	—	2.35	27.49	27.66	—	2.59	30.24
A.02.02.01.B	2.00m	m²	0.93	27.51	—	2.60	30.11	30.26	—	2.86	33.12

House Renovation Grants, Repairs and Alterations

Hutchins Priced Schedules 2021	Unit	Labour Hours	Labour Net £	Plant Net £	Materials Net £	Unit Rate Net £	Labour Gross £	Plant Gross £	Materials Gross £	Unit Rate Gross £
								(Gross rates include 10% profit)		
A.02.	**REPAIRS AND ALTERATIONS**									
A.02.02.	**EARTHWORK SUPPORT AND HARDCORE**									
A.02.02.02.	**In loose ground (four uses assumed) to opposing faces not exceeding 2.00m apart; maximum depth not exceeding**									
A.02.02.02.A	1.00m...................	m²	6.05	178.96	—	16.85	195.80	196.85	—	18.53	215.39
A.02.02.02.B	2.00m...................	m²	6.62	195.82	—	18.64	214.46	215.40	—	20.50	235.90
A.02.02.03.	**Imported hardcore compacted to receive concrete to finished thickness**									
A.02.02.03.A	100mm...................	m²	0.48	14.20	0.53	6.48	21.21	15.62	0.59	7.13	23.33
A.02.02.03.B	150mm...................	m²	0.72	21.30	0.64	9.95	31.89	23.43	0.70	10.95	35.08
A.02.02.03.C	225mm...................	m²	1.08	31.95	0.74	14.93	47.62	35.14	0.82	16.42	52.38
A.02.03.	**CONCRETE**									
A.02.03.01.	**Portland cement concrete in foundations**									
A.02.03.01.A	GEN1 mix..................	m³	6.00	177.48	—	91.39	268.87	195.23	—	100.53	295.76
A.02.03.01.B	GEN2 mix..................	m³	6.00	177.48	—	93.89	271.37	195.23	—	103.28	298.51
A.02.03.02.	**Concrete GEN1 oversite; thickness**									
A.02.03.02.A	100mm...................	m³	9.00	266.22	—	91.39	357.61	292.84	—	100.53	393.37
A.02.03.02.B	150mm...................	m³	8.50	251.43	—	91.39	342.82	276.57	—	100.53	377.10
A.02.03.03.	**Concrete GEN1 oversite; in patches not exceeding 4.00m2 in area; including jointing to existing, thickness**									
A.02.03.03.A	100mm...................	m³	14.00	414.12	—	91.39	505.51	455.53	—	100.53	556.06
A.02.03.03.B	150mm...................	m³	13.50	399.33	—	91.39	490.72	439.26	—	100.53	539.79
A.02.03.04.	**Extra oversite concrete for**									
A.02.03.04.A	neat cement slurry to receive asphalt, tiling, etc....................	m²	0.45	13.31	—	2.58	15.89	14.64	—	2.84	17.48
A.02.03.04.B	trowel to smooth surface..........	m²	0.55	16.27	—	—	16.27	17.90	—	—	17.90
A.02.03.05.	**Sprinkle surface with coarse carborundum at**									
A.02.03.05.A	1kg per m2 and lightly trowel........	m²	0.55	16.27	—	3.28	19.55	17.90	—	3.60	21.50
A.02.03.06.	**Clean existing concrete or rendered floors**									
A.02.03.06.A	treat with application of silicate of soda solution....................	m²	0.30	8.87	—	0.57	9.45	9.76	—	0.63	10.39
A.02.04.	**PRECAST CONCRETE**									
A.02.04.01.	**Reinforced concrete lintels cast in situ, including reinforcement**									
A.02.04.01.A	113mm x 150mm..............	m	0.50	26.31	—	4.27	30.58	28.95	—	4.69	33.64
A.02.04.01.B	113mm x 225mm..............	m	0.60	31.66	—	5.16	36.82	34.83	—	5.68	40.50

Hutchins Priced Schedules 2021		Unit	Labour Hours	Labour Net £	Plant Net £	Materials Net £	Unit Rate Net £	Labour Gross £	Plant Gross £	Materials Gross £	Unit Rate Gross £
								(Gross rates include 10% profit)			
A.02.	**REPAIRS AND ALTERATIONS**										
A.02.04.	**PRECAST CONCRETE**										
A.02.04.01.	**Reinforced concrete lintels cast in situ, including reinforcement**										
A.02.04.01.C	225mm x 150mm..............	m	0.87	45.71	—	7.20	52.91	50.28	—	7.92	58.21
A.02.04.01.D	225mm x 225mm..............	m	0.97	50.95	—	8.99	59.94	56.05	—	9.89	65.94
A.02.04.02.	**Precast concrete lintels**										
A.02.04.02.A	100mm x 150mm..............	m	0.30	15.83	—	23.67	39.50	17.41	—	26.03	43.45
A.02.04.02.B	100mm x 220mm..............	m	0.44	22.86	—	36.99	59.85	25.14	—	40.69	65.83
A.02.04.02.C	225mm x 140mm..............	m	0.57	29.88	—	52.14	82.02	32.87	—	57.36	90.22
A.02.04.03.	**Needle through brickwork; shore with one pair Acrow or other adjustable struts to every linear metre or part thereof (maximum span 2.70m); cut out and remove defective lintel; supply, hoist and build in reinforced concrete lintel; make good all brickwork (reclaimed facing bricks, PC sum £780 per 1000) and plaster disturbed; lintel size**										
A.02.04.03.A	225mm x 140mm; precast..........	m	3.25	170.57	6.78	60.32	237.67	187.63	7.45	66.36	261.44
A.02.04.03.B	225mm x 225mm; cast in situ (formwork used up to three times)...........	m	3.29	221.05	7.45	16.30	244.81	243.16	8.20	17.93	269.29
A.02.04.04.	**Cut away triangular area of brickwork above lintel; cut out and remove defective lintel; supply, hoist and build in reinforced concrete lintel cast in situ (formwork used up to three times); rebuild brickwork over, including facing bricks (PC sum £550 per 1000) to match existing; make good internal plaster**										
A.02.04.04.A	225mm x 225mm..............	m	5.85	393.18	—	26.62	419.80	432.50	—	29.28	461.78
A.02.04.05.	**Take out stone or concrete sill; supply and build in cast concrete sill including all making good**										
A.02.04.05.A	225mm x 75mm..............	m	0.74	38.69	—	14.12	52.81	42.55	—	15.54	58.09
A.02.04.06.	**Pier caps**										
A.02.04.06.A	300mm x 300mm x 75mm for 225mm piers......................	nr	0.67	35.12	—	14.35	49.47	38.63	—	15.78	54.42
A.02.04.06.B	400mm x 400mm x 75mm for 338mm piers......................	nr	0.87	45.71	—	15.65	61.37	50.28	—	17.22	67.50

House Renovation Grants, Repairs and Alterations

Hutchins Priced Schedules 2021	Unit	Labour Hours	Labour Net £	Plant Net £	Materials Net £	Unit Rate Net £	Labour Gross £	Plant Gross £	Materials Gross £	Unit Rate Gross £
							(Gross rates include 10% profit)			

A.02.   REPAIRS AND ALTERATIONS

A.02.05.   BREAKING UP CONCRETE STEPS AND FLOORS

A.02.05.01.   Break up and remove old concrete steps

| A.02.05.01.A | form new steps in concrete GEN1 including wrought formwork to risers and ends; surfaces of treads trowelled smooth | m³ | 19.00 | 562.02 | — | 120.68 | 682.70 | 618.22 | — | 132.75 | 750.97 |

A.02.05.02.   **Break up and remove concrete floors, paving, etc. at ground level and load into skip**

A.02.05.02.A	not exceeding 150mm	m²	2.25	66.56	—	—	66.56	73.21	—	—	73.21
A.02.05.02.B	150mm - 225mm	m²	4.00	118.32	—	—	118.32	130.15	—	—	130.15
A.02.05.02.C	225mm - 300mm	m²	6.00	177.48	—	—	177.48	195.23	—	—	195.23

A.02.05.03.   **Break up and remove reinforced concrete floors, pavings, etc. at ground level and load into skip**

A.02.05.03.A	not exceeding 150mm	m²	3.40	100.57	—	—	100.57	110.63	—	—	110.63
A.02.05.03.B	150mm - 225mm	m²	6.00	177.48	—	—	177.48	195.23	—	—	195.23
A.02.05.03.C	225mm - 300mm	m²	9.00	266.22	—	—	266.22	292.84	—	—	292.84

A.02.05.04.   **Break up concrete paving 750mm wide and 100mm thick for new wall and remove; excavate trench and part return, fill in and ram and remove remainder**

| A.02.05.04.A | make good concrete paving (foundation concrete measured separately) | m | 2.85 | 84.30 | — | 7.66 | 91.96 | 92.73 | — | 8.43 | 101.16 |

A.02.05.05.   **Hack up broken or sunken areas of concrete paving; spread and consolidate hardcore 150mm; lay new concrete to falls; joint to existing including trowel to form smooth surface**

| A.02.05.05.A | 100mm | m² | 2.15 | 63.60 | — | 20.13 | 83.73 | 69.96 | — | 22.15 | 92.11 |
| A.02.05.05.B | 150mm | m² | 2.45 | 72.47 | — | 25.18 | 97.65 | 79.72 | — | 27.70 | 107.42 |

A.02.06.   PAVINGS

A.02.06.01.   **Hack surface of existing paving of floors and grout and render in**

| A.02.06.01.A | 19mm cement mortar (1:2:5) | m² | 0.50 | 33.74 | — | 3.48 | 37.22 | 37.11 | — | 3.83 | 40.94 |

A.02.06.02.   **Hack off defective cement rendering to steps (treads and risers) and make out in**

| A.02.06.02.A | 25mm cement and sand (1:3); trowel including nosings and arises | m² | 1.26 | 84.35 | — | 3.56 | 87.91 | 92.78 | — | 3.92 | 96.70 |

Hutchins Priced Schedules 2021	Unit	Labour Hours	Labour Net £	Plant Net £	Materials Net £	Unit Rate Net £	Labour Gross £	Plant Gross £	Materials Gross £	Unit Rate Gross £
						(Gross rates include 10% profit)				
A.02.	**REPAIRS AND ALTERATIONS**									
A.02.06.	**PAVINGS**									
A.02.06.03.	**Clean and hack existing concrete surface to form key for**									
A.02.06.03.A	granolithic paving	m²	0.40	11.83	—	—	11.83	13.02	—	—	13.02
A.02.06.04.	**Roughen and grout edge of existing concrete paving to new**									
A.02.06.04.A	100mm	m	0.40	11.83	—	—	11.83	13.02	—	—	13.02
A.02.06.04.B	150mm	m	0.50	14.79	—	—	14.79	16.27	—	—	16.27
A.02.07.	**BREAKING OUT REINFORCED CONCRETE**									
A.02.07.01.	**Breaking up reinforced concrete**									
A.02.07.01.A	walls, columns, beams, suspended floors or roofs and loading into skip	m³	27.00	798.66	—	—	798.66	878.53	—	—	878.53
A.02.07.02.	**Cut holes through concrete for pipes, bars, etc.; per 25mm depth of cut; make good**									
A.02.07.02.A	area not exceeding 0.003m2	nr	0.20	5.92	—	0.71	6.63	6.51	—	0.78	7.29
A.02.07.02.B	0.003m2 - 0.023m2	nr	0.40	11.83	—	0.95	12.78	13.02	—	1.04	14.06
A.02.07.03.	**Cut holes through reinforced concrete for pipes, bars, etc.; per 25mm depth of cut; make good**									
A.02.07.03.A	area not exceeding 0.003m2	nr	0.30	8.87	—	0.71	9.59	9.76	—	0.78	10.54
A.02.07.03.B	0.003m2 - 0.023m2	nr	0.60	17.75	—	0.95	18.70	19.52	—	1.04	20.57
A.02.08.	**CONCRETE KERBS AND CHANNELS**									
A.02.08.01.	**Forming concrete GEN2 kerbs and channels including all necessary formwork but excluding excavation**									
A.02.08.01.A	average 0.047m2 sectional area	m	0.50	33.74	—	4.92	38.66	37.11	—	5.41	42.52
A.02.09.	**WORKS TO CHIMNEYS**									
A.02.09.01.	**Demolish brickwork, any height; clean sound whole bricks to**									
A.02.09.01.A	reuse and remove remainder	m³	6.37	334.02	—	—	334.02	367.42	—	—	367.42
A.02.09.02.	**Collect, clean and stack bricks for reuse; in**									
A.02.09.02.A	lime mortar	1000	9.39	492.22	—	—	492.22	541.45	—	—	541.45
A.02.09.02.B	compo mortar	1000	11.40	597.69	—	—	597.69	657.46	—	—	657.46
A.02.09.02.C	cement mortar	1000	14.76	773.51	—	—	773.51	850.86	—	—	850.86

House Renovation Grants, Repairs and Alterations

Hutchins Priced Schedules 2021		Unit	Labour Hours	Labour Net £	Plant Net £	Materials Net £	Unit Rate Net £	Labour Gross £	Plant Gross £	Materials Gross £	Unit Rate Gross £
								(Gross rates include 10% profit)			
A.02.	**REPAIRS AND ALTERATIONS**										
A.02.09.	**WORKS TO CHIMNEYS**										
A.02.09.03.	**Pull down chimney stacks, clean sound whole bricks for reuse and remove; up to 9.00m high or two storeys**										
A.02.09.03.A	lime mortar	m³	18.78	984.55	—	—	984.55	1083.01	—	—	1083.01
A.02.09.03.B	cement mortar	m³	24.15	1265.84	—	—	1265.84	1392.42	—	—	1392.42
A.02.09.04.	**Extra over last for each additional 3.00m or storey height**										
A.02.09.04.A	lime mortar	m³	4.70	246.16	—	—	246.16	270.78	—	—	270.78
A.02.09.04.B	cement mortar	m³	6.04	316.41	—	—	316.41	348.05	—	—	348.05
A.02.09.05.	**Rebuild single flue chimney in common brickwork (PC sum £450 per 1000); in cement mortar (1:3); including building in 185mm dia socketed and rebated clay flue liners, BS EN 13502:2002; up to 9.00m or two storeys high; overall plan dimensions**										
A.02.09.05.A	450mm x 450mm	m	1.90	99.81	—	56.39	156.20	109.79	—	62.03	171.82
A.02.09.05.B	675mm x 675mm	m	4.74	248.58	—	132.80	381.37	273.43	—	146.08	419.51
A.02.09.07.	**Rebuild double flue chimney in common brickwork (PC sum £450 per 1000); in cement mortar (1:3); including building in 185mm dia socketed and rebated clay flue liners, BS EN 13502:2002; up to 9.00m or two storeys high; overall plan dimensions**										
A.02.09.07.A	450mm x 750mm	m	3.39	177.60	—	101.49	279.09	195.36	—	111.64	307.00
A.02.09.07.B	675mm x 975mm	m	6.93	363.48	—	198.26	561.74	399.83	—	218.09	617.92
A.02.09.09.	**Take off loose chimney pot and reset including flaunching**										
A.02.09.09.A	up to two storeys or 9.00m high	nr	1.34	70.35	—	0.64	70.99	77.38	—	0.71	78.09
A.02.09.10.	**Take down and remove chimney pot, supply, set and flaunch new pot; up to two storeys or 9.00m high**										
A.02.09.10.A	300mm pot	nr	1.68	87.96	—	52.63	140.59	96.76	—	57.89	154.65
A.02.09.10.B	375mm pot	nr	1.71	89.69	—	64.73	154.42	98.66	—	71.20	169.86
A.02.09.10.C	450mm pot	nr	1.74	91.42	—	64.73	156.15	100.56	—	71.20	171.77
A.02.09.10.D	600mm pot	nr	1.81	94.99	—	96.00	190.99	104.48	—	105.60	210.08
A.02.09.11.	**Add to the foregoing for each additional storey or 3.00m high**										
A.02.09.11.A	300mm pot	nr	0.18	9.54	—	—	9.54	10.49	—	—	10.49

Hutchins Priced Schedules 2021		Unit	Labour Hours	Labour Net £	Plant Net £	Materials Net £	Unit Rate Net £	Labour Gross £	Plant Gross £	Materials Gross £	Unit Rate Gross £
							(Gross rates include 10% profit)				
A.02.	**REPAIRS AND ALTERATIONS**										
A.02.09.	**WORKS TO CHIMNEYS**										
A.02.09.11.	**Add to the foregoing for each additional storey or 3.00m high**										
A.02.09.11.B	375mm pot..................	nr	0.22	11.43	—	—	11.43	12.57	—	—	12.57
A.02.09.11.C	450mm pot..................	nr	0.25	13.31	—	—	13.31	14.65	—	—	14.65
A.02.09.11.D	600mm pot..................	nr	0.34	17.61	—	—	17.61	19.37	—	—	19.37
A.02.12.	**BRICKWORK REPAIRS, DPCS, SUNDRIES**										
A.02.12.01.	**Treat brick walls with silicone, or similar, damp-proof liquid**										
A.02.12.01.A	external...................	m²	0.15	7.76	—	0.06	7.82	8.53	—	0.07	8.60
A.02.12.02.	**Cut out defective brickwork and reface with new facing bricks (PC sum £550 per 1000)**										
A.02.12.02.A	cement mortar...............	m²	3.82	200.45	—	38.85	239.31	220.50	—	42.74	263.24
A.02.12.02.B	lime mortar................	m²	3.35	175.82	—	39.26	215.08	193.40	—	43.19	236.58
A.02.12.03.	**Cut out defective brickwork and reface with single facing bricks (PC sum £550 per 1000)**										
A.02.12.03.A	cement mortar...............	nr	0.34	17.61	—	0.68	18.29	19.37	—	0.75	20.12
A.02.12.03.B	lime mortar................	nr	0.23	12.27	—	0.68	12.94	13.49	—	0.75	14.24
A.02.12.04.	**Rake out mortar and repoint**										
A.02.12.04.A	perished mortar..............	m²	0.58	30.30	—	0.51	30.81	33.33	—	0.57	33.89
A.02.12.04.B	sound mortar................	m²	1.44	75.59	—	0.51	76.10	83.15	—	0.57	83.71
A.02.12.05.	**Cut out fractures in brickwork and build in new common brickwork (PC sum £450 per 1000); approximately 405mm wide, 225mm thick**										
A.02.12.05.A	cement mortar...............	m	2.41	126.54	—	26.91	153.45	139.20	—	29.60	168.79
A.02.12.05.B	lime mortar................	m	1.58	82.61	—	26.94	109.55	90.88	—	29.63	120.51
A.02.12.06.	**Take down segmental arch and rebuild in facings (PC sum £550 per 1000) 225mm high on face including centring**										
A.02.12.06.A	113mm wide soffit.............	m	4.02	210.94	—	10.05	220.99	232.03	—	11.06	243.09
A.02.12.06.B	225mm wide soffit.............	m	5.23	274.26	—	18.30	292.56	301.69	—	20.13	321.82
A.02.12.07.	**Cut opening through brick walls in cement mortar for doors, windows, etc.; including all necessary shoring and making good (common brickwork PC sum £450 per 1000, facing brickwork PC sum £550 per 1000)**										
A.02.12.07.A	half brick walls................	m²	1.34	58.72	0.36	20.09	79.17	64.59	0.39	22.10	87.09

House Renovation Grants, Repairs and Alterations

Hutchins Priced Schedules 2021	Unit	Labour Hours	Labour Net £	Plant Net £	Materials Net £	Unit Rate Net £	Labour Gross £	Plant Gross £	Materials Gross £	Unit Rate Gross £
							(Gross rates include 10% profit)			
A.02.	**REPAIRS AND ALTERATIONS**									
A.02.12.	**BRICKWORK REPAIRS, DPCS, SUNDRIES**									
A.02.12.07.	Cut opening through brick walls in cement mortar for doors, windows, etc.; including all necessary shoring and making good (common brickwork PC sum £450 per 1000, facing brickwork PC sum £550 per 1000)									
A.02.12.07.B	one brick walls	m²	2.68	117.54	0.72	36.98	155.24	129.29	0.79	40.68	170.77
A.02.12.07.C	one and a half brick walls	m²	4.02	176.26	0.99	53.99	231.24	193.88	1.09	59.39	254.36
A.02.12.07.D	two brick walls	m²	5.35	234.98	1.30	70.44	306.71	258.47	1.43	77.48	337.38
A.02.12.08.	**Common brickwork (PC sum £450 per 1000) in gauged mortar in small areas; bonding to existing**									
A.02.12.08.A	half brick	m²	1.64	86.18	—	33.56	119.74	94.80	—	36.92	131.71
A.02.12.08.B	one brick	m²	3.02	158.20	—	67.57	225.78	174.02	—	74.33	248.35
A.02.12.08.C	one and a half brick	m²	4.53	237.36	—	102.16	339.52	261.09	—	112.38	373.47
A.02.12.09.	**Damp-proof course in short lengths in existing walls including cutting out brickwork and building in with new common brickwork (PC sum £450 per 1000)**									
A.02.12.09.A	half brick wide; two course slate	m	1.24	65.11	—	7.61	72.71	71.62	—	8.37	79.98
A.02.12.09.B	half brick wide; bitumen felt	m	1.04	54.52	—	6.33	60.85	59.97	—	6.96	66.93
A.02.12.09.C	one brick wide and over; two course slate	m²	4.36	228.55	—	69.09	297.64	251.41	—	76.00	327.40
A.02.12.09.D	one brick wide and over; bitumen felt . . .	m²	3.59	188.08	—	58.99	247.07	206.89	—	64.89	271.78
A.02.12.10.	**Cut, tooth and bond new common brickwork (PC sum £450 per 1000) to existing**									
A.02.12.10.A	half brick	m	0.37	19.40	—	3.54	22.93	21.33	—	3.89	25.22
A.02.12.10.B	one brick	m	0.67	35.12	—	7.39	42.51	38.63	—	8.13	46.77
A.02.12.10.C	one and a half brick	m	0.94	49.27	—	10.93	60.20	54.20	—	12.02	66.22
A.02.12.11.	**Cut hole for pipes, brackets, fittings, etc.; through clinker concrete walls per 25mm in depth of cut; make good**									
A.02.12.11.A	area not exceeding 0.003m2	nr	0.08	4.19	—	0.24	4.43	4.61	—	0.26	4.87
A.02.12.11.B	over 0.003m2; not exceeding 0.03m2 . . .	nr	0.12	6.29	—	0.36	6.65	6.92	—	0.39	7.31
A.02.12.11.C	over 0.03m2; not exceeding 0.06m2 . . .	nr	0.17	8.81	—	0.47	9.28	9.69	—	0.52	10.21
A.02.12.12.	**Cut hole for pipes, brackets, fittings, etc.; through brickwork in lime mortar; make good**									
A.02.12.12.A	area not exceeding 0.003m2	nr	0.13	7.02	—	0.26	7.28	7.73	—	0.28	8.01
A.02.12.12.B	over 0.003m2; not exceeding 0.03m2 . . .	nr	0.20	10.59	—	0.39	10.97	11.65	—	0.42	12.07
A.02.12.12.C	over 0.03m2; not exceeding 0.06m2 . . .	nr	0.27	14.05	—	0.51	14.56	15.45	—	0.57	16.02

Hutchins Priced Schedules 2021		Unit	Labour Hours	Labour Net £	Plant Net £	Materials Net £	Unit Rate Net £	Labour Gross £	Plant Gross £	Materials Gross £	Unit Rate Gross £
							(Gross rates include 10% profit)				
A.02.	**REPAIRS AND ALTERATIONS**										
A.02.12.	**BRICKWORK REPAIRS, DPCS, SUNDRIES**										
A.02.12.13.	**Cut hole for pipes, brackets, fittings, etc.; through brickwork in cement mortar; make good**										
A.02.12.13.A	area not exceeding 0.003m2	nr	0.19	9.85	—	0.24	10.09	10.84	—	0.26	11.10
A.02.12.13.B	over 0.003m2; not exceeding 0.03m2 . . .	nr	0.32	16.88	—	0.36	17.24	18.57	—	0.39	18.96
A.02.12.13.C	over 0.03m2; not exceeding 0.06m2 . . .	nr	0.42	22.12	—	0.47	22.60	24.33	—	0.52	24.86
A.02.12.14.	**Cut horizontal chase in brickwork 112mm deep for concrete floor or landing; width**										
A.02.12.14.A	100mm	m	1.34	70.35	—	—	70.35	77.38	—	—	77.38
A.02.12.14.B	125mm	m	1.48	77.37	—	—	77.37	85.11	—	—	85.11
A.02.12.14.C	150mm	m	1.68	87.96	—	—	87.96	96.76	—	—	96.76
A.02.12.14.D	200mm	m	2.01	105.47	—	—	105.47	116.02	—	—	116.02
A.02.12.15.	**Cut horizontal chase in fair faced brickwork 112mm deep for concrete floor or landing; width**										
A.02.12.15.A	100mm	m	2.01	105.47	—	—	105.47	116.02	—	—	116.02
A.02.12.15.B	125mm	m	2.21	116.06	—	—	116.06	127.66	—	—	127.66
A.02.12.15.C	150mm	m	2.52	131.89	—	—	131.89	145.08	—	—	145.08
A.02.12.15.D	200mm	m	3.02	158.20	—	—	158.20	174.02	—	—	174.02
A.02.12.16.	**Reform brick jambs after cutting new opening in existing brickwork; in common brickwork (PC sum £450 per 1000)**										
A.02.12.16.A	half brick	m	1.34	70.35	—	4.07	74.42	77.38	—	4.48	81.86
A.02.12.16.B	one brick	m	1.68	87.96	—	8.15	96.11	96.76	—	8.96	105.72
A.02.12.16.C	one and a half brick	m	2.01	105.47	—	12.46	117.93	116.02	—	13.71	129.72
A.02.12.17.	**Reform brick jambs after cutting new opening in existing brickwork; in facing bricks (PC sum £550 per 1000)**										
A.02.12.17.A	half brick	m	1.81	94.99	—	4.91	99.90	104.48	—	5.41	109.89
A.02.12.17.B	one brick	m	2.08	109.03	—	9.83	118.86	119.94	—	10.81	130.75
A.02.12.17.C	one and a half brick	m	2.35	123.08	—	15.00	138.08	135.39	—	16.50	151.89
A.02.12.19.	**Take out and rebed door or window frame**										
A.02.12.19.A	point externally; make good internally . . .	m	0.37	19.40	—	0.39	19.78	21.33	—	0.42	21.76
A.02.12.20.	**Rake out defective pointing around door or window frame and repoint in**										
A.02.12.20.A	cement mortar	m	0.30	15.83	—	0.39	16.22	17.41	—	0.42	17.84
A.02.12.20.B	cement mortar but using mastic sealant .	m	0.40	21.07	—	0.34	21.41	23.18	—	0.37	23.55

House Renovation Grants, Repairs and Alterations

Hutchins Priced Schedules 2021	Unit	Labour Hours	Labour Net £	Plant Net £	Materials Net £	Unit Rate Net £	Labour Gross £	Plant Gross £	Materials Gross £	Unit Rate Gross £
							(Gross rates include 10% profit)			
A.02.	**REPAIRS AND ALTERATIONS**									
A.02.12.	**BRICKWORK REPAIRS, DPCS, SUNDRIES**									
A.02.12.22.	**Take out existing fireplace including surround and hearth**									
A.02.12.22.A	small iron	nr	2.68	140.70	—	—	140.70	154.76	—	—	154.76
A.02.12.22.B	large tiled	nr	3.45	181.06	—	—	181.06	199.16	—	—	199.16
A.02.12.22.C	free standing	nr	2.15	112.49	—	—	112.49	123.74	—	—	123.74
A.02.12.23.	**Take out and reset existing fireplace including surround and hearth**									
A.02.12.23.A	small iron	nr	6.71	351.63	—	3.44	355.07	386.80	—	3.78	390.58
A.02.12.23.B	large tiled	nr	8.22	430.79	—	4.86	435.65	473.87	—	5.35	479.22
A.02.12.23.C	free standing	nr	5.50	288.31	—	1.31	289.62	317.14	—	1.44	318.58
A.02.12.24.	**Fix only new fireplace including surround and hearth**									
A.02.12.24.A	small iron	nr	5.70	298.90	—	3.44	302.34	328.79	—	3.78	332.57
A.02.12.24.B	large tiled	nr	7.14	374.49	—	4.86	379.35	411.94	—	5.35	417.29
A.02.12.24.C	free standing	nr	3.15	165.23	—	1.31	166.53	181.75	—	1.44	183.19
A.02.12.25.	**Take out existing fireplace and fix only solid brick back with fine concrete behind**									
A.02.12.25.A	fix tiled surround and tiled hearth (cost of interior, surround, hearth tiles and fret not included)	nr	8.72	457.10	—	2.35	459.45	502.81	—	2.59	505.40
A.02.12.28.	**Galvanised iron air bricks built into wall as work proceeds**									
A.02.12.28.A	225mm x 75mm	nr	0.15	7.76	—	22.61	30.37	8.53	—	24.87	33.40
A.02.12.28.B	225mm x 150mm	nr	0.20	10.59	—	28.26	38.85	11.65	—	31.09	42.73
A.02.12.29.	**Terracotta air bricks built into wall as work proceeds**									
A.02.12.29.A	215mm x 65mm; square hole	nr	0.15	7.76	—	3.55	11.31	8.53	—	3.91	12.44
A.02.12.29.B	215mm x 140mm; square hole	nr	0.20	10.59	—	5.27	15.86	11.65	—	5.80	17.44
A.02.12.29.C	215mm x 215mm; square hole	nr	0.22	11.64	—	13.25	24.89	12.80	—	14.58	27.38
A.02.12.29.D	215mm x 65mm; louvred hole	nr	0.15	7.76	—	4.12	11.88	8.53	—	4.54	13.07
A.02.12.29.E	215mm x 140mm; louvred hole	nr	0.20	10.59	—	6.74	17.33	11.65	—	7.42	19.06
A.02.12.30.	**Plaster louvre type air bricks into wall as work proceeds**									
A.02.12.30.A	225mm x 75mm	nr	0.13	7.02	—	8.38	15.41	7.73	—	9.22	16.95
A.02.12.30.B	225mm x 150mm	nr	0.17	8.81	—	9.44	18.25	9.69	—	10.39	20.08
A.02.12.30.C	225mm x 225mm	nr	0.20	10.59	—	10.50	21.09	11.65	—	11.56	23.20

Hutchins Priced Schedules 2021		Unit	Labour Hours	Labour Net £	Plant Net £	Materials Net £	Unit Rate Net £	Labour Gross £	Plant Gross £	Materials Gross £	Unit Rate Gross £
								(Gross rates include 10% profit)			
A.02.	**REPAIRS AND ALTERATIONS**										
A.02.12.	**BRICKWORK REPAIRS, DPCS, SUNDRIES**										
A.02.12.31.	**Add to the foregoing air brick items for cutting through existing brick wall (any thickness); build in; make good**										
A.02.12.31.A	225mm x 75mm	nr	0.19	9.85	—	0.39	10.24	10.84	—	0.42	11.27
A.02.12.31.B	225mm x 150mm	nr	0.20	10.59	—	0.51	11.10	11.65	—	0.57	12.21
A.02.12.31.C	225mm x 225mm	nr	0.23	12.27	—	0.64	12.91	13.49	—	0.71	14.20
A.02.12.32.	**Add to the foregoing for cutting out and remove existing**										
A.02.12.32.A	225mm x 75mm	nr	0.07	3.56	—	—	3.56	3.92	—	—	3.92
A.02.12.32.B	225mm x 150mm	nr	0.08	4.19	—	—	4.19	4.61	—	—	4.61
A.02.12.32.C	225mm x 225mm	nr	0.10	5.24	—	—	5.24	5.77	—	—	5.77
A.02.14.	**DENSE AGGREGATE CONCRETE BLOCK WALLS**										
A.02.14.01.	**Dense aggregate concrete blocks in walls, etc.; in composition mortar**										
A.02.14.01.A	rough both sides 100mm slabs	m²	0.84	43.93	—	13.25	57.17	48.32	—	14.57	62.89
A.02.14.01.B	fair face one side	m²	1.24	65.11	—	13.25	78.35	71.62	—	14.57	86.19
A.02.14.01.C	fair face both sides	m²	1.34	70.35	—	13.25	83.59	77.38	—	14.57	91.95
A.02.16.	**WOODWORK REPAIRS AND REMOVALS**										
A.02.16.01.	**Remove timber and load into skip**										
A.02.16.01.A	roof timbers complete, including rafters, purlins, ceiling joists, plates and the like (measured flat on plan)	m²	0.37	10.94	—	—	10.94	12.04	—	—	12.04
A.02.16.01.B	floor construction; ground floor level	m²	0.28	8.28	—	—	8.28	9.11	—	—	9.11
A.02.16.01.C	floor construction; first floor level	m²	0.55	16.27	—	—	16.27	17.90	—	—	17.90
A.02.16.01.D	floor construction; roof level	m²	0.77	22.78	—	—	22.78	25.05	—	—	25.05
A.02.16.01.E	individual floor or roof members	m	0.30	8.87	—	—	8.87	9.76	—	—	9.76
A.02.16.01.F	extra for cutting off end flush with wall . .	nr	0.50	14.79	—	—	14.79	16.27	—	—	16.27
A.02.16.01.G	decayed or infected floor plates	m	0.40	11.83	—	—	11.83	13.02	—	—	13.02
A.02.16.01.H	tilting fillet or roll	m	0.17	5.03	—	—	5.03	5.53	—	—	5.53
A.02.16.01.I	fascia or barge board	m	0.65	19.23	—	—	19.23	21.15	—	—	21.15
A.02.16.02.	**Remove boarding, including withdrawing nails, and load into skip**										
A.02.16.02.A	softwood flooring; at ground floor level . .	m²	0.42	12.42	—	—	12.42	13.67	—	—	13.67
A.02.16.02.B	softwood flooring; at first floor level	m²	0.68	20.11	—	—	20.11	22.13	—	—	22.13
A.02.16.02.C	softwood flooring; at roof level	m²	0.80	23.66	—	—	23.66	26.03	—	—	26.03
A.02.16.02.D	softwood flooring; at gutter level	m²	0.88	26.03	—	—	26.03	28.63	—	—	28.63

House Renovation Grants, Repairs and Alterations

Hutchins Priced Schedules 2021		Unit	Labour Hours	Labour Net £	Plant Net £	Materials Net £	Unit Rate Net £	Labour Gross £	Plant Gross £	Materials Gross £	Unit Rate Gross £
								(Gross rates include 10% profit)			
A.02.	**REPAIRS AND ALTERATIONS**										
A.02.16.	**WOODWORK REPAIRS AND REMOVALS**										
A.02.16.02.	**Remove boarding, including withdrawing nails, and load into skip**										
A.02.16.02.E	chipboard flooring; at ground floor level ..	m²	0.17	5.03	—	—	5.03	5.53	—	—	5.53
A.02.16.02.F	chipboard flooring; at first floor level	m²	0.42	12.42	—	—	12.42	13.67	—	—	13.67
A.02.16.02.G	plywood flooring; at ground level	m²	0.25	7.40	—	—	7.40	8.13	—	—	8.13
A.02.16.02.H	plywood flooring; at first floor level	m²	0.48	14.20	—	—	14.20	15.62	—	—	15.62
A.02.16.03.	**Remove stud partition, softwood, including finishings both sides, and load into skip**										
A.02.16.03.A	solid.....................	m²	0.50	14.79	—	—	14.79	16.27	—	—	16.27
A.02.16.03.B	glazed; including removal of glass	m²	0.67	19.82	—	—	19.82	21.80	—	—	21.80
A.02.16.04.	**Remove wall linings, including battening behind, and load into skip**										
A.02.16.04.A	plain sheeting.................	m²	0.33	9.76	—	—	9.76	10.74	—	—	10.74
A.02.16.04.B	matchboarding...............	m²	0.45	13.31	—	—	13.31	14.64	—	—	14.64
A.02.16.05.	**Remove ceiling linings, including battening behind, and load into skip**										
A.02.16.05.A	plain sheeting.................	m²	0.50	14.79	—	—	14.79	16.27	—	—	16.27
A.02.16.05.B	matchboarding................	m²	0.67	19.82	—	—	19.82	21.80	—	—	21.80
A.02.16.06.	**Remove mouldings and load into skip**										
A.02.16.06.A	skirtings, picture rails, dado rails, architraves and the like...........	m	0.12	3.55	—	—	3.55	3.90	—	—	3.90
A.02.16.06.B	shelves, window boards and the like ...	m	0.35	10.35	—	—	10.35	11.39	—	—	11.39
A.02.16.07.	**Remove door and load into skip**										
A.02.16.07.A	single.....................	nr	0.45	13.31	—	—	13.31	14.64	—	—	14.64
A.02.16.07.B	single with frame or lining	nr	0.88	26.03	—	—	26.03	28.63	—	—	28.63
A.02.16.07.C	pair......................	nr	0.77	22.78	—	—	22.78	25.05	—	—	25.05
A.02.16.07.D	pair with frame or lining	nr	1.32	39.05	—	—	39.05	42.95	—	—	42.95
A.02.16.07.E	extra for taking out spring box	nr	0.83	24.55	—	—	24.55	27.01	—	—	27.01
A.02.16.08.	**Remove window and load into skip**										
A.02.16.08.A	casement; with frame	nr	1.32	39.05	—	—	39.05	42.95	—	—	42.95
A.02.16.08.B	double hung sash; with frame	nr	1.77	52.36	—	—	52.36	57.59	—	—	57.59
A.02.16.08.C	French with frame	pair	4.40	130.15	—	—	130.15	143.17	—	—	143.17
A.02.16.09.	**Remove staircase balustrade and load into skip**										
A.02.16.09.A	single straight flight	nr	3.85	113.88	—	—	113.88	125.27	—	—	125.27
A.02.16.09.B	dogleg flight..................	nr	5.50	162.69	—	—	162.69	178.96	—	—	178.96
A.02.16.09.C	handrail and brackets	m	0.12	3.55	—	—	3.55	3.90	—	—	3.90

Hutchins Priced Schedules 2021	Unit	Labour Hours	Labour Net £	Plant Net £	Materials Net £	Unit Rate Net £	Labour Gross £	Plant Gross £	Materials Gross £	Unit Rate Gross £
							(Gross rates include 10% profit)			
A.02.	**REPAIRS AND ALTERATIONS**									
A.02.16.	**WOODWORK REPAIRS AND REMOVALS**									
A.02.16.10.	**Remove bath fittings and load into skip**									
A.02.16.10.A	disconnected bath; including panels and frame	nr	0.45	13.31	—	—	13.31	14.64	—	—	14.64
A.02.16.11.	**Remove kitchen fittings and load into skip**									
A.02.16.11.A	wall units	nr	0.50	14.79	—	—	14.79	16.27	—	—	16.27
A.02.16.11.B	floor units	nr	0.33	9.76	—	—	9.76	10.74	—	—	10.74
A.02.16.11.C	larder units	nr	0.45	13.31	—	—	13.31	14.64	—	—	14.64
A.02.16.11.D	built-in cupboards	nr	1.55	45.85	—	—	45.85	50.43	—	—	50.43
A.02.16.11.E	pipe casings	m	0.33	9.76	—	—	9.76	10.74	—	—	10.74
A.02.16.13.	**Erect temporary hoarding comprising second-hand timber posts, rails and struts and covered with second-hand close boarding or corrugated iron sheets and dismantle on completion**									
A.02.16.13.A	1.80m high	m	1.37	92.28	—	17.07	109.35	101.51	—	18.77	120.28
A.02.16.13.B	extra for 0.75m wide door	nr	0.33	21.84	—	3.71	25.56	24.03	—	4.08	28.11
A.02.16.13.C	extra for pair of gates approximately 2.40m wide overall	nr	1.05	70.50	—	11.07	81.57	77.55	—	12.18	89.73
A.02.16.14.	**Enclose frontage to site with chestnut fencing with posts at 1.80m intervals and dismantle on completion**									
A.02.16.14.A	1.20m high	m	0.14	9.41	—	9.27	18.68	10.35	—	10.20	20.55
A.02.16.16.	**Take up defective gutter boards and bearers, supply and fix**									
A.02.16.16.A	new	m²	1.85	124.14	—	19.55	143.68	136.55	—	21.50	158.05
A.02.16.17.	**Take off defective rounded wood rolls to flats, supply and fix**									
A.02.16.17.A	new	m	0.11	7.39	—	4.75	12.14	8.13	—	5.22	13.36
A.02.16.18.	**Renew roof timbers**									
A.02.16.18.A	47mm x 100mm	m	0.08	5.04	—	3.79	8.83	5.54	—	4.17	9.72
A.02.16.18.B	47mm x 125mm	m	0.10	6.72	—	4.59	11.31	7.39	—	5.05	12.44
A.02.16.18.C	47mm x 150mm	m	0.13	8.40	—	5.50	13.90	9.24	—	6.05	15.29
A.02.16.19.	**Take down defective timber, supply and fix new**									
A.02.16.19.A	175mm x 31mm hips and ridges	m	0.20	13.44	—	4.42	17.86	14.79	—	4.86	19.65
A.02.16.19.B	150mm x 25mm fascia	m	0.20	13.44	—	3.00	16.44	14.79	—	3.30	18.08
A.02.16.19.C	225mm x 19mm soffit and bearers	m	0.30	20.16	—	6.18	26.34	22.18	—	6.80	28.98

House Renovation Grants, Repairs and Alterations

Hutchins Priced Schedules 2021	Unit	Labour Hours	Labour Net £	Plant Net £	Materials Net £	Unit Rate Net £	Labour Gross £	Plant Gross £	Materials Gross £	Unit Rate Gross £
										(Gross rates include 10% profit)
A.02. **REPAIRS AND ALTERATIONS**										
A.02.16. **WOODWORK REPAIRS AND REMOVALS**										
A.02.16.20. Take off front gate; remove defective 150mm x 150mm timber posts; grub up concrete; supply new post approximately 1.50m long; set in new concrete and rehang gate										
A.02.16.20.A treated fir post	nr	1.50	100.68	—	37.01	137.69	110.75	—	40.71	151.46
A.02.16.20.B oak post	nr	1.62	109.08	—	68.91	177.99	119.99	—	75.80	195.79
A.02.16.21. Excavate for and bolt to wood gate post										
A.02.16.21.A oak spur set in concrete	nr	1.00	67.14	—	32.56	99.70	73.86	—	35.82	109.67
A.02.16.23. Take down and remove all temporary weatherproofing, e.g. polythene sheet, hardboard, chipboard and the like; together with all associated timber work to windows and doors										
A.02.16.23.A make good all existing joinery work including withdrawing all nails	m²	0.60	40.26	—	—	40.26	44.28	—	—	44.28
A.02.16.24. Take down and remove all galvanised iron sheet covering										
A.02.16.24.A including all timber backings; make good .	m²	0.85	57.06	—	—	57.06	62.77	—	—	62.77
A.02.16.26. Temporary screens comprising										
A.02.16.26.A 47mm x 100mm framing lined both sides with building paper	m²	0.13	8.74	—	9.49	18.23	9.61	—	10.44	20.05
A.02.16.26.B 47mm x 100mm framing lined one side with 19mm matchboard	m²	0.30	20.16	—	27.00	47.16	22.18	—	29.70	51.88
A.02.16.26.C 47mm x 50mm framing lined one side with hardboard	m²	0.23	15.12	—	7.12	22.24	16.63	—	7.83	24.46
A.02.16.28. Strut up ceiling floor to ceiling and remove struts on completion										
A.02.16.28.A average 2.60m	m	0.55	36.90	3.39	—	40.28	40.59	3.73	—	44.31
A.02.16.30. Strut and support window openings; area of window										
A.02.16.30.A 1.00m2	nr	0.23	15.12	—	2.30	17.42	16.63	—	2.53	19.17
A.02.16.30.B 1.50m2	nr	0.25	16.80	—	2.69	19.49	18.48	—	2.95	21.44
A.02.16.30.C 2.00m2	nr	0.28	18.48	—	3.00	21.48	20.33	—	3.30	23.63
A.02.16.31. Remove all grease and dirt from existing flooring; remove all projecting lino nails or tacks; punch down all floor brads, resecure any loose boards, plane off and leave smooth										
A.02.16.31.A generally	m²	0.43	28.56	—	—	28.56	31.42	—	—	31.42

Hutchins Priced Schedules 2021	Unit	Labour Hours	Labour Net £	Plant Net £	Materials Net £	Unit Rate Net £	Labour Gross £	Plant Gross £	Materials Gross £	Unit Rate Gross £
						(Gross rates include 10% profit)				
A.02.	**REPAIRS AND ALTERATIONS**									
A.02.16.	**WOODWORK REPAIRS AND REMOVALS**									
A.02.16.31.	**Remove all grease and dirt from existing flooring; remove all projecting lino nails or tacks; punch down all floor brads, resecure any loose boards, plane off and leave smooth**									
A.02.16.31.B	in areas less than 1.00m2	m²	0.55	36.90	—	—	36.90	40.59	—	—	40.59
A.02.16.32.	**Take up loose floor blocks and relay in mastic**									
A.02.16.32.A	single block	nr	0.18	11.76	—	0.79	12.55	12.94	—	0.87	13.80
A.02.16.32.B	in patches up to six blocks	nr	0.11	7.39	—	1.64	9.03	8.13	—	1.80	9.93
A.02.16.33.	**Smooth hardwood floor**									
A.02.16.33.A	with electric sanding machine	m²	0.50	33.54	0.72	—	34.26	36.89	0.79	—	37.68
A.02.16.34.	**Take off existing skirting**									
A.02.16.34.A	replug grounds and refix skirting	m	0.18	12.10	—	—	12.10	13.31	—	—	13.31
A.02.16.35.	**Take off existing softwood skirting; supply and fix new**									
A.02.16.35.A	25mm x 150mm	m	0.25	16.80	—	4.82	21.62	18.48	—	5.30	23.78
A.02.16.39.	**Take up existing shrunk or worn flooring, any thickness, drawing all nails, relaying, cramping up, making up width or length with extra boarding of same thickness and clean off on completion; areas over 0.50m2**									
A.02.16.39.A	plain edge	m²	0.45	30.18	—	0.98	31.16	33.20	—	1.08	34.28
A.02.16.39.B	tongued and grooved	m²	0.55	36.90	—	1.01	37.90	40.59	—	1.11	41.69
A.02.16.40.	**Remove damaged 25mm softwood floor boards; clean joists and renew**									
A.02.16.40.A	plain edge	m²	0.43	28.56	—	14.70	43.26	31.42	—	16.17	47.59
A.02.16.40.B	tongued and grooved	m²	0.50	33.54	—	17.30	50.84	36.89	—	19.03	55.93
A.02.16.40.C	plain edge in small detached areas not exceeding 1.00m2	m²	1.10	73.80	—	14.70	88.50	81.18	—	16.17	97.35
A.02.16.40.D	tongued and grooved in detached area not exceeding 1.00m2	m²	1.35	90.60	—	17.30	107.90	99.66	—	19.03	118.69
A.02.16.40.E	plain edge over 1.00m2; not exceeding 2.50m2	m²	1.00	67.14	—	14.70	81.84	73.86	—	16.17	90.03
A.02.16.40.F	tongued and grooved over 1.00m2; not exceeding 2.50m2	m²	1.25	83.88	—	17.30	101.18	92.27	—	19.03	111.30
A.02.16.42.	**New joists and softwood floor boarding; treating joists and underside of boards with creosote substitute or other preservative**									
A.02.16.42.A	100mm x 50mm floor joists and 25mm plain edge flooring	m²	1.00	37.63	—	24.58	62.21	41.39	—	27.04	68.44

House Renovation Grants, Repairs and Alterations

Hutchins Priced Schedules 2021	Unit	Labour Hours	Labour Net £	Plant Net £	Materials Net £	Unit Rate Net £	Labour Gross £	Plant Gross £	Materials Gross £	Unit Rate Gross £
							(Gross rates include 10% profit)			
A.02.	**REPAIRS AND ALTERATIONS**									
A.02.16.	WOODWORK REPAIRS AND REMOVALS									
A.02.16.44.	**oak strip flooring pinned and glued to existing softwood floor; clean off and wax polish**									
A.02.16.44.A	14mm (PC sum £35 per m2)	m²	0.90	33.87	—	40.13	74.00	37.25	—	44.15	81.40
A.02.16.46.	**Take down door, cut 13mm off bottom edge and rehang**									
A.02.16.46.A	rehang	nr	0.80	54.04	—	—	54.04	59.44	—	—	59.44
A.02.16.47.	**Take down architraves and reduce length by 13mm and refix**									
A.02.16.47.A	one side	set	0.28	18.55	—	—	18.55	20.40	—	—	20.40
A.02.16.47.B	reduce length without removal	set	0.25	16.87	—	—	16.87	18.56	—	—	18.56
A.02.16.48.	**Take off skirting and refix**									
A.02.16.48.A	at higher level	m	0.15	10.15	—	—	10.15	11.16	—	—	11.16
A.02.16.49.	**Hardwood border to hearth**									
A.02.16.49.A	mitred	nr	0.30	20.23	—	19.30	39.53	22.25	—	21.23	43.49
A.02.16.50.	**Take down door, take out lining or frame, realign and refix**									
A.02.16.50.A	ease, adjust and rehang door; refix existing architraves; make good work disturbed	nr	2.51	168.76	—	—	168.76	185.64	—	—	185.64
A.02.16.52.	**Take down door**									
A.02.16.52.A	ease and rehang	nr	0.90	60.76	—	—	60.76	66.83	—	—	66.83
A.02.16.53.	**Take down door, ease and rehang on new butt hinges; remove lock and furniture, supply and fit new lock and furniture**									
A.02.16.53.A	rim lock	nr	1.51	101.29	—	22.06	123.34	111.41	—	24.26	135.68
A.02.16.53.B	mortice lock	nr	1.76	118.16	—	18.26	136.42	129.97	—	20.09	150.06
A.02.16.54.	**Take down door, take apart and fit new panel or rail**									
A.02.16.54.A	rehang	nr	2.41	161.98	—	3.26	165.24	178.17	—	3.59	181.76
A.02.16.55.	**Renew weatherboard**									
A.02.16.55.A	to external softwood door	nr	0.50	33.74	—	8.29	42.03	37.11	—	9.12	46.23
A.02.16.56.	**Take down, ease and adjust and rehang**									
A.02.16.56.A	casement sash	nr	0.60	40.53	—	—	40.53	44.58	—	—	44.58

Hutchins Priced Schedules 2021	Unit	Labour Hours	Labour Net £	Plant Net £	Materials Net £	Unit Rate Net £	Labour Gross £	Plant Gross £	Materials Gross £	Unit Rate Gross £
							(Gross rates include 10% profit)			

A.02.   **REPAIRS AND ALTERATIONS**

A.02.16.   **WOODWORK REPAIRS AND REMOVALS**

	Unit	Labour Hours	Labour Net £	Plant Net £	Materials Net £	Unit Rate Net £	Labour Gross £	Plant Gross £	Materials Gross £	Unit Rate Gross £
A.02.16.57. Take off defective staff and parting beads										
A.02.16.57.A to double hung sash window and renew .	nr	0.40	27.02	—	1.09	28.11	29.72	—	1.20	30.92
A.02.16.58. Take out double hung sashes										
A.02.16.58.A ease, adjust and rehang, including new cords	nr	0.68	45.57	—	2.45	48.02	50.13	—	2.69	52.82
A.02.16.59. Cut out defective glazing bars to skylights, windows, doors or greenhouses										
A.02.16.59.A renew	m	0.45	30.38	—	1.91	32.29	33.42	—	2.10	35.52
A.02.16.60. Strengthening handrail and balusters including										
A.02.16.60.A renewing defective balusters	m	0.63	42.21	—	2.94	45.15	46.43	—	3.23	49.66
A.02.16.61. Cutting out defective and worn portion of tread										
A.02.16.61.A piecing in new	nr	0.40	27.02	—	0.67	27.69	29.72	—	0.74	30.46
A.02.16.62. Take off and renew ironmongery fixed to softwood										
A.02.16.62.A strong pattern zinc placed butts 75mm . .	pair	0.67	44.90	—	1.59	46.49	49.39	—	1.75	51.13
A.02.16.62.B strong pattern zinc plated butts 100mm . .	pair	0.79	53.30	—	2.12	55.42	58.63	—	2.33	60.96
A.02.16.62.C steel washered brass butts 75mm	pair	0.67	44.90	—	3.45	48.34	49.39	—	3.79	53.18
A.02.16.62.D steel washered brass butts 100mm	pair	0.79	53.30	—	4.03	57.33	58.63	—	4.43	63.06
A.02.16.62.E brass rising butts 75mm	pair	0.81	54.37	—	7.58	61.95	59.81	—	8.34	68.15
A.02.16.62.F brass rising butts 100mm	pair	0.93	62.77	—	10.81	73.59	69.05	—	11.89	80.94
A.02.16.62.G steel tee hinges 305mm	pair	0.57	38.11	—	4.85	42.96	41.92	—	5.34	47.26
A.02.16.62.H steel tee hinges 457mm	pair	0.61	41.20	—	12.72	53.92	45.32	—	13.99	59.31
A.02.16.62.I rim lock and furniture	nr	0.79	53.30	—	22.06	75.36	58.63	—	24.26	82.89
A.02.16.62.J mortice lock and furniture	nr	0.92	62.10	—	18.26	80.37	68.31	—	20.09	88.40
A.02.16.62.K Suffolk latch	nr	0.67	44.90	—	7.06	51.96	49.39	—	7.77	57.15
A.02.16.62.L 100mm straight barrel bolt, polished chrome	nr	0.34	22.99	—	3.39	26.38	25.28	—	3.73	29.02
A.02.16.62.M 150mm straight barrel bolt, polished chrome	nr	0.39	26.35	—	4.93	31.28	28.98	—	5.42	34.40
A.02.16.62.N casement stay with two pins; 250mm . . .	nr	0.31	20.90	—	9.71	30.61	22.99	—	10.68	33.67
A.02.16.62.O casement fastener; wedge pattern	nr	0.36	24.33	—	4.19	28.52	26.76	—	4.61	31.37
A.02.16.62.P sliding sash fastener	nr	0.74	49.60	—	10.43	60.03	54.56	—	11.47	66.03
A.02.16.62.Q sash lift	nr	0.24	15.86	—	1.00	16.86	17.45	—	1.10	18.54

House Renovation Grants, Repairs and Alterations

Hutchins Priced Schedules 2021		Unit	Labour Hours	Labour Net £	Plant Net £	Materials Net £	Unit Rate Net £	Labour Gross £	Plant Gross £	Materials Gross £	Unit Rate Gross £
								(Gross rates include 10% profit)			
A.02.	**REPAIRS AND ALTERATIONS**										
A.02.17.	**SHORING**										
A.02.17.01.	**Erect temporary dead shoring to form opening using three pairs 150mm x 150mm uprights and three 225mm x 150mm needles, braces and 225mm x 225mm base plates, holing brickwork for needles, all cartage, make good; remove on completion**										
A.02.17.01.A	assumed volume of timber 0.09m3	item	34.95	2348.99	—	62.33	2411.32	2583.89	—	68.57	2652.46
A.02.17.01.B	add to or deduct for every 0.03m3 more or less than 0.09m3	item	1.00	67.14	—	2.27	69.41	73.86	—	2.49	76.35
A.02.17.02.	**Erect temporary raking shoring including rakers, wall plate, needles, holing brickwork, cartage, make good; remove on completion**										
A.02.17.02.A	assumed volume of timber 0.30m3	item	15.98	1073.81	—	20.40	1094.21	1181.20	—	22.44	1203.64
A.02.17.02.B	add to or deduct for every 0.03m3 more or less than 0.09m3	item	1.50	100.68	—	2.27	102.95	110.75	—	2.49	113.24
A.02.17.03.	**Erect temporary flying shoring including horizontal shores, struts, wall plates, posts, needles, holing brickwork, cartage, make good; remove on completion**										
A.02.17.03.A	assumed volume of timber 0.60m3	item	39.94	2684.57	—	40.80	2725.37	2953.03	—	44.88	2997.91
A.02.17.03.B	add to or deduct for every 0.03m3 more or less than 0.60m3	item	2.00	134.22	—	43.07	177.29	147.64	—	47.37	195.01
A.02.17.04.	**Erect permanent raking shoring; as described above but left in position for an indefinite period**										
A.02.17.04.A	0.30m3	item	9.99	671.16	—	43.07	714.23	738.27	—	47.37	785.65
A.02.17.04.B	add to or deduct for each 0.03m3 or timber more or less than 0.30m3	item	1.00	67.14	—	4.53	71.68	73.86	—	4.99	78.84
A.02.17.05.	**Erect permanent flying shoring; as described above but left in position for an indefinite period**										
A.02.17.05.A	0.60m3	item	26.96	1812.05	—	87.27	1899.32	1993.25	—	95.99	2089.25
A.02.17.05.B	add to or deduct for every 0.03m3 more or less than 0.60m3	item	1.35	90.60	—	8.50	99.10	99.66	—	9.35	109.01
A.02.19.	**FINISHES, REPAIRS AND RENEWALS**										
A.02.19.01.	**Remove surface finishes; floor**										
A.02.19.01.A	carpet and underlay	m²	0.13	3.85	—	—	3.85	4.23	—	—	4.23
A.02.19.01.B	linoleum sheeting	m²	0.12	3.55	—	—	3.55	3.90	—	—	3.90

House Renovation Grants, Repairs and Alterations

Hutchins Priced Schedules 2021	Unit	Labour Hours	Labour Net £	Plant Net £	Materials Net £	Unit Rate Net £	Labour Gross £	Plant Gross £	Materials Gross £	Unit Rate Gross £
							(Gross rates include 10% profit)			

A.02. REPAIRS AND ALTERATIONS

A.02.19. FINISHES, REPAIRS AND RENEWALS

A.02.19.01. Remove surface finishes; floor

A.02.19.01.C	screed....................	m²	0.50	14.79	—	—	14.79	16.27	—	—	16.27
A.02.19.01.D	granolithic and screed............	m²	0.67	19.82	—	—	19.82	21.80	—	—	21.80
A.02.19.01.E	terrazzo or ceramic tiles; screed......	m²	1.10	32.54	—	—	32.54	35.79	—	—	35.79
A.02.19.02.	**Remove surface finishes; wall**										
A.02.19.02.A	plasterboard.................	m²	0.45	13.31	—	—	13.31	14.64	—	—	14.64
A.02.19.02.B	plaster.....................	m²	0.22	6.51	—	—	6.51	7.16	—	—	7.16
A.02.19.02.C	cement rendering; pebbledashing.....	m²	0.45	13.31	—	—	13.31	14.64	—	—	14.64
A.02.19.02.D	tiling and screed..............	m²	0.55	16.27	—	—	16.27	17.90	—	—	17.90
A.02.19.03.	**Remove surface finishes; ceiling**										
A.02.19.03.A	plasterboard and skim including withdrawing nails..............	m²	0.33	9.76	—	—	9.76	10.74	—	—	10.74
A.02.19.03.B	wood lath and plaster including withdrawing nails..............	m²	0.55	16.27	—	—	16.27	17.90	—	—	17.90
A.02.19.03.C	suspended..................	m²	0.83	24.55	—	—	24.55	27.01	—	—	27.01
A.02.19.03.D	plaster moulded cornice; per 25mm girth .	m	0.17	5.03	—	—	5.03	5.53	—	—	5.53
A.02.19.04.	**Prepare surface to be sound and clean; apply two coats UniBond Super PVA Universal Adhesive and sealer to receive plaster or cement rendering**										
A.02.19.04.A	walls; existing cement and sand base over 300mm wide..............	m²	0.15	10.15	—	0.50	10.65	11.16	—	0.55	11.71
A.02.19.04.B	walls; existing glazed tile base over 300mm wide..................	m²	0.12	8.13	—	0.34	8.47	8.95	—	0.37	9.32
A.02.19.04.C	walls; existing painted base over 300mm wide.....................	m²	0.14	9.48	—	0.38	9.85	10.42	—	0.42	10.84
A.02.19.04.D	walls; existing concrete base over 300mm wide..................	m²	0.15	10.15	—	0.42	10.57	11.16	—	0.46	11.62
A.02.19.04.E	ceilings; existing cement and sand base over 300mm wide..............	m²	0.19	12.50	—	0.50	13.00	13.75	—	0.55	14.30
A.02.19.04.F	ceilings; existing painted base over 300mm wide.................	m²	0.17	11.49	—	0.38	11.87	12.64	—	0.42	13.06
A.02.19.04.G	ceilings; existing concrete base over 300mm wide.................	m²	0.19	12.50	—	0.42	12.92	13.75	—	0.46	14.21
A.02.19.05.	**Hack down defective ceiling plaster and laths; clean out old nails ready for**										
A.02.19.05.A	new laths or plasterboard..........	m²	0.43	28.70	—	—	28.70	31.57	—	—	31.57
A.02.19.06.	**Hack down defective ceiling plaster and clean laths ready for**										
A.02.19.06.A	new plaster.................	m²	0.30	20.23	—	—	20.23	22.25	—	—	22.25

House Renovation Grants, Repairs and Alterations

Hutchins Priced Schedules 2021	Unit	Labour Hours	Labour Net £	Plant Net £	Materials Net £	Unit Rate Net £	Labour Gross £	Plant Gross £	Materials Gross £	Unit Rate Gross £
							(Gross rates include 10% profit)			
A.02. **REPAIRS AND ALTERATIONS**										
A.02.19. **FINISHES, REPAIRS AND RENEWALS**										
A.02.19.07. **Take down temporary boarded linings to ceilings**										
A.02.19.07.A clean joists	m²	0.40	11.83	—	—	11.83	13.02	—	—	13.02
A.02.19.08. **Expanded metal lathing to ceiling joists and render**										
A.02.19.08.A float and set	m²	0.85	57.40	—	11.14	68.54	63.14	—	12.26	75.39
A.02.19.09. **Plaster and set on**										
A.02.19.09.A existing laths	m²	0.63	42.21	—	6.34	48.55	46.43	—	6.98	53.41
A.02.19.10. **Plaster baseboard; scrim and set ceilings with**										
A.02.19.10.A patent plaster	m²	0.50	33.74	—	3.36	37.10	37.11	—	3.70	40.81
A.02.19.12. **Hack down defective ceiling plaster and fix plaster baseboard; scrim and set with patent plaster including jointing to existing**										
A.02.19.12.A area not exceeding 1.00m2	m²	1.00	67.48	—	3.36	70.84	74.23	—	3.70	77.93
A.02.19.12.B area 1.00m2 - 4.00m2	m²	0.70	47.25	—	3.36	50.61	51.97	—	3.70	55.67
A.02.19.13. **Make good at intersection of wall and ceiling plaster after replastering**										
A.02.19.13.A wall or ceiling	m	0.28	18.55	—	0.22	18.77	20.40	—	0.24	20.64
A.02.19.13.B make good cracks in ceiling plaster	m	0.23	15.19	—	0.22	15.41	16.71	—	0.24	16.95
A.02.19.14. **Hack brick, stone or concrete walls to form key**										
A.02.19.14.A for plaster	m²	0.75	22.19	—	—	22.19	24.40	—	—	24.40
A.02.19.16. **Hack off wall plaster and rake out brick joints to form key**										
A.02.19.16.A for new plaster	m²	0.75	22.19	—	—	22.19	24.40	—	—	24.40
A.02.19.18. **Rake out joints of brickwork**										
A.02.19.18.A to form key	m²	0.40	11.83	—	—	11.83	13.02	—	—	13.02
A.02.19.19. **Dub out uneven walls to receive**										
A.02.19.19.A new plaster	m²	0.20	13.51	—	1.77	15.28	14.86	—	1.95	16.81
A.02.19.20. **Render and set**										
A.02.19.20.A brick or block walls	m²	0.43	28.70	—	5.21	33.91	31.57	—	5.73	37.30

Hutchins Priced Schedules 2021		Unit	Labour Hours	Labour Net £	Plant Net £	Materials Net £	Unit Rate Net £	Labour Gross £	Plant Gross £	Materials Gross £	Unit Rate Gross £
								(Gross rates include 10% profit)			
A.02.	**REPAIRS AND ALTERATIONS**										
A.02.19.	**FINISHES, REPAIRS AND RENEWALS**										
A.02.19.21.	**Hack down defective wall plaster in small quantities; plaster and set including jointing to existing**										
A.02.19.21.A	area not exceeding 1.00m2	m²	0.88	59.08	—	5.65	64.73	64.99	—	6.22	71.21
A.02.19.21.B	area 1.00m2 - 4.00m2	m²	0.45	30.38	—	5.65	36.03	33.42	—	6.22	39.64
A.02.19.24.	**Render and set brick walls**										
A.02.19.24.A	including jointing new to old and a small amount of dubbing out	m²	0.55	37.10	—	7.19	44.29	40.81	—	7.91	48.72
A.02.19.26.	**Make good cracks**										
A.02.19.26.A	in wall plaster	m	0.23	15.19	—	0.22	15.41	16.71	—	0.24	16.95
A.02.19.26.B	in moulded cornice; per 25mm girth of cornice	m	0.13	8.47	—	0.44	8.90	9.32	—	0.48	9.79
A.02.19.26.C	around door and window frames and repoint	m	0.20	13.51	—	0.22	13.73	14.86	—	0.24	15.10
A.02.19.28.	**Hack down and re-run plaster cornices per 25mm girth of cornice**										
A.02.19.28.A	coved	m	0.13	8.47	—	0.44	8.90	9.32	—	0.48	9.79
A.02.19.28.B	moulded	m	0.17	11.16	—	0.22	11.37	12.27	—	0.24	12.51
A.02.19.30.	**Make good plaster around pipes**										
A.02.19.30.A	small pipes	nr	0.15	10.15	—	0.44	10.58	11.16	—	0.48	11.64
A.02.19.30.B	large pipes	nr	0.17	11.16	—	0.44	11.59	12.27	—	0.48	12.75
A.02.19.32.	**White glazed wall tiles fixed with adhesive; pointed in white cement grout to wall in small quantities in repairs**										
A.02.19.32.B	150mm x 150mm x 5mm	m²	1.20	80.32	—	6.87	87.19	88.35	—	7.56	95.91
A.02.19.34.	**White glazed wall tiles bedded in cement mortar; pointed in white cement in small quantities in repairs**										
A.02.19.34.B	150mm x 150mm x 5mm	m²	2.51	168.76	—	9.34	178.10	185.64	—	10.27	195.91
A.02.19.38.	**Hack off glazed tiles to wall**										
A.02.19.38.A	in detached areas 2.00m2 - 5.00m2	m²	1.00	29.58	—	—	29.58	32.54	—	—	32.54
A.02.19.38.B	in patches 0.50m2 - 2.00m2	m²	2.50	73.95	—	—	73.95	81.35	—	—	81.35
A.02.19.38.C	in single tiles in patches up to 0.50m2 . .	nr	0.30	8.87	—	—	8.87	9.76	—	—	9.76
A.02.19.40.	**Take out broken cappings horizontal or vertical and renew**										
A.02.19.40.A	12mm x 150mm x 75mm angle	nr	0.09	6.05	—	7.61	13.66	6.65	—	8.37	15.03
A.02.19.41.	**Take out broken moulded cappings and renew**										
A.02.19.41.A	12mm x 150mm x 75mm	nr	0.10	6.72	—	1.35	8.07	7.39	—	1.49	8.88

House Renovation Grants, Repairs and Alterations

Hutchins Priced Schedules 2021	Unit	Labour Hours	Labour Net £	Plant Net £	Materials Net £	Unit Rate Net £	Labour Gross £	Plant Gross £	Materials Gross £	Unit Rate Gross £
						(Gross rates include 10% profit)				
A.02.	**REPAIRS AND ALTERATIONS**									
A.02.19.	**FINISHES, REPAIRS AND RENEWALS**									
A.02.19.42.	**Holes through wall tiling, any colour, for pipes, brackets, etc.; make good**									
A.02.19.42.A	small pipe	nr	0.28	8.28	—	—	8.28	9.11	—	—	9.11
A.02.19.42.B	large pipe	nr	0.38	11.24	—	—	11.24	12.36	—	—	12.36
A.02.19.44.	**Strip loose tiles and clean**									
A.02.19.44.A	for reuse	m²	0.40	11.83	—	—	11.83	13.02	—	—	13.02
A.02.19.45.	**Refix only salvaged tiles with**									
A.02.19.45.A	adhesive	m²	1.13	75.95	—	1.25	77.19	83.54	—	1.37	84.91
A.02.19.48.	**Provide and fix wall tiles with adhesive**									
A.02.19.48.A	150mm x 150mm x 5mm	m²	1.13	75.95	—	7.10	83.05	83.54	—	7.81	91.35
A.02.19.49.	**Hack off defective rendering to concreted areas; grout and render in cement mortar**									
A.02.19.49.A	19mm thick	m²	0.50	33.74	—	0.22	33.96	37.11	—	0.24	37.36
A.02.19.50.	**Hack off and renew cement rendered plinth including joints new to old**									
A.02.19.50.A	225mm high	m	0.15	10.15	—	0.78	10.93	11.16	—	0.85	12.02
A.02.19.52.	**Cut out cracks in rendering; make good to existing**									
A.02.19.52.A	Snowcrete or rough cast	m	0.40	27.02	—	0.82	27.84	29.72	—	0.90	30.62
A.02.19.52.B	make good to Snowcrete or rough cast to match adjacent work around reset window and door frames	m	0.24	16.20	—	0.56	16.76	17.82	—	0.61	18.43
A.02.19.54.	**Hack off defective rendering to walls; prepare for and cement render**									
A.02.19.54.A	two coats (plain face)	m²	0.50	33.74	—	3.00	36.73	37.11	—	3.29	40.41
A.02.19.56.	**Cement wash walls**									
A.02.19.56.A	one coat	m²	0.15	10.15	—	0.29	10.44	11.16	—	0.32	11.48
A.02.19.56.B	two coats	m²	0.27	18.21	—	0.38	18.59	20.04	—	0.42	20.45
A.02.19.58.	**Hack off broken cement rendering and renew rendering to**									
A.02.19.58.A	three-sided kerb to gully	nr	0.45	30.38	—	0.55	30.93	33.42	—	0.61	34.03

Hutchins Priced Schedules 2021	Unit	Labour Hours	Labour Net £	Plant Net £	Materials Net £	Unit Rate Net £	Labour Gross £	Plant Gross £	Materials Gross £	Unit Rate Gross £
								(Gross rates include 10% profit)		

A.02.   **REPAIRS AND ALTERATIONS**

A.02.19.   **FINISHES, REPAIRS AND RENEWALS**

A.02.19.60.   Hack off all loose stucco rendering; hack back brick or concrete to form key and render with cement mortar; trowel smooth including reproducing all profiles and ruled joints

	Unit	Labour Hours	Labour Net £	Plant Net £	Materials Net £	Unit Rate Net £	Labour Gross £	Plant Gross £	Materials Gross £	Unit Rate Gross £
A.02.19.60.A to match existing	m²	1.00	67.48	—	2.66	70.14	74.23	—	2.93	77.16
A.02.19.60.B in patches not exceeding 1.00m2	m²	1.26	84.35	—	2.66	87.01	92.78	—	2.93	95.71

A.02.29.   **PLUMBING REPAIRS AND RENEWALS**

A.02.29.01.   Remove gutterwork and pipework; gutterwork and supports

	Unit	Labour Hours	Labour Net £	Plant Net £	Materials Net £	Unit Rate Net £	Labour Gross £	Plant Gross £	Materials Gross £	Unit Rate Gross £
A.02.29.01.A cement	m	0.21	20.18	—	—	20.18	22.19	—	—	22.19
A.02.29.01.B PVCu	m	0.23	21.89	—	—	21.89	24.08	—	—	24.08
A.02.29.01.C cast iron	m	0.27	25.89	—	—	25.89	28.47	—	—	28.47

A.02.29.02.   Remove gutterwork and pipework; rainwater pipework and supports

	Unit	Labour Hours	Labour Net £	Plant Net £	Materials Net £	Unit Rate Net £	Labour Gross £	Plant Gross £	Materials Gross £	Unit Rate Gross £
A.02.29.02.A cement	m	0.18	17.23	—	—	17.23	18.95	—	—	18.95
A.02.29.02.B PVCu	m	0.20	19.03	—	—	19.03	20.94	—	—	20.94
A.02.29.02.C cast iron	m	0.24	23.03	—	—	23.03	25.33	—	—	25.33

A.02.29.03.   Remove gutterwork and pipework; rainwater shoe

	Unit	Labour Hours	Labour Net £	Plant Net £	Materials Net £	Unit Rate Net £	Labour Gross £	Plant Gross £	Materials Gross £	Unit Rate Gross £
A.02.29.03.A PVCu	nr	0.04	4.00	—	—	4.00	4.40	—	—	4.40
A.02.29.03.B cast iron	nr	0.06	5.81	—	—	5.81	6.39	—	—	6.39

A.02.29.04.   Remove gutterwork and pipework; rainwater head and support

	Unit	Labour Hours	Labour Net £	Plant Net £	Materials Net £	Unit Rate Net £	Labour Gross £	Plant Gross £	Materials Gross £	Unit Rate Gross £
A.02.29.04.A PVCu	nr	0.22	21.32	—	—	21.32	23.45	—	—	23.45
A.02.29.04.B cast iron	nr	0.27	25.89	—	—	25.89	28.47	—	—	28.47

A.02.29.05.   Remove gutterwork and pipework; soil and ventilation pipework and supports

	Unit	Labour Hours	Labour Net £	Plant Net £	Materials Net £	Unit Rate Net £	Labour Gross £	Plant Gross £	Materials Gross £	Unit Rate Gross £
A.02.29.05.A PVCu	m	0.36	34.55	—	—	34.55	38.00	—	—	38.00
A.02.29.05.B cast iron	m	0.41	38.54	—	—	38.54	42.40	—	—	42.40
A.02.29.05.C lead	m	0.45	43.21	—	—	43.21	47.53	—	—	47.53

A.02.29.06.   Remove gutterwork and pipework; service, waste and overflow pipework and supports

	Unit	Labour Hours	Labour Net £	Plant Net £	Materials Net £	Unit Rate Net £	Labour Gross £	Plant Gross £	Materials Gross £	Unit Rate Gross £
A.02.29.06.A PVCu	m	0.08	7.52	—	—	7.52	8.27	—	—	8.27
A.02.29.06.B copper	m	0.10	9.80	—	—	9.80	10.78	—	—	10.78
A.02.29.06.C lead	m	0.10	9.80	—	—	9.80	10.78	—	—	10.78
A.02.29.06.D galvanised steel	m	0.10	9.80	—	—	9.80	10.78	—	—	10.78

A.02.29.08.   Remove sanitary fittings including taps and trap

	Unit	Labour Hours	Labour Net £	Plant Net £	Materials Net £	Unit Rate Net £	Labour Gross £	Plant Gross £	Materials Gross £	Unit Rate Gross £
A.02.29.08.A WC suite	nr	0.25	24.17	—	—	24.17	26.59	—	—	26.59
A.02.29.08.B wash hand basin	nr	0.22	21.32	—	—	21.32	23.45	—	—	23.45

House Renovation Grants, Repairs and Alterations

Hutchins Priced Schedules 2021		Unit	Labour Hours	Labour Net £	Plant Net £	Materials Net £	Unit Rate Net £	Labour Gross £	Plant Gross £	Materials Gross £	Unit Rate Gross £
									(Gross rates include 10% profit)		
A.02.	**REPAIRS AND ALTERATIONS**										
A.02.29.	PLUMBING REPAIRS AND RENEWALS										
A.02.29.08.	Remove sanitary fittings including taps and trap										
A.02.29.08.C	bath	nr	0.32	29.98	—	—	29.98	32.98	—	—	32.98
A.02.29.08.D	sink unit	nr	0.22	21.32	—	—	21.32	23.45	—	—	23.45
A.02.29.08.E	shower	nr	0.09	8.66	—	—	8.66	9.53	—	—	9.53
A.02.29.10.	**Remove sanitary fittings including taps, trap and service and waste pipes; not exceeding 3.00m girth**										
A.02.29.10.A	WC suite	nr	0.32	29.98	—	—	29.98	32.98	—	—	32.98
A.02.29.10.B	wash hand basin	nr	0.25	24.17	—	—	24.17	26.59	—	—	26.59
A.02.29.10.C	bath	nr	0.38	35.69	—	—	35.69	39.26	—	—	39.26
A.02.29.10.D	sink unit	nr	0.25	24.17	—	—	24.17	26.59	—	—	26.59
A.02.29.10.E	shower	nr	0.13	12.66	—	—	12.66	13.92	—	—	13.92
A.02.29.11.	**Remove bathroom toilet fittings**										
A.02.29.11.A	toilet roll holder	nr	0.03	2.86	—	—	2.86	3.14	—	—	3.14
A.02.29.11.B	soap dispenser	nr	0.03	2.86	—	—	2.86	3.14	—	—	3.14
A.02.29.11.C	towel rail	nr	0.05	4.57	—	—	4.57	5.02	—	—	5.02
A.02.29.11.D	towel holder	nr	0.10	9.80	—	—	9.80	10.78	—	—	10.78
A.02.29.11.E	mirror	nr	0.10	9.80	—	—	9.80	10.78	—	—	10.78
A.02.29.12.	**Remove equipment; excluding any draining down of system**										
A.02.29.12.A	cold water tank	nr	1.60	152.56	—	—	152.56	167.81	—	—	167.81
A.02.29.12.B	hot water cylinder	nr	0.80	76.52	—	—	76.52	84.17	—	—	84.17
A.02.29.12.C	gas water heater	nr	2.66	253.25	—	—	253.25	278.57	—	—	278.57
A.02.29.12.D	gas fire	nr	1.33	126.58	—	—	126.58	139.23	—	—	139.23
A.02.29.12.E	expansion tank	nr	1.21	115.06	—	—	115.06	126.57	—	—	126.57
A.02.29.14.	**Clean out eaves and parapet gutters; remove rubbish**										
A.02.29.14.A	any height or position	m	0.08	7.52	—	—	7.52	8.27	—	—	8.27
A.02.29.16.	**Clean out rainwater pipes, stack pipes, etc.**										
A.02.29.16.A	any height or position	m	0.08	7.52	—	—	7.52	8.27	—	—	8.27
A.02.29.17.	**Take down existing gutters, clean and refix to fascia or on brackets, seal joints with putty**										
A.02.29.17.A	set to proper falls	m	0.42	40.26	—	4.23	44.48	44.28	—	4.65	48.93
A.02.29.18.	**Take down and remove existing 100mm cast iron gutters; provide and fix new gutters**										
A.02.29.18.A	half round	m	0.67	63.29	—	27.30	90.58	69.62	—	30.02	99.64

House Renovation Grants, Repairs and Alterations

Hutchins Priced Schedules 2021	Unit	Labour Hours	Labour Net £	Plant Net £	Materials Net £	Unit Rate Net £	Labour Gross £	Plant Gross £	Materials Gross £	Unit Rate Gross £
							(Gross rates include 10% profit)			

A.02.   **REPAIRS AND ALTERATIONS**

A.02.29.   **PLUMBING REPAIRS AND RENEWALS**

A.02.29.18.   Take down and remove existing 100mm cast iron gutters; provide and fix new gutters

A.02.29.18.B ogee	m	0.67	63.29	—	29.97	93.26	69.62	—	32.97	102.58

A.02.29.19.   Take down existing rainwater pipes and refix to walls

A.02.29.19.A 50mm, 65mm, or 75mm	m	0.45	43.21	—	0.47	43.68	47.53	—	0.52	48.04

A.02.29.20.   Take down and remove existing cast iron rainwater pipes and provide and fix new pipes

A.02.29.20.A 65mm	m	0.67	63.29	—	52.46	115.75	69.62	—	57.70	127.32
A.02.29.20.B rainwater shoe (not extra over)	nr	0.30	28.74	—	27.57	56.31	31.62	—	30.33	61.94

A.02.29.22.   Cut out and reform caulked lead joints in cast iron soil, vent or waste pipes

A.02.29.22.A 50mm	nr	0.42	40.26	—	4.71	44.97	44.28	—	5.18	49.46
A.02.29.22.B 75mm	nr	0.48	46.06	—	7.04	53.10	50.67	—	7.74	58.41
A.02.29.22.C 100mm	nr	0.70	66.14	—	8.64	74.78	72.76	—	9.50	82.26

A.02.29.24.   Cut and adapt existing cast iron soil pipe for a new WC by inserting a branch and bend or long junction; connect to pan trap; make good to wall

A.02.29.24.A 100mm	nr	2.90	276.28	—	98.58	374.86	303.91	—	108.44	412.34
A.02.29.24.B extra for access door	nr	—	—	—	42.40	42.40	—	—	46.64	46.64

A.02.29.42.   Renew broken stopcock box with hinged lid

A.02.29.42.A 127mm x 127mm x 76mm	nr	0.30	28.74	—	19.08	47.82	31.62	—	20.99	52.60
A.02.29.42.B 152mm x 152mm x 76mm	nr	0.33	31.69	—	26.50	58.19	34.86	—	29.15	64.01

A.02.29.44.   Galvanised water tube including bends, sockets, etc. (tees excepted); in repairs to existing work; heavy weight

A.02.29.44.A 15mm	m	0.40	37.97	—	5.16	43.14	41.77	—	5.68	47.45
A.02.29.44.B 20mm	m	0.45	43.21	—	6.68	49.88	47.53	—	7.34	54.87
A.02.29.44.C 25mm	m	0.48	46.06	—	7.94	54.00	50.67	—	8.74	59.41

A.02.29.46.   Galvanised water tube including bends, sockets, etc. (tees excepted); in repairs to existing work; medium weight

A.02.29.46.A 15mm	m	0.40	37.97	—	4.91	42.89	41.77	—	5.40	47.17
A.02.29.46.B 20mm	m	0.45	43.21	—	5.12	48.33	47.53	—	5.63	53.16
A.02.29.46.C 25mm	m	0.48	46.06	—	5.88	51.95	50.67	—	6.47	57.14

House Renovation Grants, Repairs and Alterations

Hutchins Priced Schedules 2021		Unit	Labour Hours	Labour Net £	Plant Net £	Materials Net £	Unit Rate Net £	Labour Gross £	Plant Gross £	Materials Gross £	Unit Rate Gross £
							(Gross rates include 10% profit)				
A.02.	**REPAIRS AND ALTERATIONS**										
A.02.29.	**PLUMBING REPAIRS AND RENEWALS**										
A.02.29.48.	**Cut existing iron pipe for and insert new tees**										
A.02.29.48.A	15mm	nr	0.42	40.26	—	1.92	42.18	44.28	—	2.11	46.39
A.02.29.48.B	20mm	nr	0.51	48.92	—	2.12	51.04	53.81	—	2.33	56.14
A.02.29.48.C	25mm	nr	0.61	57.58	—	2.88	60.46	63.34	—	3.17	66.51
A.02.29.50.	**Copper tubing including capillary bends, etc. (tees excepted); in repairs to existing work**										
A.02.29.50.A	15mm	m	0.36	34.55	—	3.79	38.34	38.00	—	4.17	42.17
A.02.29.50.B	22mm	m	0.42	40.26	—	6.17	46.42	44.28	—	6.78	51.07
A.02.29.50.C	28mm	m	0.48	46.06	—	8.14	54.20	50.67	—	8.95	59.62
A.02.29.52.	**Cut existing copper pipes and insert new capillary tees**										
A.02.29.52.A	15mm	nr	0.61	57.58	—	0.70	58.28	63.34	—	0.77	64.11
A.02.29.52.B	22mm	nr	0.73	69.09	—	1.99	71.09	76.00	—	2.19	78.19
A.02.29.52.C	28mm	nr	0.91	86.32	—	9.97	96.29	94.95	—	10.97	105.92
A.02.29.54.	**Polyethylene tubing including compression bends, etc. (tees excepted); in repairs to existing work**										
A.02.29.54.A	20mm	m	0.73	69.09	—	1.75	70.84	76.00	—	1.92	77.92
A.02.29.54.B	25mm	m	0.85	80.61	—	1.86	82.46	88.67	—	2.04	90.71
A.02.29.54.C	32mm	m	1.03	97.83	—	2.36	100.20	107.62	—	2.60	110.22
A.02.29.56.	**Cut existing polyethylene pipes and insert new compression tees**										
A.02.29.56.A	20mm	nr	0.91	86.32	—	4.12	90.44	94.95	—	4.54	99.49
A.02.29.56.B	25mm	nr	1.03	97.83	—	6.37	104.21	107.62	—	7.01	114.63
A.02.29.56.C	32mm	nr	1.21	115.06	—	7.88	122.94	126.57	—	8.66	135.23
A.02.29.58.	**Cover iron or copper pipes with hair felt and twine; in any position**										
A.02.29.58.A	up to 22mm dia	m	0.42	40.26	—	0.11	40.36	44.28	—	0.12	44.40
A.02.29.60.	**Take off existing bib valve; prepare iron or copper pipes and provide and fit new bib valve**										
A.02.29.60.A	13mm	nr	0.61	57.58	—	5.64	63.22	63.34	—	6.20	69.54
A.02.29.60.B	19mm	nr	0.73	69.09	—	7.05	76.14	76.00	—	7.75	83.76
A.02.29.61.	**Cut into iron or copper pipes and fit new stopcock**										
A.02.29.61.A	13mm	nr	0.79	74.80	—	5.62	80.42	82.28	—	6.18	88.46
A.02.29.61.B	19mm	nr	0.91	86.32	—	11.48	97.80	94.95	—	12.63	107.58

House Renovation Grants, Repairs and Alterations

Hutchins Priced Schedules 2021		Unit	Labour Hours	Labour Net £	Plant Net £	Materials Net £	Unit Rate Net £	Labour Gross £	Plant Gross £	Materials Gross £	Unit Rate Gross £
								(Gross rates include 10% profit)			
A.02.	**REPAIRS AND ALTERATIONS**										
A.02.29.	**PLUMBING REPAIRS AND RENEWALS**										
A.02.29.62.	**Rewasher ball valve, tap or indoor stopcock**										
A.02.29.62.A	13mm	nr	0.30	28.74	—	0.14	28.88	31.62	—	0.15	31.77
A.02.29.62.B	19mm	nr	0.36	34.55	—	0.36	34.90	38.00	—	0.39	38.39
A.02.29.64.	**Take off existing copper trap to bath, basin or sink; provide and fit new plastic trap**										
A.02.29.64.A	32mm	nr	0.73	69.09	—	16.12	85.22	76.00	—	17.73	93.74
A.02.29.64.B	38mm	nr	0.79	74.80	—	19.75	94.55	82.28	—	21.72	104.01
A.02.29.66.	**Take off stopcock to iron or copper pipes both ends; provide and fit new stopcock**										
A.02.29.66.A	13mm	nr	0.39	37.40	—	5.62	43.02	41.14	—	6.18	47.32
A.02.29.66.B	19mm	nr	0.51	48.92	—	11.48	60.40	53.81	—	12.63	66.44
A.02.29.66.C	25mm	nr	0.64	60.43	—	21.06	81.50	66.48	—	23.17	89.64
A.02.29.68.	**Take off ball valve to iron or copper pipe; provide and fit new ball valve and ball float complete**										
A.02.29.68.A	13mm	nr	0.30	28.74	—	6.99	35.73	31.62	—	7.68	39.30
A.02.29.68.B	19mm	nr	0.39	37.40	—	21.77	59.17	41.14	—	23.95	65.09
A.02.29.70.	**Supply WC suite complete (PC sum £240 per nr) and connect to existing services**										
A.02.29.70.A	copper or iron	nr	3.51	333.76	—	252.00	585.76	367.14	—	277.20	644.34
A.02.29.71.	**Supply low level WC suite (PC sum £140 per nr) and connect to existing services**										
A.02.29.71.A	copper or iron	nr	3.81	362.60	—	147.00	509.60	398.86	—	161.70	560.56
A.02.29.72.	**Remove defective WC pan and fix new (PC sum £60 per nr)**										
A.02.29.72.A	make good all connections	nr	2.36	224.41	—	63.00	287.41	246.85	—	69.30	316.15
A.02.29.74.	**Take off defective seat to pedestal pan; supply and fix new seat**										
A.02.29.74.A	plastic seat	nr	0.45	43.21	—	17.49	60.70	47.53	—	19.24	66.77
A.02.29.76.	**Take off WC seat; renew joints to flush pipe including closet and outlet connection and pan**										
A.02.29.76.A	refix seat	nr	0.70	66.14	—	1.96	68.11	72.76	—	2.16	74.92

House Renovation Grants, Repairs and Alterations

Hutchins Priced Schedules 2021	Unit	Labour Hours	Labour Net £	Plant Net £	Materials Net £	Unit Rate Net £	Labour Gross £	Plant Gross £	Materials Gross £	Unit Rate Gross £
							(Gross rates include 10% profit)			

A.02.   **REPAIRS AND ALTERATIONS**

A.02.29.   **PLUMBING REPAIRS AND RENEWALS**

Hutchins Priced Schedules 2021	Unit	Labour Hours	Labour Net £	Plant Net £	Materials Net £	Unit Rate Net £	Labour Gross £	Plant Gross £	Materials Gross £	Unit Rate Gross £
A.02.29.80. Disconnect ball valve to WC tank										
A.02.29.80.A re-washer and clean out	nr	0.88	83.46	—	0.36	83.82	91.81	—	0.39	92.20
A.02.29.81. Unscrew ball to valve; supply and fit										
A.02.29.81.A new ball	nr	0.18	17.23	—	2.23	19.45	18.95	—	2.45	21.40
A.02.29.82. Supply and fix flat back basin including taps, traps and wall brackets (PC sum £100 per nr), connect to existing services										
A.02.29.82.A iron or copper	nr	3.51	333.76	—	105.00	438.76	367.14	—	115.50	482.64
A.02.29.84. Supply and fix pedestal basin including taps, traps and wall brackets (PC sum £150 per nr); connect to existing services										
A.02.29.84.A iron or copper	nr	3.51	333.76	—	157.50	491.26	367.14	—	173.25	540.39
A.02.29.86. Supply and fix 1700mm long, white reinforced acrylic rectangular bath (PC sum £180 per nr); with waste and overflow; complete with 3/4in bath shower mixer tap with shower head and hose (PC sum £75 per nr) and trap; connect to existing services										
A.02.29.86.A iron or copper	nr	3.51	333.76	—	320.93	654.69	367.14	—	353.02	720.16
A.02.29.88. Clear blockage; flush out to										
A.02.29.88.A WC pans and traps	nr	0.76	71.95	—	—	71.95	79.14	—	—	79.14
A.02.29.88.B traps and waste pipes of baths, sinks, lavatory basins, etc.	nr	0.61	57.58	—	—	57.58	63.34	—	—	63.34
A.02.29.90. Disconnect all pipework; take out and remove galvanised steel hot water tank; provide and install copper indirect cylinder; allow for cutting holes, tank connectors, make up and all connections to existing pipework; capacity										
A.02.29.90.A 114l	nr	1.81	172.64	—	217.72	390.36	189.90	—	239.49	429.39

House Renovation Grants, Repairs and Alterations

Hutchins Priced Schedules 2021	Unit	Labour Hours	Labour Net £	Plant Net £	Materials Net £	Unit Rate Net £	Labour Gross £	Plant Gross £	Materials Gross £	Unit Rate Gross £
							(Gross rates include 10% profit)			

A.02.   **REPAIRS AND ALTERATIONS**

A.02.29.   **PLUMBING REPAIRS AND RENEWALS**

A.02.29.96.   **Turn off water supply; disconnect all pipework; take out and remove galvanised steel cold water storage tank; provide and install plastic tank complete with ball valve, lid and insulation and allow for cutting holes, tank connectors, make up and all connections to existing pipework; capacity**

	Unit	Labour Hours	Labour Net £	Plant Net £	Materials Net £	Unit Rate Net £	Labour Gross £	Plant Gross £	Materials Gross £	Unit Rate Gross £
A.02.29.96.A 182l (PC sum £140 per nr)	nr	3.81	362.60	—	199.93	562.52	398.86	—	219.92	618.78
A.02.29.98. **Clean and scour out open-top storage tanks**										
A.02.29.98.A any size and position	nr	4.48	425.89	—	—	425.89	468.47	—	—	468.47
A.02.39. **GLAZING REPAIRS AND RENEWALS**										
A.02.39.01. **Hack out all types of broken glass**										
A.02.39.01.A except plate glass	m²	2.34	88.05	—	—	88.05	96.86	—	—	96.86
A.02.39.01.B plate glass	m²	3.55	133.59	—	—	133.59	146.95	—	—	146.95
A.02.39.02. **Carefully take out all types of glass and set aside for re-use**										
A.02.39.02.A except plate glass	m²	3.23	121.54	—	—	121.54	133.70	—	—	133.70
A.02.39.02.B plate glass	m²	4.68	176.11	—	—	176.11	193.72	—	—	193.72
A.02.39.04. **Remove old putty; paint rebate one coat oil colour ready to receive**										
A.02.39.04.A new glass	m	0.16	6.02	—	0.11	6.13	6.62	—	0.12	6.75
A.02.39.05. **Glaze, sprig and putty to wood sashes; average 0.40m2**										
A.02.39.05.A 4mm clear	m²	1.10	41.39	—	31.04	72.44	45.53	—	34.15	79.68
A.02.39.05.B 4mm obscured	m²	1.10	41.39	—	47.72	89.11	45.53	—	52.49	98.02
A.02.39.05.C 7mm wired cast	m²	1.20	45.16	—	84.72	129.88	49.67	—	93.19	142.86
A.02.39.05.D add for glazing to metal sashes	m²	0.40	15.05	—	0.66	15.71	16.56	—	0.73	17.28
A.02.39.08. **Wired cast glass in roof lights in panes**										
A.02.39.08.A up to 0.70m2	m²	0.90	33.87	—	84.72	118.59	37.25	—	93.19	130.45
A.02.39.08.B exceeding 0.70m2	m²	1.00	37.63	—	84.72	122.35	41.39	—	93.19	134.59
A.02.39.10. **Float glass**										
A.02.39.10.A 6mm	m²	3.50	131.71	—	38.16	169.87	144.88	—	41.98	186.86
A.02.39.12. **Remove temporary coverings to sashes**										
A.02.39.12.A stopping nail holes, etc.	m²	1.10	41.39	—	—	41.39	45.53	—	—	45.53

House Renovation Grants, Repairs and Alterations

Hutchins Priced Schedules 2021		Unit	Labour Hours	Labour Net £	Plant Net £	Materials Net £	Unit Rate Net £	Labour Gross £	Plant Gross £	Materials Gross £	Unit Rate Gross £
								(Gross rates include 10% profit)			
A.02.	**REPAIRS AND ALTERATIONS**										
A.02.41.	**PAINTING AND DECORATING**										
A.02.41.01.	**Brush down brick walls; plaster walls, or ceilings and apply two coats lime white on brick walls**										
A.02.41.01.A	new work...................	m²	0.33	12.42	—	0.17	12.59	13.66	—	0.19	13.85
A.02.41.01.B	old work...................	m²	0.38	14.30	—	0.17	14.47	15.73	—	0.19	15.92
A.02.41.02.	**Brush down brick walls; plaster walls, or ceilings and apply two coats emulsion paint on plaster**										
A.02.41.02.A	new work...................	m²	0.46	17.31	—	0.51	17.82	19.04	—	0.56	19.61
A.02.41.02.B	old work...................	m²	0.51	19.19	—	0.51	19.70	21.11	—	0.56	21.67
A.02.41.03.	**Brush down brick walls; plaster walls, or ceiling and apply two coats emulsion paint on brick walls**										
A.02.41.03.A	new work...................	m²	0.58	21.83	—	0.77	22.59	24.01	—	0.85	24.85
A.02.41.03.B	old work...................	m²	0.63	23.71	—	0.77	24.48	26.08	—	0.85	26.92
A.02.41.04.	**Wash down plaster surfaces; fill in minor cracks, nail holes, etc. with filler; bring forward paint as necessary ready for**										
A.02.41.04.A	new decoration..............	m²	0.20	7.53	—	0.49	8.02	8.28	—	0.54	8.82
A.02.41.06.	**Prepare and apply oil colour on ceilings**										
A.02.41.06.A	one coat...................	m²	0.27	10.16	—	0.54	10.70	11.18	—	0.60	11.78
A.02.41.06.B	two coats..................	m²	0.55	20.70	—	1.13	21.82	22.77	—	1.24	24.01
A.02.41.08.	**Prepare and apply oil colour on plaster walls**										
A.02.41.08.A	one coat...................	m²	0.21	7.90	—	0.54	8.45	8.69	—	0.60	9.29
A.02.41.08.B	two coats..................	m²	0.43	16.18	—	1.13	17.31	17.80	—	1.24	19.04
A.02.41.10.	**Prepare and apply oil colour on brick walls**										
A.02.41.10.A	one coat...................	m²	0.27	10.16	—	0.86	11.02	11.18	—	0.94	12.12
A.02.41.10.B	two coats..................	m²	0.54	20.32	—	1.75	22.07	22.35	—	1.93	24.28
A.02.41.11.	**Wash down, touch up and apply two coats of oil colour to**										
A.02.41.11.A	general wood surfaces...........	m²	0.65	24.46	—	0.87	25.32	26.91	—	0.95	27.86
A.02.41.11.B	add for each extra coat applied or deduct for one coat.................	m²	0.24	9.03	—	0.48	9.51	9.93	—	0.53	10.47

Hutchins Priced Schedules 2021	Unit	Labour Hours	Labour Net £	Plant Net £	Materials Net £	Unit Rate Net £	Labour Gross £	Plant Gross £	Materials Gross £	Unit Rate Gross £
							(Gross rates include 10% profit)			
A.02.	**REPAIRS AND ALTERATIONS**									
A.02.41.	**PAINTING AND DECORATING**									
A.02.41.11.	**Wash down, touch up and apply two coats of oil colour to**									
A.02.41.11.C	window frames and sashes (measured over glass)	m²	1.00	37.63	—	0.61	38.24	41.39	—	0.67	42.06
A.02.41.11.D	add for each extra coat applied or deduct for one coat on window frames and sashes	m²	0.40	15.05	—	0.34	15.39	16.56	—	0.37	16.93
A.02.41.14.	**Wash down, touch up and apply two coats of oil colour on surfaces**									
A.02.41.14.A	not exceeding 150mm girth	m	0.14	5.27	—	0.10	5.37	5.80	—	0.11	5.91
A.02.41.14.B	150mm - 300mm girth	m	0.24	9.03	—	0.25	9.29	9.93	—	0.28	10.21
A.02.41.14.C	add for each extra cost applied or deduct for one coat not exceeding 150mm girth	m	0.07	2.63	—	0.06	2.69	2.90	—	0.06	2.96
A.02.41.14.D	add for each extra cost applied or deduct for one coat 150mm - 300mm girth	m	0.11	4.14	—	0.14	4.28	4.55	—	0.16	4.71
A.02.41.16.	**Clean down and bring forward all bare patches and apply oil colour to metal**									
A.02.41.16.A	frames and sashes (measured over glass)	m²	1.11	41.77	—	0.87	42.63	45.95	—	0.95	46.90
A.02.41.16.B	general surfaces; over 300mm girth	m²	0.69	25.96	—	0.61	26.58	28.56	—	0.67	29.23
A.02.41.16.C	not exceeding 150mm girth	m	0.30	11.29	—	0.10	11.39	12.42	—	0.11	12.53
A.02.41.16.D	150mm - 300mm girth	m	0.39	14.68	—	0.25	14.93	16.14	—	0.28	16.42
A.02.41.18.	**Clean down and apply gloss oil to fireplace jambs, stoves, mantel registers and similar**									
A.02.41.18.A	one coat	nr	1.00	37.63	—	1.44	39.07	41.39	—	1.58	42.98
A.02.41.22.	**Clean and apply gloss oil to casement stays, fasteners, bolts, rim locks and sundry similar fittings**									
A.02.41.22.A	one coat	nr	0.20	7.53	—	0.05	7.57	8.28	—	0.05	8.33
A.02.41.24.	**Prepare polished wood surfaces and repolish**									
A.02.41.24.A	existing	m²	2.10	79.02	—	0.29	79.31	86.93	—	0.32	87.25
A.02.41.24.B	handrails	m	0.60	22.58	—	0.10	22.67	24.84	—	0.11	24.94
A.02.41.26.	**Strip, body in and repolish**									
A.02.41.26.A	wood surfaces	m²	4.70	176.86	—	0.49	177.35	194.55	—	0.53	195.08
A.02.41.26.B	handrail	m	1.10	41.39	—	0.12	41.51	45.53	—	0.13	45.67
A.02.41.28.	**Prepare and wax polish**									
A.02.41.28.A	flooring	m²	0.50	18.82	—	0.29	19.11	20.70	—	0.32	21.02

House Renovation Grants, Repairs and Alterations

Hutchins Priced Schedules 2021	Unit	Labour Hours	Labour Net £	Plant Net £	Materials Net £	Unit Rate Net £	Labour Gross £	Plant Gross £	Materials Gross £	Unit Rate Gross £
							(Gross rates include 10% profit)			
A.02.	**REPAIRS AND ALTERATIONS**									
A.02.41.	PAINTING AND DECORATING									
A.02.41.32.	**Strip paper from walls or ceilings, stop, size ready for new paper**									
A.02.41.32.A	first layer	m²	0.30	11.29	—	0.21	11.50	12.42	—	0.23	12.65
A.02.41.32.B	each extra layer	m²	0.12	4.52	—	0.21	4.73	4.97	—	0.23	5.20
A.02.41.32.C	first layer varnished paper	m²	0.60	22.58	—	0.21	22.79	24.84	—	0.23	25.07
A.02.41.32.D	each extra layer varnished paper	m²	0.22	8.28	—	0.21	8.49	9.11	—	0.23	9.34
A.02.41.34.	**Cut, trim and hang paper to**									
A.02.41.34.A	walls; woodchip effect	m²	0.26	9.78	—	1.83	11.62	10.76	—	2.01	12.78
A.02.41.34.B	walls; standard (PC sum £16 per roll) . . .	m²	0.26	9.78	—	3.81	13.60	10.76	—	4.19	14.96
A.02.41.34.C	ceilings; lining	m²	0.30	11.29	—	0.46	11.75	12.42	—	0.51	12.93
A.02.41.36.	**Wash down and apply masonry paint on external rendered walls**									
A.02.41.36.A	two coats	m²	0.53	19.94	—	0.36	20.30	21.94	—	0.40	22.33
A.02.41.38.	**Wash down and apply two coats of masonry paint on external rendered walls**									
A.02.41.38.A	Sandtex Matt	m²	0.54	20.32	—	1.24	21.56	22.35	—	1.36	23.71
A.02.41.38.B	Snowcem Matt	m²	0.42	15.80	—	0.82	16.63	17.39	—	0.91	18.29
A.02.41.44.	**Wash down, touch up and apply oil colour externally to general wood surfaces**									
A.02.41.44.A	two coats external	m²	0.68	25.59	—	0.97	26.56	28.15	—	1.06	29.21
A.02.41.44.B	add for each extra coat applied external .	m²	0.27	10.16	—	0.54	10.70	11.18	—	0.59	11.77
A.02.41.46.	**Wash down, touch up and apply oil colour on window frames and sashes (measured over glass)**									
A.02.41.46.A	two coats external	m²	1.21	45.53	—	0.61	46.14	50.09	—	0.67	50.76
A.02.41.46.B	add for each extra coat applied	m²	0.43	16.18	—	0.34	16.52	17.80	—	0.37	18.17
A.02.41.50.	**Burn off paint to woodwork and prepare for priming**									
A.02.41.50.A	general surfaces	m²	0.80	30.10	—	—	30.10	33.11	—	—	33.11
A.02.41.51.	**Strip paint with**									
A.02.41.51.A	paint remover	m²	0.50	18.82	—	0.06	18.88	20.70	—	0.07	20.77
A.02.41.52.	**Prime and apply oil colour to putties after reglazing**									
A.02.41.52.A	two coats	m	0.18	6.77	—	0.11	6.89	7.45	—	0.12	7.57

Hutchins Priced Schedules 2021	Unit	Labour Hours	Labour Net £	Plant Net £	Materials Net £	Unit Rate Net £	Labour Gross £	Plant Gross £	Materials Gross £	Unit Rate Gross £
							(Gross rates include 10% profit)			
A.02.	**REPAIRS AND ALTERATIONS**									
A.02.41.	**PAINTING AND DECORATING**									
A.02.41.62.	**Clean down, wire brush and bring forward all bare patches and apply oil colour on previously painted metalwork**									
A.02.41.62.A	general surfaces; over 300mm girth	m²	0.78	29.35	—	0.97	30.32	32.29	—	1.06	33.35
A.02.41.62.B	general surfaces; not exceeding 150mm girth	m	0.39	14.68	—	0.15	14.83	16.14	—	0.17	16.31
A.02.41.62.C	general surfaces; 150mm - 300mm girth .	m	0.51	19.19	—	0.31	19.50	21.11	—	0.34	21.45
A.02.41.62.D	corrugated surfaces (measured flat)	m²	0.87	32.74	—	1.12	33.86	36.01	—	1.23	37.24
A.02.41.62.E	structural steelwork	m²	0.96	36.12	—	0.97	37.09	39.74	—	1.06	40.80
A.02.41.62.F	railings, balusters, etc. (measured flat overall)	m²	0.87	32.74	—	0.97	33.71	36.01	—	1.06	37.08
A.02.41.62.G	stairs (measured overall)	m²	0.84	31.61	—	0.97	32.58	34.77	—	1.06	35.83
A.02.41.62.H	windows glazed doors in small panes . . .	m²	1.98	74.51	—	0.66	75.17	81.96	—	0.73	82.69
A.02.41.62.I	windows glazed doors in medium panes .	m²	1.74	65.48	—	0.61	66.09	72.02	—	0.67	72.70
A.02.41.62.J	windows glazed doors in large panes . . .	m²	1.50	56.45	—	0.51	56.95	62.09	—	0.56	62.65
A.02.41.62.K	eaves gutters (inside and outside)	m	0.51	19.19	—	0.66	19.85	21.11	—	0.73	21.84
A.02.41.62.L	rainwater pipes, soil pipes, etc.	m	0.51	19.19	—	0.36	19.55	21.11	—	0.39	21.50
A.02.41.62.M	pipes bars, straps, etc.; up to 150mm girth	m	0.45	16.93	—	0.15	17.09	18.63	—	0.17	18.79
A.02.41.66.	**Treating wood surfaces with two coats preservative**									
A.02.41.66.A	sawn surfaces; creosote substitute	m²	0.40	15.05	—	0.42	15.48	16.56	—	0.47	17.02
A.02.41.66.B	wrought surfaces; creosote substitute . . .	m²	0.38	14.30	—	0.28	14.58	15.73	—	0.31	16.04
A.02.41.66.C	sawn surfaces; Cuprinol clear	m²	0.40	15.05	—	1.94	16.99	16.56	—	2.13	18.69
A.02.41.66.D	wrought surfaces; Cuprinol clear	m²	0.38	14.30	—	1.28	15.58	15.73	—	1.41	17.14
A.02.41.66.E	sawn surfaces; fungicidal and insecticidal preservative	m²	0.38	14.30	—	0.69	14.99	15.73	—	0.75	16.48
A.02.41.66.F	wrought surfaces; fungicidal and insecticidal preservative	m²	0.36	13.55	—	0.45	14.00	14.90	—	0.50	15.40
A.02.44.	**MASONRY REPAIRS AND RENEWALS**									
A.02.44.01.	**Take down masonry, clean and set aside**									
A.02.44.01.A	ashlar walling	m²	2.01	105.47	—	—	105.47	116.02	—	—	116.02
A.02.44.01.B	cornices, etc.	m	0.80	42.15	—	—	42.15	46.36	—	—	46.36
A.02.44.01.C	arches	nr	0.54	28.20	—	—	28.20	31.02	—	—	31.02
A.02.44.01.D	steps, sills, etc.	m	1.01	52.73	—	—	52.73	58.01	—	—	58.01
A.02.44.02.	**Take down masonry, clean and reset**									
A.02.44.02.A	ashlar walling	m²	5.20	272.48	—	10.94	283.42	299.73	—	12.03	311.76
A.02.44.02.B	cornices, etc.	m	2.21	116.06	—	2.77	118.83	127.66	—	3.05	130.71

House Renovation Grants, Repairs and Alterations

Hutchins Priced Schedules 2021		Unit	Labour Hours	Labour Net £	Plant Net £	Materials Net £	Unit Rate Net £	Labour Gross £	Plant Gross £	Materials Gross £	Unit Rate Gross £
							(Gross rates include 10% profit)				
A.02.	**REPAIRS AND ALTERATIONS**										
A.02.44.	**MASONRY REPAIRS AND RENEWALS**										
A.02.44.02.	**Take down masonry, clean and reset**										
A.02.44.02.C	arches	nr	1.48	77.37	—	4.89	82.26	85.11	—	5.37	90.48
A.02.44.02.D	steps, sills, etc.	m	1.34	70.35	—	3.96	74.31	77.38	—	4.36	81.74
A.02.44.03.	**Cut out decayed Portland (or similar) stone in facings of wall built in lime mortar in adjacent stones; prepare for, supply and fix new stone (PC sum £2580 per m3)**										
A.02.44.03.A	average 50mm thick; point and clean down on completion	m²	2.01	105.47	—	153.21	258.67	116.02	—	168.53	284.54
A.02.44.03.B	in separate stones	m²	2.41	126.54	—	169.35	295.89	139.20	—	186.28	325.48
A.02.44.04.	**Rake out joints and repoint**										
A.02.44.04.A	ashlar stonework	m²	0.94	49.27	—	0.66	49.94	54.20	—	0.73	54.93
A.02.44.04.B	squared rubble	m²	1.07	56.30	—	0.66	56.96	61.93	—	0.73	62.66
A.02.44.05.	**Redress face of walling, where decayed, with picked face and repoint**										
A.02.44.05.A	generally	m²	5.37	281.29	—	0.92	282.21	309.41	—	1.02	310.43
A.02.44.08.	**Take up and reset 50mm thick Yorkstone slabs; any size, in landings, hearths, cover stones, paving, etc.**										
A.02.44.08.A	in lime mortar	m²	1.34	70.35	—	3.48	73.82	77.38	—	3.82	81.21
A.02.44.08.B	in cement mortar	m²	1.61	84.40	—	3.20	87.60	92.84	—	3.52	96.36
A.02.44.10.	**Cut and form toothing in old masonry for**										
A.02.44.10.A	new brick or stone	m²	1.34	70.35	—	—	70.35	77.38	—	—	77.38
A.02.44.11.	**Take down and reset blocking courses, cornices, strings, plinths, apexes, kneelers, etc.**										
A.02.44.11.A	in lime mortar	m	0.80	42.15	—	0.64	42.79	46.36	—	0.71	47.07
A.02.44.11.B	in cement mortar	m	0.94	49.27	—	0.59	49.87	54.20	—	0.65	54.85
A.02.44.14.	**Take down and reset window sills, steps, etc. including cutting away and making good**										
A.02.44.14.A	in lime mortar	m	1.34	70.35	—	0.39	70.73	77.38	—	0.42	77.81
A.02.44.14.B	in cement mortar	m	1.48	77.37	—	0.36	77.73	85.11	—	0.39	85.50
A.02.44.15.	**Repair with granolithic (2:7) including cutting out to depth of at least 19mm; finish concrete fair and flush with original surface**										
A.02.44.15.A	treads	m²	1.74	91.42	—	2.97	94.39	100.56	—	3.26	103.83

House Renovation Grants, Repairs and Alterations

Hutchins Priced Schedules 2021	Unit	Labour Hours	Labour Net £	Plant Net £	Materials Net £	Unit Rate Net £	Labour Gross £	Plant Gross £	Materials Gross £	Unit Rate Gross £
								(Gross rates include 10% profit)		

A.02. **REPAIRS AND ALTERATIONS**

A.02.44. **MASONRY REPAIRS AND RENEWALS**

A.02.44.15. Repair with granolithic (2:7) including cutting out to depth of at least 19mm; finish concrete fair and flush with original surface

A.02.44.15.B	landings	m²	1.07	56.30	—	2.97	59.27	61.93	—	3.26	65.19

A.02.46. **ROOFING REPAIRS AND RENEWALS**

A.02.46.01. Remove coverings and load into skip; roof coverings

A.02.46.01.A	slates	m²	0.55	16.27	—	—	16.27	17.90	—	—	17.90
A.02.46.01.B	nibbed tiles	m²	0.45	13.31	—	—	13.31	14.64	—	—	14.64
A.02.46.01.C	corrugated metal sheeting	m²	0.45	13.31	—	—	13.31	14.64	—	—	14.64
A.02.46.01.D	underfelt and nails	m²	0.07	2.07	—	—	2.07	2.28	—	—	2.28
A.02.46.01.E	three layers felt	m²	0.28	8.28	—	—	8.28	9.11	—	—	9.11
A.02.46.01.F	sheet metal	m²	0.55	16.27	—	—	16.27	17.90	—	—	17.90
A.02.46.01.G	metal flashings; horizontal	m	0.22	6.51	—	—	6.51	7.16	—	—	7.16
A.02.46.01.H	metal flashings; stepped	m	0.28	8.28	—	—	8.28	9.11	—	—	9.11
A.02.46.01.I	tile or slate battens, including withdrawing nails	m²	0.10	2.96	—	—	2.96	3.25	—	—	3.25

A.02.46.02. Remove coverings, carefully handling and disposing by an approved method toxic or other special waste

A.02.46.02.A	asbestos cement sheeting	m²	1.48	77.37	—	—	77.37	85.11	—	—	85.11

A.02.46.03. Strip, clean and set aside sound slates or tiles for re-use

A.02.46.03.A	slates	m²	0.40	11.83	—	—	11.83	13.02	—	—	13.02
A.02.46.03.B	tiles	m²	0.30	8.87	—	—	8.87	9.76	—	—	9.76
A.02.46.03.C	cement slates	m²	0.40	11.83	—	—	11.83	13.02	—	—	13.02

A.02.46.04. Strip, clean and set aside for removal

A.02.46.04.A	slates	m²	0.30	8.87	—	—	8.87	9.76	—	—	9.76
A.02.46.04.B	tiles	m²	0.24	7.10	—	—	7.10	7.81	—	—	7.81
A.02.46.04.C	cement slates	m²	0.30	8.87	—	—	8.87	9.76	—	—	9.76

A.02.46.06. Renew tile battens to 100mm gauge

A.02.46.06.A	area not exceeding 8.50m2	m²	0.23	11.95	—	5.07	17.02	13.15	—	5.57	18.72
A.02.46.06.B	over 8.50m2	m²	0.17	9.12	—	5.07	14.19	10.03	—	5.57	15.61

A.02.46.08. Renew slate battens to 205mm gauge

A.02.46.08.A	area not exceeding 8.50m2	m²	0.17	9.12	—	3.40	12.52	10.03	—	3.74	13.77
A.02.46.08.B	over 8.50m2	m²	0.11	5.98	—	3.40	9.37	6.57	—	3.74	10.31

A.02.46.10. Renew roofing felt

A.02.46.10.A	generally	m²	0.09	4.93	—	3.05	7.97	5.42	—	3.35	8.77

House Renovation Grants, Repairs and Alterations

Hutchins Priced Schedules 2021		Unit	Labour Hours	Labour Net £	Plant Net £	Materials Net £	Unit Rate Net £	Labour Gross £	Plant Gross £	Materials Gross £	Unit Rate Gross £
							(Gross rates include 10% profit)				
A.02.	**REPAIRS AND ALTERATIONS**										
A.02.46.	**ROOFING REPAIRS AND RENEWALS**										
A.02.46.12.	**Take off and renew single slates including clips; first slate**										
A.02.46.12.A	400mm x 200mm	nr	0.36	18.66	—	1.92	20.58	20.53	—	2.11	22.64
A.02.46.12.B	500mm x 250mm	nr	0.38	19.71	—	4.96	24.67	21.68	—	5.46	27.14
A.02.46.12.C	600mm x 300mm	nr	0.39	20.44	—	9.93	30.37	22.49	—	10.92	33.41
A.02.46.14.	**Take off and renew second and subsequent slates; up to 30nr**										
A.02.46.14.A	400mm x 200mm	nr	0.15	7.76	—	1.92	9.68	8.53	—	2.11	10.64
A.02.46.14.B	500mm x 250mm	nr	0.15	8.07	—	4.96	13.03	8.88	—	5.46	14.34
A.02.46.14.C	600mm x 300mm	nr	0.16	8.49	—	9.93	18.42	9.34	—	10.92	20.26
A.02.46.16.	**Take off and renew single tiles; first tile**										
A.02.46.16.A	clay (PC sum £600 per 1000)	nr	0.24	12.69	—	0.60	13.29	13.95	—	0.66	14.61
A.02.46.16.B	concrete	nr	0.24	12.69	—	0.32	13.01	13.95	—	0.35	14.31
A.02.46.18.	**renew second and subsequent tiles; up to 50nr**										
A.02.46.18.A	clay (PC sum £600 per 1000)	nr	0.15	7.76	—	0.60	8.36	8.53	—	0.66	9.19
A.02.46.18.B	concrete	nr	0.15	7.76	—	0.32	8.08	8.53	—	0.35	8.89
A.02.46.21.	**Cement mortar angle fillet to slate or tile roof at abutment to walls or chimney stacks**										
A.02.46.21.A	1nr - 3nr	m	0.27	14.05	—	0.36	14.40	15.45	—	0.39	15.84
A.02.46.22.	**Strip roof slating, sort slates and reslate roof using 50% existing slates**										
A.02.46.22.A	400mm x 200mm	m²	0.77	40.47	—	13.43	53.90	44.52	—	14.78	59.29
A.02.46.22.B	500mm x 250mm	m²	0.54	28.10	—	19.84	47.94	30.91	—	21.82	52.73
A.02.46.22.C	600mm x 300mm	m²	0.34	17.61	—	29.78	47.40	19.37	—	32.76	52.14
A.02.46.24.	**Strip 267mm x 165mm plain roof tiles and retile using 50% existing tiles**										
A.02.46.24.A	clay (PC sum for new £600 per 1000) . . .	m²	1.14	59.76	—	21.00	80.76	65.73	—	23.10	88.83
A.02.46.24.B	concrete	m²	1.14	59.76	—	11.20	70.96	65.73	—	12.32	78.05
A.02.46.26.	**Take up old lead; any position or weight**										
A.02.46.26.A	set aside	m²	0.24	23.03	—	—	23.03	25.33	—	—	25.33
A.02.46.28.	**Take up and relay lead to boarded flats**										
A.02.46.28.A	including dressing over rolls and drips with new bossed ends, etc.	m²	2.60	247.44	—	—	247.44	272.19	—	—	272.19

House Renovation Grants, Repairs and Alterations

Hutchins Priced Schedules 2021		Unit	Labour Hours	Labour Net £	Plant Net £	Materials Net £	Unit Rate Net £	Labour Gross £	Plant Gross £	Materials Gross £	Unit Rate Gross £
							(Gross rates include 10% profit)				
A.02.	**REPAIRS AND ALTERATIONS**										
A.02.46.	**ROOFING REPAIRS AND RENEWALS**										
A.02.46.29.	**Redress lead flashings**										
A.02.46.29.A	rewedge and repoint	m	0.48	46.06	—	—	46.06	50.67	—	—	50.67
A.02.46.30.	**Take up old zinc and remove from site**										
A.02.46.30.A	in any position	m²	0.61	57.58	—	—	57.58	63.34	—	—	63.34
A.02.46.32.	**Take up existing zinc and supply and lay new zinc including rolls, laps, etc.**										
A.02.46.32.A	0.4mm thick (10G)	m²	2.42	230.22	—	25.87	256.09	253.24	—	28.46	281.70
A.02.46.32.B	0.6mm thick (12G)	m²	2.72	258.96	—	33.68	292.64	284.85	—	37.05	321.91
A.02.46.32.C	0.8mm thick (14G)	m²	3.02	287.79	—	45.60	333.40	316.57	—	50.16	366.74
A.02.46.34.	**Prepare zinc roofs and supply and lay bitumenised fabric**										
A.02.46.34.A	bedded in bitumen, apply two coats of bitumen paint	m²	1.27	66.78	—	6.84	73.63	73.46	—	7.53	80.99
A.02.46.36.	**Seal crack in holes in zinc or asphalt flats and apply bitumen paint**										
A.02.46.36.A	two coats	m²	0.60	31.66	—	1.67	33.33	34.83	—	1.83	36.66
A.02.46.38.	**Clean and treat defective sheet zinc**										
A.02.46.38.A	Zinc repair compound	m²	0.74	38.69	—	3.76	42.45	42.55	—	4.14	46.69
A.02.46.39.	**Remove slates, take out defective box gutter linings or valleys and renew wood linings; line with zinc and replace slates**										
A.02.46.39.A	0.6mm thick (12G)	m²	4.84	460.43	—	51.81	512.24	506.48	—	56.99	563.46
A.02.46.40.	**Labour and transport up to 5 miles each way; set up and remove ladders**										
A.02.46.40.A	up to two storeys high	job	2.00	59.16	4.60	—	63.76	65.08	5.06	—	70.14
A.02.46.40.B	up to four storeys high	job	2.50	73.95	4.60	—	78.55	81.35	5.06	—	86.41
A.02.46.40.C	extra over last two items if access is difficult, e.g. house in a terrace with no side entrance	job	2.00	59.16	4.60	—	63.76	65.08	5.06	—	70.14
A.02.46.42.	**Remove waterproofing finishes and load into skip**										
A.02.46.42.A	asphalt paving	m²	0.67	19.82	—	—	19.82	21.80	—	—	21.80
A.02.46.42.B	asphalt roofing	m²	1.10	32.54	—	—	32.54	35.79	—	—	35.79
A.02.46.42.C	asphalt skirting	m	0.17	5.03	—	—	5.03	5.53	—	—	5.53

House Renovation Grants, Repairs and Alterations

Hutchins Priced Schedules 2021	Unit	Labour Hours	Labour Net £	Plant Net £	Materials Net £	Unit Rate Net £	Labour Gross £	Plant Gross £	Materials Gross £	Unit Rate Gross £
							(Gross rates include 10% profit)			
A.02.	**REPAIRS AND ALTERATIONS**									
A.02.50.	**ROOFING REPAIRS BY SPECIALISTS**									
A.02.50.01.	**Clean, prepare and apply bitumen roofing compound to**									
A.02.50.01.A	sound surfaces, asphalt, concrete, felt, slate or tile	m²	0.55	16.27	—	6.47	22.74	17.90	—	7.11	25.01
A.02.50.02.	**Clean, prepare and apply bitumen roofing compound and fungicide solution**									
A.02.50.02.A	to surfaces likely to support fungal or algae growth, asphalt, concrete, felt, slate or tile	m²	0.60	17.75	—	6.74	24.48	19.52	—	7.41	26.93
A.02.52.	**DRAINAGE REPAIRS AND RENEWALS**									
A.02.52.01.	**Break up 100mm concrete paving and hardcore for excavation of drain trench**									
A.02.52.01.A	not exceeding 600mm wide	m	1.10	32.54	—	—	32.54	35.79	—	—	35.79
A.02.52.02.	**Hand excavate trench for 100mm drain including backfilling and carting away remainder**									
A.02.52.02.A	average 0.50m deep	m	1.44	42.60	—	—	42.60	46.85	—	—	46.85
A.02.52.02.B	average 1.00m deep	m	2.90	85.78	—	—	85.78	94.36	—	—	94.36
A.02.52.02.C	average 1.50m deep	m	5.16	152.63	—	—	152.63	167.90	—	—	167.90
A.02.52.03.	**100mm concrete bed and haunch to**									
A.02.52.03.A	pipe	m	1.10	32.54	—	16.54	49.08	35.79	—	18.19	53.98
A.02.52.04.	**Hepworth SuperSleve clayware pipes; laid and jointed in short lengths**									
A.02.52.04.A	nominal dia 100mm	m	0.25	16.87	—	10.40	27.26	18.56	—	11.43	29.99
A.02.52.05.	**Make good 100mm concrete paving and hardcore under, after drainwork**									
A.02.52.05.A	average 600mm wide	m	1.30	38.45	—	6.09	44.55	42.30	—	6.70	49.00
A.02.52.06.	**Cut into brick side of exposed manhole and concrete benching for new branch drain and three quarter section channel; make good brickwork and benching**									
A.02.52.06.A	100mm	nr	1.51	101.29	—	60.43	161.72	111.41	—	66.47	177.89
A.02.52.07.	**Glazed clayware gully and grid 150mm x 150mm x 100mm with 100mm outlet join to drain**									
A.02.52.07.A	including necessary excavation and concrete bed	nr	1.21	80.99	—	91.66	172.65	89.09	—	100.83	189.91

House Renovation Grants, Repairs and Alterations

Hutchins Priced Schedules 2021	Unit	Labour Hours	Labour Net £	Plant Net £	Materials Net £	Unit Rate Net £	Labour Gross £	Plant Gross £	Materials Gross £	Unit Rate Gross £
										(Gross rates include 10% profit)
A.02. **REPAIRS AND ALTERATIONS**										
A.02.52. **DRAINAGE REPAIRS AND RENEWALS**										
A.02.52.08. **Concrete kerb 100mm outlet joint to drain**										
A.02.52.08.A including necessary formwork	nr	0.63	42.21	—	5.57	47.78	46.43	—	6.13	52.56
A.02.52.09. **Demolish kerb, disconnect and remove gully, supply and connect new gully and grid including work to concrete bed, new grid, rendering three sides and remake connection after rodding**										
A.02.52.09.A 150mm x 150mm x 100mm	nr	3.01	202.50	—	94.86	297.37	222.75	—	104.35	327.10
A.02.52.10. **Break up defective kerb-surround to manhole and reform in**										
A.02.52.10.A fine concrete splayed and rendered	nr	0.78	52.29	—	2.68	54.97	57.52	—	2.95	60.47
A.02.52.12. **Cut through external 225mm brick wall and 150mm concrete floor and connect to trap of pan at one end and existing pipe at the other end**										
A.02.52.12.A 102mm bend	nr	3.77	253.11	—	43.43	296.54	278.42	—	47.78	326.20
A.02.52.14. **Form new manhole on line of existing drain with 150mm concrete base and 225mm class B engineering brick wall (PC sum £500 per 1000); cover and frame; cut away existing drain within manhole to form channel, provide and set two three-quarter section channels and form concrete benching and cement render walls**										
A.02.52.14.A 0.90m deep	nr	18.08	1215.02	—	261.28	1476.31	1336.52	—	287.41	1623.94
A.02.52.14.B add for every extra 300mm depth in excess of 0.90m	nr	4.52	303.72	—	53.38	357.10	334.09	—	58.72	392.81
A.02.52.15. **Take up existing manhole cover and frame; provide, bed and seal new cover and frame; 600 x 450mm**										
A.02.52.15.A light duty cast iron; A15 solid top	nr	0.55	37.10	—	67.31	104.41	40.81	—	74.04	114.85
A.02.52.15.B medium duty cast iron; B125 solid top . .	nr	0.65	43.89	—	115.93	159.82	48.28	—	127.53	175.80
A.02.52.16. **Excavate for stoppage in 100mm drain; cut out and renew two lengths of pipe; fill in and test; reform paving**										
A.02.52.16.A assumed depth 0.90m	nr	5.02	337.53	—	52.99	390.52	371.28	—	58.29	429.57

House Renovation Grants, Repairs and Alterations

Hutchins Priced Schedules 2021	Unit	Labour Hours	Labour Net £	Plant Net £	Materials Net £	Unit Rate Net £	Labour Gross £	Plant Gross £	Materials Gross £	Unit Rate Gross £
							(Gross rates include 10% profit)			
A.02. **REPAIRS AND ALTERATIONS**										
A.02.52. **DRAINAGE REPAIRS AND RENEWALS**										
A.02.52.18. **Unstop gullies**										
A.02.52.18.A remove silt, clean and flush with disinfectant	nr	0.58	38.85	—	—	38.85	42.73	—	—	42.73
A.02.52.20. **Clear soil drains; rod and flush in sections**										
A.02.52.20.A average length 25.00m	nr	2.89	194.04	—	—	194.04	213.44	—	—	213.44
A.02.52.21. **Unstop and smoke test**										
A.02.52.21.A cast iron soil pipe	nr	2.11	141.75	—	—	141.75	155.92	—	—	155.92
A.02.52.22. **Take up manhole cover; clean, sand out and rebed in grease**										
A.02.52.22.A frame channels	nr	0.55	37.10	—	—	37.10	40.81	—	—	40.81
A.02.52.24. **Clayware channel to gully set in concrete with brick kerb rendered in cement mortar**										
A.02.52.24.A 450mm long	nr	1.00	67.48	—	14.13	81.61	74.23	—	15.54	89.77
A.02.52.24.B supply and fit new gully grid	nr	0.13	8.47	—	8.09	16.56	9.32	—	8.90	18.21
A.02.52.26. **Open up manhole; break out brickwork and benching to main channel; provide for and insert one three quarter channel bend and join to existing channel; reform benching; make good all work disturbed and refix manhole cover**										
A.02.52.26.A 100mm	nr	5.02	337.53	—	35.91	373.44	371.28	—	39.50	410.78
A.02.52.28. **Open up manhole 0.90m x 0.90m inside on plan and 1.27m deep; break out all channels and branches; break out one brick side of manhole at one end and extend manhole**										
A.02.52.28.A by 150mm x 0.90m inside with 150mm concrete at bottom, one brick side in stock bricks and 150mm concrete cover. Build in two pipes 150mm and one pipe 100mm, provide and insert straight main channel 150mm and six half branch channel bends 100mm and reform benching. Make good all work disturbed and refix manhole cover	nr	12.05	810.01	—	432.11	1242.13	891.02	—	475.32	1366.34
A.02.52.28.B extra over excavation for breaking out brick manhole 0.80m x 0.70m inside on plan and 0.90m deep to invert	nr	1.51	101.29	—	—	101.29	111.41	—	—	111.41
A.02.52.28.C add or deduct for every 300mm more or less than 0.90m deep to invert	nr	0.50	33.74	—	—	33.74	37.11	—	—	37.11

Hutchins Priced Schedules 2021		Unit	Labour Hours	Labour Net £	Plant Net £	Materials Net £	Unit Rate Net £	Labour Gross £	Plant Gross £	Materials Gross £	Unit Rate Gross £
								(Gross rates include 10% profit)			
A.02.	**REPAIRS AND ALTERATIONS**										
A.02.52.	**DRAINAGE REPAIRS AND RENEWALS**										
A.02.52.30.	**Seal open ends of disused clayware pipes with concrete plugs 300mm long**										
A.02.52.30.A	100mm dia.	nr	0.30	20.23	—	0.26	20.49	22.25	—	0.29	22.54
A.02.52.30.B	150mm dia.	nr	0.40	27.02	—	0.61	27.63	29.72	—	0.67	30.39
A.02.52.30.C	225mm dia.	nr	0.55	37.10	—	2.09	39.19	40.81	—	2.30	43.11

Drainage, Sewerage and Public Works

Hutchins Priced Schedules 2021		Unit	Labour Hours	Labour Net £	Plant Net £	Materials Net £	Unit Rate Net £	Labour Gross £	Plant Gross £	Materials Gross £	Unit Rate Gross £
								(Gross rates include 10% profit)			
B.01.	**DRAINAGE AND SEWERAGE**										
B.01.01.	**EXCAVATING TRENCHES**										
B.01.01.01.	Excavate trenches in firm soil; 450mm wide; earthwork support (four uses assumed); grade bottom; backfill; compact; dispose of surplus; hand labour; average depth										
B.01.01.01.A	0.50m	m	1.31	38.75	—	0.38	39.13	42.62	—	0.41	43.04
B.01.01.01.B	0.75m	m	1.97	58.27	—	0.53	58.80	64.10	—	0.58	64.68
B.01.01.01.C	1.00m	m	2.63	77.80	—	0.67	78.47	85.57	—	0.74	86.32
B.01.01.01.D	1.25m	m	3.92	115.95	—	0.85	116.80	127.55	—	0.94	128.48
B.01.01.01.E	1.50m	m	4.69	138.73	—	1.05	139.78	152.60	—	1.16	153.76
B.01.01.01.F	1.75m	m	5.48	162.10	—	1.20	163.30	178.31	—	1.32	179.63
B.01.01.01.G	2.00m	m	6.26	185.17	—	1.35	186.52	203.69	—	1.49	205.17
B.01.01.01.H	2.25m	m	8.52	252.02	—	1.50	253.52	277.22	—	1.65	278.87
B.01.01.01.I	2.50m	m	9.46	279.83	—	1.75	281.58	307.81	—	1.93	309.74
B.01.01.01.J	2.75m	m	10.42	308.22	—	1.88	310.10	339.05	—	2.06	341.11
B.01.01.01.K	3.00m	m	11.35	335.73	—	2.00	337.73	369.31	—	2.20	371.51
B.01.01.02.	Excavate trenches in firm soil; 600mm wide earthwork support (four uses assumed); grade bottom; compact; dispose of surplus; hand labour; average depth										
B.01.01.02.A	0.50m	m	1.75	51.77	—	0.42	52.19	56.94	—	0.47	57.41
B.01.01.02.B	0.75m	m	2.63	77.80	—	0.67	78.47	85.57	—	0.74	86.32
B.01.01.02.C	1.00m	m	3.50	103.53	—	0.90	104.43	113.88	—	0.99	114.87
B.01.01.02.D	1.25m	m	5.22	154.41	—	1.15	155.56	169.85	—	1.27	171.11
B.01.01.02.E	1.50m	m	6.26	185.17	—	1.35	186.52	203.69	—	1.49	205.17
B.01.01.02.F	1.75m	m	7.30	215.93	—	1.58	217.51	237.53	—	1.73	239.26
B.01.01.02.G	2.00m	m	8.34	246.70	—	1.80	248.50	271.37	—	1.98	273.35
B.01.01.02.H	2.25m	m	11.35	335.73	—	2.00	337.73	369.31	—	2.20	371.51
B.01.01.02.I	2.50m	m	12.62	373.30	—	2.25	375.55	410.63	—	2.48	413.11
B.01.01.02.J	2.75m	m	13.88	410.57	—	2.48	413.05	451.63	—	2.72	454.35
B.01.01.02.K	3.00m	m	15.14	447.84	—	2.73	450.57	492.63	—	3.00	495.62

Drainage, Sewerage and Public Works

Hutchins Priced Schedules 2021		Unit	Labour Hours	Labour Net £	Plant Net £	Materials Net £	Unit Rate Net £	Labour Gross £	Plant Gross £	Materials Gross £	Unit Rate Gross £
								(Gross rates include 10% profit)			
B.01.	**DRAINAGE AND SEWERAGE**										
B.01.01.	**EXCAVATING TRENCHES**										
B.01.01.03.	Excavate trenches in firm soil; 750mm wide earthwork support (four uses assumed); grade bottom; compact; dispose of surplus; hand labour; average depth										
B.01.01.03.A	0.50m	m	2.18	64.48	—	0.60	65.09	70.93	—	0.66	71.59
B.01.01.03.B	0.75m	m	3.29	97.32	—	0.85	98.17	107.05	—	0.94	107.99
B.01.01.03.C	1.00m	m	4.37	129.26	—	1.15	130.42	142.19	—	1.27	143.46
B.01.01.03.D	1.25m	m	6.54	193.45	—	1.45	194.90	212.80	—	1.60	214.40
B.01.01.03.E	1.50m	m	7.82	231.32	—	1.75	233.07	254.45	—	1.93	256.37
B.01.01.03.F	1.75m	m	9.13	270.07	—	1.95	272.02	297.07	—	2.15	299.22
B.01.01.03.G	2.00m	m	10.44	308.82	—	2.25	311.07	339.70	—	2.48	342.17
B.01.01.03.H	2.25m	m	14.21	420.33	—	2.55	422.88	462.36	—	2.81	465.17
B.01.01.03.I	2.50m	m	15.78	466.77	—	2.85	469.62	513.45	—	3.14	516.59
B.01.01.03.J	2.75m	m	17.35	513.21	—	3.08	516.29	564.53	—	3.39	567.92
B.01.01.03.K	3.00m	m	18.94	560.25	—	3.38	563.62	616.27	—	3.71	619.98
B.01.01.04.	Excavate trenches in firm soil; 450mm wide earthwork support (four uses assumed); grade bottom; compact; dispose of surplus; machine excavation; average depth										
B.01.01.04.A	0.50m	m	0.17	5.03	6.61	0.38	12.01	5.53	7.27	0.41	13.21
B.01.01.04.B	0.75m	m	0.24	7.10	9.34	0.53	16.97	7.81	10.28	0.58	18.67
B.01.01.04.C	1.00m	m	0.34	10.06	13.22	0.67	23.95	11.06	14.54	0.74	26.34
B.01.01.04.D	1.25m	m	0.47	13.90	18.30	0.85	33.05	15.29	20.13	0.94	36.36
B.01.01.04.E	1.50m	m	0.55	16.27	21.43	1.05	38.75	17.90	23.57	1.16	42.62
B.01.01.04.F	1.75m	m	0.65	19.23	25.34	1.20	45.77	21.15	27.87	1.32	50.34
B.01.01.04.G	2.00m	m	0.74	21.89	28.82	1.35	52.06	24.08	31.70	1.49	57.26
B.01.01.04.H	2.25m	m	0.97	28.69	33.90	1.50	64.09	31.56	37.29	1.65	70.50
B.01.01.04.I	2.50m	m	1.07	31.65	38.94	1.75	72.35	34.82	42.84	1.93	79.58
B.01.01.04.J	2.75m	m	1.18	34.90	42.07	1.88	78.85	38.39	46.28	2.06	86.74
B.01.01.04.K	3.00m	m	1.28	37.86	45.98	2.00	85.85	41.65	50.58	2.20	94.43
B.01.01.05.	Excavate trenches in firm soil; 600mm wide earthwork support (four uses assumed); grade bottom; compact; dispose of surplus; machine excavation; average depth										
B.01.01.05.A	0.50m	m	0.23	6.80	8.95	0.42	16.18	7.48	9.85	0.47	17.80
B.01.01.05.B	0.75m	m	0.34	10.06	13.22	0.67	23.95	11.06	14.54	0.74	26.34
B.01.01.05.C	1.00m	m	0.44	13.02	17.13	0.90	31.04	14.32	18.84	0.99	34.15
B.01.01.05.D	1.25m	m	0.62	18.34	24.16	1.15	43.65	20.17	26.58	1.27	48.02
B.01.01.05.E	1.50m	m	0.74	21.89	28.82	1.35	52.06	24.08	31.70	1.49	57.26
B.01.01.05.F	1.75m	m	0.88	26.03	32.34	1.58	59.94	28.63	35.57	1.73	65.94
B.01.01.05.G	2.00m	m	1.00	29.58	37.03	1.80	68.41	32.54	40.73	1.98	75.25
B.01.01.05.H	2.25m	m	1.28	37.86	45.98	2.00	85.85	41.65	50.58	2.20	94.43

Hutchins Priced Schedules 2021		Unit	Labour Hours	Labour Net £	Plant Net £	Materials Net £	Unit Rate Net £	Labour Gross £	Plant Gross £	Materials Gross £	Unit Rate Gross £
								(Gross rates include 10% profit)			
B.01.	**DRAINAGE AND SEWERAGE**										
B.01.01.	**EXCAVATING TRENCHES**										
B.01.01.05.	Excavate trenches in firm soil; 600mm wide earthwork support (four uses assumed); grade bottom; compact; dispose of surplus; machine excavation; average depth										
B.01.01.05.I	2.50m	m	1.43	42.30	51.81	2.25	96.36	46.53	56.99	2.48	105.99
B.01.01.05.J	2.75m	m	1.57	46.44	56.50	2.48	105.42	51.08	62.15	2.72	115.96
B.01.01.05.K	3.00m	m	1.70	50.29	62.33	2.73	115.34	55.31	68.56	3.00	126.87
B.01.01.06.	Excavate trenches in firm soil; 750mm wide earthwork support (four uses assumed); grade bottom; compact; dispose of surplus; machine excavation; average depth										
B.01.01.06.A	0.50m	m	0.26	7.69	10.13	0.60	18.42	8.46	11.14	0.66	20.26
B.01.01.06.B	0.75m	m	0.40	11.83	15.60	0.85	28.28	13.02	17.16	0.94	31.11
B.01.01.06.C	1.00m	m	0.53	15.68	20.64	1.15	37.47	17.25	22.71	1.27	41.22
B.01.01.06.D	1.25m	m	0.74	21.89	28.82	1.45	52.16	24.08	31.70	1.60	57.37
B.01.01.06.E	1.50m	m	0.90	26.62	35.07	1.75	63.45	29.28	38.58	1.93	69.79
B.01.01.06.F	1.75m	m	1.04	30.76	40.51	1.95	73.22	33.84	44.56	2.15	80.54
B.01.01.06.G	2.00m	m	1.19	35.20	46.33	2.25	83.79	38.72	50.97	2.48	92.16
B.01.01.06.H	2.25m	m	1.55	45.85	56.50	2.55	104.90	50.43	62.15	2.81	115.39
B.01.01.06.I	2.50m	m	1.72	50.88	62.33	2.85	116.06	55.97	68.56	3.14	127.66
B.01.01.06.J	2.75m	m	1.88	55.61	68.19	3.08	126.88	61.17	75.01	3.39	139.57
B.01.01.06.K	3.00m	m	2.05	60.64	75.97	3.38	139.99	66.70	83.57	3.71	153.99
B.01.02.	**PIPE BEDS AND COVERINGS**										
B.01.02.01.	Sand; 50mm beds; width										
B.01.02.01.A	450mm	m	0.11	3.25	—	1.27	4.52	3.58	—	1.39	4.97
B.01.02.01.B	525mm	m	0.12	3.55	—	1.60	5.15	3.90	—	1.76	5.66
B.01.02.01.C	600mm	m	0.14	4.14	—	1.81	5.95	4.56	—	1.99	6.55
B.01.02.01.D	750mm	m	0.17	5.03	—	2.23	7.26	5.53	—	2.46	7.99
B.01.02.02.	Granular material; DTp Type 1; 50mm beds; width										
B.01.02.02.A	450mm	m	0.11	3.25	—	0.99	4.24	3.58	—	1.08	4.66
B.01.02.02.B	525mm	m	0.12	3.55	—	1.50	5.05	3.90	—	1.65	5.56
B.01.02.02.C	600mm	m	0.14	4.14	—	1.74	5.88	4.56	—	1.91	6.47
B.01.02.02.D	750mm	m	0.17	5.03	—	1.97	7.00	5.53	—	2.17	7.70
B.01.02.03.	Granular material; DTp Type 1; 100mm beds; width										
B.01.02.03.A	450mm	m	0.20	5.92	—	2.11	8.03	6.51	—	2.32	8.83
B.01.02.03.B	525mm	m	0.24	7.10	—	2.49	9.59	7.81	—	2.74	10.55
B.01.02.03.C	600mm	m	0.28	8.28	—	3.00	11.29	9.11	—	3.30	12.42
B.01.02.03.D	750mm	m	0.35	10.35	—	3.99	14.34	11.39	—	4.39	15.78

Drainage, Sewerage and Public Works

Hutchins Priced Schedules 2021	Unit	Labour Hours	Labour Net £	Plant Net £	Materials Net £	Unit Rate Net £	Labour Gross £	Plant Gross £	Materials Gross £	Unit Rate Gross £
							(Gross rates include 10% profit)			
B.01.	**DRAINAGE AND SEWERAGE**									
B.01.02.	PIPE BEDS AND COVERINGS									
B.01.02.04.	**Granular material; DTp Type 1; 150mm beds; width**									
B.01.02.04.A	450mm....................	m	0.31	9.17	—	3.47	12.64	10.09	—	3.82	13.91
B.01.02.04.B	525mm....................	m	0.36	10.65	—	3.99	14.64	11.71	—	4.39	16.10
B.01.02.04.C	600mm....................	m	0.42	12.42	—	4.46	16.88	13.67	—	4.91	18.57
B.01.02.04.D	750mm....................	m	0.53	15.68	—	5.49	21.17	17.25	—	6.04	23.29
B.01.02.05.	**Granular material; DTp Type 1; 100mm beds; filling to half height of pipe; width**									
B.01.02.05.A	450mm to 100mm pipe...........	m	0.32	9.47	—	3.00	12.47	10.41	—	3.30	13.72
B.01.02.05.B	525mm to 150mm pipe...........	m	0.42	12.42	—	3.99	16.41	13.67	—	4.39	18.05
B.01.02.05.C	600mm to 225mm pipe...........	m	0.53	15.68	—	4.98	20.65	17.25	—	5.47	22.72
B.01.02.05.D	750mm to 300mm pipe...........	m	0.73	21.59	—	6.95	28.54	23.75	—	7.64	31.39
B.01.02.06.	**Granular material; DTp Type 1; 150mm beds; filling to half height of pipe; width**									
B.01.02.06.A	450mm to 100mm pipe...........	m	0.43	12.72	—	4.46	17.18	13.99	—	4.91	18.90
B.01.02.06.B	525mm to 150mm pipe...........	m	0.54	15.97	—	5.49	21.47	17.57	—	6.04	23.61
B.01.02.06.C	600mm to 225mm pipe...........	m	0.66	19.52	—	6.48	26.00	21.48	—	7.13	28.60
B.01.02.06.D	750mm to 300mm pipe...........	m	0.91	26.92	—	8.97	35.88	29.61	—	9.86	39.47
B.01.02.07.	**Granular material; DTp Type 1; bed and covering; thickness**									
B.01.02.07.A	450mm x 350mm thick to 100mm pipe..	m	0.70	20.71	—	7.46	28.17	22.78	—	8.21	30.99
B.01.02.07.B	450mm x 450mm thick to 100mm pipe..	m	0.90	26.62	—	9.95	36.57	29.28	—	10.95	40.23
B.01.02.07.C	525mm x 400mm thick to 150mm pipe..	m	0.89	26.33	—	9.95	36.28	28.96	—	10.95	39.91
B.01.02.07.D	525mm x 500mm thick to 150mm pipe..	m	1.14	33.72	—	12.44	46.16	37.09	—	13.68	50.78
B.01.02.07.E	600mm x 475mm thick to 225mm pipe..	m	1.14	33.72	—	12.44	46.16	37.09	—	13.68	50.78
B.01.02.07.F	600mm x 575mm thick to 225mm pipe..	m	1.42	42.00	—	15.44	57.45	46.20	—	16.99	63.19
B.01.02.07.G	750mm x 550mm thick to 300mm pipe..	m	1.58	46.74	—	17.41	64.15	51.41	—	19.16	70.57
B.01.02.07.H	750mm x 650mm thick to 300mm pipe..	m	1.91	56.50	—	20.89	77.39	62.15	—	22.98	85.12
B.01.02.09.	**Plain concrete mix GEN1 - 40mm aggregate; 100mm beds; width**									
B.01.02.09.A	450mm....................	m	0.29	8.58	—	4.61	13.19	9.44	—	5.07	14.51
B.01.02.09.B	525mm....................	m	0.34	10.06	—	5.31	15.37	11.06	—	5.84	16.90
B.01.02.09.C	600mm....................	m	0.38	11.24	—	6.09	17.33	12.36	—	6.70	19.07
B.01.02.09.D	750mm....................	m	0.48	14.20	—	7.66	21.86	15.62	—	8.43	24.04
B.01.02.10.	**Plain concrete mix GEN1 - 40mm aggregate; 150mm beds; width**									
B.01.02.10.A	450mm....................	m	0.43	12.72	—	6.79	19.51	13.99	—	7.47	21.46
B.01.02.10.B	525mm....................	m	0.50	14.79	—	8.01	22.80	16.27	—	8.81	25.08

Drainage, Sewerage and Public Works

Hutchins Priced Schedules 2021		Unit	Labour Hours	Labour Net £	Plant Net £	Materials Net £	Unit Rate Net £	Labour Gross £	Plant Gross £	Materials Gross £	Unit Rate Gross £
								(Gross rates include 10% profit)			
B.01.	**DRAINAGE AND SEWERAGE**										
B.01.02.	**PIPE BEDS AND COVERINGS**										
B.01.02.10.	Plain concrete mix GEN1 - 40mm aggregate; 150mm beds; width										
B.01.02.10.C	600mm....................	m	0.58	17.16	—	9.14	26.30	18.87	—	10.05	28.93
B.01.02.10.D	750mm....................	m	0.72	21.30	—	11.40	32.70	23.43	—	12.54	35.97
B.01.02.11.	**Plain concrete mix GEN1 - 40mm aggregate; 100mm beds; filling to half height of pipe; width**										
B.01.02.11.A	450mm to 100mm pipe..........	m	0.44	13.02	—	6.79	19.80	14.32	—	7.47	21.78
B.01.02.11.B	525mm to 150mm pipe..........	m	0.56	16.56	—	8.79	25.36	18.22	—	9.67	27.89
B.01.02.11.C	600mm to 225mm pipe..........	m	0.72	21.30	—	8.79	30.09	23.43	—	9.67	33.10
B.01.02.11.D	750mm to 300mm pipe..........	m	1.01	29.88	—	11.23	41.10	32.86	—	12.35	45.21
B.01.02.12.	**Plain concrete mix GEN1 - 40mm aggregate; 150mm beds; filling to half height of pipe; width**										
B.01.02.12.A	450mm to 100mm pipe..........	m	0.59	17.45	—	7.22	24.68	19.20	—	7.95	27.14
B.01.02.12.B	525mm to 150mm pipe..........	m	0.74	21.89	—	8.53	30.42	24.08	—	9.38	33.46
B.01.02.12.C	600mm to 225mm pipe..........	m	0.91	26.92	—	11.75	38.67	29.61	—	12.93	42.54
B.01.02.12.D	750mm to 300mm pipe..........	m	1.25	36.98	—	15.41	52.38	40.67	—	16.95	57.62
B.01.02.13.	**Plain concrete mix GEN1 - 40mm aggregate; bed and covering; 450mm wide**										
B.01.02.13.A	350mm thick to 100mm pipe........	m	0.96	28.40	—	15.93	44.33	31.24	—	17.52	48.76
B.01.02.13.B	450mm thick to 100mm pipe........	m	1.24	36.68	—	20.54	57.22	40.35	—	22.60	62.94
B.01.02.14.	**Plain concrete mix GEN1 - 40mm aggregate; bed and covering; 525mm wide**										
B.01.02.14.A	400mm thick to 150mm pipe........	m	1.22	36.09	—	21.24	57.33	39.70	—	23.36	63.06
B.01.02.14.B	500mm thick to 150mm pipe........	m	1.56	46.14	—	26.63	72.78	50.76	—	29.30	80.06
B.01.02.15.	**Plain concrete mix GEN1 - 40mm aggregate; bed and covering; 600mm wide**										
B.01.02.15.A	475mm thick to 225mm pipe........	m	1.56	46.14	—	26.72	72.87	50.76	—	29.39	80.15
B.01.02.15.B	575mm thick to 225mm pipe........	m	1.94	57.39	—	33.25	90.63	63.12	—	36.57	99.70
B.01.02.16.	**Plain concrete mix GEN1 - 40mm aggregate; bed and covering; 750mm wide**										
B.01.02.16.A	550mm thick to 300mm pipe........	m	2.17	64.19	—	37.86	102.05	70.61	—	41.65	112.26
B.01.02.16.B	650mm thick to 300mm pipe........	m	2.62	77.50	—	45.17	122.67	85.25	—	49.69	134.94

Drainage, Sewerage and Public Works

Hutchins Priced Schedules 2021	Unit	Labour Hours	Labour Net £	Plant Net £	Materials Net £	Unit Rate Net £	Labour Gross £	Plant Gross £	Materials Gross £	Unit Rate Gross £
							(Gross rates include 10% profit)			
B.01.	**DRAINAGE AND SEWERAGE**									
B.01.04.	BREAKING UP PAVED SURFACES									
B.01.04.01.	**Break up paving with compressed air equipment for trenches 600mm wide; average thickness**									
B.01.04.01.A	75mm tarmacadam	m	0.43	12.72	1.43	—	14.15	13.99	1.57	—	15.56
B.01.04.01.B	150mm plain concrete	m	1.35	39.93	4.41	—	44.35	43.93	4.86	—	48.78
B.01.04.01.C	150mm reinforced concrete	m	1.83	54.13	8.76	—	62.89	59.54	9.63	—	69.18
B.01.05.	**REINSTATE PAVED SURFACES**									
B.01.05.01.	**Reinstate paving; average 600mm wide; 100mm hardcore bed; paving average thickness**									
B.01.05.01.A	75mm tarmacadam	m²	0.42	12.42	—	6.72	19.14	13.67	—	7.39	21.06
B.01.05.01.B	150mm concrete	m²	0.85	25.14	—	33.39	58.53	27.66	—	36.73	64.39
B.01.06.	**SUPERSLEVE DRAIN PIPES AND FITTINGS**									
B.01.06.01.	**Push-fit polypropylene flexible couplings; for underground drainage; in trenches; 100mm nominal dia pipes**									
B.01.06.01.A	in runs exceeding 3.00m long	m	0.11	7.46	—	10.40	17.86	8.21	—	11.43	19.64
B.01.06.01.B	in runs not exceeding 3.00m long	m	0.14	9.48	—	10.40	19.87	10.42	—	11.43	21.86
B.01.06.01.C	extra for bend	nr	0.15	10.15	—	14.24	24.38	11.16	—	15.66	26.82
B.01.06.01.E	extra for junction	nr	0.13	8.47	—	30.77	39.24	9.32	—	33.85	43.16
B.01.06.02.	**Push-fit polypropylene flexible couplings; for underground drainage; in trenches; 150mm nominal dia pipes**									
B.01.06.02.A	in runs exceeding 3.00m long	m	0.14	9.48	—	19.22	28.69	10.42	—	21.14	31.56
B.01.06.02.B	in runs not exceeding 3.00m long	m	0.19	12.84	—	19.22	32.05	14.12	—	21.14	35.26
B.01.06.02.C	extra for bend	nr	0.17	11.16	—	28.29	39.45	12.27	—	31.12	43.39
B.01.06.02.E	extra for junction	nr	0.15	10.15	—	41.56	51.71	11.16	—	45.72	56.88
B.01.06.03.	**Push-fit polypropylene flexible couplings; for underground drainage; in trenches; 225mm nominal dia pipes**									
B.01.06.03.A	in runs exceeding 3.00m long	m	0.18	12.03	—	62.11	74.14	13.23	—	68.32	81.55
B.01.06.03.B	in runs not exceeding 3.00m long	m	0.26	17.34	—	62.11	79.45	19.07	—	68.32	87.39
B.01.06.03.C	extra for bend	nr	0.18	12.23	—	142.85	155.08	13.46	—	157.14	170.59
B.01.06.03.E	extra for junction	nr	0.18	12.17	—	256.16	268.32	13.38	—	281.78	295.16

Drainage, Sewerage and Public Works

Hutchins Priced Schedules 2021		Unit	Labour Hours	Labour Net £	Plant Net £	Materials Net £	Unit Rate Net £	Labour Gross £	Plant Gross £	Materials Gross £	Unit Rate Gross £
								(Gross rates include 10% profit)			
B.01.	**DRAINAGE AND SEWERAGE**										
B.01.06.	**SUPERSLEVE DRAIN PIPES AND FITTINGS**										
B.01.06.04.	**Push-fit polypropylene flexible couplings; for underground drainage; in trenches; 300mm nominal dia pipes**										
B.01.06.04.A	in runs exceeding 3.00m long	m	0.23	15.26	—	95.08	110.33	16.78	—	104.59	121.37
B.01.06.04.B	in runs not exceeding 3.00m long	m	0.35	23.39	—	95.08	118.47	25.73	—	104.59	130.31
B.01.06.04.C	extra for bend	nr	0.20	13.37	—	271.29	284.66	14.71	—	298.42	313.13
B.01.06.04.E	extra for junction	nr	0.22	14.52	—	543.06	557.58	15.97	—	597.37	613.34
B.01.07.	**SUPERSLEVE ACCESSORIES**										
B.01.07.01.	**Jointing to drains; excavation and concrete surrounds**										
B.01.07.01.A	square gulley; trapped; square grid; 100mm outlet	nr	0.75	50.61	—	103.54	154.15	55.67	—	113.90	169.57
B.01.07.01.B	square gulley; trapped; horizontal back inlet to small waste pipe; square grid; 100mm outlet	nr	0.90	60.76	—	88.70	149.46	66.83	—	97.57	164.40
B.01.07.01.C	square gulley; trapped; horizontal back inlet to large waste pipe; square grid; 100mm outlet	nr	0.90	60.76	—	57.65	118.41	66.83	—	63.42	130.25
B.01.07.01.D	square hopper; integral vertical back inlet; trapped; square sealing plate; 100mm outlet	nr	0.90	60.76	—	88.70	149.46	66.83	—	97.57	164.40
B.01.07.01.E	square hopper; integral vertical back inlet; untrapped; square sealing plate; 100mm outlet	nr	0.90	60.76	—	95.28	156.04	66.83	—	104.81	171.64
B.01.07.01.F	access gulley; integral vertical back inlet; rodding eye; stopper; plastic grid; hinged grate and frame	nr	0.90	60.76	—	95.28	156.04	66.83	—	104.81	171.64
B.01.07.01.H	polypropylene inspection chamber; 475mm dia; 595mm deep; five 100/110mm inlets; base; ductile iron cover and frame	nr	1.00	67.48	—	450.44	517.92	74.23	—	495.48	569.71
B.01.07.01.I	polypropylene inspection chamber; 475mm dia; 1030mm deep; five 150/160mm inlets; base; ductile iron cover and frame	nr	1.00	67.48	—	577.10	644.57	74.23	—	634.81	709.03
B.01.07.01.J	polypropylene inspection chamber; 475mm dia; 1030mm deep; three 150/160mm inlets (and two stoppered); base; ductile iron cover and frame	nr	1.11	74.27	—	577.10	651.36	81.69	—	634.81	716.50
B.01.10.	**PVCU UNDERGROUND DRAINAGE PIPES AND FITTINGS**										
B.01.10.01.	**OsmaDrain system; in trenches; 110mm nominal dia pipe**										
B.01.10.01.A	in runs exceeding 3.00m long	m	0.12	8.13	—	13.19	21.32	8.95	—	14.51	23.45
B.01.10.01.B	in runs not exceeding 3.00m long	m	0.14	9.48	—	13.19	22.66	10.42	—	14.51	24.93

Drainage, Sewerage and Public Works

Hutchins Priced Schedules 2021		Unit	Labour Hours	Labour Net £	Plant Net £	Materials Net £	Unit Rate Net £	Labour Gross £	Plant Gross £	Materials Gross £	Unit Rate Gross £
								(Gross rates include 10% profit)			
B.01.	**DRAINAGE AND SEWERAGE**										
B.01.10.	PVCU UNDERGROUND DRAINAGE PIPES AND FITTINGS										
B.01.10.01.	OsmaDrain system; in trenches; 110mm nominal dia pipe										
B.01.10.01.C	extra for short radius bend; single socket .	nr	0.14	9.48	—	35.01	44.49	10.42	—	38.51	48.94
B.01.10.01.D	extra for short radius bend; double socket	nr	0.14	9.48	—	41.07	50.54	10.42	—	45.17	55.60
B.01.10.01.E	extra for long radius bend	nr	0.14	9.48	—	76.94	86.42	10.42	—	84.64	95.06
B.01.10.01.F	extra for equal junction	nr	0.12	8.13	—	49.45	57.58	8.95	—	54.39	63.34
B.01.10.02.	OsmaDrain system; in trenches; 160mm nominal dia pipe										
B.01.10.02.A	in runs exceeding 3.00m long	m	0.15	9.81	—	30.25	40.06	10.79	—	33.28	44.07
B.01.10.02.B	in runs not exceeding 3.00m long	m	0.17	11.16	—	30.25	41.41	12.27	—	33.28	45.55
B.01.10.02.C	extra for short radius bend; single socket .	nr	0.18	11.83	—	89.38	101.21	13.01	—	98.32	111.33
B.01.10.02.D	extra for short radius bend; double socket	nr	0.18	11.83	—	97.83	109.66	13.01	—	107.61	120.62
B.01.10.02.E	extra for long radius bend	nr	0.18	11.83	—	208.07	219.90	13.01	—	228.87	241.89
B.01.10.02.F	extra for equal junction	nr	0.17	11.16	—	161.40	172.55	12.27	—	177.54	189.81
B.01.10.02.G	extra for unequal junction	nr	0.17	11.16	—	123.24	134.40	12.27	—	135.56	147.83
B.01.10.03.	PVCu accessories; jointing to drains; excavation and concrete surrounds										
B.01.10.03.A	110mm rodding point	nr	0.36	24.33	—	126.18	150.51	26.76	—	138.80	165.56
B.01.10.03.C	access gulley; round; sealing rings; grid and access plug; 110mm outlet	nr	0.60	40.53	—	63.37	103.89	44.58	—	69.70	114.28
B.01.10.03.D	inspection chamber; 315mm dia base with stoppered inlets; 110mm outlet, shaft, cover and frame; up to 600mm deep	nr	1.04	69.90	—	270.21	340.11	76.89	—	297.23	374.12
B.01.10.04.	PVCu slotted pipes and fittings; 110mm nominal pipe sizes										
B.01.10.04.A	in runs exceeding 3.00m long	m	0.20	5.92	—	20.58	26.50	6.51	—	22.64	29.15
B.01.10.04.B	in runs not exceeding 3.00m long	m	0.25	7.40	—	20.58	27.98	8.13	—	22.64	30.77
B.01.12.	**CAST IRON DRAIN PIPES AND FITTINGS**										
B.01.12.01.	Spigot and socket caulked lead joints; in trenches; 100mm pipes										
B.01.12.01.A	laid straight	m	0.56	52.91	—	45.26	98.17	58.21	—	49.79	108.00
B.01.12.01.B	in runs not exceeding 3.00m long	m	0.76	71.95	—	45.26	117.21	79.14	—	49.79	128.93
B.01.12.01.C	extra for bend; short radius	nr	0.56	52.91	—	36.73	89.65	58.21	—	40.40	98.61
B.01.12.01.D	extra for bend; long radius	nr	0.56	52.91	—	56.81	109.72	58.21	—	62.49	120.69
B.01.12.01.E	extra for branch; single	nr	0.76	71.95	—	53.99	125.94	79.14	—	59.39	138.53
B.01.12.01.F	extra for branch; double	nr	0.94	89.17	—	82.60	171.78	98.09	—	90.87	188.96

Hutchins Priced Schedules 2021	Unit	Labour Hours	Labour Net £	Plant Net £	Materials Net £	Unit Rate Net £	Labour Gross £	Plant Gross £	Materials Gross £	Unit Rate Gross £
								(Gross rates include 10% profit)		

B.01.   **DRAINAGE AND SEWERAGE**

B.01.12.   **CAST IRON DRAIN PIPES AND FITTINGS**

B.01.12.01.   Spigot and socket caulked lead joints; in trenches; 100mm pipes

B.01.12.01.G	extra for drain connector; large socket for clayware; 305mm long	nr	0.56	52.91	—	40.22	93.13	58.21	—	44.24	102.44
B.01.12.01.H	extra for drain connector; large socket for WC; 300mm long	nr	0.42	39.69	—	36.88	76.56	43.65	—	40.57	84.22

B.01.12.02.   Spigot and socket caulked lead joints; in trenches; 150mm pipes

B.01.12.02.A	laid straight	m	0.70	66.14	—	88.07	154.21	72.76	—	96.88	169.64
B.01.12.02.B	in runs not exceeding 3.00m long	m	0.94	89.17	—	88.07	177.25	98.09	—	96.88	194.97
B.01.12.02.C	extra for bend; short radius	nr	0.70	66.14	—	63.33	129.47	72.76	—	69.66	142.42
B.01.12.02.D	extra for bend; medium radius	nr	0.70	66.14	—	209.64	275.78	72.76	—	230.61	303.36
B.01.12.02.E	extra for branch; single	nr	0.94	89.17	—	121.57	210.74	98.09	—	133.72	231.81
B.01.12.02.F	extra for branch; double	nr	1.18	112.21	—	282.49	394.70	123.43	—	310.74	434.16
B.01.12.02.G	extra for; drain connector; large socket for clayware; 305mm long	nr	0.70	66.14	—	70.48	136.62	72.76	—	77.53	150.28

B.01.12.03.   Timesaver rainwater shoe; horizontal or vertical inlet; setting on and bedding in GEN1 concrete

B.01.12.03.A	100mm	nr	0.52	49.49	—	300.88	350.36	54.44	—	330.96	385.40

B.01.12.04.   Yard gulley; bedding on and setting in GEN1 concrete

B.01.12.04.A	deans; trapped; galvanised sediment pan; 267mm round heavy grating; 100mm outlet	nr	2.24	212.90	—	503.72	716.62	234.18	—	554.10	788.28
B.01.12.04.B	garage; trapless; galvanised sediment pan; 267mm round heavy grating; 100mm outlet	nr	2.09	198.52	—	1029.19	1227.71	218.38	—	1132.11	1350.48
B.01.12.04.C	garage; trapped; rodding eye; galvanised perforated sediment pan; stopper; 267mm round heavy grating; 100mm outlet	nr	2.78	264.76	—	1088.66	1353.43	291.24	—	1197.53	1488.77
B.01.12.04.D	square top; trapped; galvanised sediment pan; 255mm x 255mm square grating; 100mm outlet	nr	2.75	261.81	—	423.95	685.76	287.99	—	466.34	754.34

B.01.14.   **CAST IRON TIMESAVER PIPES AND FITTINGS**

B.01.14.01.   Drainage Castings Timesaver System; mechanical coupling joints; in trenches; 100mm pipes

B.01.14.01.A	laid straight	m	0.35	33.40	—	52.63	86.03	36.75	—	57.89	94.64
B.01.14.01.B	in runs not exceeding 3.00m long	m	0.47	44.92	—	52.63	97.55	49.41	—	57.89	107.30
B.01.14.01.C	extra for bend; medium radius	nr	0.42	39.69	—	105.74	145.42	43.65	—	116.31	159.96

Drainage, Sewerage and Public Works

Hutchins Priced Schedules 2021		Unit	Labour Hours	Labour Net £	Plant Net £	Materials Net £	Unit Rate Net £	Labour Gross £	Plant Gross £	Materials Gross £	Unit Rate Gross £
								(Gross rates include 10% profit)			
B.01.	**DRAINAGE AND SEWERAGE**										
B.01.14.	**CAST IRON TIMESAVER PIPES AND FITTINGS**										
B.01.14.01.	**Drainage Castings Timesaver System; mechanical coupling joints; in trenches; 100mm pipes**										
B.01.14.01.D	extra for bend; long radius	nr	0.42	39.69	—	105.74	145.42	43.65	—	116.31	159.96
B.01.14.01.E	extra for branch; single - plain 100mm x 100mm	nr	0.52	49.49	—	130.20	179.69	54.44	—	143.22	197.66
B.01.14.01.G	extra for branch double - plain 100mm x 100mm	nr	0.67	63.29	—	141.79	205.07	69.62	—	155.96	225.58
B.01.14.01.I	extra for transitional pipe; socket for WC .	nr	0.35	33.40	—	40.22	73.62	36.75	—	44.24	80.98
B.01.14.01.J	extra for transitional pipe; socket for clayware	nr	0.24	23.03	—	35.06	58.10	25.33	—	38.57	63.91
B.01.14.02.	**Drainage Castings Timesaver System; mechanical coupling joints; in trenches 150mm pipes**										
B.01.14.02.A	laid straight	m	0.42	39.69	—	82.37	122.06	43.65	—	90.61	134.26
B.01.14.02.B	in runs not exceeding 3.00m long	m	0.57	54.06	—	82.37	136.43	59.46	—	90.61	150.07
B.01.14.02.C	extra for bend; medium radius	nr	0.49	46.63	—	117.03	163.67	51.30	—	128.74	180.03
B.01.14.02.D	extra for bend; long radius	nr	0.49	46.63	—	82.37	129.01	51.30	—	90.61	141.91
B.01.14.02.E	extra for branch; single	nr	0.59	56.44	—	113.07	169.51	62.08	—	124.38	186.46
B.01.14.02.H	extra for transitional pipe; socket for clayware	nr	0.56	52.91	—	41.34	94.25	58.21	—	45.47	103.68
B.01.14.02.I	extra for isolated Timesaver joint	nr	0.29	27.60	—	35.06	62.66	30.36	—	38.57	68.93
B.01.14.03.	**Timesaver rainwater shoes; horizontal or vertical inlet; setting on and bedding in GEN1 concrete**										
B.01.14.03.A	100mm	nr	0.35	33.40	—	323.56	356.96	36.75	—	355.91	392.66
B.01.14.04.	**Yard gulley; bedding on and setting in GEN1 concrete**										
B.01.14.04.A	deans; trapped; galvanised sediment pan; 267mm round heavy grating; 100mm outlet	nr	2.03	192.81	—	523.27	716.09	212.10	—	575.60	787.70
B.01.14.04.B	garage; trapless; galvanised sediment pan; 267mm round heavy grating; 100mm outlet	nr	1.88	178.44	—	1048.74	1227.18	196.29	—	1153.61	1349.90
B.01.14.04.C	garage; trapped; rodding eye; galvanised perforated sediment pan; stopper; 267mm round heavy grating; 100mm outlet	nr	2.09	198.52	—	1111.08	1309.61	218.38	—	1222.19	1440.57
B.01.14.04.D	square top; trapped; galvanised sediment pan; 255mm x 255mm square grating; 100mm outlet	nr	2.06	195.67	—	443.50	639.17	215.24	—	487.85	703.08
B.01.14.05.	**Inspection chambers; bolted flat cover; bedding in cement mortar (1:3)**										
B.01.14.05.B	100mm x 100mm; one branch each side .	nr	0.73	69.09	—	288.70	357.80	76.00	—	317.57	393.58

Drainage, Sewerage and Public Works

Hutchins Priced Schedules 2021	Unit	Labour Hours	Labour Net £	Plant Net £	Materials Net £	Unit Rate Net £	Labour Gross £	Plant Gross £	Materials Gross £	Unit Rate Gross £
										(Gross rates include 10% profit)
B.01.	**DRAINAGE AND SEWERAGE**									
B.01.14.	**CAST IRON TIMESAVER PIPES AND FITTINGS**									
B.01.14.05.	**Inspection chambers; bolted flat cover; bedding in cement mortar (1:3)**									
B.01.14.05.E	150mm x 100mm; one branch each side .	nr	1.00	94.98	—	373.22	468.20	104.48	—	410.54	515.02
B.01.14.05.F	150mm x 150mm; one branch each side .	nr	1.37	130.57	—	468.67	599.24	143.63	—	515.54	659.17
B.01.16.	**HEPWORTH DRAIN PIPES AND FITTINGS**									
B.01.16.02.	**Butt jointed; in trenches; 100mm nominal dia pipes**									
B.01.16.02.A	in runs exceeding 3.00m long	m	0.29	8.58	—	10.40	18.97	9.44	—	11.43	20.87
B.01.16.02.B	in runs not exceeding 3.00m long	m	0.34	10.06	—	10.40	20.45	11.06	—	11.43	22.50
B.01.16.02.C	extra for junction	nr	0.24	7.10	—	30.77	37.87	7.81	—	33.85	41.66
B.01.16.03.	**Butt jointed; in trenches; 150mm nominal dia pipes**									
B.01.16.03.A	in runs exceeding 3.00m long	m	0.37	10.94	—	19.22	30.16	12.04	—	21.14	33.18
B.01.16.03.B	in runs not exceeding 3.00m long	m	0.42	12.42	—	19.22	31.64	13.67	—	21.14	34.80
B.01.16.03.C	extra for junction	nr	0.29	8.58	—	37.86	46.44	9.44	—	41.65	51.09
B.01.16.04.	**Butt jointed; in trenches; 225mm nominal dia pipes**									
B.01.16.04.A	in runs exceeding 3.00m long	m	0.57	16.86	—	49.10	65.96	18.55	—	54.01	72.55
B.01.16.04.B	in runs not exceeding 3.00m long	m	0.62	18.34	—	49.10	67.44	20.17	—	54.01	74.18
B.01.16.04.C	extra for junction	nr	0.43	12.72	—	256.16	268.88	13.99	—	281.78	295.77
B.01.18.	**HEPLINE PERFORATED PIPES AND FITTINGS**									
B.01.18.01.	**Dry push-fit flexible integral polyethylene sleeve joints; in trenches; 100mm nominal dia pipes**									
B.01.18.01.A	in runs exceeding 3.00m long	m	0.21	6.21	—	15.95	22.16	6.83	—	17.54	24.38
B.01.18.01.B	in runs not exceeding 3.00m long	m	0.26	7.69	—	15.95	23.64	8.46	—	17.54	26.00
B.01.18.01.C	extra for bends	nr	0.29	8.58	—	14.24	22.81	9.44	—	15.66	25.10
B.01.18.01.D	extra for junction	nr	0.24	7.10	—	30.77	37.87	7.81	—	33.85	41.66
B.01.18.02.	**Dry push-fit flexible integral polyethylene sleeve joints; in trenches; 150mm nominal dia pipes**									
B.01.18.02.A	in runs exceeding 3.00m long	m	0.28	8.28	—	28.98	37.26	9.11	—	31.88	40.99
B.01.18.02.B	in runs not exceeding 3.00m long	m	0.38	11.24	—	28.98	40.22	12.36	—	31.88	44.24
B.01.18.02.C	extra for bend	nr	0.32	9.47	—	28.29	37.76	10.41	—	31.12	41.53
B.01.18.02.D	extra for junction	nr	0.29	8.58	—	41.56	50.14	9.44	—	45.72	55.15

Drainage, Sewerage and Public Works

Hutchins Priced Schedules 2021	Unit	Labour Hours	Labour Net £	Plant Net £	Materials Net £	Unit Rate Net £	Labour Gross £	Plant Gross £	Materials Gross £	Unit Rate Gross £
										(Gross rates include 10% profit)
B.01.	**DRAINAGE AND SEWERAGE**									
B.01.18.	**HEPLINE PERFORATED PIPES AND FITTINGS**									
B.01.18.03.	**Dry push-fit flexible integral polyethylene sleeve joints; in trenches; 225mm nominal dia pipes**									
B.01.18.03.A	in runs exceeding 3.00m long	m	0.45	13.31	—	61.32	74.63	14.64	—	67.45	82.09
B.01.18.03.B	in runs not exceeding 3.00m long	m	0.51	15.09	—	61.32	76.41	16.59	—	67.45	84.05
B.01.18.03.C	extra for bend	nr	0.50	14.79	—	142.85	157.64	16.27	—	157.14	173.41
B.01.18.03.D	extra for junction	nr	0.40	11.83	—	256.16	267.99	13.02	—	281.78	294.79
B.01.22.	**POROUS CONCRETE DRAIN PIPES**									
B.01.22.03.	**Rolling ring seal joints; in trenches; 225mm nominal dia pipes**									
B.01.22.03.A	in runs exceeding 3.00m long	m	0.80	23.66	—	29.67	53.34	26.03	—	32.64	58.67
B.01.22.03.B	in runs not exceeding 3.00m long	m	0.83	24.55	—	29.67	54.22	27.01	—	32.64	59.65
B.01.22.04.	**Rolling ring seal joints; in trenches; 300mm nominal dia pipes**									
B.01.22.04.A	in runs exceeding 3.00m long	m	0.96	28.40	—	31.95	60.35	31.24	—	35.15	66.38
B.01.22.04.B	in runs not exceeding 3.00m long	m	1.00	29.46	—	31.95	61.41	32.41	—	35.15	67.55
B.01.22.05.	**Rolling ring seal joints; in trenches; 450mm nominal dia pipes**									
B.01.22.05.A	in runs exceeding 3.00m long	m	1.28	37.86	—	46.70	84.57	41.65	—	51.37	93.02
B.01.22.05.B	in runs not exceeding 3.00m long	m	1.33	39.28	—	46.70	85.99	43.21	—	51.37	94.58
B.01.22.06.	**Rolling ring seal joints; in trenches; 600mm nominal dia pipes**									
B.01.22.06.A	in runs exceeding 3.00m long	m	1.60	47.33	—	76.91	124.24	52.06	—	84.60	136.66
B.01.22.06.B	in runs not exceeding 3.00m long	m	1.66	49.10	—	76.91	126.02	54.01	—	84.60	138.62
B.01.24.	**MANHOLES, ETC.**									
B.01.24.01.	**Excavate for manholes and soakaways; dispose of surplus; hand labour; maximum depth not exceeding**									
B.01.24.01.A	1.00m	m³	5.45	161.21	—	—	161.21	177.33	—	—	177.33
B.01.24.01.B	2.00m	m³	6.58	194.64	—	—	194.64	214.10	—	—	214.10
B.01.24.02.	**Excavated material; part backfill**									
B.01.24.02.A	remainder wheel and deposit	m³	2.50	73.95	—	—	73.95	81.35	—	—	81.35
B.01.24.03.	**Fill hardcore**									
B.01.24.03.A	into soakaways	m³	0.50	14.79	—	49.29	64.08	16.27	—	54.22	70.48
B.01.24.04.	**Earthwork support (four uses assumed) to sides of excavation**									
B.01.24.04.A	firm ground; depth not exceeding 1.00m .	m²	0.64	18.93	—	2.35	21.28	20.82	—	2.59	23.41
B.01.24.04.B	firm ground; depth not exceeding 2.00m .	m²	0.70	20.71	—	2.60	23.31	22.78	—	2.86	25.64

Drainage, Sewerage and Public Works

Hutchins Priced Schedules 2021		Unit	Labour Hours	Labour Net £	Plant Net £	Materials Net £	Unit Rate Net £	Labour Gross £	Plant Gross £	Materials Gross £	Unit Rate Gross £
										(Gross rates include 10% profit)	
B.01.	**DRAINAGE AND SEWERAGE**										
B.01.24.	**MANHOLES, ETC.**										
B.01.24.04.	Earthwork support (four uses assumed) to sides of excavation										
B.01.24.04.C	loose ground; depth not exceeding 1.00m................	m²	4.98	147.31	—	16.85	164.15	162.04	—	18.53	180.57
B.01.24.04.D	loose ground; depth not exceeding 2.00m................	m²	5.45	161.12	—	18.64	179.76	177.23	—	20.50	197.74
B.01.24.05.	Hire of skip; delivery to site; removing when full; disposal of contents; payment of tipping charges; size of skip										
B.01.24.05.A	4.50m3................	m³	—	—	43.60	—	43.60	—	47.96	—	47.96
B.01.24.06.	Base; concrete GEN1										
B.01.24.06.A	100mm thick................	m³	0.75	22.19	—	91.39	113.58	24.40	—	100.53	124.93
B.01.24.06.B	150mm thick................	m³	1.15	34.02	—	91.39	125.41	37.42	—	100.53	137.95
B.01.24.07.	Benching; concrete GEN1 to steep slopes to channels and branches; finished with 13mm cement mortar (1:3) trowelled smooth; average thickness										
B.01.24.07.A	225mm................	m²	2.65	78.39	—	22.25	100.64	86.23	—	24.47	110.70
B.01.24.07.B	300mm................	m²	3.31	97.91	—	29.65	127.56	107.70	—	32.61	140.31
B.01.24.08.	Fine concrete splayed kerb around manhole frame										
B.01.24.08.A	600mm x 450mm................	nr	1.40	41.41	—	1.83	43.24	45.55	—	2.01	47.56
B.01.24.09.	Reinforced suspended cover slab; including fabric reinforcement										
B.01.24.09.A	100mm - 150mm thick............	m³	7.76	229.54	—	93.85	323.39	252.49	—	103.23	355.73
B.01.24.10.	Common brickwork (PC sum £450 per 1000) in cement mortar (1:3) manhole walls										
B.01.24.10.A	half brick................	m²	1.00	67.14	—	32.43	99.57	73.86	—	35.67	109.53
B.01.24.10.B	one brick................	m²	1.63	109.69	—	65.71	175.40	120.66	—	72.28	192.94
B.01.24.11.	Extra over common brickwork for fair face and flush pointing as the work proceeds										
B.01.24.11.A	stretcher bond................	m²	0.28	18.89	—	—	18.89	20.77	—	—	20.77
B.01.24.11.B	Flemish bond................	m²	0.30	20.23	—	—	20.23	22.25	—	—	22.25
B.01.24.12.	Oversail common brickwork at top of manhole										
B.01.24.12.A	one course................	m	0.20	13.51	—	—	13.51	14.86	—	—	14.86
B.01.24.12.B	two courses................	m	0.40	27.02	—	—	27.02	29.72	—	—	29.72
B.01.24.12.C	three courses................	m	0.60	40.53	—	—	40.53	44.58	—	—	44.58

Drainage, Sewerage and Public Works

Hutchins Priced Schedules 2021		Unit	Labour Hours	Labour Net £	Plant Net £	Materials Net £	Unit Rate Net £	Labour Gross £	Plant Gross £	Materials Gross £	Unit Rate Gross £
								(Gross rates include 10% profit)			
B.01.	**DRAINAGE AND SEWERAGE**										
B.01.24.	MANHOLES, ETC.										
B.01.24.13.	**Building in end of pipe to half brick wall; common bricks**										
B.01.24.13.A	100mm pipe	nr	0.09	5.71	—	—	5.71	6.28	—	—	6.28
B.01.24.13.B	150mm pipe	nr	0.10	6.72	—	—	6.72	7.39	—	—	7.39
B.01.24.13.C	225mm pipe	nr	0.11	7.46	—	—	7.46	8.21	—	—	8.21
B.01.24.15.	**Building in end of pipe to one brick wall; common bricks**										
B.01.24.15.A	100mm pipe	nr	0.19	10.17	—	—	10.17	11.19	—	—	11.19
B.01.24.15.B	150mm pipe	nr	0.23	12.27	—	—	12.27	13.49	—	—	13.49
B.01.24.15.C	225mm pipe	nr	0.25	13.31	—	—	13.31	14.65	—	—	14.65
B.01.24.16.	**Class B engineering bricks (PC sum £500 per 1000) in cement mortar (1:3) manhole walls**										
B.01.24.16.A	half brick	m²	1.03	69.50	—	35.25	104.75	76.44	—	38.78	115.22
B.01.24.16.B	one brick	m²	1.76	118.16	—	71.20	189.35	129.97	—	78.31	208.29
B.01.24.17.	**Extra over class B engineering brickwork for fair face and flush pointing as the work proceeds**										
B.01.24.17.A	stretcher bond	m²	0.28	18.89	—	—	18.89	20.77	—	—	20.77
B.01.24.17.B	Flemish bond	m²	0.30	20.23	—	—	20.23	22.25	—	—	22.25
B.01.24.18.	**Oversail class B engineering brickwork at top of manhole**										
B.01.24.18.A	one course	m	0.20	13.51	—	—	13.51	14.86	—	—	14.86
B.01.24.18.B	two courses	m	0.40	27.02	—	—	27.02	29.72	—	—	29.72
B.01.24.18.C	three courses	m	0.60	40.53	—	—	40.53	44.58	—	—	44.58
B.01.24.19.	**Building in end of pipe to half brick wall; class B engineering bricks**										
B.01.24.19.A	100mm pipe	nr	0.10	6.72	—	—	6.72	7.39	—	—	7.39
B.01.24.19.B	150mm pipe	nr	0.12	7.80	—	—	7.80	8.58	—	—	8.58
B.01.24.19.C	225mm pipe	nr	0.12	8.13	—	—	8.13	8.95	—	—	8.95
B.01.24.20.	**Building in end of pipe to one brick wall; class B engineering bricks**										
B.01.24.20.A	100mm pipe	nr	0.16	10.48	—	—	10.48	11.53	—	—	11.53
B.01.24.20.B	150mm pipe	nr	0.19	12.50	—	—	12.50	13.75	—	—	13.75
B.01.24.20.C	225mm pipe	nr	0.21	14.18	—	—	14.18	15.60	—	—	15.60
B.01.24.21.	**Step irons; galvanised general purpose pattern; building into brickwork**										
B.01.24.21.A	115mm tails	nr	0.06	3.70	—	3.55	7.25	4.07	—	3.91	7.97

Drainage, Sewerage and Public Works

Hutchins Priced Schedules 2021		Unit	Labour Hours	Labour Net £	Plant Net £	Materials Net £	Unit Rate Net £	Labour Gross £	Plant Gross £	Materials Gross £	Unit Rate Gross £
							(Gross rates include 10% profit)				
B.01.	**DRAINAGE AND SEWERAGE**										
B.01.24.	**MANHOLES, ETC.**										
B.01.24.21.	**Step irons; galvanised general purpose pattern; building into brickwork**										
B.01.24.21.B	230mm tails	nr	0.07	4.37	—	5.79	10.16	4.81	—	6.37	11.17
B.01.24.22.	**Vitrified clay channel; set and jointed in cement mortar (1:3)**										
B.01.24.22.A	100mm half round straight main channel; 600mm long	nr	0.20	13.51	—	9.73	23.23	14.86	—	10.70	25.56
B.01.24.22.B	150mm half round straight main channel; 600mm long	nr	0.28	18.55	—	15.99	34.54	20.40	—	17.59	38.00
B.01.24.22.C	100mm half round main channel bend . .	nr	0.21	14.18	—	13.14	27.32	15.60	—	14.45	30.05
B.01.24.22.D	150mm half round main channel bend . .	nr	0.30	20.23	—	22.30	42.53	22.25	—	24.53	46.78
B.01.24.22.E	150mm - 100mm half round straight taper main channel	nr	0.30	20.23	—	31.35	51.58	22.25	—	34.48	56.73
B.01.24.22.F	150mm - 100mm half round taper main channel bend	nr	0.30	20.23	—	93.60	113.83	22.25	—	102.96	125.21
B.01.24.22.G	100mm half section branch channel bend	nr	0.21	14.18	—	31.14	45.32	15.60	—	34.25	49.85
B.01.24.22.H	150mm half section branch channel bend	nr	0.29	19.56	—	50.77	70.33	21.51	—	55.85	77.36
B.01.24.22.I	100mm three quarter section branch channel bend	nr	0.21	14.18	—	34.20	48.38	15.60	—	37.62	53.22
B.01.24.22.J	150mm three quarter section branch channel bend	nr	0.29	19.56	—	57.08	76.64	21.51	—	62.79	84.30
B.01.24.23.	**Covers and frames; bedded and flaunched in cement mortar (1:3); cover sealed in grease and sand**										
B.01.24.23.A	medium duty cast iron; B125 solid top manhole cover and frame; 600 x 450mm .	nr	0.57	38.11	—	116.88	154.99	41.92	—	128.57	170.49
B.01.24.23.B	light duty cast iron; A15 solid top manhole cover and frame; 600 x 450mm .	nr	0.33	22.25	—	68.26	90.51	24.47	—	75.09	99.56
B.01.24.24.	**Intercepting trap; vitrified clay; with stopper; cement joints to channel and pipe; bedding and surrounding with 150mm concrete GEN1**										
B.01.24.24.A	100mm	nr	0.53	35.76	—	177.69	213.45	39.33	—	195.46	234.79
B.01.24.24.B	150mm	nr	0.73	48.93	—	254.48	303.40	53.82	—	279.92	333.74
B.01.24.24.C	225mm	nr	1.00	67.48	—	771.48	838.96	74.23	—	848.63	922.86
B.01.24.25.	**Sewer connection; hand excavation; searching for existing 225mm vitrified clay live sewer pipe; breaking into; inserting and connecting new 100mm vitrified clay saddle junction; make good; backfill; consolidate; depth**										
B.01.24.25.A	2.00m	nr	10.81	726.67	—	30.20	756.87	799.34	—	33.22	832.56
B.01.24.25.B	3.00m	nr	13.47	905.52	—	30.20	935.72	996.07	—	33.22	1029.29

Drainage, Sewerage and Public Works

Hutchins Priced Schedules 2021	Unit	Labour Hours	Labour Net £	Plant Net £	Materials Net £	Unit Rate Net £	Labour Gross £	Plant Gross £	Materials Gross £	Unit Rate Gross £
										(Gross rates include 10% profit)

(Gross rates include 10% profit)

B.01.   **DRAINAGE AND SEWERAGE**

B.01.24.   **MANHOLES, ETC.**

B.01.24.26.   **Inspection chamber; 0.60m x 0.45m x 0.90m deep. All excavation work; concrete base 150mm; one brick walls in common brickwork (PC sum £450 per 1000) in cement mortar, rendered internally; one straight channel and two branch channels with concrete benching; building in ends of 100mm drain pipes; manhole cover and frame**

Item	Description	Unit	Labour Hours	Labour Net £	Plant Net £	Materials Net £	Unit Rate Net £	Labour Gross £	Plant Gross £	Materials Gross £	Unit Rate Gross £
B.01.24.26.A	medium duty cast iron; B125 solid top; 600 x 450mm	nr	14.56	978.78	—	406.03	1384.81	1076.66	—	446.64	1523.29
B.01.24.26.B	extra for class B engineering bricks (PC sum £500 per 1000) in lieu of commons (PC sum £450 per 1000)	nr	0.45	30.38	—	18.00	48.38	33.42	—	19.80	53.22
B.01.24.26.C	add or deduct for each 150mm depth in excess of or less than 0.90m in the foregoing inspection chamber if built in common bricks (PC sum £450 per 1000)	nr	1.76	118.16	—	61.11	179.26	129.97	—	67.22	197.19
B.01.24.26.D	add or deduct for each 150mm depth in excess of or less than 0.90m in the foregoing inspection chamber if built in class B engineering bricks (PC sum £500 per 1000)	nr	2.01	135.02	—	66.96	201.98	148.53	—	73.65	222.18

B.01.24.27.   **Rendering; 13mm cement and sand**

Item	Description	Unit	Labour Hours	Labour Net £	Plant Net £	Materials Net £	Unit Rate Net £	Labour Gross £	Plant Gross £	Materials Gross £	Unit Rate Gross £
B.01.24.27.A	to brick walls; internally	m²	0.40	27.02	—	1.90	28.92	29.72	—	2.09	31.81

B.01.24.28.   **Interceptor chamber; 0.75m x 0.60m x 1.20m deep. All excavation work; concrete base 150mm; one brick walls in common brickwork (PC sum £450 per 1000) in cement mortar, rendered internally; one straight channel and concrete benching; 100mm interceptor trap and one end of 100mm drain pipe built in; brickwork at top corbelled over for, and fitted with, manhole cover and frame**

Item	Description	Unit	Labour Hours	Labour Net £	Plant Net £	Materials Net £	Unit Rate Net £	Labour Gross £	Plant Gross £	Materials Gross £	Unit Rate Gross £
B.01.24.28.A	medium duty cast iron; B125 solid top; 600 x 450mm	nr	22.10	1485.00	—	681.97	2166.97	1633.51	—	750.16	2383.67
B.01.24.28.B	extra class B engineering bricks (PC sum £500 per 1000) in lieu of commons	nr	0.75	50.61	—	28.40	79.01	55.67	—	31.24	86.91
B.01.24.28.C	add or deduct for each 150mm depth in excess of or less than 1.20m in the foregoing interceptor chamber built in common bricks (PC sum £450 per 1000)	nr	2.11	141.75	—	73.39	215.13	155.92	—	80.72	236.64

Drainage, Sewerage and Public Works

Hutchins Priced Schedules 2021	Unit	Labour Hours	Labour Net £	Plant Net £	Materials Net £	Unit Rate Net £	Labour Gross £	Plant Gross £	Materials Gross £	Unit Rate Gross £
								(Gross rates include 10% profit)		

B.01. DRAINAGE AND SEWERAGE

B.01.24. MANHOLES, ETC.

B.01.24.28. Interceptor chamber; 0.75m x 0.60m x 1.20m deep. All excavation work; concrete base 150mm; one brick walls in common brickwork (PC sum £450 per 1000) in cement mortar, rendered internally; one straight channel and concrete benching; 100mm interceptor trap and one end of 100mm drain pipe built in; brickwork at top corbelled over for, and fitted with, manhole cover and frame

B.01.24.28.D	add or deduct for each 150mm depth in excess of or less than 1.20m in the foregoing interceptor chamber built in class B engineering bricks (PC sum £500 per 1000)	nr	2.41	161.98	—	80.49	242.46	178.17	—	88.53	266.71

B.01.26. INSPECTION CHAMBERS

B.01.26.01. Shallow inspection chamber; 250mm dia; place in excavation; invert depth

| B.01.26.01.A | 600mm .... | nr | 0.83 | 56.05 | — | 234.07 | 290.12 | 61.66 | — | 257.48 | 319.13 |

B.01.26.02. Single seal circular ductile iron cover and frame

| B.01.26.02.A | 475mm dia.................. | nr | 0.29 | 19.56 | — | 137.13 | 156.69 | 21.51 | — | 150.85 | 172.36 |

B.01.26.03. Universal inspection chamber; 475mm dia; placing in excavation; invert depth

B.01.26.03.A	500mm........................	nr	0.88	59.08	—	321.83	380.91	64.99	—	354.02	419.00
B.01.26.03.B	730mm........................	nr	1.05	70.23	—	424.13	494.37	77.26	—	466.55	543.81
B.01.26.03.C	960mm........................	nr	1.21	80.99	—	448.49	529.48	89.09	—	493.34	582.43

B.01.28. PIPE TRENCHES IN SOFT SOIL

B.01.28.01. Excavate trenches; by machine; grade bottoms; backfill; compact; dispose of surplus excavated material; earthwork support measured separately

B.01.28.01.A	0.60m width x 0.90m depth	m	1.45	42.89	3.91	—	46.80	47.18	4.30	—	51.48
B.01.28.01.B	0.60m width x 1.20m depth	m	1.80	53.24	5.08	—	58.33	58.57	5.59	—	64.16
B.01.28.01.C	0.60m width x 1.50m depth	m	2.30	68.03	6.26	—	74.29	74.84	6.88	—	81.72
B.01.28.01.D	0.75m width x 0.90m depth	m	1.65	48.81	4.65	—	53.46	53.69	5.12	—	58.81
B.01.28.01.E	0.75m width x 1.20m depth	m	2.30	68.03	6.26	—	74.29	74.84	6.88	—	81.72
B.01.28.01.F	0.75m width x 1.50m depth	m	2.70	79.87	7.78	—	87.65	87.85	8.56	—	96.41
B.01.28.01.G	0.75m width x 1.80m depth	m	3.50	103.53	9.34	—	112.87	113.88	10.28	—	124.16
B.01.28.01.H	0.75m width x 2.10m depth	m	3.65	107.97	11.30	—	119.27	118.76	12.43	—	131.19
B.01.28.01.I	0.75m width x 2.40m depth	m	4.40	130.15	12.47	—	142.62	143.17	13.72	—	156.89
B.01.28.01.J	0.75m width x 2.70m depth	m	5.00	147.90	15.17	—	163.07	162.69	16.69	—	179.38
B.01.28.01.K	0.75m width x 3.00m depth	m	5.90	174.52	16.77	—	191.30	191.97	18.45	—	210.43
B.01.28.01.L	0.90m width x 1.80m depth	m	3.65	107.97	11.30	—	119.27	118.76	12.43	—	131.19
B.01.28.01.M	0.90m width x 2.10m depth	m	4.00	118.32	13.22	—	131.54	130.15	14.54	—	144.69
B.01.28.01.N	0.90m width x 2.40m depth	m	5.00	147.90	15.17	—	163.07	162.69	16.69	—	179.38
B.01.28.01.O	0.90m width x 2.70m depth	m	5.40	159.73	16.77	—	176.51	175.71	18.45	—	194.16
B.01.28.01.P	0.90m width x 3.00m depth	m	6.70	198.19	18.69	—	216.88	218.00	20.56	—	238.56

Drainage, Sewerage and Public Works

Hutchins Priced Schedules 2021		Unit	Labour Hours	Labour Net £	Plant Net £	Materials Net £	Unit Rate Net £	Labour Gross £	Plant Gross £	Materials Gross £	Unit Rate Gross £
										(Gross rates include 10% profit)	
B.01.	**DRAINAGE AND SEWERAGE**										
B.01.29.	**PIPE TRENCHES IN CLAY OR COMPACT GRAVEL**										
B.01.29.01.	Excavate trenches; by machine; grade bottoms; backfill; compact; dispose of surplus excavated material; earthwork support measured separately										
B.01.29.01.A	0.60m width x 0.90m depth	m	1.55	45.85	4.65	—	50.50	50.43	5.12	—	55.55
B.01.29.01.B	0.60m width x 1.20m depth	m	1.90	56.20	6.26	—	62.46	61.82	6.88	—	68.70
B.01.29.01.C	0.60m width x 1.50m depth	m	2.40	70.99	7.78	—	78.77	78.09	8.56	—	86.65
B.01.29.01.D	0.75m width x 0.90m depth	m	1.75	51.77	5.08	—	56.85	56.94	5.59	—	62.53
B.01.29.01.E	0.75m width x 1.20m depth	m	2.40	70.99	7.39	—	78.38	78.09	8.13	—	86.22
B.01.29.01.F	0.75m width x 1.50m depth	m	2.80	82.82	9.34	—	92.17	91.11	10.28	—	101.39
B.01.29.01.G	0.75m width x 1.80m depth	m	3.40	100.57	11.30	—	111.87	110.63	12.43	—	123.06
B.01.29.01.H	0.75m width x 2.10m depth	m	3.80	112.40	13.22	—	125.62	123.64	14.54	—	138.18
B.01.29.01.I	0.75m width x 2.40m depth	m	4.60	136.07	15.17	—	151.24	149.67	16.69	—	166.36
B.01.29.01.J	0.75m width x 2.70m depth	m	5.30	156.77	16.77	—	173.55	172.45	18.45	—	190.90
B.01.29.01.K	0.75m width x 3.00m depth	m	6.50	192.27	18.30	—	210.57	211.50	20.13	—	231.63
B.01.29.01.L	0.90m width x 1.80m depth	m	3.80	112.40	13.22	—	125.62	123.64	14.54	—	138.18
B.01.29.01.M	0.90m width x 2.10m depth	m	4.15	122.76	15.99	—	138.75	135.03	17.59	—	152.62
B.01.29.01.N	0.90m width x 2.40m depth	m	5.30	156.77	17.95	—	174.72	172.45	19.74	—	192.19
B.01.29.01.O	0.90m width x 2.70m depth	m	5.60	165.65	20.25	—	185.90	182.21	22.28	—	204.49
B.01.29.01.P	0.90m width x 3.00m depth	m	7.25	214.46	22.60	—	237.05	235.90	24.86	—	260.76
B.01.30.	**PIPE TRENCHES IN CHALK**										
B.01.30.01.	Excavate trenches; by machine; grade bottoms; backfill; compact; dispose of surplus excavated material; earthwork support measured separately										
B.01.30.01.A	0.60m width x 0.90m depth	m	1.80	53.24	5.43	—	58.68	58.57	5.98	—	64.55
B.01.30.01.B	0.60m width x 1.20m depth	m	2.75	81.35	7.04	—	88.38	89.48	7.74	—	97.22
B.01.30.01.C	0.60m width x 1.50m depth	m	3.20	94.66	8.95	—	103.61	104.12	9.85	—	113.97
B.01.30.01.D	0.75m width x 0.90m depth	m	2.10	62.12	6.61	—	68.73	68.33	7.27	—	75.60
B.01.30.01.E	0.75m width x 1.20m depth	m	3.20	94.66	8.95	—	103.61	104.12	9.85	—	113.97
B.01.30.01.F	0.75m width x 1.50m depth	m	4.35	128.67	10.52	—	139.19	141.54	11.57	—	153.11
B.01.30.01.G	0.75m width x 1.80m depth	m	4.80	141.98	13.22	—	155.20	156.18	14.54	—	170.72
B.01.30.01.H	0.75m width x 2.10m depth	m	5.10	150.86	15.60	—	166.46	165.94	17.16	—	183.10
B.01.30.01.I	0.75m width x 2.40m depth	m	5.90	174.52	17.52	—	192.04	191.97	19.27	—	211.24
B.01.30.01.J	0.75m width x 2.70m depth	m	7.00	207.06	19.47	—	226.53	227.77	21.42	—	249.18
B.01.30.01.K	0.75m width x 3.00m depth	m	8.70	257.35	21.43	—	278.77	283.08	23.57	—	306.65
B.01.30.01.L	0.90m width x 1.80m depth	m	5.10	150.86	15.60	—	166.46	165.94	17.16	—	183.10

Hutchins Priced Schedules 2021	Unit	Labour Hours	Labour Net £	Plant Net £	Materials Net £	Unit Rate Net £	Labour Gross £	Plant Gross £	Materials Gross £	Unit Rate Gross £
							(Gross rates include 10% profit)			

B.01.   **DRAINAGE AND SEWERAGE**

B.01.30.   **PIPE TRENCHES IN CHALK**

B.01.30.01.   Excavate trenches; by machine; grade bottoms; backfill; compact; dispose of surplus excavated material; earthwork support measured separately

	Unit	Labour Hours	Labour Net £	Plant Net £	Materials Net £	Unit Rate Net £	Labour Gross £	Plant Gross £	Materials Gross £	Unit Rate Gross £
B.01.30.01.M 0.90m width x 2.10m depth	m	5.90	174.52	18.69	—	193.21	191.97	20.56	—	212.53
B.01.30.01.N 0.90m width x 2.40m depth	m	7.00	207.06	21.04	—	228.10	227.77	23.14	—	250.91
B.01.30.01.O 0.90m width x 2.70m depth	m	7.75	229.25	23.38	—	252.63	252.17	25.72	—	277.89
B.01.30.01.P 0.90m width x 3.00m depth	m	9.00	266.22	26.08	—	292.30	292.84	28.69	—	321.53
B.01.32. **BREAK UP PAVED SURFACES**										
B.01.32.01. **Extra over trench excavation for breaking up with compressed air equipment**										
B.01.32.01.A 75mm tarmacadam	m²	0.43	12.72	1.40	—	14.12	13.99	1.54	—	15.53
B.01.32.01.B 150mm plain concrete	m²	1.35	39.93	4.41	—	44.35	43.93	4.86	—	48.78
B.01.32.01.C 150mm reinforced concrete	m²	1.83	54.13	5.97	—	60.10	59.54	6.57	—	66.11
B.01.33. **REINSTATE PAVED SURFACES**										
B.01.33.01. **Reinstate paving; 100mm hardcore bed**										
B.01.33.01.A 75mm tarmacadam	m²	0.42	12.42	—	29.72	42.15	13.67	—	32.70	46.36
B.01.33.02. **Reinstate paving; 150mm hardcore bed**										
B.01.33.02.A 150mm plain concrete	m²	0.85	25.14	—	26.45	51.59	27.66	—	29.10	56.75
B.01.34. **OPEN EARTHWORK SUPPORT**										
B.01.34.01. **Four uses assumed; both sides of trench measured; average depth of trench**										
B.01.34.01.A 0.90m	m	0.22	6.51	—	2.80	9.31	7.16	—	3.08	10.24
B.01.34.01.B 1.20m	m	0.30	8.87	—	3.58	12.45	9.76	—	3.94	13.70
B.01.34.01.C 1.50m	m	0.40	11.83	—	4.43	16.26	13.02	—	4.87	17.89
B.01.34.01.D 1.80m	m	0.45	13.31	—	5.20	18.51	14.64	—	5.72	20.37
B.01.34.01.E 2.10m	m	0.55	16.27	—	6.43	22.70	17.90	—	7.07	24.97
B.01.34.01.F 2.40m	m	0.70	20.71	—	7.25	27.96	22.78	—	7.98	30.76
B.01.34.01.G 2.70m	m	0.85	25.14	—	8.01	33.15	27.66	—	8.81	36.46
B.01.34.01.H 3.00m	m	1.00	29.58	—	8.83	38.41	32.54	—	9.71	42.25
B.01.36. **CLOSE EARTHWORK SUPPORT**										
B.01.36.01. **Four uses assumed; both sides of trench measured; average depth of trench**										
B.01.36.01.A 0.90m	m	1.65	48.81	—	8.38	57.19	53.69	—	9.22	62.91
B.01.36.01.B 1.20m	m	2.20	65.08	—	10.83	75.91	71.58	—	11.92	83.50
B.01.36.01.C 1.50m	m	2.90	85.78	—	13.21	98.99	94.36	—	14.53	108.89
B.01.36.01.D 1.80m	m	3.50	103.53	—	15.66	119.19	113.88	—	17.23	131.11

Drainage, Sewerage and Public Works

Hutchins Priced Schedules 2021		Unit	Labour Hours	Labour Net £	Plant Net £	Materials Net £	Unit Rate Net £	Labour Gross £	Plant Gross £	Materials Gross £	Unit Rate Gross £
										(Gross rates include 10% profit)	
B.01.	**DRAINAGE AND SEWERAGE**										
B.01.36.	**CLOSE EARTHWORK SUPPORT**										
B.01.36.01.	Four uses assumed; both sides of trench measured; average depth of trench										
B.01.36.01.E	2.10m	m	4.10	121.28	—	19.29	140.57	133.41	—	21.22	154.62
B.01.36.01.F	2.40m	m	5.20	153.82	—	22.36	176.18	169.20	—	24.60	193.80
B.01.36.01.G	2.70m	m	6.20	183.40	—	24.09	207.49	201.74	—	26.50	228.24
B.01.36.01.H	3.00m	m	7.25	214.46	—	26.42	240.87	235.90	—	29.06	264.96
B.01.38.	**PIPE BEDS AND COVERINGS**										
B.01.38.01.	Plain concrete mix GEN1; 150mm bed; width										
B.01.38.01.A	400mm	m	0.43	12.72	—	5.83	18.55	13.99	—	6.41	20.41
B.01.38.01.B	450mm	m	0.50	14.79	—	6.53	21.32	16.27	—	7.18	23.45
B.01.38.01.C	525mm	m	0.58	17.16	—	7.66	24.82	18.87	—	8.43	27.30
B.01.38.01.D	600mm	m	0.72	21.30	—	8.70	30.00	23.43	—	9.57	33.00
B.01.38.01.E	675mm	m	0.85	25.14	—	9.75	34.89	27.66	—	10.72	38.38
B.01.38.01.F	750mm	m	1.00	29.58	—	10.88	40.46	32.54	—	11.97	44.51
B.01.38.01.G	825mm	m	1.14	33.72	—	12.01	45.73	37.09	—	13.21	50.31
B.01.38.01.H	900mm	m	1.28	37.86	—	13.14	51.01	41.65	—	14.46	56.11
B.01.38.02.	Plain concrete mix GEN1; 150mm bed and filling to half height of pipe; width										
B.01.38.02.A	400mm to 100mm pipe	m	0.59	17.45	—	7.31	24.76	19.20	—	8.04	27.24
B.01.38.02.B	450mm to 150mm pipe	m	0.74	21.89	—	8.88	30.77	24.08	—	9.77	33.84
B.01.38.02.C	525mm to 225mm pipe	m	0.91	26.92	—	10.53	37.45	29.61	—	11.59	41.19
B.01.38.02.D	600mm to 300mm pipe	m	1.25	36.98	—	13.40	50.38	40.67	—	14.74	55.42
B.01.38.02.E	675mm to 375mm pipe	m	1.37	40.52	—	15.84	56.37	44.58	—	17.43	62.00
B.01.38.02.F	750mm to 450mm pipe	m	1.63	48.22	—	18.45	66.67	53.04	—	20.30	73.33
B.01.38.02.G	825mm to 525mm pipe	m	1.90	56.20	—	21.32	77.53	61.82	—	23.46	85.28
B.01.38.02.H	900mm to 600mm pipe	m	2.06	60.93	—	24.11	85.04	67.03	—	26.52	93.55
B.01.38.04.	Plain concrete mix GEN1; 150mm bed and covering; width										
B.01.38.04.A	400mm to 100mm pipe	m	1.24	36.68	—	14.45	51.13	40.35	—	15.89	56.24
B.01.38.04.B	450mm to 150mm pipe	m	1.56	46.14	—	17.58	63.73	50.76	—	19.34	70.10
B.01.38.04.C	525mm to 225mm pipe	m	1.94	57.39	—	20.98	78.36	63.12	—	23.07	86.20
B.01.38.04.D	600mm to 300mm pipe	m	2.62	77.50	—	26.63	104.13	85.25	—	29.30	114.55
B.01.38.04.E	675mm to 375mm pipe	m	3.44	101.76	—	31.68	133.44	111.93	—	34.85	146.78
B.01.38.04.F	750mm to 450mm pipe	m	5.00	147.90	—	36.90	184.80	162.69	—	40.60	203.29
B.01.38.04.G	825mm to 525mm pipe	m	7.58	224.22	—	42.74	266.95	246.64	—	47.01	293.65
B.01.38.04.H	900mm to 600mm pipe	m	9.77	289.00	—	48.05	337.04	317.90	—	52.85	370.75

Drainage, Sewerage and Public Works

						(Gross rates include 10% profit)				
B.01.	**DRAINAGE AND SEWERAGE**									
B.01.40.	OSMA ULTRARIB FOUL AND SURFACE WATER DRAINAGE SYSTEM									
B.01.40.01.	**For sewers; in trenches; 150mm nominal dia pipes**									
B.01.40.01.A	in runs exceeding 3.00m long	m	0.25	7.40	—	14.00	21.39	8.13	—	15.40	23.53
B.01.40.01.B	extra for bend	nr	0.22	6.51	—	42.57	49.08	7.16	—	46.83	53.98
B.01.40.01.C	extra for junction	nr	0.29	8.58	—	87.62	96.20	9.44	—	96.38	105.82
B.01.40.01.D	extra for adaptor to clay spigot	nr	0.13	3.85	—	83.98	87.83	4.23	—	92.38	96.61
B.01.40.02.	**For sewers; in trenches; 225mm nominal dia pipes**									
B.01.40.02.A	in runs exceeding 3.00m long	m	0.29	8.58	—	40.69	49.27	9.44	—	44.76	54.19
B.01.40.02.B	extra for bend	nr	0.26	7.69	—	179.06	186.75	8.46	—	196.96	205.42
B.01.40.02.C	extra for junction	nr	0.35	10.35	—	249.43	259.78	11.39	—	274.37	285.76
B.01.40.02.D	extra for adaptor	nr	0.17	5.03	—	80.96	85.99	5.53	—	89.06	94.59
B.01.40.03.	**For sewers; in trenches; 300mm nominal dia pipes**									
B.01.40.03.A	in runs exceeding 3.00m long	m	0.42	12.42	—	62.91	75.33	13.67	—	69.20	82.86
B.01.40.03.B	extra for bend	nr	0.37	10.94	—	325.16	336.10	12.04	—	357.67	369.71
B.01.40.03.C	extra for junction	nr	0.48	14.20	—	509.82	524.02	15.62	—	560.80	576.42
B.01.40.03.D	extra for adaptor	nr	0.18	5.32	—	113.05	118.37	5.86	—	124.35	130.21
B.01.42.	**SUPERSLEVE DRAIN PIPES AND FITTINGS**									
B.01.42.01.	**Push-fit flexible socket joints; for sewerage; in trenches; 100mm nominal dia pipes**									
B.01.42.01.A	in runs exceeding 3.00m long	m	0.13	8.47	—	10.40	18.86	9.32	—	11.43	20.75
B.01.42.01.B	in runs not exceeding 3.00m long	m	0.17	11.16	—	10.40	21.55	12.27	—	11.43	23.71
B.01.42.01.C	extra for bend	nr	0.15	10.15	—	14.24	24.38	11.16	—	15.66	26.82
B.01.42.01.E	extra for junction	nr	0.13	8.47	—	30.77	39.24	9.32	—	33.85	43.16
B.01.42.02.	**Push-fit flexible socket joints; for sewerage; in trenches; 150mm nominal dia pipes**									
B.01.42.02.A	in runs exceeding 3.00m long	m	0.18	11.83	—	19.22	31.04	13.01	—	21.14	34.15
B.01.42.02.B	in runs not exceeding 3.00m long	m	0.23	15.19	—	19.22	34.40	16.71	—	21.14	37.84
B.01.42.02.C	extra for bend	nr	0.24	15.86	—	28.29	44.15	17.45	—	31.12	48.57
B.01.42.02.E	extra for junction	nr	0.15	10.15	—	41.56	51.71	11.16	—	45.72	56.88
B.01.42.03.	**Push-fit flexible socket joints; for sewerage; in trenches; 225mm nominal dia pipes**									
B.01.42.03.A	in runs exceeding 3.00m long	m	0.20	13.51	—	62.11	75.62	14.86	—	68.32	83.18

Drainage, Sewerage and Public Works

Hutchins Priced Schedules 2021		Unit	Labour Hours	Labour Net £	Plant Net £	Materials Net £	Unit Rate Net £	Labour Gross £	Plant Gross £	Materials Gross £	Unit Rate Gross £
							(Gross rates include 10% profit)				
B.01.	**DRAINAGE AND SEWERAGE**										
B.01.42.	SUPERSLEVE DRAIN PIPES AND FITTINGS										
B.01.42.03.	Push-fit flexible socket joints; for sewerage; in trenches; 225mm nominal dia pipes										
B.01.42.03.B	in runs not exceeding 3.00m long	m	0.25	16.87	—	62.11	78.98	18.56	—	68.32	86.87
B.01.42.03.C	extra for bend	nr	0.25	16.87	—	142.85	159.72	18.56	—	157.14	175.69
B.01.42.03.E	extra for junction	nr	0.22	14.85	—	256.16	271.01	16.34	—	281.78	298.11
B.01.42.04.	Push-fit flexible socket joints; for sewerage; in trenches; 300mm nominal dia pipes										
B.01.42.04.A	in runs exceeding 3.00m long	m	0.31	20.90	—	95.08	115.98	22.99	—	104.59	127.58
B.01.42.04.B	in runs not exceeding 3.00m long	m	0.39	26.35	—	95.08	121.42	28.98	—	104.59	133.57
B.01.42.04.C	extra for bend	nr	0.39	26.35	—	271.29	297.63	28.98	—	298.42	327.40
B.01.42.04.E	extra for junction	nr	0.37	25.00	—	543.06	568.06	27.50	—	597.37	624.87
B.01.44.	CONCRETE CYLINDRICAL DRAIN PIPES AND FITTINGS; CLASS 120										
B.01.44.02.	Flexible joints; for drainage and sewerage; in trenches; 225mm nominal dia pipes										
B.01.44.02.A	in runs exceeding 3.00m long	m	0.56	19.95	—	21.20	41.15	21.94	—	23.32	45.26
B.01.44.02.B	in runs not exceeding 3.00m long	m	0.75	26.61	—	21.20	47.81	29.27	—	23.32	52.59
B.01.44.02.C	extra for bend	nr	0.28	9.89	—	211.93	221.82	10.88	—	233.13	244.01
B.01.44.02.D	extra for junction 150mm	nr	0.18	6.33	—	222.04	228.37	6.96	—	244.25	251.21
B.01.44.03.	Flexible joints; for drainage and sewerage; in trenches; 300mm nominal dia pipes										
B.01.44.03.A	in runs exceeding 3.00m long	m	0.75	26.60	—	22.83	49.43	29.26	—	25.11	54.37
B.01.44.03.B	in runs not exceeding 3.00m long	m	1.00	35.48	—	22.83	58.31	39.03	—	25.11	64.14
B.01.44.03.C	extra for bend	nr	0.36	12.66	—	228.25	240.91	13.93	—	251.07	265.00
B.01.44.03.D	extra for junction 150mm	nr	0.23	8.09	—	159.78	167.87	8.90	—	175.76	184.66
B.01.44.04.	Flexible joints; for drainage and sewerage; in trenches; 375mm nominal dia pipes										
B.01.44.04.A	in runs exceeding 3.00m long	m	0.93	33.23	—	28.26	61.49	36.56	—	31.08	67.64
B.01.44.04.B	in runs not exceeding 3.00m long	m	1.25	44.36	—	28.26	72.61	48.79	—	31.08	79.87
B.01.44.04.C	extra for bend	nr	0.50	17.73	—	282.51	300.24	19.50	—	310.76	330.26
B.01.44.04.D	extra for junction 150mm	nr	0.32	11.36	—	197.76	209.11	12.49	—	217.53	230.03
B.01.44.05.	Flexible joints; for drainage and sewerage; in trenches; 450mm nominal dia pipes										
B.01.44.05.A	in runs exceeding 3.00m long	m	1.12	39.89	—	33.36	73.25	43.88	—	36.69	80.58
B.01.44.05.B	in runs not exceeding 3.00m long	m	1.50	53.23	—	33.36	86.58	58.55	—	36.69	95.24
B.01.44.05.C	extra for bend	nr	0.61	21.28	—	333.62	354.89	23.40	—	366.98	390.38

Hutchins Priced Schedules 2021	Unit	Labour Hours	Labour Net £	Plant Net £	Materials Net £	Unit Rate Net £	Labour Gross £	Plant Gross £	Materials Gross £	Unit Rate Gross £
										(Gross rates include 10% profit)
B.01.	**DRAINAGE AND SEWERAGE**									
B.01.44.	**CONCRETE CYLINDRICAL DRAIN PIPES AND FITTINGS; CLASS 120**									
B.01.44.05.	**Flexible joints; for drainage and sewerage; in trenches; 450mm nominal dia pipes**									
B.01.44.05.D	extra for junction 150mm	nr	0.39	13.61	—	233.53	247.14	14.97	—	256.88	271.86
B.01.44.06.	**Flexible joints; for drainage and sewerage; in trenches; 525mm nominal dia pipes**									
B.01.44.06.A	in runs exceeding 3.00m long	m	1.31	46.55	—	43.76	90.31	51.21	—	48.14	99.35
B.01.44.06.B	in runs not exceeding 3.00m long	m	1.75	62.09	—	43.76	105.86	68.30	—	48.14	116.44
B.01.44.06.C	extra for bend	nr	0.71	24.83	—	437.66	462.49	27.31	—	481.43	508.74
B.01.44.06.D	extra for junction 150mm	nr	0.45	15.90	—	306.37	322.27	17.49	—	337.01	354.49
B.01.44.07.	**Flexible joints; for drainage and sewerage; in trenches; 600mm nominal dia pipes**									
B.01.44.07.A	in runs exceeding 3.00m long	m	1.49	53.18	—	54.94	108.11	58.50	—	60.43	118.93
B.01.44.07.B	in runs not exceeding 3.00m long	m	2.00	70.97	—	54.94	125.91	78.07	—	60.43	138.50
B.01.44.07.C	extra for bend	nr	0.90	31.96	—	549.37	581.33	35.16	—	604.31	639.47
B.01.44.07.D	extra for junction 150mm	nr	0.52	18.15	—	384.56	402.71	19.96	—	423.02	442.98
B.01.48.	**PRECAST CONCRETE CIRCULAR MANHOLE RINGS AND ACCESSORIES**									
B.01.48.01.	**Bedding, jointing and pointing in cement mortar (1:3) on prepared bed; ring dia**									
B.01.48.01.B	900mm	m	4.02	269.98	—	93.99	363.97	296.98	—	103.39	400.37
B.01.48.01.C	1050mm	m	5.40	362.80	—	132.50	495.30	399.08	—	145.75	544.83
B.01.48.01.D	1200mm	m	6.78	455.62	—	152.82	608.44	501.18	—	168.10	669.28
B.01.48.02.	**Cover slabs; heavy duty; reinforced; 600mm access opening; ring dia**									
B.01.48.02.B	900mm	nr	1.56	104.65	—	87.55	192.19	115.11	—	96.30	211.41
B.01.48.02.C	1050mm	nr	1.81	121.52	—	92.84	214.36	133.67	—	102.13	235.79
B.01.48.02.D	1200mm	nr	2.06	138.39	—	112.72	251.10	152.22	—	123.99	276.21
B.01.48.03.	**Reducing slab; heavy duty; reinforced**									
B.01.48.03.A	1200mm - 1050mm dia	nr	0.53	35.29	8.43	147.92	191.64	38.81	9.27	162.72	210.80
B.01.50.	**ROAD GULLIES, GRATINGS, COVERS AND FRAMES**									
B.01.50.01.	**Vitrified clay road gullies including excavation; 150mm concrete bed and surround and jointing to drain**									
B.01.50.01.A	300mm x 600mm deep; 100mm outlet . .	nr	2.18	146.79	—	229.76	376.54	161.47	—	252.73	414.20
B.01.50.01.B	450mm x 900mm deep; 150mm outlet . .	nr	4.32	290.28	—	380.26	670.54	319.31	—	418.28	737.59

Drainage, Sewerage and Public Works

Hutchins Priced Schedules 2021		Unit	Labour Hours	Labour Net £	Plant Net £	Materials Net £	Unit Rate Net £	Labour Gross £	Plant Gross £	Materials Gross £	Unit Rate Gross £
								(Gross rates include 10% profit)			
B.01.	**DRAINAGE AND SEWERAGE**										
B.01.50.	ROAD GULLIES, GRATINGS, COVERS AND FRAMES										
B.01.50.02.	Cast iron hinged roadway gratings and frames; bedding on one course of class B engineering brickwork (PC sum £500 per 1000) built brick-on-edge										
B.01.50.02.A	385mm x 317mm	nr	1.06	70.91	—	54.72	125.63	78.00	—	60.19	138.19
B.01.50.02.B	500mm x 350mm	nr	1.11	74.27	—	81.83	156.09	81.69	—	90.01	171.70
B.01.50.03.	Access cover and frame; B125; medium duty; circular single seal solid top; bedding frame in cement and sand (1:3) and the cover in grease and sand										
B.01.50.03.A	450mm dia	nr	0.95	63.85	—	62.99	126.84	70.23	—	69.29	139.52
B.01.50.04.	Access cover and frame; C250 heavy duty; circular single seal solid top; bedding frame in cement and sand (1:3) and the cover in grease and sand										
B.01.50.04.A	600mm x 600mm	nr	1.26	84.35	—	125.91	210.26	92.78	—	138.50	231.29
B.02.	**DRAINAGE AND SEWERAGE REPAIRS AND ALTERATIONS**										
B.02.01.	REPAIRS										
B.02.01.01.	Break up concrete paving and hardcore under 600mm wide for excavation drain trench										
B.02.01.01.A	100mm thick	m	1.10	32.54	—	—	32.54	35.79	—	—	35.79
B.02.01.02.	Hand excavate trench for 100mm drain including backfilling; cart away remainder; including necessary earthwork support (four uses assumed)										
B.02.01.02.A	average 0.50m deep	m	1.44	42.60	—	3.88	46.47	46.85	—	4.27	51.12
B.02.01.02.B	average 1.00m deep	m	2.90	85.78	—	7.75	93.54	94.36	—	8.53	102.89
B.02.01.02.C	average 1.50m deep	m	5.16	152.63	—	11.63	164.27	167.90	—	12.80	180.69
B.02.01.02.D	100mm concrete bed and haunch to pipe	m	1.10	32.54	—	7.22	39.76	35.79	—	7.95	43.74
B.02.01.04.	Hepworth SuperSleve clayware pipes; laid and jointed in short lengths										
B.02.01.04.A	100mm	m	0.25	16.87	—	10.40	27.26	18.56	—	11.43	29.99
B.02.01.05.	Make good 100mm concrete paving and hardcore under, after drainwork										
B.02.01.05.A	average 600mm wide	m	1.30	38.45	—	6.09	44.55	42.30	—	6.70	49.00

Hutchins Priced Schedules 2021	Unit	Labour Hours	Labour Net £	Plant Net £	Materials Net £	Unit Rate Net £	Labour Gross £	Plant Gross £	Materials Gross £	Unit Rate Gross £
									(Gross rates include 10% profit)	
B.02.	**DRAINAGE AND SEWERAGE REPAIRS AND ALTERATIONS**									
B.02.01.	**REPAIRS**									
B.02.01.06.	Cut into brick side of exposed manhole and concrete benching for new branch drain and three quarter section channel; make good common brickwork (PC sum £450 per 1000) and benching									
B.02.01.06.A	100mm	nr	1.51	101.29	—	67.18	168.47	111.41	—	73.90	185.32
B.02.01.08.	Glazed clayware gully and grid with 100mm outlet joint to drain; including necessary excavation and concrete bed									
B.02.01.08.A	150mm x 150mm x 100mm	nr	1.21	80.99	—	89.66	170.65	89.09	—	98.62	187.71
B.02.01.09.	Concrete kerb around glazed clayware gully; including necessary formwork									
B.02.01.09.A	100mm	nr	0.63	42.21	—	5.05	47.26	46.43	—	5.55	51.98
B.02.01.10.	Demolish kerb; disconnect and remove gully; supply and connect new gully and grid including work to concrete bed; new kerb; render three sides and remake connection after rodding									
B.02.01.10.A	150mm x 150mm x 100mm	nr	3.01	202.50	—	93.04	295.55	222.75	—	102.35	325.10
B.02.01.11.	Break up defective kerb surround to manhole and reform in									
B.02.01.11.A	fine concrete splayed and rendered	nr	0.78	52.29	—	6.00	58.29	57.52	—	6.60	64.12
B.02.01.12.	Cut through external brick wall and 150mm concrete floor; and connect 102mm bend to trap of pan at one end existing pipe at the other end									
B.02.01.12.A	225mm brick wall	nr	3.77	253.11	—	22.45	275.56	278.42	—	24.69	303.12
B.02.01.14.	Form new manhole on line of existing drain with 150mm concrete base, 225mm common brick walls (PC sum £450 per 1000), access cover and frame; B125; medium duty; circular single seal solid top 600 x 450mm; cut away existing drain within manhole to form channel; provide and set 2nr three-quarter section channels; form concrete benching and cement render walls									
B.02.01.14.A	0.90m deep	nr	18.08	1215.02	—	385.44	1600.46	1336.52	—	423.98	1760.51
B.02.01.14.B	add for every extra 300mm depth in excess depth in excess of 0.90m	nr	4.52	303.72	—	53.63	357.35	334.09	—	59.00	393.09

Drainage, Sewerage and Public Works

Hutchins Priced Schedules 2021	Unit	Labour Hours	Labour Net £	Plant Net £	Materials Net £	Unit Rate Net £	Labour Gross £	Plant Gross £	Materials Gross £	Unit Rate Gross £
									(Gross rates include 10% profit)	
B.02.	**DRAINAGE AND SEWERAGE REPAIRS AND ALTERATIONS**									
B.02.01.	**REPAIRS**									
B.02.01.16.	**Take up existing manhole cover and frame, and provide, bed and seal new cover and frame; 600mm x 450mm**									
B.02.01.16.A	light duty cast iron; A15 solid top	nr	0.55	37.10	—	67.67	104.77	40.81	—	74.43	115.24
B.02.01.16.B	medium duty cast iron; B125 solid top . .	nr	0.65	43.89	—	116.29	160.18	48.28	—	127.92	176.19
B.02.01.18.	**Excavate for stoppage in 100mm drain; cut out and renew two lengths of pipe; fill in and test; reform paving**									
B.02.01.18.A	assumed depth 0.90m	nr	5.02	337.53	—	35.88	373.41	371.28	—	39.46	410.75
B.02.01.19.	**Unstop gullies; remove silt**									
B.02.01.19.A	clean and flush with disinfectant	nr	0.58	38.85	—	—	38.85	42.73	—	—	42.73
B.02.01.21.	**Clear drains; rod and flush in sections**									
B.02.01.21.A	average length 25.00m	nr	2.89	194.04	—	—	194.04	213.44	—	—	213.44
B.02.01.22.	**Unstop and smoke test soil pipe**									
B.02.01.22.A	cast iron	nr	2.11	141.75	—	—	141.75	155.92	—	—	155.92
B.02.01.23.	**Take up manhole cover; clean and sand out frame channels**									
B.02.01.23.A	rebed in grease	nr	0.55	37.10	—	—	37.10	40.81	—	—	40.81
B.02.01.24.	**Clayware channel to gully set in concrete with brick kerb rendered in cement mortar**									
B.02.01.24.A	450mm long	nr	1.00	67.48	—	14.62	82.10	74.23	—	16.08	90.31
B.02.01.24.B	supply and fit new gully grid	nr	0.13	8.47	—	8.09	16.56	9.32	—	8.90	18.21
B.02.01.26.	**Open up manhole; break out brickwork and benching to main channel; provide for and insert one three quarter channel bend 100mm and join to existing channel; reform benching**									
B.02.01.26.A	make good all work disturbed and refix manhole cover	nr	5.02	337.53	—	42.74	380.27	371.28	—	47.02	418.30

Hutchins Priced Schedules 2021	Unit	Labour Hours	Labour Net £	Plant Net £	Materials Net £	Unit Rate Net £	Labour Gross £	Plant Gross £	Materials Gross £	Unit Rate Gross £
										(Gross rates include 10% profit)

B.02.   **DRAINAGE AND SEWERAGE REPAIRS AND ALTERATIONS**

B.02.01.   **REPAIRS**

B.02.01.28.   Open up manhole, 0.90m x 0.90m inside on plan; break out all channels and branches; break out one brick side of manhole at one end and extend manhole by 150mm x 0.90m inside with 150mm concrete at bottom; one brick side in stock bricks and 150mm concrete cover; build in two pipes 150mm and one pipe 100mm; provide and insert straight main channel 150mm and six half branch channel bends 100mm and reform benching; make good all work disturbed and refix manhole cover (common brickwork PC sum £450 per 1000)

B.02.01.28.A 1.27m deep.................	nr	12.05	810.01	—	432.11	1242.13	891.02	—	475.32	1366.34
B.02.01.28.B extra over excavation for breaking out brick manhole 0.80m x 0.70m inside on plan and 0.90m deep to invert.......	nr	1.51	101.29	—	—	101.29	111.41	—	—	111.41
B.02.01.28.C add or deduct for every 300mm more or less than 0.90m deep to invert.......	nr	0.50	33.74	—	—	33.74	37.11	—	—	37.11

B.02.01.30.   Seal open ends of disused clayware pipes with concrete plugs; 300mm long

B.02.01.30.A 100mm dia...................	nr	0.30	20.23	—	0.26	20.49	22.25	—	0.29	22.54
B.02.01.30.B 150mm dia...................	nr	0.40	27.02	—	0.61	27.63	29.72	—	0.67	30.39
B.02.01.30.C 225mm dia...................	nr	0.55	37.10	—	2.09	39.19	40.81	—	2.30	43.11

B.03.   **ROADS, FOOTWAYS, KERBS, WATER MAINS, CABLE LAYING AND PILING**

B.03.01.   **ROADS**

B.03.01.01.   Cut up turves; wheel away and stack

B.03.01.01.A not exceeding 100mm...........	m²	0.40	11.83	—	—	11.83	13.02	—	—	13.02

B.03.01.02.   Take turves off stack, re-lay, level and roll

B.03.01.02.A not exceeding 100m............	m²	0.55	16.27	—	—	16.27	17.90	—	—	17.90

B.03.01.03.   Excavate oversite average 150mm deep; wheel and deposit

B.03.01.03.A by hand....................	m²	0.38	11.24	—	—	11.24	12.36	—	—	12.36
B.03.01.03.B by machine.................	m²	—	—	0.78	—	0.78	—	0.86	—	0.86

Drainage, Sewerage and Public Works

Hutchins Priced Schedules 2021	Unit	Labour Hours	Labour Net £	Plant Net £	Materials Net £	Unit Rate Net £	Labour Gross £	Plant Gross £	Materials Gross £	Unit Rate Gross £
								(Gross rates include 10% profit)		

B.03. ROADS, FOOTWAYS, KERBS, WATER MAINS, CABLE LAYING AND PILING

B.03.01. ROADS

B.03.01.04. Hand excavation to reduce levels average 150mm deep; remove excavated material by dumper 100m; spread and level

	Unit	Labour Hours	Labour Net £	Plant Net £	Materials Net £	Unit Rate Net £	Labour Gross £	Plant Gross £	Materials Gross £	Unit Rate Gross £
B.03.01.04.A loose soil	m²	0.55	16.27	0.21	—	16.48	17.90	0.23	—	18.13
B.03.01.04.B firm soil or sand	m²	0.66	19.52	0.21	—	19.73	21.48	0.23	—	21.71
B.03.01.04.C light clay, compact soil or gravel	m²	0.69	20.41	0.21	—	20.62	22.45	0.23	—	22.68
B.03.01.04.D stiff heavy clay	m²	0.92	27.21	0.21	—	27.43	29.93	0.23	—	30.17
B.03.01.04.E soft chalk	m²	1.38	40.82	0.21	—	41.03	44.90	0.23	—	45.14

B.03.01.05. Hand excavation to reduce levels average 225mm deep; remove excavated material by dumper 100m; and deposit

	Unit	Labour Hours	Labour Net £	Plant Net £	Materials Net £	Unit Rate Net £	Labour Gross £	Plant Gross £	Materials Gross £	Unit Rate Gross £
B.03.01.05.A loose soil	m²	0.65	19.23	0.35	—	19.58	21.15	0.39	—	21.54
B.03.01.05.B firm soil or sand	m²	0.78	23.07	0.35	—	23.42	25.38	0.39	—	25.77
B.03.01.05.C light clay, compact soil or gravel	m²	0.81	23.96	0.35	—	24.31	26.36	0.39	—	26.74
B.03.01.05.D stiff heavy clay	m²	1.08	31.95	0.35	—	32.30	35.14	0.39	—	35.53
B.03.01.05.E soft chalk	m²	1.63	48.22	0.35	—	48.57	53.04	0.39	—	53.42

B.03.01.06. Machine excavation to reduce level; load excavated material into

	Unit	Labour Hours	Labour Net £	Plant Net £	Materials Net £	Unit Rate Net £	Labour Gross £	Plant Gross £	Materials Gross £	Unit Rate Gross £
B.03.01.06.A lorries	m³	1.00	29.58	10.13	—	39.71	32.54	11.14	—	43.68

B.03.01.07. Hand excavation to reduce levels; remove excavated material by dumper 400m and deposit

	Unit	Labour Hours	Labour Net £	Plant Net £	Materials Net £	Unit Rate Net £	Labour Gross £	Plant Gross £	Materials Gross £	Unit Rate Gross £
B.03.01.07.A loose soil	m³	3.00	88.74	5.62	—	94.36	97.61	6.18	—	103.79
B.03.01.07.B firm soil or sand	m³	3.60	106.49	5.62	—	112.11	117.14	6.18	—	123.32
B.03.01.07.C light clay, compact soil or gravel	m³	3.75	110.93	5.62	—	116.54	122.02	6.18	—	128.20
B.03.01.07.D stiff heavy clay	m³	5.00	147.90	5.62	—	153.52	162.69	6.18	—	168.87
B.03.01.07.E soft chalk	m³	7.50	221.85	5.62	—	227.47	244.04	6.18	—	250.22

B.03.01.08. Consolidate

	Unit	Labour Hours	Labour Net £	Plant Net £	Materials Net £	Unit Rate Net £	Labour Gross £	Plant Gross £	Materials Gross £	Unit Rate Gross £
B.03.01.08.A formation	m²	0.10	2.96	—	—	2.96	3.25	—	—	3.25

B.03.01.09. Roll formation

	Unit	Labour Hours	Labour Net £	Plant Net £	Materials Net £	Unit Rate Net £	Labour Gross £	Plant Gross £	Materials Gross £	Unit Rate Gross £
B.03.01.09.B 8t roller	m²	0.06	1.80	1.20	—	3.00	1.98	1.32	—	3.30

B.03.01.10. Break up 100mm tarmacadam paving

	Unit	Labour Hours	Labour Net £	Plant Net £	Materials Net £	Unit Rate Net £	Labour Gross £	Plant Gross £	Materials Gross £	Unit Rate Gross £
B.03.01.10.A by hand	m²	1.30	38.45	—	—	38.45	42.30	—	—	42.30
B.03.01.10.B by two tool compressor	m²	0.20	5.92	0.90	—	6.82	6.51	0.99	—	7.50

Hutchins Priced Schedules 2021		Unit	Labour Hours	Labour Net £	Plant Net £	Materials Net £	Unit Rate Net £	Labour Gross £	Plant Gross £	Materials Gross £	Unit Rate Gross £
							(Gross rates include 10% profit)				
B.03.	**ROADS, FOOTWAYS, KERBS, WATER MAINS, CABLE LAYING AND PILING**										
B.03.01.	**ROADS**										
B.03.01.11.	**Break up 100mm tarmacadam paving and hardcore bed**										
B.03.01.11.A	by hand	m²	1.50	44.37	—	—	44.37	48.81	—	—	48.81
B.03.01.11.B	by two tool compressor	m²	0.30	8.87	1.26	—	10.14	9.76	1.39	—	11.15
B.03.01.13.	**Break up 150mm surface concrete**										
B.03.01.13.A	by hand	m²	2.50	73.95	—	—	73.95	81.35	—	—	81.35
B.03.01.13.B	by two tool compressor	m²	0.40	11.83	1.33	—	13.16	13.02	1.47	—	14.48
B.03.01.14.	**Break up 150mm surface concrete and 100mm hardcore bed**										
B.03.01.14.A	by hand	m²	3.00	88.74	—	—	88.74	97.61	—	—	97.61
B.03.01.14.B	by two tool compressor	m²	0.50	14.79	2.27	—	17.06	16.27	2.50	—	18.77
B.03.01.15.	**Break up 225mm surface concrete**										
B.03.01.15.A	by hand	m²	4.00	118.32	—	—	118.32	130.15	—	—	130.15
B.03.01.15.B	by two tool compressor	m²	0.60	17.75	3.28	—	21.03	19.52	3.61	—	23.13
B.03.01.16.	**Break up 300mm surface concrete**										
B.03.01.16.A	by hand	m²	6.00	177.48	—	—	177.48	195.23	—	—	195.23
B.03.01.16.B	by two tool compressor	m²	1.00	29.58	3.79	—	33.37	32.54	4.17	—	36.70
B.03.01.18.	**Break up 150mm reinforced surface concrete**										
B.03.01.18.A	by hand	m²	3.75	110.93	—	—	110.93	122.02	—	—	122.02
B.03.01.18.B	by two tool compressor	m²	0.60	17.75	3.28	—	21.03	19.52	3.61	—	23.13
B.03.01.20.	**Break up 225mm reinforced surface concrete**										
B.03.01.20.A	by hand	m²	6.00	177.48	—	—	177.48	195.23	—	—	195.23
B.03.01.20.B	by two tool compressor	m²	0.90	26.62	4.91	—	31.53	29.28	5.40	—	34.68
B.03.01.21.	**Break up 300mm reinforced surface concrete**										
B.03.01.21.A	by hand	m²	9.00	266.22	—	—	266.22	292.84	—	—	292.84
B.03.01.21.B	by two tool compressor	m²	1.50	44.37	6.53	—	50.90	48.81	7.18	—	55.99
B.03.01.22.	**Level off slightly uneven areas by bulldozer and remove arisings; distance**										
B.03.01.22.A	20.00m	m²	0.25	7.40	1.96	—	9.35	8.13	2.15	—	10.29
B.03.01.24.	**Take up granite sets**										
B.03.01.24.A	set aside for reuse	m²	0.60	17.75	—	—	17.75	19.52	—	—	19.52
B.03.01.25.	**Clean granite sets; relay on**										
B.03.01.25.A	25mm cement and sand bed and grout joints	m²	2.70	79.87	—	4.98	84.85	87.85	—	5.48	93.33

Drainage, Sewerage and Public Works

Hutchins Priced Schedules 2021	Unit	Labour Hours	Labour Net £	Plant Net £	Materials Net £	Unit Rate Net £	Labour Gross £	Plant Gross £	Materials Gross £	Unit Rate Gross £
							(Gross rates include 10% profit)			
B.03.	**ROADS, FOOTWAYS, KERBS, WATER MAINS, CABLE LAYING AND PILING**									
B.03.01.	**ROADS**									
B.03.01.26.	**Hand packed boulder stone pitching blinded with ashes or fine ballast and rolled with heavy roller**									
B.03.01.26.A	225mm thick	m²	0.90	26.65	7.66	19.81	54.12	29.32	8.43	21.79	59.54
B.03.01.28.	**Furnace clinker hardcore spread levelled and rolled; thickness**									
B.03.01.28.A	100mm	m²	0.20	5.92	—	6.62	12.53	6.51	—	7.28	13.79
B.03.01.28.B	150mm	m²	0.25	7.40	—	9.95	17.35	8.13	—	10.95	19.08
B.03.01.29.	**Brick hardcore bed spread levelled and rolled; thickness**									
B.03.01.29.A	100mm	m²	0.36	10.65	—	6.62	17.27	11.71	—	7.28	18.99
B.03.01.29.B	150mm	m²	0.54	15.97	—	9.95	25.92	17.57	—	10.95	28.52
B.03.01.29.C	225mm	m²	0.81	23.96	—	14.88	38.84	26.36	—	16.37	42.72
B.03.01.30.	**Blind hardcore with**									
B.03.01.30.A	ashes	m²	0.16	4.73	—	9.10	13.83	5.21	—	10.01	15.22
B.03.01.32.	**Bitumen macadam 100mm work to roads in two coats; 70mm thick base course; 30mm thick wearing course of 10mm graded limestone aggregate; grit sprayed on**									
B.03.01.32.A	generally	m²	0.60	17.75	—	44.52	62.27	19.52	—	48.97	68.49
B.03.01.34.	**Polythene building sheets; medium grade laying on**									
B.03.01.34.A	hardcore or ashes including laps	m²	0.05	1.48	—	0.83	2.31	1.63	—	0.91	2.54
B.03.01.36.	**Blinding layer**									
B.03.01.36.A	50mm concrete GEN1	m²	0.35	10.35	—	5.31	15.66	11.39	—	5.84	17.23
B.03.01.38.	**Surface concrete GEN2 spread and levelled to falls and cambers; tamped around reinforcement (measured separately); thickness**									
B.03.01.38.A	125mm	m²	0.75	22.19	—	13.06	35.24	24.40	—	14.36	38.76
B.03.01.38.B	150mm	m²	0.85	25.14	—	14.93	40.08	27.66	—	16.43	44.08
B.03.01.38.C	175mm	m²	1.00	29.58	—	17.44	47.02	32.54	—	19.18	51.72
B.03.01.38.D	225mm	m²	1.15	34.02	—	22.36	56.37	37.42	—	24.59	62.01
B.03.01.39.	**Fabric reinforcement lapped in surface concrete; ref**									
B.03.01.39.A	C283 weighing 2.61kg/m2	m²	0.06	4.03	—	3.83	7.87	4.44	—	4.22	8.65
B.03.01.39.B	C385 weighing 3.41kg/m2	m²	0.08	5.04	—	5.01	10.05	5.54	—	5.51	11.05

Drainage, Sewerage and Public Works

Hutchins Priced Schedules 2021		Unit	Labour Hours	Labour Net £	Plant Net £	Materials Net £	Unit Rate Net £	Labour Gross £	Plant Gross £	Materials Gross £	Unit Rate Gross £
								(Gross rates include 10% profit)			
B.03.	**ROADS, FOOTWAYS, KERBS, WATER MAINS, CABLE LAYING AND PILING**										
B.03.01.	**ROADS**										
B.03.01.39.	**Fabric reinforcement lapped in surface concrete; ref**										
B.03.01.39.C	C503 weighing 4.34kg/m2	m²	0.10	6.72	—	6.38	13.11	7.39	—	7.02	14.42
B.03.01.39.D	C636 weighing 5.55kg/m2	m²	0.13	8.47	—	8.16	16.63	9.32	—	8.97	18.29
B.03.01.40.	**Expansion joint**										
B.03.01.40.A	20mm x 175mm high	m	0.25	7.40	—	3.79	11.19	8.13	—	4.17	12.31
B.03.01.41.	**Running top edge of expansion joint with**										
B.03.01.41.A	bitumen	m	0.18	5.32	—	0.20	5.52	5.86	—	0.22	6.07
B.03.01.42.	**Extra over surface concrete for forming channel**										
B.03.01.42.A	300mm wide including formwork	m	0.25	7.40	—	0.71	8.10	8.13	—	0.78	8.91
B.03.01.44.	**Treating surface of concrete with silicate of soda**										
B.03.01.44.A	two coats	m²	0.12	3.55	—	0.76	4.31	3.90	—	0.83	4.74
B.03.01.44.B	three coats	m²	0.20	5.92	—	1.14	7.05	6.51	—	1.25	7.76
B.03.01.46.	**Formwork to edge of surface concrete; straight; height not exceeding**										
B.03.01.46.A	125mm	m	0.19	7.15	—	0.24	7.39	7.86	—	0.26	8.12
B.03.01.46.B	150mm	m	0.23	8.65	—	0.28	8.93	9.52	—	0.30	9.82
B.03.01.46.C	175mm	m	0.27	10.16	—	0.32	10.48	11.18	—	0.35	11.52
B.03.01.46.D	225mm	m	0.34	12.79	—	0.35	13.15	14.07	—	0.39	14.46
B.03.01.47.	**Formwork to edge of surface concrete; curved; height not exceeding**										
B.03.01.47.A	125mm	m	0.38	14.30	—	0.48	14.78	15.73	—	0.53	16.26
B.03.01.47.B	150mm	m	0.46	17.31	—	0.55	17.86	19.04	—	0.61	19.65
B.03.01.47.C	175mm	m	0.54	20.32	—	0.57	20.89	22.35	—	0.63	22.98
B.03.01.47.D	225mm	m	0.68	25.59	—	0.84	26.43	28.15	—	0.92	29.07
B.03.04.	**FOOTWAYS**										
B.03.04.01.	**Take up precast concrete paving slabs; rebed in**										
B.03.04.01.A	lime mortar (1:4), grout in cement mortar (1:3)	m²	1.20	35.50	—	5.52	41.02	39.05	—	6.07	45.12
B.03.04.02.	**Precast concrete paving slabs 50mm thick; bedding in**										
B.03.04.02.A	lime mortar (1:4), grout in cement mortar (1:3)	m²	0.65	19.23	—	23.39	42.62	21.15	—	25.73	46.88

Drainage, Sewerage and Public Works

Hutchins Priced Schedules 2021		Unit	Labour Hours	Labour Net £	Plant Net £	Materials Net £	Unit Rate Net £	Labour Gross £	Plant Gross £	Materials Gross £	Unit Rate Gross £
							(Gross rates include 10% profit)				
B.03.	**ROADS, FOOTWAYS, KERBS, WATER MAINS, CABLE LAYING AND PILING**										
B.03.04.	**FOOTWAYS**										
B.03.04.03.	**Cutting on paving slabs**										
B.03.04.03.A	raking	m	0.63	18.64	—	—	18.64	20.50	—	—	20.50
B.03.04.03.B	curved	m	1.10	32.54	—	—	32.54	35.79	—	—	35.79
B.03.04.04.	**Gravel paving; two coats; rolled; thickness**										
B.03.04.04.A	50mm	m²	0.30	8.87	0.43	2.49	11.80	9.76	0.48	2.74	12.98
B.03.04.04.B	63mm	m²	0.35	10.35	0.54	3.24	14.13	11.39	0.60	3.56	15.55
B.03.04.05.	**Fine clinker ash bed; 75mm thick**										
B.03.04.05.A	spread, levelled and rolled	m²	0.30	8.87	0.43	16.43	25.74	9.76	0.48	18.07	28.31
B.03.04.06.	**Sand bed; 25mm thick**										
B.03.04.06.A	spread, levelled and rolled	m²	0.10	2.96	0.22	1.06	4.23	3.25	0.24	1.16	4.66
B.03.04.08.	**Bitumen macadam; 65mm work to footway**										
B.03.04.08.A	two coats; 45mm base course; 20mm wearing course of 6mm medium graded limestone aggregate; 14mm chippings sprinkled and rolled into wearing course .	m²	0.60	17.78	7.66	33.39	58.83	19.56	8.43	36.73	64.71
B.03.04.10.	**Concrete GEN2 paving; thickness**										
B.03.04.10.A	100mm	m²	0.60	17.75	—	9.93	27.67	19.52	—	10.92	30.44
B.03.04.10.B	150mm	m²	0.85	25.14	—	14.93	40.08	27.66	—	16.43	44.08
B.03.04.11.	**Extra for marking out**										
B.03.04.11.A	concrete paving in panels	m²	0.30	8.87	—	—	8.87	9.76	—	—	9.76
B.03.04.12.	**Precast concrete edging; bedded and pointed in cement mortar (1:3) on and including 150mm x 75mm concrete foundation and haunching**										
B.03.04.12.A	50mm x 150mm flat top	m	0.40	11.83	—	6.95	18.78	13.02	—	7.64	20.66
B.03.04.12.B	50mm x 150mm bullnosed	m	0.40	11.83	—	8.94	20.78	13.02	—	9.84	22.85
B.03.04.13.	**Treated softwood edging; 150mm x 38mm; staked at**										
B.03.04.13.A	1050mm centres	m	0.30	8.87	—	4.62	13.49	9.76	—	5.08	14.84
B.03.05.	**KERBS**										
B.03.05.01.	**Hand excavation; 300mm x 300mm foundation trench**										
B.03.05.01.A	consolidate	m	0.20	5.92	—	—	5.92	6.51	—	—	6.51

Drainage, Sewerage and Public Works

Hutchins Priced Schedules 2021	Unit	Labour Hours	Labour Net £	Plant Net £	Materials Net £	Unit Rate Net £	Labour Gross £	Plant Gross £	Materials Gross £	Unit Rate Gross £
										(Gross rates include 10% profit)
B.03.	**ROADS, FOOTWAYS, KERBS, WATER MAINS, CABLE LAYING AND PILING**									
B.03.05.	**KERBS**									
B.03.05.02.	**Concrete GEN1 bed and haunching to kerb**									
B.03.05.02.A	300mm high overall	m	0.33	9.76	—	4.61	14.37	10.74	—	5.07	15.81
B.03.05.03.	**Precast concrete kerb 150mm x 305mm; bedded, jointed and pointed in cement mortar (1:3); haunched both sides with concrete**									
B.03.05.03.A	straight	m	0.35	23.66	—	15.42	39.07	26.02	—	16.96	42.98
B.03.05.03.B	curved	m	0.70	47.25	—	15.42	62.66	51.97	—	16.96	68.93
B.03.05.04.	**Precast concrete channel 125mm x 255mm; bedded, jointed and pointed in cement mortar (1:3); haunched both sides with concrete**									
B.03.05.04.A	straight	m	0.25	16.87	—	11.73	28.60	18.56	—	12.90	31.46
B.03.05.04.B	curved	m	0.50	33.74	—	11.73	45.47	37.11	—	12.90	50.02
B.03.05.05.	**Precast concrete quadrant; bedded, jointed and pointed in cement mortar (1:3); haunched with concrete**									
B.03.05.05.A	305mm radius x 255mm deep	nr	0.28	18.55	—	31.62	50.17	20.40	—	34.78	55.19
B.03.05.05.B	455mm radius x 255mm deep	nr	0.33	21.91	—	41.12	63.03	24.10	—	45.23	69.33
B.03.06.	**WATER MAINS**									
B.03.06.01.	**Excavation of trenches to receive pipes 0.60m wide x 1.00m deep; grading bottom; earthwork support; backfilling; compacting; disposal of surplus excavated material; by hand**									
B.03.06.01.B	firm soil	m	2.63	77.80	—	0.90	78.70	85.57	—	0.99	86.57
B.03.06.02.	**Excavation of trenches to receive pipes 0.60m wide x 1.00m deep; grading bottom; earthwork support; backfilling; compacting; disposal of surplus excavated material; by machine**									
B.03.06.02.B	firm soil	m	0.44	13.02	17.13	0.90	31.04	14.32	18.84	0.99	34.15
B.03.06.04.	**Extra over excavating tarmacadam road surfacing**									
B.03.06.04.A	replace with existing consolidated broken tarmacadam as temporary surfacing . . .	m	0.70	20.71	—	—	20.71	22.78	—	—	22.78
B.03.06.04.B	remove existing tarmacadam and reinstate with new tarmacadam	m	1.10	32.54	5.83	28.98	67.34	35.79	6.41	31.88	74.08

Drainage, Sewerage and Public Works

Hutchins Priced Schedules 2021	Unit	Labour Hours	Labour Net £	Plant Net £	Materials Net £	Unit Rate Net £	Labour Gross £	Plant Gross £	Materials Gross £	Unit Rate Gross £
							(Gross rates include 10% profit)			

B.03. ROADS, FOOTWAYS, KERBS, WATER MAINS, CABLE LAYING AND PILING

B.03.06. WATER MAINS

B.03.06.06. Break up concrete paving by drill; replace with existing broken concrete; consolidate as temporary surfacing

	Unit	Labour Hours	Labour Net £	Plant Net £	Materials Net £	Unit Rate Net £	Labour Gross £	Plant Gross £	Materials Gross £	Unit Rate Gross £
B.03.06.06.A 150mm	m	1.00	29.58	—	—	29.58	32.54	—	—	32.54

B.03.06.10. Hand excavation for hydrant pit beyond extent of pipe trench

	Unit	Labour Hours	Labour Net £	Plant Net £	Materials Net £	Unit Rate Net £	Labour Gross £	Plant Gross £	Materials Gross £	Unit Rate Gross £
B.03.06.10.A part backfill and dispose of surplus excavated material	nr	1.50	44.37	—	—	44.37	48.81	—	—	48.81

B.03.06.11. Class B engineering bricks (PC sum £500 per 1000); walls of hydrant pit; half brick thick

	Unit	Labour Hours	Labour Net £	Plant Net £	Materials Net £	Unit Rate Net £	Labour Gross £	Plant Gross £	Materials Gross £	Unit Rate Gross £
B.03.06.11.A laid dry	nr	1.00	67.48	—	45.50	112.98	74.23	—	50.05	124.28
B.03.06.11.B cement mortar (1:3)	nr	2.51	168.76	—	49.30	218.06	185.64	—	54.23	239.87

B.03.06.12. Concrete surround to surface box

	Unit	Labour Hours	Labour Net £	Plant Net £	Materials Net £	Unit Rate Net £	Labour Gross £	Plant Gross £	Materials Gross £	Unit Rate Gross £
B.03.06.12.A GEN1 mix	nr	0.45	13.31	—	7.40	20.71	14.64	—	8.14	22.78

B.03.06.14. Surface box; cast iron opening size 380mm x 230mm; 100mm deep with drop-in lid and chain; bedded and flaunched in cement mortar (1:3)

	Unit	Labour Hours	Labour Net £	Plant Net £	Materials Net £	Unit Rate Net £	Labour Gross £	Plant Gross £	Materials Gross £	Unit Rate Gross £
B.03.06.14.A 420mm x 255mm overall	nr	0.50	33.74	—	40.72	74.46	37.11	—	44.79	81.90
B.03.06.14.B temporary reinstatement	nr	1.00	29.58	—	2.35	31.93	32.54	—	2.59	35.12
B.03.06.14.C cost per pit complete	nr	5.96	289.76	—	99.68	389.44	318.74	—	109.64	428.39

B.03.07. CABLE LAYING FOR ELECTRICITY AUTHORITIES

B.03.07.01. Excavation of trenches to receive cables; grading bottom; compacting; part backfill and dispose of surplus excavated material; by hand

	Unit	Labour Hours	Labour Net £	Plant Net £	Materials Net £	Unit Rate Net £	Labour Gross £	Plant Gross £	Materials Gross £	Unit Rate Gross £
B.03.07.01.A 300mm wide x 525mm deep	m	0.75	22.19	—	—	22.19	24.40	—	—	24.40
B.03.07.01.B 373mm wide x 675mm deep	m	1.30	38.45	—	—	38.45	42.30	—	—	42.30

B.03.07.02. Excavation of trenches to receive cables; grading bottom; compacting; part backfill and dispose of surplus excavated material; by machine

	Unit	Labour Hours	Labour Net £	Plant Net £	Materials Net £	Unit Rate Net £	Labour Gross £	Plant Gross £	Materials Gross £	Unit Rate Gross £
B.03.07.02.A 300mm wide x 525mm deep	m	0.40	11.83	1.52	—	13.36	13.02	1.68	—	14.69
B.03.07.02.B 373mm wide x 675mm deep	m	0.70	20.71	2.35	—	23.05	22.78	2.58	—	25.36

B.03.07.03. Break up 50mm tarmacadam paving and 75mm hardcore bed

	Unit	Labour Hours	Labour Net £	Plant Net £	Materials Net £	Unit Rate Net £	Labour Gross £	Plant Gross £	Materials Gross £	Unit Rate Gross £
B.03.07.03.A 450mm wide	m	0.60	17.75	0.97	—	18.72	19.52	1.07	—	20.59

Hutchins Priced Schedules 2021		Unit	Labour Hours	Labour Net £	Plant Net £	Materials Net £	Unit Rate Net £	Labour Gross £	Plant Gross £	Materials Gross £	Unit Rate Gross £
								(Gross rates include 10% profit)			
B.03.	**ROADS, FOOTWAYS, KERBS, WATER MAINS, CABLE LAYING AND PILING**										
B.03.07.	**CABLE LAYING FOR ELECTRICITY AUTHORITIES**										
B.03.07.04.	**Temporarily re-lay paving with existing broken tarmacadam and consolidate**										
B.03.07.04.A	450mm wide...............	m	0.30	8.87	27.60	—	36.47	9.76	30.36	—	40.12
B.03.07.05.	**Take up paving slabs and ash bed**										
B.03.07.05.A	450mm wide...............	m	0.40	11.83	—	—	11.83	13.02	—	—	13.02
B.03.07.06.	**Temporarily re-lay paving with**										
B.03.07.06.A	existing slabs on ash bed..........	m	0.45	13.31	—	5.11	18.42	14.64	—	5.62	20.26
B.03.07.08.	**Take up granite sett paving**										
B.03.07.08.A	450mm wide...............	m	0.55	16.27	—	—	16.27	17.90	—	—	17.90
B.03.07.09.	**Temporarily re-lay paving with existing setts**										
B.03.07.09.A	450mm wide...............	m	0.40	11.83	—	—	11.83	13.02	—	—	13.02
B.03.07.10.	**Break up 75mm tarmacadam road and 100mm hardcore bed**										
B.03.07.10.A	450mm wide...............	m	0.70	20.71	1.37	—	22.07	22.78	1.50	—	24.28
B.03.07.12.	**Temporarily re-lay paving with existing broken tarmacadam and consolidate**										
B.03.07.12.A	450mm wide...............	m	0.40	11.83	38.34	—	50.17	13.02	42.17	—	55.19
B.03.07.13.	**Break up 150mm concrete paving and 150mm hardcore bed**										
B.03.07.13.A	450mm wide...............	m	1.00	29.58	7.46	—	37.04	32.54	8.21	—	40.74
B.03.07.14.	**Temporarily re-lay paving with existing broken concrete and consolidate**										
B.03.07.14.A	450mm wide...............	m	0.60	17.75	27.60	—	45.35	19.52	30.36	—	49.88
B.03.07.16.	**Labour unwinding and laying cable direct in trench**										
B.03.07.16.A	25mm..................	m	0.10	2.96	—	—	2.96	3.25	—	—	3.25
B.03.07.16.B	31mm - 50mm.............	m	0.12	3.55	—	—	3.55	3.90	—	—	3.90
B.03.07.16.C	50mm - 63mm.............	m	0.14	4.14	—	—	4.14	4.56	—	—	4.56
B.03.07.16.D	63mm - 75mm.............	m	0.20	5.92	—	—	5.92	6.51	—	—	6.51
B.03.07.18.	**Labour unwinding and drawing cable through ducts in trench**										
B.03.07.18.A	25mm..................	m	0.22	6.51	—	—	6.51	7.16	—	—	7.16
B.03.07.18.B	31mm - 50mm.............	m	0.25	7.40	—	—	7.40	8.13	—	—	8.13
B.03.07.18.C	50mm - 63mm.............	m	0.30	8.87	—	—	8.87	9.76	—	—	9.76
B.03.07.18.D	63mm - 75mm.............	m	0.40	11.83	—	—	11.83	13.02	—	—	13.02
B.03.07.19.	**Labour laying tile cable covers**										
B.03.07.19.A	300mm x 100mm.............	m	0.30	8.87	—	—	8.87	9.76	—	—	9.76

Drainage, Sewerage and Public Works

Hutchins Priced Schedules 2021		Unit	Labour Hours	Labour Net £	Plant Net £	Materials Net £	Unit Rate Net £	Labour Gross £	Plant Gross £	Materials Gross £	Unit Rate Gross £
								(Gross rates include 10% profit)			
B.03.	**ROADS, FOOTWAYS, KERBS, WATER MAINS, CABLE LAYING AND PILING**										
B.03.07.	**CABLE LAYING FOR ELECTRICITY AUTHORITIES**										
B.03.07.19.	**Labour laying tile cable covers**										
B.03.07.19.B	225mm x 150mm	m	0.35	10.35	—	—	10.35	11.39	—	—	11.39
B.03.07.20.	**Labour laying concrete cable covers**										
B.03.07.20.A	900mm x 175mm	m	0.40	11.83	—	—	11.83	13.02	—	—	13.02
B.03.07.20.B	900mm x 225mm	m	0.50	14.79	—	—	14.79	16.27	—	—	16.27
B.03.07.21.	**Labour laying stoneware ducts**										
B.03.07.21.A	100mm	m	0.25	7.40	—	—	7.40	8.13	—	—	8.13

Hutchins Priced Schedules 2021	Unit	Labour Hours	Labour Net £	Plant Net £	Materials Net £	Unit Rate Net £	Labour Gross £	Plant Gross £	Materials Gross £	Unit Rate Gross £
										(Gross rates include 10% profit)
C.01.	**SEEDING, TURFING AND PLANTING**									
C.01.01.	**PREPARATORY ITEMS**									
C.01.01.01.	**Temporarily enclose site with chestnut fencing; up to 20 times used**									
C.01.01.01.A	1.35m high; each use	m	0.25	7.40	—	0.91	8.31	8.13	—	1.00	9.14
C.01.01.02.	**Cut down hedge; grub up roots; burn or deposit in skip; height**									
C.01.01.02.A	600mm	m	1.97	58.27	—	—	58.27	64.10	—	—	64.10
C.01.01.02.B	900mm	m	2.63	77.80	—	—	77.80	85.57	—	—	85.57
C.01.01.02.C	1200mm	m	3.12	92.29	—	—	92.29	101.52	—	—	101.52
C.01.01.02.D	1500mm	m	4.27	126.31	—	—	126.31	138.94	—	—	138.94
C.01.01.02.E	1800mm	m	5.74	169.79	—	—	169.79	186.77	—	—	186.77
C.01.01.03.	**Cut down tree; lop off branches; grub up roots; burn or deposit in skip; fill hole with excavated material; up to**									
C.01.01.03.A	450mm girth; 140mm dia	nr	16.00	473.28	—	—	473.28	520.61	—	—	520.61
C.01.01.03.B	900mm girth; 290mm dia	nr	28.00	828.24	—	—	828.24	911.06	—	—	911.06
C.01.01.03.C	1350mm girth; 430mm dia	nr	42.00	1242.36	—	—	1242.36	1366.60	—	—	1366.60
C.01.01.03.D	1800mm girth; 570mm dia	nr	56.00	1656.48	—	—	1656.48	1822.13	—	—	1822.13
C.01.01.03.E	2250mm girth; 720mm dia	nr	69.00	2041.02	—	—	2041.02	2245.12	—	—	2245.12
C.01.01.03.F	2700mm girth; 860mm dia	nr	81.00	2395.98	—	—	2395.98	2635.58	—	—	2635.58
C.01.01.03.G	3150mm girth; 1000mm dia	nr	92.00	2721.36	—	—	2721.36	2993.50	—	—	2993.50
C.01.01.03.H	3600mm girth; 1150mm dia	nr	102.00	3017.16	—	—	3017.16	3318.88	—	—	3318.88
C.01.01.04.	**Clear site of bushes, scrub and undergrowth; cutting down small trees; grub up roots; burn or deposit in skip; average height**									
C.01.01.04.A	1.50m	m²	0.30	8.87	—	—	8.87	9.76	—	—	9.76
C.01.02.	**SEEDING**									
C.01.02.01.	**Ground preparation**									
C.01.02.01.A	clear site of rubbish	m²	0.08	2.37	—	—	2.37	2.60	—	—	2.60
C.01.02.01.B	strip site of surface vegetation	m²	0.07	2.07	—	—	2.07	2.28	—	—	2.28
C.01.02.01.C	cultivate 150mm deep; remove stones and vegetable matter	m²	0.12	3.55	—	—	3.55	3.90	—	—	3.90
C.01.02.01.D	fill into barrows; wheel up to 20.00m; deposit in skip	m²	0.17	5.03	—	—	5.03	5.53	—	—	5.53

Landscaping

	Unit	Labour Hours	Labour Net £	Plant Net £	Materials Net £	Unit Rate Net £	Labour Gross £	Plant Gross £	Materials Gross £	Unit Rate Gross £
						(Gross rates include 10% profit)				
C.01.	**SEEDING, TURFING AND PLANTING**									
C.01.02.	**SEEDING**									
C.01.02.01.	**Ground preparation**									
C.01.02.01.E	hire 4.50m3 skip; delivery to site; remove when full; dispose of contents; pay tipping charges	m²	—	—	6.89	—	6.89	—	7.58	—	7.58
C.01.02.01.F	grade	m²	0.08	2.37	—	—	2.37	2.60	—	—	2.60
C.01.02.01.G	standard loam 150mm deep; spread and level	m²	0.41	12.13	—	7.64	19.77	13.34	—	8.40	21.74
C.01.02.01.H	fork and rake	m²	0.06	1.77	—	—	1.77	1.95	—	—	1.95
C.01.02.01.I	dress with bonemeal; lightly rake in	m²	0.02	0.59	—	0.23	0.82	0.65	—	0.26	0.91
C.01.02.02.	**Sowing and maintenance**									
C.01.02.02.A	seed with seed mixture	m²	0.10	2.96	—	0.32	3.28	3.25	—	0.35	3.61
C.01.02.02.B	twice roll	m²	0.07	2.07	—	—	2.07	2.28	—	—	2.28
C.01.02.02.C	scythe to top	m²	0.01	0.30	—	—	0.30	0.33	—	—	0.33
C.01.02.02.D	scythe to reduce	m²	0.01	0.30	—	—	0.30	0.33	—	—	0.33
C.01.02.02.E	keep grass mown to height of 50mm during contract	m²	0.07	2.07	—	—	2.07	2.28	—	—	2.28
C.01.02.02.F	keep area free of stones exceeding 12mm	m²	0.07	2.07	—	—	2.07	2.28	—	—	2.28
C.01.02.02.G	roll in two directions (maintenance)	m²	0.06	1.77	—	—	1.77	1.95	—	—	1.95
C.01.02.02.H	scythe to top (maintenance)	m²	0.06	1.77	—	—	1.77	1.95	—	—	1.95
C.01.02.02.I	scythe to reduce (maintenance)	m²	0.06	1.77	—	—	1.77	1.95	—	—	1.95
C.01.02.02.J	twice box mow (maintenance)	m²	0.08	2.37	—	—	2.37	2.60	—	—	2.60
C.01.02.02.K	dress with fish manure (maintenance)	m²	0.01	0.30	—	0.13	0.43	0.33	—	0.15	0.47
C.01.02.02.L	water as necessary (maintenance)	m²	0.06	1.77	—	—	1.77	1.95	—	—	1.95
C.01.03.	**TURFING**									
C.01.03.01.	**Ground preparation**									
C.01.03.01.A	clear site of rubbish	m²	0.08	2.37	—	—	2.37	2.60	—	—	2.60
C.01.03.01.B	strip site of surface vegetation	m²	0.07	2.07	—	—	2.07	2.28	—	—	2.28
C.01.03.01.C	cultivate 150mm deep; remove stones and vegetable matter	m²	0.12	3.55	—	—	3.55	3.90	—	—	3.90
C.01.03.01.D	fill into barrows; wheel up to 20.00m deposit in skip	m²	0.17	5.03	—	—	5.03	5.53	—	—	5.53
C.01.03.01.E	hire 4.50m3 skip; delivery to site; remove when full; dispose of contents; pay tipping charges	m²	—	—	6.89	—	6.89	—	7.58	—	7.58
C.01.03.01.F	grade	m²	0.08	2.37	—	—	2.37	2.60	—	—	2.60
C.01.03.01.G	standard loam 100mm deep; spread and level	m²	0.27	7.99	—	5.08	13.07	8.79	—	5.59	14.37
C.01.03.01.H	fork and rake	m²	0.06	1.77	—	—	1.77	1.95	—	—	1.95
C.01.03.01.I	dress with bonemeal; lightly raked in	m²	0.02	0.59	—	0.17	0.76	0.65	—	0.18	0.83

Landscaping

Hutchins Priced Schedules 2021		Unit	Labour Hours	Labour Net £	Plant Net £	Materials Net £	Unit Rate Net £	Labour Gross £	Plant Gross £	Materials Gross £	Unit Rate Gross £
									(Gross rates include 10% profit)		
C.01.	**SEEDING, TURFING AND PLANTING**										
C.01.03.	TURFING										
C.01.03.02.	**Laying and maintenance**										
C.01.03.02.A	25mm turves and laying (PC sum £2.40 per m2)	m²	0.33	9.76	—	2.52	12.28	10.74	—	2.77	13.51
C.01.03.02.B	twice roll	m²	0.07	2.07	—	—	2.07	2.28	—	—	2.28
C.01.03.02.C	scythe to top	m²	0.01	0.30	—	—	0.30	0.33	—	—	0.33
C.01.03.02.D	scythe to reduce	m²	0.01	0.30	—	—	0.30	0.33	—	—	0.33
C.01.03.02.E	keep grass mown to height of 50mm during contract	m²	0.07	2.07	—	—	2.07	2.28	—	—	2.28
C.01.03.02.F	roll in two directions	m²	0.06	1.77	—	—	1.77	1.95	—	—	1.95
C.01.03.02.G	scythe to top (maintenance)	m²	0.06	1.77	—	—	1.77	1.95	—	—	1.95
C.01.03.02.H	scythe to reduce (maintenance)	m²	0.06	1.77	—	—	1.77	1.95	—	—	1.95
C.01.03.02.I	twice box mow (maintenance)	m²	0.08	2.37	—	—	2.37	2.60	—	—	2.60
C.01.03.02.J	top dress with fine sifted soil brushed into joints (maintenance)	m²	0.06	1.77	—	0.84	2.61	1.95	—	0.92	2.88
C.01.03.02.K	dress with fish manure (maintenance)	m²	0.01	0.30	—	0.13	0.43	0.33	—	0.15	0.47
C.01.03.02.L	water as necessary (maintenance)	m²	0.06	1.77	—	—	1.77	1.95	—	—	1.95
C.01.04.	**SHRUB PLANTING**										
C.01.04.01.	**Preparation**										
C.01.04.01.A	clear site of rubbish	m²	0.08	2.37	—	—	2.37	2.60	—	—	2.60
C.01.04.01.B	excavation to reduce levels 300mm deep	m²	1.08	31.95	—	—	31.95	35.14	—	—	35.14
C.01.04.01.C	fill into barrows; wheel up to 20.00m; deposit in skip	m²	0.33	9.76	—	—	9.76	10.74	—	—	10.74
C.01.04.01.D	hire 4.50m3 skip; delivery to site; remove when full; dispose of contents; pay tipping charges	m²	—	—	13.78	—	13.78	—	15.16	—	15.16
C.01.04.01.E	cultivate 150mm deep; remove stones and vegetable matter	m²	0.08	2.37	—	—	2.37	2.60	—	—	2.60
C.01.04.01.F	manure 100mm deep; worked into subsoil	m²	0.17	5.03	—	0.49	5.52	5.53	—	0.54	6.07
C.01.04.01.G	standard loam; 375mm deep	m²	1.00	29.58	—	17.40	46.98	32.54	—	19.14	51.68
C.01.04.01.H	spread topsoil to even levels and camber where necessary	m²	0.05	1.48	—	—	1.48	1.63	—	—	1.63
C.01.04.02.	**Preparation to existing areas**										
C.01.04.02.A	clear beds of rubbish; debris; vegetation	m²	0.08	2.37	—	—	2.37	2.60	—	—	2.60
C.01.04.02.B	cultivate 150mm deep; remove stones and vegetable matter	m²	0.08	2.37	—	—	2.37	2.60	—	—	2.60
C.01.04.02.C	rake to even levels and camber where necessary	m²	0.05	1.48	—	—	1.48	1.63	—	—	1.63

Landscaping

	Unit	Labour Hours	Labour Net £	Plant Net £	Materials Net £	Unit Rate Net £	Labour Gross £	Plant Gross £	Materials Gross £	Unit Rate Gross £
							(Gross rates include 10% profit)			
C.01.	**SEEDING, TURFING AND PLANTING**									
C.01.04.	**SHRUB PLANTING**									
C.01.04.03.	**Planting operations**									
C.01.04.03.A	clear weed growth and rubbish before planting	m²	0.08	2.37	—	—	2.37	2.60	—	—	2.60
C.01.04.03.B	peat 50mm deep; fork into upper 200mm of topsoil	m²	0.25	7.40	—	6.57	13.96	8.13	—	7.22	15.36
C.01.04.03.C	bulb planting (PC sum £0.25 per nr) in beds	nr	0.04	1.18	—	0.26	1.45	1.30	—	0.29	1.59
C.01.04.03.D	bulb planting (PC sum £0.25 per nr) in grassed areas with dibber; topped up with loose topsoil	nr	0.09	2.66	—	0.26	2.92	2.93	—	0.29	3.22
C.01.04.03.E	ground cover shrubs (PC sum £10 per nr)	nr	0.22	6.51	—	10.50	17.01	7.16	—	11.55	18.71
C.01.04.03.F	hedging plants (PC sum £3.90 per nr)	nr	0.27	7.99	—	4.10	12.08	8.79	—	4.50	13.29
C.01.04.03.G	plants; open ground grown (PC sum £10 per nr)	nr	0.28	8.28	—	10.50	18.78	9.11	—	11.55	20.66
C.01.04.03.H	plants; container grown (PC sum £14 per nr)	nr	0.30	8.87	—	14.70	23.57	9.76	—	16.17	25.93
C.01.04.03.I	water at time of planting	m²	0.15	4.44	—	—	4.44	4.88	—	—	4.88
C.01.04.03.J	Vitax Q4 fertiliser	m²	0.02	0.59	—	0.27	0.86	0.65	—	0.29	0.94
C.01.04.03.K	cultivation, lightly, after planting	m²	0.08	2.37	—	—	2.37	2.60	—	—	2.60
C.01.04.04.	**Maintenance**									
C.01.04.04.A	remove weed growth and rubbish; light cultivation; per visit	m²	0.17	5.03	—	—	5.03	5.53	—	—	5.53
C.01.04.04.B	water; per visit	m²	0.05	1.48	—	—	1.48	1.63	—	—	1.63
C.01.04.04.C	mulching; spent mushroom compost; spread over area 80mm deep; per visit	m²	0.13	3.85	—	9.94	13.79	4.23	—	10.94	15.17
C.01.04.04.D	prune shrubs; per visit	nr	0.07	2.07	—	—	2.07	2.28	—	—	2.28
C.01.04.04.E	prune roses; per visit	nr	0.12	3.55	—	—	3.55	3.90	—	—	3.90
C.01.05.	**CLIMBER PLANTING**									
C.01.05.01.	**Preparation**									
C.01.05.01.A	clear rubbish, debris and vegetation from position	nr	0.08	2.37	—	—	2.37	2.60	—	—	2.60
C.01.05.01.B	excavate position 900mm x 230mm x 450mm deep	nr	0.47	13.90	—	—	13.90	15.29	—	—	15.29
C.01.05.01.C	fill into barrows; wheel up to 20.00m; deposit in skip	nr	0.19	5.62	—	—	5.62	6.18	—	—	6.18
C.01.05.01.D	hire 4.50m3 skip; delivery to site; remove when full; dispose of contents; pay tipping charges	nr	—	—	4.27	—	4.27	—	4.70	—	4.70
C.01.05.01.E	cultivate bottom of position to depth of 150mm	nr	0.03	0.89	—	—	0.89	0.98	—	—	0.98

Landscaping

Hutchins Priced Schedules 2021		Unit	Labour Hours	Labour Net £	Plant Net £	Materials Net £	Unit Rate Net £	Labour Gross £	Plant Gross £	Materials Gross £	Unit Rate Gross £
								(Gross rates include 10% profit)			
C.01.	**SEEDING, TURFING AND PLANTING**										
C.01.05.	**CLIMBER PLANTING**										
C.01.05.01.	**Preparation**										
C.01.05.01.F	manure in bottom of position; 100mm deep	nr	0.17	5.03	—	0.18	5.21	5.53	—	0.19	5.73
C.01.05.01.G	blended loam topsoil; 500mm deep	nr	0.71	21.00	—	4.68	25.68	23.10	—	5.15	28.25
C.01.05.01.H	spread topsoil to even levels and camber	nr	0.03	0.89	—	—	0.89	0.98	—	—	0.98
C.01.05.02.	**Planting operations**										
C.01.05.02.A	clear weed growth and rubbish before planting	nr	0.05	1.48	—	—	1.48	1.63	—	—	1.63
C.01.05.02.B	peat 50mm deep; fork into upper 200mm of topsoil	nr	0.22	6.51	—	2.20	8.71	7.16	—	2.42	9.58
C.01.05.02.C	climber plant; open ground grown (PC sum £11 per nr)	nr	0.28	8.28	—	11.55	19.83	9.11	—	12.71	21.82
C.01.05.02.D	climber plant; container grown (PC sum £15 per nr)	nr	0.30	8.87	—	15.75	24.62	9.76	—	17.33	27.09
C.01.05.02.E	water at time of planting	nr	0.05	1.48	—	—	1.48	1.63	—	—	1.63
C.01.05.02.F	Vitax Q4 fertiliser	nr	0.02	0.59	—	0.07	0.66	0.65	—	0.07	0.72
C.01.05.02.G	cultivation; lightly; after planting	nr	0.05	1.48	—	—	1.48	1.63	—	—	1.63
C.01.05.03.	**Maintenance**										
C.01.05.03.A	remove weed growth and rubbish; light cultivation; per visit	nr	0.10	2.96	—	—	2.96	3.25	—	—	3.25
C.01.05.03.B	water; per visit	nr	0.01	0.30	—	—	0.30	0.33	—	—	0.33
C.01.05.03.C	mulching; spent mushroom compost; spread over area 80mm deep; per visit	nr	0.07	2.07	—	9.09	11.16	2.28	—	10.00	12.27
C.01.05.03.D	prune climber; per visit	nr	0.10	2.96	—	—	2.96	3.25	—	—	3.25
C.01.06.	**TREE PLANTING**										
C.01.06.02.	**Preparation**										
C.01.06.02.A	clear rubbish, debris and vegetation from position; size 300mm x 300mm x 300mm; for transplants, whips and feathered trees	nr	0.01	0.30	—	—	0.30	0.33	—	—	0.33
C.01.06.02.B	clear rubbish; debris and vegetation from position size 900mm x 900mm x 600mm for light standard and selected standard trees	nr	0.07	2.07	—	—	2.07	2.28	—	—	2.28
C.01.06.02.C	clear rubbish; debris and vegetation from position size 1200mm x 1200mm x 1000mm for heavy and extra heavy standard trees	nr	0.12	3.55	—	—	3.55	3.90	—	—	3.90
C.01.06.02.D	excavate to form tree pit size 300mm x 300mm x 300mm	nr	0.14	4.14	—	—	4.14	4.56	—	—	4.56
C.01.06.02.E	excavate to form tree pit size 900mm x 900mm x 600mm	nr	2.43	71.88	—	—	71.88	79.07	—	—	79.07

Landscaping

Hutchins Priced Schedules 2021	Unit	Labour Hours	Labour Net £	Plant Net £	Materials Net £	Unit Rate Net £	Labour Gross £	Plant Gross £	Materials Gross £	Unit Rate Gross £
							(Gross rates include 10% profit)			

C.01.   **SEEDING, TURFING AND PLANTING**

C.01.06.   **TREE PLANTING**

C.01.06.02.   **Preparation**

C.01.06.02.F	excavate to form tree pit size 1200mm x 1200mm x 1000mm	nr	7.20	212.98	—	—	212.98	234.27	—	—	234.27
C.01.06.02.G	fill into barrows; wheel up to 20.00m; deposit in skip; tree pit size 300mm x 300mm x 300mm	nr	0.03	0.89	—	—	0.89	0.98	—	—	0.98
C.01.06.02.H	fill into barrows; wheel up to 20.00m; deposit in skip; tree pit size 900mm x 900mm x 600mm	nr	0.54	15.97	—	—	15.97	17.57	—	—	17.57
C.01.06.02.I	fill into barrows; wheel up to 20.00m; deposit in skip; tree pit size 1200mm x 1200mm x 1000mm	nr	1.58	46.74	—	—	46.74	51.41	—	—	51.41
C.01.06.02.J	hire 4.50m3 skip; delivery to site; remove when full; dispose of contents; pay tipping charges; 300mm x 300mm x 300mm	nr	—	—	1.22	—	1.22	—	1.34	—	1.34
C.01.06.02.K	hire 4.50m3 skip; delivery to site; remove when full; dispose of contents; pay tipping charges; 900mm x 900mm x 600mm	nr	—	—	22.24	—	22.24	—	24.46	—	24.46
C.01.06.02.L	hire 4.50m3 skip; delivery to site; remove when full; dispose of contents; pay tipping charges; 1200mm x 1200mm x 1000mm	nr	—	—	65.92	—	65.92	—	72.52	—	72.52
C.01.06.02.M	cultivate 150mm deep to bottom of pit; size 300mm x 300mm	nr	0.01	0.30	—	—	0.30	0.33	—	—	0.33
C.01.06.02.N	cultivate 150mm deep to bottom of pit; size 900mm x 900mm	nr	0.07	2.07	—	—	2.07	2.28	—	—	2.28
C.01.06.02.O	cultivate 150mm deep to bottom of pit; size 1200mm x 1200mm	nr	0.12	3.55	—	—	3.55	3.90	—	—	3.90
C.01.06.02.P	manure 100mm deep to bottom of pit; size 300mm x 300mm	nr	0.02	0.59	—	0.07	0.66	0.65	—	0.07	0.73
C.01.06.02.Q	manure 100mm deep to bottom of pit; size 900mm x 900mm	nr	0.13	3.85	—	0.60	4.45	4.23	—	0.66	4.89
C.01.06.02.R	manure 100mm deep to bottom of pit; size 1200mm x 1200mm	nr	0.23	6.80	—	1.07	7.88	7.48	—	1.18	8.66
C.01.06.02.S	imported fibrous loam to pit; size 300mm x 300mm x 300mm	nr	0.07	2.07	—	1.28	3.35	2.28	—	1.41	3.69
C.01.06.02.T	imported fibrous loam to pit; size 900mm x 900mm x 600mm	nr	1.26	37.27	—	20.76	58.03	41.00	—	22.84	63.83
C.01.06.02.U	imported fibrous loam to pit; size 1200mm x 1200mm x 1000mm	nr	3.73	110.33	—	61.04	171.37	121.37	—	67.14	188.51
C.01.06.02.V	finish soil to even levels to pit; size 300mm x 300mm	nr	0.01	0.30	—	—	0.30	0.33	—	—	0.33
C.01.06.02.W	finish soil to even levels to pit; size 900mm x 900mm	nr	0.03	0.89	—	—	0.89	0.98	—	—	0.98

Landscaping

Hutchins Priced Schedules 2021	Unit	Labour Hours	Labour Net £	Plant Net £	Materials Net £	Unit Rate Net £	Labour Gross £	Plant Gross £	Materials Gross £	Unit Rate Gross £
										(Gross rates include 10% profit)
C.01.	**SEEDING, TURFING AND PLANTING**									
C.01.06.	**TREE PLANTING**									
C.01.06.02.	**Preparation**									
C.01.06.02.X	finish soil to even levels to pit; size 1200mm x 1200mm	nr	0.04	1.18	—	—	1.18	1.30	—	—	1.30
C.01.06.04.	**Planting operations**									
C.01.06.04.A	clear weed growth and rubbish before planting to pit; size 300mm x 300m	nr	0.01	0.30	—	—	0.30	0.33	—	—	0.33
C.01.06.04.B	clear weed growth and rubbish before planting to pit; size 900mm x 900m	nr	0.09	2.66	—	—	2.66	2.93	—	—	2.93
C.01.06.04.C	clear weed growth and rubbish before planting to pit; size 1200mm x 1200mm .	nr	0.16	4.73	—	—	4.73	5.21	—	—	5.21
C.01.06.04.D	peat 50mm deep forked into upper 200mm of topsoil to pit; size 300mm x 300mm.	nr	0.02	0.59	—	0.74	1.33	0.65	—	0.82	1.47
C.01.06.04.E	peat 50mm deep forked into upper 200mm of topsoil to pit; size 900mm x 900mm.	nr	0.16	4.73	—	6.57	11.30	5.21	—	7.22	12.43
C.01.06.04.F	peat 50mm deep forked into upper 200mm of topsoil to pit; size 1200mm x 1200mm.	nr	0.29	8.58	—	10.19	18.77	9.44	—	11.21	20.65
C.01.06.04.G	tree planting; transplants (PC sum £0.60 per nr)	nr	0.10	2.96	—	0.63	3.59	3.25	—	0.69	3.95
C.01.06.04.H	tree planting; whips (PC sum £10 per nr) .	nr	0.12	3.55	—	10.60	14.15	3.90	—	11.66	15.56
C.01.06.04.I	tree planting; feathered (PC sum £20 per nr)	nr	0.15	4.44	—	21.20	25.64	4.88	—	23.32	28.20
C.01.06.04.J	tree planting; light standard (PC sum £50 per nr)	nr	0.19	5.62	—	53.00	58.62	6.18	—	58.30	64.48
C.01.06.04.K	tree planting; standard (PC sum £50 per nr)	nr	0.20	5.92	—	53.00	58.92	6.51	—	58.30	64.81
C.01.06.04.L	tree planting; selected standard (PC sum £50 per nr)	nr	0.22	6.51	—	53.00	59.51	7.16	—	58.30	65.46
C.01.06.04.M	tree planting; heavy standard (PC sum £65 per nr)	nr	0.24	7.10	—	68.25	75.35	7.81	—	75.08	82.88
C.01.06.04.N	tree planting; extra heavy standard (PC sum £150 per nr)	nr	0.27	7.99	—	159.00	166.99	8.79	—	174.90	183.69
C.01.06.04.O	water at time of planting to pit; size 300mm x 300mm	nr	0.01	0.30	—	—	0.30	0.33	—	—	0.33
C.01.06.04.P	water at time of planting to pit; size 900mm x 900mm	nr	0.05	1.48	—	—	1.48	1.63	—	—	1.63
C.01.06.04.Q	water at time of planting to pit; size 1200mm x 1200mm	nr	0.08	2.37	—	—	2.37	2.60	—	—	2.60
C.01.06.04.R	Vitax Q4 fertiliser to pit; size 300mm x 300mm.	nr	0.01	0.30	—	0.13	0.43	0.33	—	0.15	0.47
C.01.06.04.S	Vitax Q4 fertiliser to pit; size 900mm x 900mm.	nr	0.01	0.30	—	1.00	1.30	0.33	—	1.10	1.43

Landscaping

Hutchins Priced Schedules 2021		Unit	Labour Hours	Labour Net £	Plant Net £	Materials Net £	Unit Rate Net £	Labour Gross £	Plant Gross £	Materials Gross £	Unit Rate Gross £
								(Gross rates include 10% profit)			
C.01.	**SEEDING, TURFING AND PLANTING**										
C.01.06.	TREE PLANTING										
C.01.06.04.	Planting operations										
C.01.06.04.T	Vitax Q4 fertiliser to pit; size 1200mm x 1200mm...................	nr	0.02	0.59	—	1.67	2.26	0.65	—	1.83	2.48
C.01.06.04.U	lightly cultivate after planting to pit; size 300mm x 300mm............	nr	0.01	0.30	—	—	0.30	0.33	—	—	0.33
C.01.06.04.V	lightly cultivate after planting to pit; size 900mm x 900mm............	nr	0.03	0.89	—	—	0.89	0.98	—	—	0.98
C.01.06.04.W	lightly cultivate after planting to pit; size 1200mm x 1200mm..........	nr	0.04	1.18	—	—	1.18	1.30	—	—	1.30
C.01.06.04.X	turf (PC sum £2.40 per m2) around tree position after planting to form opening 600mm x 600mm to pit; size 900mm x 900mm...................	nr	0.25	7.40	—	1.27	8.67	8.13	—	1.40	9.53
C.01.06.04.Y	turf (PC sum £2.40 per m2) position after planting to form opening 600mm x 600mm to pit; size 1200mm x 1200mm ..	nr	0.45	13.31	—	2.42	15.73	14.64	—	2.66	17.30
C.01.06.05.	**Maintenance**										
C.01.06.05.A	remove weed growth and rubbish; light cultivation; per visit; to pit; size 300mm x 300mm...................	nr	0.01	0.30	—	—	0.30	0.33	—	—	0.33
C.01.06.05.B	remove weed growth and rubbish; light cultivation; per visit; to pit; size 900mm x 900mm...................	nr	0.06	1.77	—	—	1.77	1.95	—	—	1.95
C.01.06.05.C	remove weed growth and rubbish; light cultivation; per visit; to pit; size 1200mm x 1200mm...................	nr	0.11	3.25	—	—	3.25	3.58	—	—	3.58
C.01.06.05.D	watering; per visit; to pit; size 300mm x 300mm...................	nr	0.01	0.30	—	—	0.30	0.33	—	—	0.33
C.01.06.05.E	watering; per visit; to pit; size 900mm x 900mm...................	nr	0.04	1.18	—	—	1.18	1.30	—	—	1.30
C.01.06.05.F	watering; per visit; to pit; size 1200mm x 1200mm...................	nr	0.07	2.07	—	—	2.07	2.28	—	—	2.28
C.01.06.05.G	mulching; spent mushroom compost; spread 80mm deep; per visit; over pit; size 300mm x 300mm.............	nr	0.01	0.30	—	2.89	3.18	0.33	—	3.18	3.50
C.01.06.05.H	mulching; spent mushroom compost; spread 80mm deep; per visit; over pit; size 900mm x 900mm............	nr	0.10	2.96	—	15.29	18.25	3.25	—	16.82	20.07
C.01.06.05.I	mulching; spent mushroom compost; spread 80mm deep; per visit; over pit; size 1200mm x 1200mm..........	nr	0.17	5.03	—	27.16	32.18	5.53	—	29.87	35.40
C.01.06.05.J	prune trees; per visit.............	nr	0.17	5.03	—	—	5.03	5.53	—	—	5.53
C.01.06.06.	**Sundry items**										
C.01.06.06.A	spiral tree guards for protection against rabbit damage with bamboo cane.....	nr	0.03	0.89	—	1.39	2.27	0.98	—	1.53	2.50

Hutchins Priced Schedules 2021	Unit	Labour Hours	Labour Net £	Plant Net £	Materials Net £	Unit Rate Net £	Labour Gross £	Plant Gross £	Materials Gross £	Unit Rate Gross £
									(Gross rates include 10% profit)	
C.01.	**SEEDING, TURFING AND PLANTING**									
C.01.06.	**TREE PLANTING**									
C.01.06.06.	**Sundry items**									
C.01.06.06.B	tree spat mulch mats	nr	0.08	2.37	—	2.07	4.43	2.60	—	2.27	4.88
C.01.06.06.C	rot proof, chemical resistant plastic mesh with tree stake	m	0.30	8.87	—	3.35	12.22	9.76	—	3.68	13.44
C.01.06.06.D	tubular tree shelter with tree stake	nr	0.08	2.37	—	2.14	4.51	2.60	—	2.35	4.96
C.01.06.06.E	tree stake	nr	0.33	9.76	—	0.83	10.59	10.74	—	0.91	11.65
C.01.06.06.F	tree ties (3nr) with tree stake	nr	0.15	4.44	—	1.81	6.25	4.88	—	1.99	6.87
C.01.06.06.G	steel weld mesh tree guard with tree stake	nr	1.58	46.74	—	21.97	68.71	51.41	—	24.17	75.58
C.02.	**PATHS, PAVING AND WALLS**									
C.02.01.	**PATHS AND PAVING**									
C.02.01.01.	**Surface excavation**									
C.02.01.01.A	150mm deep; remove soil up to 18.00m; deposit; trim and consolidate new surface	m²	0.50	14.79	—	—	14.79	16.27	—	—	16.27
C.02.01.01.B	fill into barrows; wheel up to 20m deposit in skip	m²	0.15	4.44	—	—	4.44	4.88	—	—	4.88
C.02.01.01.C	hire 4.50m3 skip; delivery to site; remove when full; dispose of contents; pay tipping charges	m²	—	—	6.89	—	6.89	—	7.58	—	7.58
C.02.01.02.	**Ash or fine clinker; spread, levelled and rolled**									
C.02.01.02.A	50mm thick	m²	0.25	7.40	—	9.10	16.50	8.13	—	10.01	18.15
C.02.01.02.B	75mm thick	m²	0.30	8.87	—	13.76	22.64	9.76	—	15.14	24.90
C.02.01.03.	**Brick hardcore 100mm; spread, levelled watered, rammed, blinded with ashes and rolled; cambered surface**									
C.02.01.03.A	finished to receive surfacing material . . .	m²	0.35	10.35	—	17.39	27.75	11.39	—	19.13	30.52
C.02.01.04.	**Brick hardcore 75mm; spread, levelled, watered, rammed and blinded with ashes; 100mm concrete paving**									
C.02.01.04.A	to falls; in bays; formwork; expansion joints; spade finish to falls	m²	1.20	35.50	—	29.74	65.23	39.05	—	32.71	71.76
C.02.01.05.	**Crazy paving; broken precast concrete slabs; mortar bed; pointing**									
C.02.01.05.A	50mm	m²	0.57	29.88	—	21.07	50.95	32.87	—	23.18	56.05
C.02.01.06.	**Precast concrete slabs 50mm; natural finish; mortar bed; pointing**									
C.02.01.06.A	600mm x 600mm	m²	0.33	21.91	—	24.10	46.01	24.10	—	26.51	50.61
C.02.01.06.B	600mm x 750mm	m²	0.28	18.55	—	23.86	42.41	20.40	—	26.24	46.65

Landscaping

Hutchins Priced Schedules 2021

Hutchins Priced Schedules 2021	Unit	Labour Hours	Labour Net £	Plant Net £	Materials Net £	Unit Rate Net £	Labour Gross £	Plant Gross £	Materials Gross £	Unit Rate Gross £
							(Gross rates include 10% profit)			
C.02.	**PATHS, PAVING AND WALLS**									
C.02.01.	**PATHS AND PAVING**									
C.02.01.06.	**Precast concrete slabs 50mm; natural finish; mortar bed; pointing**									
C.02.01.06.C	600mm x 900mm	m²	0.25	16.87	—	21.55	38.42	18.56	—	23.70	42.26
C.02.01.07.	**Precast concrete slabs 50mm; 25mm sand bed**									
C.02.01.07.A	spread; levelled; consolidated 600mm x 600mm	m²	0.60	40.53	—	22.15	62.68	44.58	—	24.36	68.94
C.02.01.07.B	75mm fine clinker; spread; levelled consolidated 600mm x 600mm	m²	0.75	50.61	—	35.91	86.52	55.67	—	39.50	95.17
C.02.01.08.	**Precast concrete edging to paths; bedding; pointing in cement mortar; haunched with concrete**									
C.02.01.08.A	50mm x 150mm; flat top	m	0.20	13.51	—	9.70	23.21	14.86	—	10.67	25.53
C.02.01.08.B	50mm x 150mm; bullnosed	m	0.20	13.51	—	11.38	24.89	14.86	—	12.51	27.37
C.02.01.09.	**Site-mixed (1:3:6) concrete path; slightly cambered; formwork; trowelled smooth**									
C.02.01.09.A	75mm	m²	0.45	30.38	0.96	9.44	40.78	33.42	1.05	10.38	44.85
C.02.01.09.B	100mm	m²	0.55	37.10	1.17	12.55	50.82	40.81	1.29	13.80	55.90
C.02.01.09.C	extra for marking out concrete in square or crazy pattern	m²	0.15	10.15	—	—	10.15	11.16	—	—	11.16
C.02.01.10.	**Formwork to edge of path**									
C.02.01.10.A	75mm high (four uses assumed)	m	0.12	4.33	—	0.37	4.70	4.76	—	0.41	5.17
C.02.01.10.B	100mm high (four uses assumed)	m	0.13	4.78	—	0.45	5.23	5.26	—	0.49	5.75
C.02.01.10.C	150mm high (four uses assumed)	m	0.16	6.06	—	0.61	6.67	6.66	—	0.67	7.33
C.02.01.10.D	75mm high, left in	m	0.10	3.76	—	1.40	5.16	4.14	—	1.54	5.68
C.02.01.10.E	100mm high, left in	m	0.11	4.14	—	1.72	5.86	4.55	—	1.90	6.45
C.02.01.10.F	150mm high, left in	m	0.14	5.27	—	2.37	7.64	5.80	—	2.61	8.40
C.02.01.11.	**Paving; paviors laid to flats, falls, cross falls, slopes not exceeding 15 deg from horizontal**									
C.02.01.11.A	brick paviors (PC sum £1135 per 1000); 75mm thick; bedding and jointing in cement mortar (1:3); laid stretcher bond .	m²	1.39	93.15	—	52.36	145.52	102.47	—	57.60	160.07
C.02.01.11.B	brick paviors; (PC sum £1135 per 1000); 75mm thick; bedding and jointing in cement mortar (1:3); laid in herringbone bond	m²	1.50	100.55	—	52.36	152.91	110.60	—	57.60	168.20
C.02.01.11.C	Keyblok concrete block paviors; 200mm x 100mm; 60mm thick; red colour; laid on 50mm screeded bed; compact and vibrate with hand operated vibrating plate; laid flat in herringbone bond	m²	0.75	50.61	3.55	24.95	79.11	55.67	3.90	27.45	87.02

Hutchins Priced Schedules 2021	Unit	Labour Hours	Labour Net £	Plant Net £	Materials Net £	Unit Rate Net £	Labour Gross £	Plant Gross £	Materials Gross £	Unit Rate Gross £
							(Gross rates include 10% profit)			

C.02. PATHS, PAVING AND WALLS

C.02.01. PATHS AND PAVING

C.02.01.11. Paving; paviors laid to flats, falls, cross falls, slopes not exceeding 15 deg from horizontal

C.02.01.11.D Keyblok concrete block paviors; 200mm x 100mm; 80mm thick; red colour; laid on 50mm screeded bed; compact and vibrate with hand operated vibrating plate; laid flat in herringbone bond	m²	0.85	57.40	3.55	30.83	91.78	63.14	3.90	33.91	100.95
C.02.01.12. Kerb; brick on flat; brick paviors; (PC sum £1135 per 1000); bedding and jointing; flush pointing; straight										
C.02.01.12.A 102.5mm wide x 75mm high	m	0.14	9.48	—	6.27	15.74	10.42	—	6.90	17.32
C.02.01.12.B 215mm wide x 75mm high	m	0.25	16.87	—	12.66	29.52	18.56	—	13.92	32.48
C.02.01.13. Treated sawn timber edging; nailed to and including 47mm x 50mm treated and pointed stakes driven in at 1.50m centres (one side only measured)										
C.02.01.13.A 25mm x 150mm	m	0.15	10.15	—	4.45	14.60	11.16	—	4.90	16.06
C.02.02. WALLS										
C.02.02.01. Excavate trench; garden wall foundations; part backfill; part wheel and load into skip										
C.02.02.01.A 375mm x 300mm deep	m	0.56	16.56	—	—	16.56	18.22	—	—	18.22
C.02.02.01.B 375mm x 450mm deep	m	0.80	23.66	—	—	23.66	26.03	—	—	26.03
C.02.02.02. Site-mixed (1:3:6) concrete in trench										
C.02.02.02.A 375mm wide x 150mm thick	m	0.60	17.75	1.27	6.61	25.63	19.52	1.40	7.28	28.20
C.02.02.03. Common bricks wall (PC sum £450 per 1000)										
C.02.02.03.A half brick thick; pointed both sides; brick-on-edge coping	m²	1.31	87.78	—	31.89	119.66	96.55	—	35.08	131.63
C.02.02.03.B one brick thick; pointed both sides; brick-on-edge coping	m²	2.26	151.89	—	65.51	217.41	167.08	—	72.07	239.15
C.02.02.03.C half brick thick; joints raked out; rendered in cement mortar both sides	m²	1.96	131.60	—	36.88	168.48	144.76	—	40.57	185.32
C.02.02.03.D one brick thick; joints raked out; rendered in cement mortar both sides	m²	2.81	188.99	—	70.51	259.50	207.89	—	77.56	285.45
C.02.02.04. Brick wall in reclaimed stocks (PC sum £780 per 1000); brick-on-edge coping										
C.02.02.04.A one brick thick; pointed both sides	m²	2.26	151.89	—	107.97	259.87	167.08	—	118.77	285.85
C.02.02.04.B one brick thick; rendered in cement mortar both sides	m²	2.81	188.99	—	113.74	302.73	207.89	—	125.11	333.00

Landscaping

(Gross rates include 10% profit)

	Unit	Labour Hours	Labour Net £	Plant Net £	Materials Net £	Unit Rate Net £	Labour Gross £	Plant Gross £	Materials Gross £	Unit Rate Gross £
C.02.	**PATHS, PAVING AND WALLS**									
C.02.02.	**WALLS**									
C.02.02.05.	**Brick wall in facing bricks (PC sum £550 per 1000); brick-on-edge coping**									
C.02.02.05.A	half brick thick; pointed both sides	m²	1.31	87.78	—	38.49	126.26	96.55	—	42.34	138.89
C.02.02.05.B	one brick thick; pointed both sides	m²	2.26	151.89	—	78.61	230.51	167.08	—	86.48	253.56
C.02.02.06.	**Double course tile creasing**									
C.02.02.06.A	brick-on-flat coping (PC sum £550 per 1000); cement mortar fillets	m	0.58	38.85	—	10.39	49.23	42.73	—	11.42	54.16
C.02.02.07.	**Coping; flat; twice throated; precast concrete**									
C.02.02.07.A	600mm x 300mm x 50mm	m	0.60	40.53	—	15.55	56.08	44.58	—	17.10	61.69
C.02.02.08.	**Metal coping**									
C.02.02.08.A	angle iron at ends	nr	0.04	2.69	—	4.70	7.38	2.96	—	5.17	8.12
C.02.02.09.	**Excavate pit; part backfill; part wheel and load into skip**									
C.02.02.09.A	525mm x 525mm x 300mm deep for 225mm x 225mm brick pier	nr	0.68	20.11	—	—	20.11	22.13	—	—	22.13
C.02.02.09.B	650mm x 650mm x 300mm deep for 338mm x 338mm brick pier	nr	0.93	27.51	—	—	27.51	30.26	—	—	30.26
C.02.02.10.	**Site-mixed (1:3:6) concrete in pits**									
C.02.02.10.A	525mm x 525mm x 150mm	nr	0.30	8.87	0.64	4.81	14.32	9.76	0.70	5.29	15.75
C.02.02.10.B	650mm x 650mm x 150mm	nr	0.40	11.83	0.85	7.42	20.10	13.02	0.93	8.16	22.11
C.02.02.11.	**Brick piers; 225mm x 225mm; pointed on all faces**									
C.02.02.11.A	common brickwork (PC sum £450 per 1000)	m	0.90	60.76	—	15.04	75.80	66.83	—	16.55	83.38
C.02.02.11.B	facing bricks (PC sum £550 per 1000) . .	m	1.06	70.91	—	18.04	88.95	78.00	—	19.85	97.85
C.02.02.12.	**Pier cap; weathered four ways; precast concrete**									
C.02.02.12.A	305mm x 305mm	nr	0.50	33.74	—	7.67	41.41	37.11	—	8.44	45.55
C.02.02.14.	**Brick piers; 338mm x 338mm; pointed on all faces**									
C.02.02.14.A	common brickwork (PC sum £450 per 1000)	m	1.51	101.29	—	30.67	131.95	111.41	—	33.74	145.15
C.02.02.14.B	facing bricks (PC sum £550 per 1000) . .	m	1.76	118.16	—	39.01	157.17	129.97	—	42.92	172.89
C.02.02.15.	**Pier cap; weathered four ways; precast concrete**									
C.02.02.15.A	420mm x 420mm	nr	0.65	43.89	—	14.54	58.43	48.28	—	15.99	64.27

Hutchins Priced Schedules 2021		Unit	Labour Hours	Labour Net £	Plant Net £	Materials Net £	Unit Rate Net £	Labour Gross £	Plant Gross £	Materials Gross £	Unit Rate Gross £
								(Gross rates include 10% profit)			
C.02.	**PATHS, PAVING AND WALLS**										
C.02.02.	**WALLS**										
C.02.02.16.	**Flint stone (PC sum £215 per m3) walling built in cement lime mortar and pointed both sides**										
C.02.02.16.A	225mm thick	m²	3.14	210.97	—	85.17	296.14	232.07	—	93.69	325.76
C.02.02.17.	**Flint stone (PC sum £215 per m3) cavity walling comprising 100mm flint facework; to backing of**										
C.02.02.17.A	half brick in commons (PC sum £450 per 1000); inner skin in cement lime mortar and pointed one side	m²	2.47	165.74	—	78.18	243.92	182.31	—	86.00	268.32
C.02.02.18.	**Marshalls Superscreen walling; 90mm precast concrete blocks; bedded in gauged mortar (1:1:6); jointing with flush joints both sides**										
C.02.02.18.A	290mm x 290mm open	m²	1.11	74.27	—	52.84	127.11	81.69	—	58.13	139.82
C.02.02.18.B	290mm x 290mm solid	m²	1.11	74.27	—	63.79	138.06	81.69	—	70.17	151.87
C.02.02.18.C	pilaster; 190mm x 194mm x 194mm; pointed all round	m	1.11	74.27	—	29.28	103.55	81.69	—	32.21	113.90
C.02.02.18.D	add for reinforcing pilasters over three courses high; 4nr - 12mm dia mild steel reinforcing rods set in concrete foundations (measured separately); placing pilaster blocks over rods; filling centre void with cement mortar packed around rods	m	0.55	37.10	—	6.03	43.13	40.81	—	6.63	47.44
C.02.02.18.E	pilaster cap 194mm x 194mm x 51mm . .	nr	0.28	18.89	—	4.73	23.62	20.77	—	5.21	25.98
C.02.02.20.	**Marshalite reconstructed walling; pitch faced stones and jumpers; 220mm x 100mm x 65mm blocks; bedded and pointed in cement mortar (1:3)**										
C.02.02.20.A	ash multi	m²	1.21	80.99	—	67.83	148.82	89.09	—	74.61	163.70
C.02.02.20.B	buff	m²	1.21	80.99	—	67.83	148.82	89.09	—	74.61	163.70
C.02.02.22.	**Marshalite reconstructed walling; rustic faced stones and jumpers; 220mm x 100mm x 65mm blocks; bedded and pointed in cement mortar (1:3)**										
C.02.02.22.A	ash multi	m²	1.21	80.99	—	69.45	150.44	89.09	—	76.39	165.48
C.02.02.22.B	buff	m²	1.21	80.99	—	69.45	150.44	89.09	—	76.39	165.48

Landscaping

Hutchins Priced Schedules 2021		Unit	Labour Hours	Labour Net £	Plant Net £	Materials Net £	Unit Rate Net £	Labour Gross £	Plant Gross £	Materials Gross £	Unit Rate Gross £
								(Gross rates include 10% profit)			
C.02.	**PATHS, PAVING AND WALLS**										
C.02.02.	**WALLS**										
C.02.02.24.	Marshalite reconstructed coping; Saxon textured split faced stone; 600mm x 136mm x 50mm bedded and pointed in cement mortar (1:3)										
C.02.02.24.B	buff	m	2.00	134.69	—	12.52	147.21	148.16	—	13.77	161.93
C.02.02.26.	**Marshalite reconstructed attached pier; pitch faced stone and jumpers; 330mm x 330mm; bedded and pointed in cement mortar (1:3)**										
C.02.02.26.A	ash multi	m	1.51	101.29	—	59.73	161.02	111.41	—	65.70	177.12
C.02.02.26.B	buff	m	1.51	101.29	—	59.73	161.02	111.41	—	65.70	177.12
C.02.02.28.	**Marshalite reconstructed attached pier; pitch faced stone and jumpers; 410mm x 410mm; bedded and pointed in cement mortar (1:3)**										
C.02.02.28.A	ash multi	m	2.01	135.02	—	58.22	193.25	148.53	—	64.05	212.57
C.02.02.28.B	buff	m	2.01	135.02	—	58.22	193.25	148.53	—	64.05	212.57
C.02.02.30.	**Marshalls precast pillar cap; 280mm x 280mm x 63-32mm; bedded and pointed in cement mortar (1:3)**										
C.02.02.30.A	off-white	nr	0.50	33.74	—	11.80	45.54	37.11	—	12.98	50.09
C.02.02.32.	**Marshalls precast pillar cap; 380mm x 380mm x 76-32mm; bedded and pointed in cement mortar (1:3)**										
C.02.02.32.A	off-white	nr	0.75	50.61	—	12.13	62.74	55.67	—	13.34	69.01
C.02.02.33.	**Marshalls precast pillar cap; 460mm x 460mm x 90-38mm; bedded and pointed in cement mortar (1:3)**										
C.02.02.33.A	off-white	nr	1.13	75.95	—	12.47	88.42	83.54	—	13.72	97.26
C.03.	**FENCING AND GATES**										
C.03.01.	**PANEL FENCING**										
C.03.01.01.	**Lapped panels; 1830mm wide; natural waney edged timber; overlapped; weathered capping strip; waney edged timber sandwiched between five pairs planed battens; timber posts; treated; including excavating holes, setting posts, backfilling with concrete; height**										
C.03.01.01.A	900mm	m	0.57	38.11	—	34.03	72.14	41.92	—	37.43	79.35
C.03.01.01.B	1200mm	m	0.64	42.88	—	35.24	78.12	47.17	—	38.76	85.93
C.03.01.01.C	1500mm	m	0.71	47.58	—	36.40	83.98	52.34	—	40.04	92.38
C.03.01.01.D	1830mm	m	0.78	52.29	—	40.45	92.74	57.52	—	44.50	102.02

Landscaping

Hutchins Priced Schedules 2021		Unit	Labour Hours	Labour Net £	Plant Net £	Materials Net £	Unit Rate Net £	Labour Gross £	Plant Gross £	Materials Gross £	Unit Rate Gross £
								(Gross rates include 10% profit)			
C.03.	**FENCING AND GATES**										
C.03.01.	PANEL FENCING										
C.03.01.02.	Lapped panels; 1830mm wide; natural waney edged timber; overlapped; weathered capping strip; waney edged timber sandwiched between five pairs planed battens treated; concrete gravel board; H section posts; including excavating holes, setting posts, backfilling; height										
C.03.01.02.A	1200mm....................	m	0.50	33.74	—	50.89	84.63	37.11	—	55.98	93.09
C.03.01.02.B	1500mm....................	m	0.53	35.76	—	51.79	87.55	39.33	—	56.97	96.31
C.03.01.02.C	1830mm....................	m	0.57	38.11	—	54.10	92.21	41.92	—	59.52	101.43
C.03.01.05.	Featherboard panels; 1830mm wide; three horizontal rails; weathered; clad with 100mm featherboard pales; timber posts; treated; chamfered post cap; timber gravel board including excavating, setting post and backfilling with concrete; height										
C.03.01.05.B	1200mm....................	m	0.68	45.90	—	36.71	82.61	50.49	—	40.38	90.87
C.03.01.05.C	1500mm....................	m	0.76	50.95	—	39.80	90.74	56.04	—	43.78	99.82
C.03.01.05.D	1800mm....................	m	0.83	55.72	—	41.68	97.40	61.29	—	45.85	107.14
C.03.01.06.	Featherboard panels; 1830mm wide; three horizontal rails; weathered; clad with 100mm featherboard pales; treated concrete gravel board; concrete H section posts; including excavating, setting posts and backfilling with concrete; height										
C.03.01.06.B	1200mm....................	m	0.52	34.75	—	50.06	84.80	38.22	—	55.06	93.29
C.03.01.06.C	1500mm....................	m	0.56	37.44	—	52.89	90.33	41.18	—	58.18	99.36
C.03.01.06.D	1800mm....................	m	0.60	40.19	—	55.20	95.40	44.21	—	60.73	104.94
C.03.01.07.	Trellis panels; square top; 1800mm wide 38mm x 19mm softwood framing; weathered capping; 19mm x 19mm softwood horizontal and vertical infill at 150mm centres; timber posts; chamfered post cap; timber gravel board; treated; height										
C.03.01.07.A	1200mm....................	m	0.41	27.69	—	33.39	61.08	30.46	—	36.73	67.19
C.03.01.07.B	1500mm....................	m	0.55	37.10	—	35.46	72.55	40.81	—	39.00	79.81

Landscaping

| | | | | | | | | | (Gross rates include 10% profit) | |

C.03.	**FENCING AND GATES**									
C.03.01.	**PANEL FENCING**									
C.03.01.09.	**Trellis panels; square top; 1800mm wide; 38mm x 19mm softwood framing; weathered capping; 19mm x 19mm softwood horizontal and vertical infill at 150mm centres; treated; concrete gravel board; concrete H section posts; height**									
C.03.01.09.A	1200mm	m	0.50	33.40	—	39.03	72.43	36.74	—	42.93	79.68
C.03.01.09.B	1500mm	m	0.64	42.88	—	41.09	83.97	47.17	—	45.19	92.36
C.03.03.	**GATES**									
C.03.03.01.	**Lapped panel gate; 910mm wide; galvanised ring latch; heavy hinges; hanging**									
C.03.03.01.A	915mm height	nr	0.13	8.47	—	59.17	67.64	9.32	—	65.09	74.40
C.03.03.01.B	1220mm height	nr	0.25	16.87	—	63.59	80.46	18.56	—	69.95	88.51
C.03.03.01.C	1524mm height	nr	0.38	25.34	—	66.24	91.58	27.87	—	72.86	100.74
C.03.03.01.D	1829mm height	nr	0.48	32.06	—	72.42	104.48	35.27	—	79.66	114.93
C.03.03.03.	**Featherboard panel gate; 910mm wide; supplied with fittings; heavy hinges; hanging**									
C.03.03.03.B	1508mm height	nr	0.50	33.74	—	68.00	101.74	37.11	—	74.80	111.91
C.03.03.03.C	1813mm height	nr	0.60	40.53	—	70.65	111.18	44.58	—	77.71	122.29
C.03.05.	**WOOD PRESERVATIVES**									
C.03.05.01.	**One coat timber preservative on wrought timber**									
C.03.05.01.A	general surfaces; not exceeding 150mm girth	m	0.04	1.51	—	0.03	1.53	1.66	—	0.03	1.68
C.03.05.01.B	150mm - 300mm girth	m	0.09	3.39	—	0.06	3.45	3.73	—	0.07	3.79
C.03.05.01.C	over 300mm girth	m²	0.19	7.15	—	0.20	7.35	7.86	—	0.22	8.08
C.03.05.02.	**One coat timber preservative on sawn timber**									
C.03.05.02.A	general surfaces; not exceeding 150mm girth	m	0.05	1.88	—	0.08	1.96	2.07	—	0.08	2.15
C.03.05.02.B	150mm - 300mm girth	m	0.10	3.76	—	0.15	3.91	4.14	—	0.17	4.31
C.03.05.02.C	over 300mm girth	m²	0.20	7.53	—	0.51	8.03	8.28	—	0.56	8.83
C.03.05.03.	**One coat fungicidal and insecticidal preservative on wrought timber**									
C.03.05.03.A	general surfaces; not exceeding 150mm girth	m	0.04	1.51	—	0.03	1.53	1.66	—	0.03	1.68

Landscaping

| | | | | | | | | | | (Gross rates include 10% profit) |

C.03. **FENCING AND GATES**

C.03.05. **WOOD PRESERVATIVES**

C.03.05.03. One coat fungicidal and insecticidal preservative on wrought timber

		Unit	Labour Hours	Labour Net £	Plant Net £	Materials Net £	Unit Rate Net £	Labour Gross £	Plant Gross £	Materials Gross £	Unit Rate Gross £
C.03.05.03.B	150mm - 300mm girth	m	0.07	2.63	—	0.05	2.69	2.90	—	0.06	2.95
C.03.05.03.C	over 300mm girth	m²	0.18	6.77	—	0.14	6.92	7.45	—	0.16	7.61
C.03.05.04.	**One coat fungicidal and insecticidal preservative on sawn timber**										
C.03.05.04.A	general surfaces; not exceeding 150mm girth	m	0.05	1.88	—	0.04	1.92	2.07	—	0.04	2.11
C.03.05.04.B	150mm - 300mm girth	m	0.08	3.01	—	0.06	3.08	3.31	—	0.07	3.38
C.03.05.04.C	over 300mm girth	m²	0.19	7.15	—	0.22	7.37	7.86	—	0.24	8.11
C.03.05.05.	**One coat Cuprinol preservative on wrought timber**										
C.03.05.05.A	general surfaces; not exceeding 150mm girth	m	0.04	1.51	—	0.11	1.61	1.66	—	0.12	1.78
C.03.05.05.B	150mm - 300mm girth	m	0.08	3.01	—	0.18	3.19	3.31	—	0.20	3.51
C.03.05.05.C	over 300mm girth	m²	0.19	7.15	—	0.62	7.77	7.86	—	0.68	8.55
C.03.05.06.	**One coat Cuprinol preservative on sawn timber**										
C.03.05.06.A	general surfaces; not exceeding 150mm girth	m	0.05	1.88	—	0.18	2.06	2.07	—	0.20	2.27
C.03.05.06.B	150mm - 300mm girth	m	0.09	3.39	—	0.37	3.75	3.73	—	0.40	4.13
C.03.05.06.C	over 300mm girth	m²	0.20	7.53	—	1.02	8.55	8.28	—	1.13	9.41
C.04.	**GENERAL FENCING**										
C.04.01.	**POST AND WIRE FENCING**										
C.04.01.01.	**Prosure Mesh Panel Fencing System; 5mm wire x 200mm x 50mm open mesh galvanised and PPC green; horizontal vee profiles at 70mm from top, then 900mm then 700mm**										
C.04.01.01.A	height 1730m	m	0.31	20.50	—	13.59	34.09	22.55	—	14.95	37.49
C.04.01.01.B	60mm x 60mm x 2400mm SHS green post; set in concrete; including brackets, bolts and caps	nr	0.44	29.71	—	30.57	60.28	32.68	—	33.63	66.31
C.04.01.01.C	extra for mesh gate; 1800m x 1200m wide single leaf green; including 5mm wire x 200mm x 50mm infill and 80mm x 80mm posts; 2400mm length set in concrete	nr	3.25	218.43	—	502.59	721.02	240.27	—	552.85	793.12
C.04.01.01.D	extra for mesh gate; 1800m x 3000m wide double leaf green; including 5mm wire x 200mm x 50mm infill and 80mm x 80mm posts; 2750mm length set in concrete	nr	6.45	433.50	—	983.07	1416.57	476.85	—	1081.37	1558.22

Landscaping

Hutchins Priced Schedules 2021		Unit	Labour Hours	Labour Net £	Plant Net £	Materials Net £	Unit Rate Net £	Labour Gross £	Plant Gross £	Materials Gross £	Unit Rate Gross £
							(Gross rates include 10% profit)				
C.04.	**GENERAL FENCING**										
C.04.01.	**POST AND WIRE FENCING**										
C.04.01.02.	**Prosure Mesh Panel Fencing System; 5mm wire x 200mm x 50mm open mesh galvanised and PPC green; horizontal beams at 70mm from top, then 1100mm, then 1100mm**										
C.04.01.02.A	height 2400mm	m	0.44	29.71	—	17.65	47.36	32.68	—	19.42	52.10
C.04.01.02.B	60mm x 60mm x 3100mm SHS green post; set in concrete; including steel clips, anti-vandal bolts and caps	nr	0.60	40.33	—	38.56	78.89	44.36	—	42.42	86.78
C.04.01.02.C	extra for mesh gate; 2400m x 1200m wide single leaf green; including 5mm wire x 200mm x 50mm infill and 80mm x 80mm posts; 3150mm length set in concrete	nr	3.90	262.12	—	526.61	788.73	288.33	—	579.27	867.60
C.04.01.02.D	extra for mesh gate; 2400m x 3000m wide double leaf green; including 5mm wire x 200mm x 50mm infill and 80mm x 80mm posts; 3150mm length set in concrete	nr	7.74	520.21	—	1009.09	1529.30	572.23	—	1110.00	1682.23
C.04.01.03.	**Fencing of five 4mm line wires and 125mm x 125mm to 75mm x 75mm x 1670mm reinforced concrete tapered posts at 3000mm centres; set 600mm deep into ground in concrete; height**										
C.04.01.03.A	1000mm	m	0.28	18.89	—	8.31	27.19	20.77	—	9.14	29.91
C.04.01.03.B	extra for 125mm x 125mm end straining posts with 100mm x 75mm strut set in concrete	nr	0.44	29.71	—	54.09	83.79	32.68	—	59.49	92.17
C.04.01.03.C	extra for corner straining post with two struts set in concrete	nr	0.66	44.56	—	61.81	106.37	49.02	—	67.99	117.01
C.04.01.04.	**Fencing of seven 4mm line wires and 125mm x 125mm to 75mm x 2070mm reinforced concrete tapered posts at 3000mm at centres; set 600mm deep into ground in concrete; height**										
C.04.01.04.A	1400mm	m	0.31	20.50	—	9.80	30.30	22.55	—	10.78	33.33
C.04.01.04.B	extra for 125mm x 125mm end straining post with 100mm x 75mm strut set in concrete	nr	0.44	29.71	—	54.09	83.80	32.68	—	59.49	92.17
C.04.01.04.C	extra for corner straining post with two struts set in concrete	nr	0.66	44.36	—	61.81	106.17	48.80	—	67.99	116.79

Landscaping

Hutchins Priced Schedules 2021	Unit	Labour Hours	Labour Net £	Plant Net £	Materials Net £	Unit Rate Net £	Labour Gross £	Plant Gross £	Materials Gross £	Unit Rate Gross £
							(Gross rates include 10% profit)			

C.04.   GENERAL FENCING

C.04.02.   GALVANISED CHAIN-LINK FENCING

C.04.02.01.   Fencing of 50mm mesh x 3mm chain link two 3.15mm line wires and 40mm x 40mm x 1500mm steel angle posts at 3000mm centres; set 600mm deep into ground in concrete; height

	Unit	Labour Hours	Labour Net £	Plant Net £	Materials Net £	Unit Rate Net £	Labour Gross £	Plant Gross £	Materials Gross £	Unit Rate Gross £
C.04.02.01.A 900mm	m	0.32	21.57	—	10.44	32.02	23.73	—	11.49	35.22
C.04.02.01.B extra for 50mm x 50mm angle end straining post with 40mm x 40mm steel angle strut; each bent over at bottom and set in concrete	nr	0.49	33.07	—	60.66	93.73	36.37	—	66.72	103.10
C.04.02.01.C extra for corner straining post with two struts set in concrete	nr	0.74	49.60	—	73.52	123.12	54.56	—	80.87	135.43

C.04.02.02.   Fencing of 50mm mesh x 3mm chain link three 3.15mm line wires and 45mm x 45mm x 2000mm steel angle posts at 3000mm centres; set 600mm deep into ground in concrete; height

	Unit	Labour Hours	Labour Net £	Plant Net £	Materials Net £	Unit Rate Net £	Labour Gross £	Plant Gross £	Materials Gross £	Unit Rate Gross £
C.04.02.02.A 1400mm	m	0.38	25.34	—	14.23	39.57	27.87	—	15.65	43.53
C.04.02.02.B extra for 50mm x 50mm angle end straining post with 45mm x 45mm steel angle strut; each bent over at bottom and set in concrete	nr	0.58	38.85	—	53.93	92.77	42.73	—	59.32	102.05
C.04.02.02.C extra for corner straining post with two struts set in concrete	nr	0.86	58.07	—	67.98	126.04	63.88	—	74.77	138.65

C.04.02.03.   Fencing of 50mm mesh x 3mm chain link, three 3.15mm line wires and 45mm x 45mm x 2600mm steel angle posts at 3000mm centres; set 760mm deep into ground in concrete; height

	Unit	Labour Hours	Labour Net £	Plant Net £	Materials Net £	Unit Rate Net £	Labour Gross £	Plant Gross £	Materials Gross £	Unit Rate Gross £
C.04.02.03.A 1800mm	m	0.48	32.40	—	24.41	56.80	35.63	—	26.85	62.49
C.04.02.03.B extra for 60mm x 60mm angle end straining post with 45mm x 45mm steel angle strut; each bent over at bottom and set in concrete	nr	0.74	49.94	—	66.08	116.02	54.93	—	72.69	127.62
C.04.02.03.C extra for corner straining post with two struts set in concrete	nr	1.11	74.60	—	108.17	182.78	82.06	—	118.99	201.05

C.04.02.04.   Fencing of 50mm mesh x 3mm chain link, two 3.15mm line wires and 120mm x 120mm to 75mm x 75mm x 1600mm reinforced concrete tapered posts at 3000mm centres; set 600mm deep into ground in concrete; height

	Unit	Labour Hours	Labour Net £	Plant Net £	Materials Net £	Unit Rate Net £	Labour Gross £	Plant Gross £	Materials Gross £	Unit Rate Gross £
C.04.02.04.A 900mm	m	0.30	20.23	—	14.93	35.16	22.25	—	16.43	38.68

Landscaping

	Unit	Labour Hours	Labour Net £	Plant Net £	Materials Net £	Unit Rate Net £	Labour Gross £	Plant Gross £	Materials Gross £	Unit Rate Gross £
							(Gross rates include 10% profit)			
C.04.	**GENERAL FENCING**									
C.04.02.	**GALVANISED CHAIN-LINK FENCING**									
C.04.02.04.	Fencing of 50mm mesh x 3mm chain link, two 3.15mm line wires and 120mm x 120mm to 75mm x 75mm x 1600mm reinforced concrete tapered posts at 3000mm centres; set 600mm deep into ground in concrete; height									
C.04.02.04.B	extra for 125mm x 125mm end straining post with 100mm x 75mm strut set in concrete	nr	0.60	40.53	—	46.93	87.46	44.58	—	51.63	96.21
C.04.02.04.C	extra for corner straining post with two struts set in concrete	nr	0.90	60.76	—	66.80	127.56	66.83	—	73.48	140.31
C.04.02.05.	Fencing of 50mm mesh x 3mm chain link, three 3.15mm line wires and 125mm x 125mm to 75mm x 75mm x 2070mm reinforced concrete tapered posts at 3000mm centres; set 600mm deep into ground in concrete; height									
C.04.02.05.A	1400mm	m	0.38	25.34	—	15.00	40.34	27.87	—	16.50	44.37
C.04.02.05.B	extra for 125mm x 125mm end straining post with 100mm x 75mm strut set in concrete	nr	0.71	47.58	—	60.11	107.70	52.34	—	66.12	118.47
C.04.02.05.C	extra for corner straining post with two struts set in concrete	nr	1.07	71.91	—	84.68	156.60	79.11	—	93.15	172.26
C.04.02.06.	Fencing of 50mm mesh x 3mm chain link, three 3.15mm line wires and 125mm x 125mm to 75mm x 75mm x 2630mm reinforced concrete tapered posts at 3000mm centres; set 760mm deep into ground in concrete; height									
C.04.02.06.A	1800mm	m	0.47	31.72	—	18.53	50.25	34.90	—	20.38	55.28
C.04.02.06.B	extra for 125mm x 125mm end straining post with 100mm x 75mm strut set in concrete	nr	0.88	59.41	—	76.66	136.07	65.36	—	84.32	149.68
C.04.02.06.C	extra for corner straining post with two struts set in concrete	nr	1.33	89.12	—	106.99	196.11	98.03	—	117.68	215.72
C.04.02.07.	Security fencing of 50mm mesh x 3mm chain link, three 3.15mm line wires; three rows of barbed wire and 125mm x 125mm to 75mm x 75mm x 3000mm reinforced concrete tapered posts, with cranked top, at 3000mm centres; set 760mm deep into ground in concrete; height									
C.04.02.07.A	1820mm	m	0.48	32.40	—	19.82	52.22	35.63	—	21.80	57.44
C.04.02.07.B	extra for 125mm x 125mm end straining post with 100mm x 75mm strut set in concrete	nr	0.88	59.41	—	76.66	136.07	65.36	—	84.32	149.68
C.04.02.07.C	extra for corner straining post with two struts set in concrete	nr	1.33	89.12	—	81.88	171.00	98.03	—	90.07	188.10

Hutchins Priced Schedules 2021	Unit	Labour Hours	Labour Net £	Plant Net £	Materials Net £	Unit Rate Net £	Labour Gross £	Plant Gross £	Materials Gross £	Unit Rate Gross £
								(Gross rates include 10% profit)		
C.04. **GENERAL FENCING**										
C.04.04. **CLEFT-PALE FENCING**										
C.04.04.01. Fencing of chestnut pales 90mm apart; two lines of binding wire and 63mm approximate dia x 1675mm chestnut posts at 2280mm centres; driven 600mm into ground; height										
C.04.04.01.A 1066mm..................	m	0.32	21.57	—	7.63	29.20	23.73	—	8.39	32.12
C.04.04.01.B extra for 76mm to 101mm approximate dia straining post with strut spiked to post	nr	0.50	33.74	—	9.03	42.77	37.11	—	9.93	47.05
C.04.04.01.C extra for corner straining post with two struts spiked to post.............	nr	0.75	50.61	—	12.29	62.89	55.67	—	13.51	69.18
C.04.06. **GALVANISED STEEL PALISADE FENCING**										
C.04.06.01. Fencing of triple pointed corrugated pales (1.9kg/m) fixed with 6mm rivets to two 50mm x 40mm x 6mm horizontal angle rails; bottom rail fitted with two support feet per bay; set into ground in concrete and with 102mm x 44mm RSJ posts at 2750mm centres; set 760mm deep into ground in concrete; height										
C.04.06.01.A 1800mm..................	m	0.67	44.90	—	42.61	87.51	49.39	—	46.87	96.26
C.04.06.01.B 2400mm..................	m	0.75	50.61	—	55.07	105.68	55.67	—	60.58	116.25
C.04.06.02. Fencing of triple pointed corrugated pales (2.42kg/m) fixed with 8mm rivets to two 50mm x 40mm x 6mm horizontal angle rails; bottom rail fitted with two support feet per bay; set into ground in concrete and with 102mm x 44mm RSJ posts at 2750mm centres; set 760mm deep into ground in concrete; height										
C.04.06.02.A 1800mm..................	m	0.75	50.61	—	55.38	105.99	55.67	—	60.92	116.59
C.04.06.02.B 2400mm..................	m	0.88	59.08	—	61.08	120.16	64.99	—	67.19	132.18

Landscaping

Hutchins Priced Schedules 2021	Unit	Labour Hours	Labour Net £	Plant Net £	Materials Net £	Unit Rate Net £	Labour Gross £	Plant Gross £	Materials Gross £	Unit Rate Gross £
							(Gross rates include 10% profit)			
C.04.	**GENERAL FENCING**									
C.04.08.	TREATED SOFTWOOD CLOSE-BOARDED FENCING									
C.04.08.01.	Fencing of 14mm to 7mm x 100mm feather edged boarding; two 75mm x 75mm arris rails; 100mm x 100mm x 1600mm sawn posts at 3000mm centres; set 600mm deep into ground in concrete; 22mm x 150mm gravel board with 47mm x 50mm centre stumps driven 600mm into ground; 25mm x 65mm counter rail and 38mm x 65mm weather capping; height									
C.04.08.01.A	1000mm..................	m	0.73	48.93	—	28.10	77.03	53.82	—	30.91	84.73
C.04.08.02.	Fencing of 14mm to 7mm x 100mm feather edged boarding; three 75mm x 75mm arris rails; 100mm x 125mm x 2350mm sawn posts at 3000mm centres set 760mm deep into ground in concrete; 22mm x 150mm gravel board with 47mm x 50mm centre stumps driven 600mm into ground; 25mm x 65mm counter rail and 38mm x 65mm weather capping; height									
C.04.08.02.A	1600mm..................	m	1.03	69.16	—	39.27	108.43	76.07	—	43.20	119.27
C.04.08.03.	Fencing of 1830mm x 910mm interwoven panels; of 6mm x 75mm woven slats; 19mm x 38mm framing and weathered capping; nailed between 75mm x 75mm x 1520mm capped posts at 1900mm centres; set 600mm deep into ground in concrete; height									
C.04.08.03.A	910mm..................	m	0.45	30.38	—	19.64	50.02	33.42	—	21.60	55.02
C.04.08.04.	Fencing of 1830mm x 1830mm interwoven panels; of 6mm x 75mm woven slats; 19mm x 38mm framing and weathered capping; nailed between 75mm x 75mm x 2590mm capped posts at 1900mm centres; set 760mm deep into ground in concrete; height									
C.04.08.04.A	1830mm..................	m	0.60	40.53	—	24.87	65.39	44.58	—	27.35	71.93

Hutchins Priced Schedules 2021	Unit	Labour Hours	Labour Net £	Plant Net £	Materials Net £	Unit Rate Net £	Labour Gross £	Plant Gross £	Materials Gross £	Unit Rate Gross £
									(Gross rates include 10% profit)	

C.04. GENERAL FENCING

C.04.08. TREATED SOFTWOOD CLOSE-BOARDED FENCING

C.04.08.05. Fencing of 1830mm x 900mm waney edged panels; 16mm x 38mm framing and weathered capping; nailed between 75mm x 75mm x 1520mm capped posts at 1900mm centres; set 600mm into ground set in concrete; height

C.04.08.05.A	900mm.....................	m	0.45	30.38	—	24.08	54.46	33.42	—	26.49	59.91

C.04.08.06. Fencing of 1830mm x 1800mm waney edge panels; 16mm x 38mm framing and weathered capping; nailed between 75mm x 75mm x 2590mm capped posts at 1900mm centres; set 760mm into ground set in concrete; height

C.04.08.06.A	1800mm....................	m	0.60	40.53	—	28.19	68.72	44.58	—	31.01	75.59

C.04.09. GATES

C.04.09.01. 42mm outside dia primed tubular steel single leaf gates filled in with chain link and with two 150mm x 150mm reinforced concrete gate posts; each with strut and set of fittings; including setting posts and struts in concrete and hanging gates

C.04.09.01.A	1000mm x 900mm.............	nr	2.39	160.30	—	330.12	490.41	176.33	—	363.13	539.46
C.04.09.01.B	1000mm x 1400mm............	nr	2.51	168.76	—	390.26	559.03	185.64	—	429.29	614.93
C.04.09.01.C	1000mm x 1800mm............	nr	2.64	177.17	—	455.34	632.51	194.88	—	500.88	695.76

C.04.09.02. 48mm outside dia primed tubular steel single leaf gates filled in with chain link and with two 150mm x 150mm reinforced concrete gate posts; each with strut and set of fittings; including setting posts and struts in concrete and hanging gates

C.04.09.02.A	3000mm x 900mm.............	nr	3.77	253.11	—	489.25	742.36	278.42	—	538.17	816.59
C.04.09.02.B	3000mm x 1400mm............	nr	4.02	269.98	—	607.32	877.30	296.98	—	668.05	965.03

C.04.09.03. Galvanised steel single leaf gates; to match corrugated pale (1.9kg/m) fencing; with two 127mm x 76mm RSJ gate posts including setting posts in concrete and hanging gates

C.04.09.03.A	1000mm x 1800mm............	nr	3.77	253.11	—	384.60	637.72	278.42	—	423.06	701.49
C.04.09.03.B	1000mm x 2400mm............	nr	4.14	278.45	—	466.43	744.88	306.30	—	513.07	819.37

C.04.09.04. Galvanised steel double leaf gates; to match corrugated pale (1.9kg/m) fencing; with two 125mm x 125mm SHS gate posts including setting posts in concrete and hanging gates

C.04.09.04.A	3000mm x 1800mm............	nr	9.04	607.51	—	987.84	1595.35	668.26	—	1086.63	1754.89

Landscaping

Hutchins Priced Schedules 2021		Unit	Labour Hours	Labour Net £	Plant Net £	Materials Net £	Unit Rate Net £	Labour Gross £	Plant Gross £	Materials Gross £	Unit Rate Gross £
							(Gross rates include 10% profit)				
C.04.	**GENERAL FENCING**										
C.04.09.	GATES										
C.04.09.05.	Treated sawn softwood close-boarded gates, to match close-boarded fencing; complete with all necessary hanging and closing fittings; including hanging gates										
C.04.09.05.A	3000mm x 1800mm	nr	2.76	185.63	—	325.15	510.79	204.20	—	357.67	561.86
C.05.	**DECKING**										
C.05.01.	HARDWOOD										
C.05.01.01.	Standard duty; natural decking board; nominal size 100mm x 25mm; on flat ground with 100mm dia, 600mm long treated decking support posts; 400mm out of ground and 200mm in ground secured with Postcrete; 50mm x 100mm treated decking joists laid at 500mm centres; area										
C.05.01.01.A	2.5m x 2.5m	m²	1.45	54.56	—	84.72	139.28	60.02	—	93.19	153.21
C.05.01.01.B	2.5m x 5.0m	m²	1.30	48.92	—	77.72	126.64	53.81	—	85.50	139.31
C.05.01.01.C	2.5m x 7.5m	m²	1.25	47.04	—	76.17	123.21	51.74	—	83.79	135.53
C.05.01.01.D	5.0m x 5.0m	m²	1.20	45.16	—	70.10	115.25	49.67	—	77.10	126.78
C.05.01.01.E	extra for fascia board; supply only	m	—	—	—	2.43	2.43	—	—	2.67	2.67
C.05.01.02.	Heavy duty; natural decking board; nominal size 150mm x 32mm; on flat ground with 100mm dia, 600mm long treated decking support posts; 400mm out of ground and 200mm in ground secured with Postcrete; 50mm x 100mm treated decking joists laid at 500mm centres; area										
C.05.01.02.A	2.5m x 2.5m	m²	1.55	58.33	—	112.07	170.40	64.16	—	123.28	187.44
C.05.01.02.B	2.5m x 5.0m	m²	1.40	52.68	—	104.57	157.25	57.95	—	115.02	172.97
C.05.01.02.C	2.5m x 7.5m	m²	1.35	50.80	—	102.85	153.65	55.88	—	113.13	169.01
C.05.01.02.D	5.0m x 5.0m	m²	1.30	48.92	—	96.68	145.60	53.81	—	106.35	160.16
C.05.01.02.E	extra for fascia board; supply only	m	—	—	—	2.43	2.43	—	—	2.67	2.67

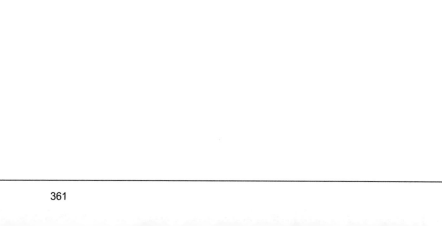

Detailed table of contents

Detailed table of contents

Detailed table of contents

Detailed table of contents

Memoranda

Memoranda

Unit equivalents

	Metric unit	Abbreviation	Imperial equivalent	
Length	1 millimetre	1 mm	0.039	in
	1 metre	1 m	{ 3.281	ft
			{ 1.094	yd
	1 kilometre	1 km	0.621	mile
Area	1 square millimetre	1 mm^2	0.001 55	in^2
	1 square metre	1 m^2	{10.764	ft^2
			{ 1.196	yd^2
	1 hectare	1 ha	{11 960	yd^2
			{ 2.471	acre
Volume	1 cubic metre	1 m^3	{35.315	ft^3
			{ 1.308	yd^3
	1 litre	1 l	{ 1.760	pint
			{ 0.220	UK gal
	5 litres	5 l	1.100	UK gal
Weights	1 kilogramme	1 kg	2.205	lb
	50 kilogrammes	50 kg	{110.231	lb
			{ 0.984	cwt
	1 tonne	1 t	0.984	ton
	1 kilogramme per metre	1 kg/m	{ 0.672	lb/ft
			{ 2.016	lb/yd
	1 kilogramme per square metre	1 kg/m^2	{ 0.205	lb/ft^2
			{ 1.843	lb/yd^2
	1 kilogramme per cubic metre	1 kg/m^3	{ 0.062	lb/ft^3
			{ 1.686	lb/yd^3
Force	1 newton	1N	0.225	lbf
	1 kilonewton	1 kN	{224.809	lbf
			{0.100 36	tonf
Pressure	1 newton per square millimetre	1 N/mm^2	145.038	lbf/in^2
			{ 0.145	lbf/in^2
	1 kilonewton per square metre	1 kN/m^2	{ 20.885	lbf/ft^2
			{ 0.009	tonf/ft^2
Energy (work, heat)	1 kilojoule	1 kJ	0.948	Btu

Unit equivalents *(continued)*

	Imperial unit	Abbreviation	Metric equivalent	
Length	1 inch	1 in	25.400	mm
	1 foot	1 ft	304.800	mm
	1 yard	1 yd	0.914	m
	1 mile	1 mile	1.609	km
Area	1 square inch	1 in^2	645.160	mm
	1 square foot	1 ft^2	0.093	m^2
	1 square yard	1 yd^2	0.836	m^2
	1 acre	1 acre	{4 046.856	m^2
			{0.405	ha
Volume	1 cubic inch	1 in^3	16 387.064	mm^3
	1 cubic foot	1 ft^3	0.028	m^3
	1 cubic yard	1 yd^3	0.765	m^3
	1 pint	1 pint	0.568	litre
	1 UK gallon	1 UK gal	4.546	litre
Weight	1 pound	1 lb	0.454	kg
	1 hundredweight	1 cwt	50.802	kg
	1 ton	1 ton	1.016	tonne
	1 pound per foot	1 lb/ft	1.488	kg/m
	1 pound per yard	1 lb/yd	0.496	kg/m
	1 pound per square foot	1 lb/ft^2	4.882	kg/m^2
	1 pound per square yard	1 lb/yd^2	0.542	kg/m^2
	1 pound per cubic foot	1 lb/ft^3	16.018	kg/m^3
	1 pound per cubic yard	1 lb/yd^3	0.593	kg/m^3
Force	1 pound force	1 lbf	4.448	N
	1 ton force	1 tonf	9.964	kN
Pressure	1 pound force per square inch	1 lbf/in^2	{ 0.007	N/mm^2
			{ 6.895	kN/m^2
	1 pound force per square foot	1 lbf/ft^2	0.048	kN/m^2
	1 ton force per square foot	1 tonf/ft^2	107.252	kN/m^2
Energy (work, space heat)	1 British thermal unit	1 Btu	1.055	kJ

Memoranda

Conversion tables

Length

mm	in
1	0.039
2	0.079
3	0.118
4	0.157
5	0.197
6	0.236
7	0.276
8	0.315
9	0.354
10	0.394
100	3.937
1 000	39.370

in	mm
$\frac{1}{16}$	1.59
$\frac{1}{8}$	3.18
$\frac{1}{4}$	6.35
$\frac{3}{8}$	9.53
$\frac{1}{2}$	12.70
$\frac{5}{8}$	15.88
$\frac{3}{4}$	19.05
$\frac{7}{8}$	22.23
1	25.40
2	50.80
3	76.20
4	101.60
5	127.00
6	152.40
7	177.80
8	203.20
9	228.60
10	254.00
11	279.40
12	304.80

m	ft
1	3.281
2	6.562
3	9.843
4	13.123
5	16.404
6	19.685
7	22.966
8	26.247
9	29.528
10	32.808
100	328.084
1 000	3 280.840

ft	m
1	0.305
2	0.610
3	0.914
4	1.219
5	1.524
6	1.829
7	2.134
8	2.438
9	2.743
10	3.048
100	30.480
1 000	304.800

m	yd
1	1.094
2	2.187
3	3.281
4	4.374
5	5.468
6	6.562
7	7.655
8	8.749
9	9.843
10	10.936
100	109.361
1 000	1 093.613

yd	m
1	0.914
2	1.829
3	2.743
4	3.658
5	4.572
6	5.486
7	6.401
8	7.315
9	8.230
10	9.144
100	91.440
1 000	914.400

km	mile
1	0.621
2	1.243
3	1.864
4	2.485
5	3.107
6	3.728
7	4.350
8	4.971
9	5.592
10	6.214
100	62.137
1 000	621.371

mile	km
1	1.609
2	3.219
3	4.828
4	6.437
5	8.047
6	9.656
7	11.265
8	12.875
9	14.484
10	16.093
100	160.934
1 000	1 609.344

Conversion tables *(continued)*

Area

mm²	in²	in²	mm²	m²	ft²	ft²	m²
1	0.001 55	1	645	1	10.764	1	0.093
2	0.003 10	2	1 290	2	21.528	2	0.186
3	0.004 65	3	1 935	3	32.292	3	0.279
4	0.006 20	4	2 581	4	43.056	4	0.372
5	0.007 75	5	3 226	5	53.820	5	0.465
6	0.009 30	6	3 871	6	64.583	6	0.557
7	0.010 85	7	4 516	7	75.347	7	0.650
8	0.012 40	8	5 161	8	86.111	8	0.743
9	0.013 95	9	5 806	9	96.875	9	0.836
10	0.015 50	10	6 452	10	107.639	10	0.929
100	0.155 00	100	64 516	100	1 076.392	100	9.290
1 000	1.550 00	1 000	645 160	1 000	10 763.915	1 000	92.903

m²	yd²	yd²	m²	ha	acre	acre	ha
1	1.196	1	0.836	1	2.471	1	0.405
2	2.392	2	1.672	2	4.942	2	0.809
3	3.588	3	2.508	3	7.413	3	1.214
4	4.784	4	3.345	4	9.884	4	1.619
5	5.980	5	4.181	5	12.355	5	2.023
6	7.176	6	5.017	6	14.826	6	2.428
7	8.372	7	5.853	7	17.297	7	2.833
8	9.568	8	6.689	8	19.768	8	3.237
9	10.764	9	7.525	9	22.239	9	3.642
10	11.960	10	8.361	10	24.711	10	4.047
100	119.599	100	83.613	100	247.105	100	40.469
1 000	1 195.991	1 000	836.127	1 000	2 471.052	1 000	404.686

Volume

m³	ft³	ft³	m³	m³	yard³	yard³	m³
1	35.315	1	0.028	1	1.308	1	0.765
2	70.629	2	0.057	2	2.616	2	1.529
3	105.944	3	0.085	3	3.924	3	2.294
4	141.259	4	0.113	4	5.232	4	3.058
5	176.574	5	0.142	5	6.540	5	3.823
6	211.888	6	0.170	6	7.848	6	4.587
7	247.203	7	0.198	7	9.156	7	5.352
8	282.518	8	0.227	8	10.464	8	6.116
9	317.833	9	0.255	9	11.772	9	6.881
10	353.147	10	0.283	10	13.080	10	7.646
100	3 531.472	100	2.832	100	130.795	100	76.455
1 000	35 314.725	1 000	28.317	1 000	1 307.953	1 000	764.554

Memoranda

Conversion tables *(continued)*

Volume (continued)

litre	pint	pint	litre	litre	UK gal	UK gal	litre
1	1.760	1	0.568	1	0.220	1	4.546
2	3.520	2	1.137	2	0.440	2	9.092
3	5.279	3	1.705	3	0.660	3	13.638
4	7.039	4	2.273	4	0.880	4	18.184
5	8.799	5	2.841	5	1.100	5	22.730
6	10.559	6	3.410	6	1.320	6	27.277
7	12.318	7	3.978	7	1.540	7	31.823
8	14.078	8	4.546	8	1.760	8	36.369
9	15.838	9	5.114	9	1.980	9	40.915
10	17.598	10	5.683	10	2.200	10	45.461
100	175.975	100	56.826	100	21.997	100	454.609
1 000	1 759.754	1 000	568.261	1 000	219.969	1 000	4 546.090

Weight

kg	lb	lb	kg	tonne	cwt	cwt	tonne
1	2.205	1	0.454	1	19.684	1	0.051
2	4.409	2	0.907	2	39.368	2	0.102
3	6.614	3	1.361	3	59.052	3	0.152
4	8.818	4	1.814	4	78.736	4	0.203
5	11.023	5	2.268	5	98.420	5	0.254
6	13.228	6	2.722	6	118.104	6	0.305
7	15.432	7	3.175	7	137.788	7	0.356
8	17.637	8	3.629	8	157.473	8	0.406
9	19.842	9	4.082	9	177.157	9	0.457
10	22.046	10	4.536	10	196.841	10	0.508
100	220.462	100	45.359	100	1 968.413	100	5.080
1 000	2 204.623	1 000	453.592	1 000	19 684.131	1 000	50.802

tonne	ton	ton	tonne	kg/m	lb/ft	lb/ft	kg/m
1	0.984	1	1.016	1	0.672	1	1.488
2	1.968	2	2.032	2	1.344	2	2.976
3	2.953	3	3.048	3	2.016	3	4.464
4	3.937	4	4.064	4	2.688	4	5.953
5	4.921	5	5.080	5	3.360	5	7.441
6	5.905	6	6.096	6	4.032	6	8.929
7	6.889	7	7.112	7	4.704	7	10.417
8	7.874	8	8.128	8	5.376	8	11.905
9	8.858	9	9.144	9	6.048	9	13.393
10	9.842	10	10.161	10	6.720	10	14.882
100	98.421	100	101.605	100	67.197	100	148.816
1 000	984.207	1 000	1 1016.047	1 000	671.969	1 000	1 488.164

Conversion tables (continued)

Weight (continued)

kg/m	lb/yd
1	2.016
2	4.032
3	6.048
4	8.064
5	10.080
6	12.095
7	14.111
8	16.127
9	18.143
10	20.159
100	201.591
1 000	2 015.907

kg/m²	lb/yd²
1	1.843
2	3.687
3	5.530
4	7.373
5	9.217
6	11.060
7	12.903
8	14.747
9	16.590
10	18.433
100	184.334
1 000	1 843.344

kg/m³	lb/yd³
1	1.686
2	3.371
3	5.057
4	6.742
5	8.428
6	10.113
7	11.799
8	13.484
9	15.170
10	16.856
100	168.555
1 000	1 685.555

lb/yd	kg/m
1	0.496
2	0.992
3	1.488
4	1.984
5	2.480
6	2.976
7	3.472
8	3.968
9	4.464
10	4.961
100	49.605
1 000	496.055

lb/yd²	kg/m²
1	0.542
2	1.085
3	1.627
4	2.170
5	2.712
6	3.255
7	3.797
8	4.340
9	4.882
10	5.425
100	54.249
1 000	542.492

lb/yd³	kg/m³
1	0.593
2	1.187
3	1.780
4	2.373
5	2.966
6	3.560
7	4.153
8	4.746
9	5.339
10	5.933
100	59.328
1 000	593.276

kg/m²	lb/ft²
1	0.205
2	0.410
3	0.614
4	0.819
5	1.024
6	1.229
7	1.434
8	1.639
9	1.843
10	2.048
100	20.482
1 000	204.816

kg/m³	lb/ft³
1	0.062
2	0.125
3	0.187
4	0.250
5	0.312
6	0.375
7	0.437
8	0.499
9	0.562
10	0.624
100	6.243
1 000	62.428

lb/ft²	kg/m²
1	4.882
2	9.765
3	14.647
4	19.530
5	24.412
6	29.295
7	34.177
8	39.059
9	43.942
10	48.824
100	488.243
1 000	4 882.430

lb/ft³	kg/m³
1	16.018
2	32.037
3	48.055
4	64.074
5	80.092
6	96.111
7	112.129
8	128.148
9	144.166
10	160.185
100	1 601.849
1 000	16 018.490

Memoranda

Conversion tables *(continued)*

Force

N	lbf		lbf	N		kN	lbf		lbf	kN
1	0.225		1	4.448		1	224.809		1	0.004
2	0.450		2	8.896		2	449.618		2	0.009
3	0.674		3	13.345		3	674.427		3	0.013
4	0.899		4	17.793		4	899.236		4	0.018
5	1.124		5	22.241		5	1 124.045		5	0.022
6	1.349		6	26.689		6	1 348.854		6	0.027
7	1.574		7	31.138		7	1 573.663		7	0.031
8	1.798		8	35.586		8	1 798.472		8	0.036
9	2.023		9	40.034		9	2 023.281		9	0.040
10	2.248		10	44.482		10	2 248.090		10	0.044
100	22.481		100	444.822		100	22 480.902		100	0.445
1 000	224.809		1 000	4 448.220		1 000	224 809.025		1 000	4.448

kN	tonf		tonf	kN
1	0.100		1	9.964
2	0.201		2	19.928
3	0.301		3	29.892
4	0.401		4	39.856
5	0.502		5	49.820
6	0.602		6	59.784
7	0.703		7	69.748
8	0.803		8	79.712
9	0.903		9	89.676
10	1.004		10	99.640
100	10.036		100	996.401
1 000	100.361		1 000	9 964.013

Pressure

N/mm^2	lbf/in^2		lbf/in^2	N/mm^2		kN/m^2	lbf/ft^2		lbf/ft^2	kN/m^2
1	145.038		1	0.007		1	20.885		1	0.048
2	290.076		2	0.014		2	41.771		2	0.096
3	145.113		3	0.021		3	62.656		3	0.144
4	580.151		4	0.028		4	83.542		4	0.192
5	725.189		5	0.034		5	104.427		5	0.239
6	870.227		6	0.041		6	125.313		6	0.287
7	1 015.265		7	0.048		7	146.198		7	0.335
8	1 160.302		8	0.055		8	167.083		8	0.383
9	1 305.340		9	0.062		9	187.969		9	0.431
10	1 450.378		10	0.069		10	208.854		10	0.479
100	14 503.779		100	0.689		100	2 088.544		100	4.788
1 000	145 037.790		1 000	6.895		1 000	20 885.442		1 000	47.880

Conversion tables *(continued)*

Pressure (continued)

kN/m^2	tonf/ft^2	tonf/ft^2	kN/m^2
1	0.009	1	107.252
2	0.019	2	214.504
3	0.028	3	321.755
4	0.037	4	429.007
5	0.047	5	536.259
6	0.056	6	643.511
7	0.065	7	750.762
8	0.075	8	858.014
9	0.084	9	965.266
10	0.093	10	1 072.518
100	0.932	100	10 725.174
1 000	9.324	1 000	107 251.740

Energy

kJ	Btu	Btu	kJ
1	0.948	1	1.055
2	1.896	2	2.110
3	2.843	3	3.165
4	3.791	4	4.220
5	4.739	5	5.275
6	5.687	6	6.330
7	6.635	7	7.385
8	7.583	8	8.440
9	8.530	9	9.496
10	9.478	10	10.551
100	94.781	100	105.506
1 000	947.813	1 000	1 055.060

Temperature

°C	°F	°F	°C
0	32	0	−17.78
10	50	5	−15.00
20	68	10	−12.22
30	86	20	−6.67
40	104	32	0.00
50	122	40	4.44
60	140	50	10.00
70	158	100	37.78
80	176	150	65.56
90	194	200	93.33
100	212	212	100.00

Memoranda

Excavation and earthwork

Bearing capacities

Type of ground	Approximate bearing capacity kN/m²
Unweathered rock	over 1 000
Solid chalk	600
Compact gravel	600
Loose gravel	up to 200
Compact sand	300
Loose sand	up to 100
Hard boulder clay	300 to 600
Stiff clay	150 to 300
Firm clay	75 to 150
Soft clay	up to 75

"Bulking" of excavated materials

Type of material	Approximate bulk of 1 m³ after excavation m³
Unweathered rock	1.75
Solid chalk	1.75
Compact gravel	1.25
Loose gravel	1.10
Compact sand	1.20
Loose sand	1.05
Hard boulder clay	1.50
Stiff clay	1.40
Firm clay	1.25
Soft clay	1.25
Subsoil	1.25
Topsoil and loam	1.25

Concrete Work

Concrete mixing materials

Based on: (1) cement $1\,440\ kg/m^3$
          (2) fine aggregate/sand $1\,600\ kg/m^3$ dry ($1\,260\ kg/m^3$ moist)
          (3) one third bulking of fine aggregate/sand from dry to moist
          (4) coarse aggregate $1\,500\ kg/m^3$
          (5) net quantities – allowance for waste to be added.

– per 2 × 25 kg bags of cement

Mix	Cement	Fine aggregate/Sand		Coarse aggregate
		dry or	(moist)	
– by volume	m^3	m^3	m^3	m^3
1:3:6	0.035	0.105	(0.140)	0.210
1:2:4	0.035	0.070	(0.093)	0.140
1:1½:3	0.035	0.053	(0.070)	0.105
– by weight	kg	kg	kg	kg
1:3:6	50	168	(176)	315
1:2:4	50	112	(117)	210
1:1½:3	50	85	(88)	158

– per cubic metre of concrete

Mix	Cement	Fine aggregate/Sand		Coarse aggregate
		dry or	(moist)	
– by volume	m^3	m^3	m^3	m^3
1:3:6	0.150	0.450	(0.600)	0.900
1:2:4	0.214	0.429	(0.572)	0.857
1:1½:3	0.273	0.409	(0.545)	0.818
– by weight	kg	kg	kg	kg
1:3:6	216	720	(756)	1 350
1:2:4	308	686	(721)	1 286
1:1½:3	393	654	(687)	1 227

Memoranda

Concrete work *(continued)*

Steel bar reinforcement

Diameter mm	Nominal weight kg/m	Length m/tonne	Sectional area mm^2
6	0.222	4 505	28.3
8	0.395	2 532	50.3
10	0.616	1 623	78.5
12	0.888	1 126	113.1
16	1.579	633	201.1
20	2.466	406	314.2
25	3.854	259	490.9
32	6.313	158	804.2
40	9.864	101	1 256.6
50	15.413	65	1 963.5

Steel fabric reinforcement

BS 4483 reference	Nominal weight kg/m^2	Mesh dimensions		Wire diameters	
		Main mm	Cross mm	Main mm	Cross mm
A 393	6.16	200	200	10	10
A 252	3.95	200	200	8	8
A 193	3.02	200	200	7	7
A 142	2.22	200	200	6	6
A 98	1.54	200	200	5	5
B 1131	10.90	100	200	12	8
B 785	8.14	100	200	10	8
B 503	5.93	100	200	8	8
B 385	4.53	100	200	7	7
B 283	3.73	100	200	6	7
B 196	3.05	100	200	5	7
C 785	6.72	100	400	10	6
C 636	5.55	80–130	400	8–10	6
C 503	4.34	100	400	8	5
C 385	3.41	100	400	7	5
C 283	2.61	100	400	6	5
D 98	1.54	200	200	5	5
D 49	0.77	100	100	2.5	2.5

Concrete Work *(continued)*

Striking times for formwork

Minimum time in days between placing concrete and striking formwork – days when frost occurs should be added.

| | "Normal" weather – about 15°C | | "Cold" weather – about 5°C | |
	Ordinary Portland cement	Rapid-hardening cement	Ordinary Portland cement	Rapid-hardening cement
	Days	Days	Days	Days
Sides of beams, columns or walls	2	1	7	5
Soffits of slabs – props left in position	4	3	10	7
Props to soffits of slabs	11	10	17	14
Soffits of beams – props left in position	7	5	14	10
Props to soffits of beams	14	12	21	17

Memoranda

Brickwork and blockwork

Bricks and mortar quantities per square metre

Based on:
(1) 215 × 102.5 × 65 mm work size bricks
(2) 10 mm mortar joints
(3) only one snapped header obtained from one facing brick
(4) net quantities - allowance for waste to be added according to the nature of the work and the type of bricks to be used.

	Common bricks	Facing bricks	Mortar for solid (wirecut) bricks	Mortar for single frogged bricks
	No/m^2	No/m^2	m^3/m^2	m^3/m^2
Unfaced walls				
102.5 mm	59.3	–	0.018	0.022
215 mm	118.5	–	0.045	0.054
327.5 mm	177.8	–	0.073	0.086
Faced walls				
102.5 mm in Stretcher bond	–	59.3	0.018	0.022
102.5 mm in English bond with snapped headers	–	88.9	0.020	0.025
102.5 mm in Flemish bond with snapped headers	–	79.0	0.019	0.024
Walls in English bond faced one side				
215 mm	29.6	88.9	0.045	0.054
327.5 mm	88.9	88.9	0.073	0.086
Walls in English bond faced both sides				
215 mm	–	118.5	0.045	0.054
327.5 mm	–	177.8	0.073	0.086
Walls in Flemish bond faced one side				
215 mm	39.5	79.0	0.045	0.054
327.5 mm	98.8	79.0	0.074	0.088
Walls in Flemish bond faced both sides				
215 mm	–	118.5	0.045	0.054
327.5 mm	19.8	158.0	0.074	0.088

Brickwork and Blockwork *(continued)*

Blocks and mortar quantities per square metre

Based on: (1)   440 × 215 mm work size blocks
           (2)   10 mm mortar joints
           (3)   net quantities – allowance for waste to be added according to the nature of the work and the type of blocks to be used.

Wall thickness	Blocks No/m^2	Mortar m^3/m^2
60 mm	9.9	0.004
75 mm	9.9	0.005
90 mm	9.9	0.006
100 mm	9.9	0.007
140 mm	9.9	0.009
190 mm	9.9	0.013
215 mm	9.9	0.014
215 mm of 100 mm blocks laid flat	20.2	0.024
215 mm of 140 mm blocks laid flat	14.8	0.019

Mortar mixing materials
Based on: (1)   cement 1 440 kg/m^3
           (2)   hydrated lime 500 kg/m^3
           (3)   sand 1 600 kg/m^3 dry (1 260 kg/m^3 moist)
           (4)   one third bulking of sand from dry to moist
           (5)   net quantities – allowance for waste to be added.

– per 2 × 25 kg bags of cement

Mix	Cement	Lime	Sand	
– by volume m^3	m^3	m^3	dry m^3	or (moist)
1:1	0.035	–	0.035	(0.047)
1:2	0.035	–	0.070	(0.093)
1:3	0.035	–	0.105	(0.140)
1:4	0.035	–	0.140	(0.187)
1:6	0.035	–	0.210	(0.280)
1:1:5	0.035	0.035	0.175	(0.233)
1:1:6	0.035	0.035	0.210	(0.280)
1:2:9	0.035	0.070	0.315	(0.420)

Memoranda

Brickwork and Blockwork *(continued)*
Mortar mixing materials
– per 2 × 25 kg bags of cement

	Mix	Cement	Lime	Sand dry	or	(moist)
– by weight		kg	kg	kg		kg
	1:1	50	–	56		(59)
	1:2	50	–	112		(117)
	1:3	50	–	168		(176)
	1:4	50	–	224		(236)
	1:6	50	–	336		(353)
	1:1:5	50	18	280		(294)
	1:1:6	50	18	336		(353)
	1:2:9	50	35	504		(529)

– per cubic metre of mortar

	Mix	Cement	Lime	Sand dry	or	(moist)
– by volume		m³	m³	m³		m³
	1:1	0.725	–	0.725		(0.967)
	1:2	0.467	–	0.933		(1.244)
	1:3	0.338	–	1.012		(1.349)
	1:4	0.260	–	1.040		(1.387)
	1:6	0.179	–	1.071		(1.428)
	1:1:5	0.196	0.196	0.983		(1.311)
	1:1:6	0.169	0.169	1.012		(1.349)
	1:2:9	0.113	0.225	1.012		(1.349)
– by weight		kg	kg	kg		kg
	1:1	1 044	–	1 160		(1 218)
	1:2	672	–	1 493		(1 567)
	1:3	487	–	1 619		(1 700)
	1:4	374	–	1 664		(1 748)
	1:6	258	–	1 714		(1 799)
	1:1:5	282	98	1 573		(1 652)
	1:1:6	243	85	1 619		(1 700)
	1:2:9	163	113	1 619		(1 700)

Roofing
Slating and tiling quantities per square metre

Based on net quantities – allowance for waste to be added according to the nature of the work and the type and dimensions of slates, tiles and battens to be used.

Notes: (1) lap – for traditional double lap slating/tiling – length of end cover of one slate/tile over the next but one slate/tile beneath
– for single lap tiling – length of end cover of one tile over the tile beneath
(2) gauge – batten centres spacing and length of slate/tile exposed
(3) nails – quantities for slating, tiling and battening can be calculated from the tables according to the type of slates/tiles, section of battens and centres spacing of the rafters or joists.

Centre-nailed slates

Size mm	Lap mm	Gauge mm	Slates No/m^2	Battens m/m^2
610 × 305	76	267	12.3	3.7
610 × 305	100	255	12.9	3.9
600 × 300	76	262	12.7	3.8
600 × 300	100	250	13.3	4.0
510 × 255	76	217	18.1	4.6
510 × 255	100	205	19.1	4.9
500 × 250	76	212	18.9	4.7
500 × 250	100	200	20.0	5.0
405 × 205	75	165	29.6	6.1
405 × 205	95	155	31.5	6.5
400 × 200	70	165	30.3	6.1
400 × 200	90	155	32.3	6.5

Plain tiles

Size mm	Lap mm	Gauge mm	Tiles No/m^2	Battens m/m^2
265 × 165	38	114	53.2	8.8
265 × 165	65	100	60.6	10.0

Single lap tiles

Size mm	Cover width mm	Lap mm	Gauge mm	Tiles No/m^2	Battens m/m^2
430 × 380	343	75	355	8.2	2.8
430 × 380	343	100	330	8.8	3.0
420 × 332	300	75	345	9.7	2.9
420 × 332	300	100	320	10.4	3.1
413 × 330	292	75	338	10.1	3.0
413 × 330	292	100	313	10.9	3.2
380 × 230	200	75	305	16.4	3.3
380 × 230	200	100	280	17.9	3.6

Memoranda

Roofing *(continued)*

Approximate numbers of round lost head nails per kilogramme

Aluminium Length × Shank		Copper Length × Shank		Steel Length × Shank	
mm	No/kg	mm	No/kg	mm	No/kg
75 × 3.75	448	65 × 3.75	178	75 × 3.75	160
65 × 3.35	672	3.35	194	65 × 3.35	240
60 × 3.35	756	50 × 3.35	292	3.00	270
50 × 3.35	860	3.00	308	60 × 3.35	270
3.00	1 008	40 × 2.65	474	3.00	330
40 × 2.65	1 390	2.36	554	50 × 3.00	360
2.36	2 128			2.65	420
				40 × 2.36	760

Approximate numbers of extra large head felt nails per kilogramme

Aluminium Length × Shank		Copper Length × Shank		Steel Length × Shank	
mm	No/kg	mm	No/kg	mm	No/kg
25 × 3.35	1 296	25 × 3.35	440	40 × 3.00	350
3.00	1 636	3.00	517	30 × 3.00	420
20 × 3.35	1 848	20 × 3.35	544	25 × 3.00	485
3.00	2 130	3.00	627	20 × 3.00	580
15 × 3.35	1 840	15 × 3.00	691	15 × 3.00	650
3.00	2 283	13 × 3.00	880	13 × 3.00	780

Roofing (continued)

Approximate number of clout, slate or tile nails per kilogramme

Aluminium

Length × Shank mm	No/kg
65 × 3.75	504
60 × 3.75	550
3.35	680
50 × 3.75	644
3.35	812
3.00	952
45 × 3.35	924
3.00	1 060
40 × 3.35	980
3.00	1 200
2.65	1 596
2.36	1 960
30 × 3.00	1 512
2.65	1 848
2.36	2 324
2.00	3 000
25 × 3.35	1 540
3.00	1 750
2.65	2 282
2.00	3 800
20 × 3.00	2 300
2.65	2 898

Copper

Length × Shank mm	No/kg
65 × 3.75	170
3.35	195
50 × 3.35	241
3.00	276
2.65	327
45 × 3.35	308
3.00	366
2.65	456
40 × 3.35	335
3.00	3.98
2.65	460
2.36	553
30 × 3.35	448
3.00	550
2.65	621
2.36	748
25 × 2.65	740
20 × 2.65	920

Steel

Length × Shank mm	No/kg
100 × 4.50	75
90 × 4.50	85
75 × 3.75	150
65 × 3.75	180
50 × 3.75	230
3.35	290
3.00	340
2.65	430
45 × 3.35	330
2.65	4.60
40 × 3.35	350
2.65	570
2.36	700
30 × 3.00	540
2.65	660
2.36	830
25 × 2.65	815
20 × 2.65	1 035
15 × 2.36	1 540
2.00	2 380

Approximate number of tile pegs per kilogramme

Aluminium

Length × Shank mm	No/kg
40 × 5.00	450
4.50	490
30 × 5.00	545
4.50	600

Steel

Length × Shank mm	No/kg
40 × 6.00	88
30 × 6.00	106

Memoranda

Roofing *(continued)*

Sheet metal roofing measurement allowances

In the absence of any directions to the contrary, allowances made in calculating the areas of the finished surface of sheet metalwork are as follows:

Item		Lead	Zinc, aluminium and copper
a.	For drips not exceeding 50 mm high	0.18 m	0.15 m
b.	For cross-welts	0.08 m	0.08 m
c.	For wood-cored rolls not exceeding 50 mm high	0.25 m	0.15 m
d.	For standing seams and welted seams	0.10 m	0.08 m
e.	For edge-welts	0.03 m	0.03 m
f.	For vertical upstands	0.15 m	0.15 m
g.	For sloping upstands and for laps in pitched roofing	0.30 m for 30°	0.30 m for 30°
		0.35 m for 25°	0.35 m for 25°
		0.45 m for 20°	0.45 m for 20°
		0.60 m for 15°	0.60 m for 15°

Sheet metal thicknesses etc.

Sheet lead Thickness	BS code	Colour marking	Sheet zinc Thickness	
mm			mm	ZG
1.32	3	Green	0.45	9
1.80	4	Blue	0.65	12
2.24	5	Red	0.80	14
2.65	6	Black		
3.15	7	White		
3.55	8	Orange		

Sheet aluminium		Sheet copper	
Thickness		Thickness	
mm	SWG	mm	SWG
0.60	23	0.45	26
0.80	21	0.55	24
		0.70	22

Woodwork – Lengths of sawn softwood per cubic metre

mm		m/m³	mm		m/m³	mm		m/m³
16 ×	16	3 906	25 ×	175	229	50 ×	50	400
	19	3 289		200	200		63	317
	22	2 841		225	178		75	267
	25	2 500		250	160		100	200
	32	1 953		300	133		125	160
	38	1 645					150	133
	44	1 420	32 ×	32	977		175	114
	50	1 250		38	822		200	100
	63	992		44	710		225	89
	75	833		50	625		250	80
	100	625		63	496		300	67
	125	500		75	417			
	150	417		100	313	63 ×	63	252
				125	250		75	212
19 ×	19	2 770		150	208		100	159
	22	2 392		175	179		125	127
	25	2 105		200	156		150	106
	32	1 645		225	139		175	91
	38	1 385		250	125		200	79
	44	1 196		300	104		225	71
	50	1 053					250	63
	63	835	38 ×	38	693		300	53
	75	702		44	598			
	100	526		50	526	75 ×	75	178
	125	421		63	418		100	133
	150	351		75	351		125	107
				100	263		150	89
22 ×	22	2 066		125	211		175	76
	25	1 818		150	175		200	67
	32	1 420		175	150		225	59
	38	1 196		200	132		250	53
	44	1 033		225	117		300	44
	50	909		250	105			
	63	722		300	88	100 ×	100	100
	75	606					150	67
	100	455	44 ×	44	517		200	50
	125	364		50	455		250	40
	150	303		63	361		300	33
				75	303			
25 ×	25	1 600		100	227	150 ×	150	44
	32	1 250		125	182		200	33
	38	1 053		150	152		250	27
	44	909		175	130		300	22
	50	800		200	114			
	63	635		225	101	200 ×	200	25
	75	533		250	91			
	100	400		300	76	250 ×	250	16
	125	320						
	150	267				300 ×	300	11

Memoranda

Woodwork *(continued)*

Lengths of boarding required per square metre

Based on net quantities – allowance for waste to be added according to the nature of the work and the type of boarding to be used.

Effective width mm	m/m²	Effective width mm	m/m²
75	13.33	150	6.67
100	10.00	175	5.71
125	8.00	200	5.00

Standard lengths of sawn softwood

m	m	m	m	m	m	m
1.8	2.1	3.0	4.2	5.1	6.0	7.2
	2.4	3.3	4.5	5.4	6.3	
	2.7	3.6	4.8	5.7	6.6	
		3.9			6.9	

Woodwork (continued)

Approximate number of steel wire nails per kilogramme

Round plain head nails
Length × Shank

mm	No/kg
150 × 6.00	29
125 × 5.60	42
5.00	53
115 × 5.00	57
100 × 5.00	66
4.50	77
4.00	88
90 × 4.00	106
75 × 4.00	121
3.75	154
3.35	194
65 × 3.35	230
3.00	275
2.65	350
60 × 3.35	255
3.00	310
2.65	385
50 × 3.35	290
3.00	340
2.65	440
2.36	550
45 × 2.65	510
2.36	640
40 × 2.65	575
2.36	750
2.00	970
30 × 2.36	840
2.00	1 170
25 × 2.00	1 430
1.80	1 720
1.60	2 210
20 × 1.60	2 710

Annular ringed shank nails
Length × Shank

mm	No/kg
100 × 5.00	66
75 × 3.75	154
65 × 3.35	230
60 × 3.35	255
50 × 3.35	290
3.00	340
2.65	440
45 × 2.65	510
40 × 2.65	575
2.36	750
30 × 2.36	840
25 × 2.00	1 430
20 × 2.00	1 900

**Oval brad or lost
Lhead nails**
Length × Shank

mm	No/kg
150 × 7.10 × 5.00	31
125 × 6.70 × 4.50	44
100 × 6.00 × 4.00	64
75 × 5.00 × 3.35	125
65 × 4.00 × 2.65	230
60 × 3.75 × 2.36	340
50 × 3.35 × 2.00	470
40 × 2.65 × 1.60	940
30 × 2.65 × 1.60	1 480
25 × 2.00 × 1.25	2 530

Round lost head nails
Length × Shank

mm	No/kg
75 × 3.75	160
65 × 3.35	240
3.00	270
60 × 3.35	270
3.00	330
50 × 3.00	360
2.65	420
40 × 2.36	760

Panel pins
Length × Shank

mm	No/kg
50 × 2.00	770
40 × 1.60	1 590
30 × 1.60	1 900
25 × 1.60	2 340
1.40	3 090
20 × 1.60	3 140
1.40	3 970
15 × 1.25	6 400

Memoranda

Woodwork *(continued)*

Wood screw sizes

Wood screw gauge	mm	in	in	Wood screw gauge	mm	in	in
0	1.52	0.060	$\frac{1}{16}$	8	4.17	0.164	
1	1.78	0.070		9	4.52	0.178	
2	2.08	0.082		10	4.88	0.192	$\frac{3}{16}+$
3	2.39	0.094	$\frac{3}{32}+$	12	5.59	0.220	$\frac{7}{32}+$
4	2.74	0.108		14	6.30	0.248	$\frac{1}{4}$
5	3.10	0.122	$\frac{1}{8}$	16	7.01	0.276	$\frac{9}{32}$
6	3.45	0.136		18	7.72	0.304	$\frac{5}{16}$
7	3.81	0.150	$\frac{5}{32}$	20	8.43	0.332	$\frac{11}{32}-$

Structural steelwork

Serial sizes and weights of structural steel sections

Universal beams		Universal beams		Universal beams		Universal columns	
mm	kg/m	mm	kg/m	mm	kg/m	mm	kg/m
914 × 419	388	457 × 152	82	203 × 133	30	356 × 406	634
	343		74		25		551
914 × 305	289		67	203 × 102	23		467
	253		60	178 × 102	19		393
	224		52	152 × 89	16		340
	201	406 × 178	74	127 × 76	13		287
838 × 292	226		67				235
	194		60	Joists		356 × 368	202
	176		54	mm	kg/m		177
762 × 267	197	406 × 140	46	254 × 203	81.85		153
	173		39	254 × 114	37.20		129
	147	356 × 171	67	203 × 152	52.09	305 × 305	283
686 × 254	170		57				240
	152		51	152 × 127	37.20		198
	140		45				158
	125	356 × 127	39	127 × 114	29.76		137
610 × 305	238		33		26.79		118
	179	305 × 165	54	127 × 76	16.37		97
	149		46			254 × 254	167
610 × 229	140		40	114 × 114	26.79		132
	125	305 × 127	48	102 × 102	23.07		107
	113		42	102 × 44	7.44		89
	101		37				73
533 × 210	122	305 × 102	33	89 × 89	19.35	203 × 203	86
	109		28				71
	101		25	76 × 76	14.67		60
	92	254 × 146	43		12.65		52
	82		37				46
457 × 191	98		31			152 × 152	37
	89	254 × 102	28				30
	82		25				23
	74		22				
	67						

Structural steelwork *(continued)*

Serial sizes and weights of structural steel sections

Channels

mm	kg/m
432 × 102	65.54
381 × 102	55.10
305 × 102	46.18
305 × 89	41.69
254 × 89	35.74
254 × 76	28.29
229 × 89	32.76
229 × 76	26.06
203 × 89	29.78
203 × 76	23.82
178 × 89	26.81
178 × 76	20.84
152 × 89	23.84
152 × 76	17.88
127 × 64	14.90
102 × 51	10.42
76 × 38	6.70

Tees cut from universal beams

mm	kg/m
305 × 457	127
	112
	101
292 × 419	113
	97
	88
267 × 381	99
	87
	74
254 × 343	85
	76
	70
	63
305 × 305	119
	90
	75
229 × 305	70
	63
	57
	51
210 × 267	61
	55
	51
	46
	41
191 × 229	49
	45
	41
	37
	34

Tees cut from universal beams *(continued)*

mm	kg/m
152 × 229	41
	37
	34
	30
	26
178 × 203	37
	34
	30
	27
140 × 203	23
	20
171 × 178	34
	29
	26
	23
127 × 178	20
	17
165 × 152	27
	23
	20
127 × 152	24
	21
	19
102 × 152	17
	14
	13
146 × 127	22
	19
	16
102 × 127	14
	13
	11
133 × 102	15
	13

Tees cut from universal columns

mm	kg/m
406 × 178	118
368 × 178	101
	89
	77
	65
305 × 152	79
	69
	59
	49
254 × 127	66
	54
	45
	37
203 × 102	43
	36
	30
	26
	23
152 × 76	19
	15
	12

Rolled tees

mm	kg/m
51 × 51	6.92
	4.76
44 × 44	4.11
	3.14

Structural steelwork *(continued)*

Serial sizes and weights of structural steel sections

Equal angles			Equal angles *(continued)*			Unequal angles			Circular hollow sections *(continued)*			Circular hollow sections *(continued)*		
mm		kg/m	mm		kg/m	mm		kg/m	mm		kg/m	mm		kg/m
250 × 250 × 35		128	50 × 50 × 8		5.82	125 × 75 × 12		17.8	60.3 × 3.2		4.51	244.5 × 6.3		37.0
	32	118		6	4.47		10	15.0		4	5.55		8	46.7
	28	104		5	3.77		8	12.2		5	6.82		10	57.8
	25	93.6		4	3.06	100 × 75 × 12		15.4	76.1 × 3.2		5.75		12.5	71.5
200 × 200 × 24		71.1		3	2.33		10	13.0		4	7.11		16	90.2
	20	59.9	45 × 45 × 6		4.00		8	10.6		5	8.77		20	111.0
	18	54.2		5	3.38	100 × 65 × 10		12.3	88.9 × 3.2		6.76	273 × 6.3		41.4
	16	48.5		4	2.74		8	9.94		4	8.38		8	52.3
150 × 150 × 18		40.1		3	2.09		7	8.77		5	10.3		10	64.9
	15	33.8	40 × 40 × 6		3.52	80 × 60 × 8		8.34	114.3 × 3.6		9.83		12.5	80.3
	12	27.3		5	2.97		7	7.36		5	13.5		16	101
	10	23.0		4	2.42		6	6.37		6.3	16.8		20	125
120 × 120 × 15		26.6		3	1.84	75 × 50 × 8		7.39	139.7 × 5		16.6		25	153
	12	21.6	30 × 30 × 5		2.18		6	5.65		6.3	20.7	323.9 × 8		62.3
	10	18.2		4	1.78	65 × 50 × 8		6.75		8	26.0		10	77.4
	8	14.7		3	1.36		6	5.16		10	32.0		12.5	96.0
100 × 100 × 15		21.9	25 × 25 × 5		1.77		5	4.35	168.3 × 5		20.1		16	121
	12	17.8		4	1.45	60 × 30 × 6		3.99		6.3	25.2		20	150
	8	12.2		3	1.11		5	3.37		8	31.6		25	184
90 × 90 × 12		15.9				40 × 25 × 4		1.93		10	29.0	355.6 × 8		68.6
	10	13.4							193.7 × 5.4		25.1		10	85.2
	8	10.9	**Unequal angles**			**Circular hollow sections**				6.3	29.1		12.5	106
	7	9.61	mm		kg/m	mm		kg/m		8	36.6		16	154
	6	8.30	200 × 150 × 18		47.1	21.3 × 3.2		1.43		10	45.3		20	166
80 × 80 × 10		11.9		15	39.6	26.9 × 3.2		1.87		12.5	55.9		25	204
	8	9.63		12	32.0	33.7 × 2.6		1.99		16	70.1	406.4 × 10		97.8
	6	7.34	200 × 100 × 15		33.7		3.2	2.41	219.1 × 6.3		33.1		12.5	121
70 × 70 × 10		10.3		12	27.3		4	2.93		8	41.6		16	154
	8	8.36		10	23.0	42.4 × 2.6		2.55		10	51.6		20	191
	6	6.38	150 × 90 × 15		26.6		3.2	3.09		12.5	63.7		25	235
60 × 60 × 10		8.69		12	21.6		4	3.79		16	80.1		32	295
	8	7.09		10	18.2	48.3 × 3.2		3.56		20	98.2	457 × 10		110
	6	5.42	150 × 75 × 15		24.8		4	4.37					12.5	137
	5	4.57		12	20.2		5	5.34					16	174
				10	17.0								20	216

Memoranda

Structural steelwork *(continued)*

Serial sizes and weights of structural steel sections

Circular hollow sections *(continued)*	
mm	kg/m
25	266
32	335
40	411

Rectangular hollow sections	
mm	kg/m
50 × 30 × 2.6	3.03
3.2	3.66
60 × 40 × 3.2	4.66
4	5.72
80 × 40 × 3.2	5.67
4	6.97
90 × 50 × 3.6	7.46
5	10.1
100 × 50 × 3.2	7.18
4	8.86
5	10.9
100 × 60 × 3.6	8.59
5	11.7
6.3	14.4
120 × 60 × 3.6	9.72
5	13.3
6.3	16.4
120 × 80 × 5	14.8
6.3	18.4
8	22.9
10	27.9
150 × 100 × 5	18.7
6.3	23.3
8	29.1
10	35.7
160 × 80 × 5	18.0
6.3	22.3
8	27.9
10	34.2

Rectangular hollow sections *(continued)*	
mm	kg/m
200 × 100 × 5	22.7
6.3	28.3
8	35.4
10	43.6
12.5	53.4
16	66.4
250 × 150 × 6.3	38.2
8	48.0
10	59.3
12.5	73.0
16	91.5
300 × 200 × 6.3	48.1
8	60.5
10	75.0
12.5	92.6
16	117
400 × 200 × 10	90.7
12.5	112
450 × 250 × 10	106
12.5	132
16	167

Square hollow sections	
mm	kg/m
20 × 20 × 2	1.12
2.6	1.39
30 × 30 × 2.6	2.21
3.2	2.65
40 × 40 × 2.6	3.03
3.2	3.66
4	4.46
50 × 50 × 3.2	4.66
4	5.72
5	6.97

Square hollow sections *(continued)*	
mm	kg/m
60 × 60 × 3.2	5.67
4	6.97
5	8.54
70 × 70 × 3.6	7.46
5	10.1
80 × 80 × 3.6	8.59
5	11.7
6.3	14.4
90 × 90 × 3.6	9.72
5	13.3
6.3	16.4
100 × 100 × 4	12.0
5	14.8
6.3	18.4
8	22.9
10	27.9
120 × 120 × 5	18.0
6.3	22.3
8	27.9
10	34.2
150 × 150 × 5	22.7
6.3	28.3
8	35.4
10	43.6
12.5	53.4
16	66.4
180 × 180 × 6.3	34.2
8	43.0
10	53.0
12.5	65.2
16	81.4

Square hollow sections *(continued)*	
mm	kg/m
200 × 200 × 6.3	38.2
8	48.0
10	59.3
12.5	73.0
16	91.5
250 × 250 × 6.3	48.1
8	60.5
10	75.0
12.5	92.6
16	117
300 × 300 × 10	90.7
12.5	112
16	142
350 × 350 × 10	106
12.5	132
16	167
400 × 400 × 10	122
12.5	152

Metalwork

Gauges and thicknesses

Imperial standard wire gauge					Birmingham gauge			
SWG	mm	in	in		BG	mm	in	in
30	0.315	0.0124			30	0.312	0.0123	
29	0.345	0.0136			29	0.353	0.0139	
28	0.367	0.0148	$\frac{1}{64}-$		28	0.397	0.015625	$\frac{1}{64}$
27	0.417	0.0164	$\frac{1}{64}+$		27	0.443	0.01745	
26	0.457	0.0180			26	0.498	0.01961	
25	0.508	0.020			25	0.560	0.02204	
24	0.559	0.022			24	0.629	0.02476	
23	0.610	0.024			23	0.707	0.02782	
22	0.711	0.028	$\frac{1}{32}-$		22	0.794	0.03125	$\frac{1}{32}$
21	0.813	0.032	$\frac{1}{32}+$		21	0.886	0.0349	
20	0.914	0.036			20	0.996	0.0392	
19	1.016	0.040			19	1.118	0.0440	
18	1.219	0.048			18	1.257	0.0495	
17	1.422	0.056	$\frac{1}{16}-$		17	1.412	0.0556	
16	1.626	0.064	$\frac{1}{16}+$		16	1.588	0.0625	$\frac{1}{16}$
15	1.829	0.072			15	1.775	0.0699	
14	2.032	0.080			14	1.994	0.0785	
13	2.337	0.092	$\frac{3}{32}-$		13	2.240	0.0882	
12	2.642	0.104			12	2.517	0.0991	
11	2.946	0.116			11	2.827	0.1113	
10	3.251	0.128	$\frac{1}{8}+$		10	3.175	0.1250	$\frac{1}{8}$
9	3.658	0.144			9	3.551	0.1398	
8	4.064	0.160	$\frac{5}{32}+$		8	3.988	0.1570	$\frac{5}{32}+$
7	4.470	0.176			7	4.481	0.1764	
6	4.877	0.192	$\frac{3}{16}+$		6	5.032	0.1981	$\frac{3}{16}+$
5	5.385	0.212			5	5.652	0.2225	
4	5.893	0.232			4	6.350	0.250	$\frac{1}{4}$
3	6.401	0.252	$\frac{1}{4}+$		3	7.122	0.2804	
2	7.010	0.276			2	7.993	0.3147	
1	7.620	0.300	$\frac{5}{16}-$		1	8.971	0.3532	$\frac{11}{32}+$

Memoranda

Plumbing and engineering installations

Roof drainage

	Roof area drained by:			
	Level gutter		Gutter falling 1/600	
Gutter and rainwater pipe sizes	Outlet one end m^2	Centre outlet m^2	Outlet one end m^2	Centre outlet m^2
76 mm half round gutter and 50 mm rainwater pipe	15	30	20	40
100 mm half round gutter and 68 mm rainwater pipe	38	75	50	100
150 mm half round gutter and 110 mm rainwater pipe	100	200	130	260
115 mm square section gutter and 65 mm square rainwater pipe	50	100	65	130

Jointing materials per pipe joint

Based on net quantities – allowance for waste to be added.

Pipe size	Cast iron soil pipes		Cast iron drain pipes	
	Lead	Yarn	Lead	Yarn
mm	kg/joint	kg/joint	kg/joint	kg/joint
50	0.65	0.07	–	–
75	1.10	0.10	–	–
100	1.85	0.13	2.55	0.13
150	–	–	4.25	0.20

Memoranda

Floor, wall and ceiling finishings

Approximate coverage of plasters per tonne

Based on net quantities – allowance for waste to be added.

Type		2 mm m²/tonne	3 mm m²/tonne	Coat thickness 5 mm m²/tonne	8 mm m²/tonne	11 mm m²/tonne
Gypsum ("Thistle")						
	undercoat	–	–	–	160	115
	finish	400	260	165	–	–
Lightweight ("Carlite")						
	browning	–	–	–	–	140
	bonding coat	–	–	–	135	90
	finish	450	–	–	–	–

Approximate coverage of renderings per cubic metre

Based on net quantities – allowance for waste to be added.

Background	6 mm m²/m³	10 mm m²/m³	Coat thickness 13 mm m²/m³	16 mm m²/m³	20 mm m²/m³
Blockwork (no grooved or keyed faces)	109	72	58	48	40
Brickwork (grooved face or joints raked out)	87	62	51	43	36
Stone rubble work	72	54	46	40	33

For mortar mixing materials, see "Brickwork and blockwork".

Memoranda

Glazing

Putty quantities per square metre of glazing

Based on: (1) 4.5 metres puttying to wood per kilogramme of putty
              (2) 3.5 metres puttying to metal per kilogramme of putty
              (3) Net quantities – allowance for waste to be added.

Pane proportions

(length × height)		up to 0.10 m^2 Putty kg/m^2	0.10–0.50 m^2 Putty kg/m^2	0.50–1.00 m^2 Putty kg/m^2	over 1.00 m^2 Putty kg/m^2
1 × 1	to wood	3.95	1.62	1.03	0.73
2 × 1	to wood	4.21	1.72	1.09	0.77
3 × 1	to wood	4.56	1.87	1.19	0.74
4 × 1	to wood	4.96	2.03	1.28	0.91
5 × 1	to wood	5.33	2.18	1.38	0.97
1 × 1	to metal	5.08	2.09	1.32	0.93
2 × 1	to metal	5.42	2.21	1.40	0.99
3 × 1	to metal	5.86	2.41	1.52	1.08
4 × 1	to metal	6.38	2.61	1.64	1.17
5 × 1	to metal	6.86	2.80	1.77	1.25

Painting and decorating

Coverage of paints

The following schedule of average coverage figures in respect of painting work is the 1974 revision of the schedule compiled and approved for the guidance of commercial organisations and professional bodies when assessing the values of materials in painting work by the Paint and Painting Industries; Liaison Committee (constituent bodies:- British Decorators Association, National Federation of Painting and Decorating Contractors, Paintmakers Association of Great Britain and Scottish Decorators Federation) whose permission to publish is hereby acknowledged.

Schedule of average coverage of paints

In this revision a range of spreading capacities is given. Figures are in square metres per litre, except for oil-bound water paint and cement-based paint which are in square metres per kilogram.

For comparative purposes figures are given for a single coat, but users are recommended to allow manufacturers' recommendations as to when to use single or multicoat systems.

It is emphasised that the figures quoted in the schedule are practical figures for brush application, achieved in scale painting work and take into account losses and wastage. They are not optimum figures based upon ideal conditions of surface, nor minimum figures reflecting the reverse of these conditions.

There will be instances when the figures indicated by paint manufacturers in their literature will be higher than those shown in the schedule. The Committee realise that under ideal conditions of application, and depending on such factors as the skill of the applicator and the type and quality of the product, better covering figures can be achieved.

The figures given are for application by brush and to appropriate systems on each surface. They are given for guidance and are qualified to allow for variation depending on certain factors.

Notes:

† Aluminium primer/sealer is normally used over "bitumen" painted surfaces.
* The texture of roughcast, Tyrolean and pebbledash can vary markedly and thus there can be significant variations in the coverage of paints applied to such surfaces. The figures given are thought to be typical but under some circumstances much lower coverages will be obtained.

In many instances the coverages achieved will be affected by the suction and texture of the backing; for example the suction and texture of brickwork can vary to such an extent that coverages outside those quoted may on occasions be obtained.

It is necessary to take these factors into account when using this table.

Memoranda

Painting and decorating *(continued)*

Schedule of average coverage of paints in square metres

Surfaces	Finishing plaster	Wood floated rendering	Smooth concrete/ cement	Fair faced brickwork	Block-work	*Rough-cast/ pebbledash	Hard-board	Soft fibre insulating board
Coating per litre								
Water thinned primer/undercoat								
as primer	13–15	–	–	–	–	–	10–12	7–10
as undercoat	–	–	–	–	–	–	–	10–12
Plaster primer (including building board)	9–11	8–12	9–11	7–9	5–7	2–4	8–10	7–9
Alkali resistant primer	7–11	6–8	7–11	6–8	4–6	2–4	–	–
External wall primer sealer	6–8	6–7	6–8	5–7	4–6	2–4	–	–
Undercoat	11–14	7–9	7–9	6–8	6–8	3–4	11–14	10–12
Gloss finish	11–14	8–10	8–10	7–9	6–8	–	11–14	10–12
Oil-based thixo-tropic finish	Figures should be obtained from individual manufacturers							
Eggshell/semi-gloss finish (oil-based)	11–14	9–11	11–14	8–10	7–9	–	10–13	10–12
Emulsion paint								
standard	12–15	8–12	11–14	8–12	6–10	2–4	12–15	8-10
contract	10–12	7–11	10–12	7–10	5–9	2–4	10–12	7-9
Glossy emulsion	Figures should be obtained from individual manufacturers							
Heavy textured coating	2–4	2–4	2–4	2–4	2–4	–	2–4	2–4
Masonry paint	5–7	4–6	5–7	4–6	3–5	2–4	–	–
per kilogram								
Oil-bound water paint	7–9	4–6	7–9	4–6	5–7	–	–	4–6
Cement-based paint	–	4–6	6–7	3–6	3–6	2–3	–	–

Painting and decorating *(continued)*

Schedule of average coverage of paints in square metres *(continued)*

Surfaces / Coating per litre	Fire retardant fibre insulating board	Smooth paper faced board	Hard asbestos sheet	Structural steelwork	Metal sheeting	Joinery	Smooth primed	Smooth under-coated surfaces
Wood primer (oil-based)	–	–	–	–	–	8–11	–	–
Water thinned primer/undercoat								
as primer	–	8–11	7–10	–	–	10–14	–	–
as undercoat	–	10–12	–	–	–	12–15	12–15	–
Aluminium sealer								
spirit-based	–	–	–	–	–	7–9	–	–
oil-based	–	–	–	–	9–13	9–13	–	–
Metal primer								
conventional	–	–	–	7–10	10–13	–	–	–
specialised			Figures should be obtained from individual manufacturers					
Plaster primer (including building board)	8–10	10–12	10–12	–	–	–	–	–
Alkali resistant primer	–	–	8–10	–	–	–	–	–
External wall primer sealer	–	–	6–8	–	–	–	–	–
Undercoat	10–12	11–14	10–12	10–12	10–12	10–12	11–14	–
Gloss finish	10–12	11–14	10–12	10–12	10–12	10–12	11–14	11–14
Oil-based thixo-tropic finish			Figures should be obtained from individual manufacturers					
Eggshell/semi-gloss finish (oil-based)	10–12	11–14	10–12	10–12	10–12	10–12	11–14	11–14
Emulsion paint								
standard	8–10	12–15	10–12	–	–	10–12	12–15	12–15
contract	–	10–12	8–10	–	–	10–12	10–12	10–12
Glossy emulsion			Figures should be obtained from individual manufacturers					
Heavy textured coating	2–4	2–4	2–4	2–4	2–4	2–4	2–4	2–4
Masonry paint	–	–	5-7	–	–	–	6-8	6-8
per kilogram								
Oil-bound water paint	–	7–9	7–9	–	–	–	7–9	–
Cement-based paint	–	–	4–6	–	–	–	–	–

Memoranda

Drainage

Bed and surround material per metre of pipe

Based on net quantities – allowance to be added for waste and for excess trench widths and/or depths

Pipe size	100 mm	150 mm	225 mm	300 mm
Bed width	450 mm	525 mm	600 mm	750 mm
Material	m³/m	m³/m	m³/m	m³/m
50 mm bed only	0.023	0.026	0.030	0.038
100 mm bed only	0.045	0.053	0.060	0.075
150 mm bed only	0.068	0.079	0.090	0.113
100 mm bed and benching or side filling to half height of pipe	0.070	0.089	0.113	0.158
150 mm bed and benching or side filling to half height of pipe	0.093	0.116	0.143	0.196
100 mm bed and side filling to full height of pipe	0.095	0.126	0.165	0.241
150 mm bed and side filling to full height of pipe	0.117	0.152	0.195	0.279
100 mm bed, side filling and 100 mm covering	0.140	0.179	0.225	0.316
150 mm bed, side filling and 150 mm covering	0.185	0.231	0.285	0.391

Approximate average weights

Aggregate	– coarse		1,500	kg/m^3	94 lb/ft^3
	– fine	– dry	1,600	kg/m^3	100 lb/ft^3
		– moist	1,260	kg/m^3	79 lb/ft^3
Asphalt	– 20 mm		49	kg/m^2	10 lb/ft^2
	– 25 mm		61	kg/m^2	13 lb/ft^2
	– 30 mm		73	kg/m^2	15 lb/ft^2
Bitumen macadam			2,000	kg/m^3	125 lb/ft^3
Blockwork	– natural aggregate	– 75 mm	160	kg/m^2	33 lb/ft^2
		– 100 mm	215	kg/m^2	44 lb/ft^2
		– 140 mm	300	kg/m^2	61 lb/ft^2
	– lightweight aggregate	– 75 mm	60	kg/m^2	12 lb/ft^2
		– 100 mm	80	kg/m^2	16 lb/ft^2
		– 140 mm	112	kg/m^2	23 lb/ft^2
Brickwork	– 102.5 mm		220	kg/m^2	45 lb/ft^2
	– 215 mm		465	kg/m^2	95 lb/ft^2
	– 327.5 mm		710	kg/m^2	145 lb/ft^2
Cement			1,440	kg/m^3	90 lb/ft^3
Concrete	– plain		2,300	kg/m^3	144 lb/ft^3
	– reinforced		2,400	kg/m^3	150 lb/ft^3
	– no-fines		1,760	kg/m^3	110 lb/ft^3
Lime	– hydrated		500	kg/m^3	31 lb/ft^3
Plasterboard	– 9.5 mm		8	kg/m^2	1.6 lb/ft^2
	– 12.7 mm		11	kg/m^2	2.3 lb/ft^2
	– 19 mm		16	kg/m^2	3.3 lb/ft^2
Plastering	– gypsum	– 5 mm	6.8	kg/m^2	1.4 lb/ft^2
		– 13 mm	22	kg/m^2	4.6 lb/ft^2
	– lightweight	– 10 mm	9.3	kg/m^2	1.9 lb/ft^2
		– 13 mm	10	kg/m^2	2 lb/ft^2
Roofing	– aluminium 0.80 mm		2.2	kg/m^2	0.5 lb/ft^2
	– copper 0.55 mm		5	kg/m^2	1 lb/ft^2
	– lead 2.24 mm		25.4	kg/m^2	5.2 lb/ft^2
	– mineral surfaced three layer felt		12	kg/m^2	2.5 lb/ft^2
	– plain tile	– clay	65	kg/m^2	13 lb/ft^2
		– concrete	80	kg/m^2	16 lb/ft^2
	– slate	– asbestos – cement	20	kg/m^2	4 lb/ft^2
		– natural	30	kg/m^2	6 lb/ft^2
	– zinc 0.65 mm		4.6	kg/m^2	0.9 lb/ft^2
Sand	– dry		1,600	kg/m^3	100 lb/ft^3
	– moist		1,260	kg/m^3	79 lb/ft^3

Memoranda

Approximate average weights *(continued)*

Screeding	– cement and sand	– 20 mm	46	kg/m²	9	lb/ft²
		– 35 mm	80	kg/m²	16	lb/ft²
		– 50 mm	115	kg/m²	24	lb/ft²
Soil	– compact		1,840	kg/m²	115	lb/ft²
Steel reinforcement – see earlier page under "Concrete work"						
Stone	– natural		2,400	kg/m³	150	lb/ft³
	– reconstructed		2,250	kg/m³	140	lb/ft³
Tarmacadam			2,000	kg/m³	125	lb/ft³
Timber	– softwood		480	kg/m³	30	lb/ft³
	– hardwood		720	kg/m³	45	lb/ft³
Water			1,000	kg/m³	62.4	lb/ft³